PRAISE FOR *FALL AND RISE*

"Remarkable and groundbreaking. . . . *Fall and Rise* is an ambitious undertaking, setting out to be an exhaustive, prismatic chronicle of 9/11. Impressively, Zuckoff pulls it off. . . . A vital and unforgettable account."

—*Washington Post*

"A riveting, harrowing, moment-by-moment narrative history of 9/11."

—*Boston Globe*

"*Fall and Rise* is a detailed, graphic, heartbreaking, and definitive portrait of that fateful day."

—*USA Today*

"A minute-to-minute, suspenseful, heart-wrenching and inspirational narrative, *Fall and Rise* should endure as a prose poem memorial to a day like no other."

—*Minneapolis Star Tribune*

"Inspiring, depressing, heartbreaking, and simply astonishing. . . . Zuckoff delivers a master class in long-form historical journalism."

—*Sydney Morning Herald*

"This book derives its power from its focus on individuals in the main unknown to the larger world. . . . With journalistic rigor, Zuckoff acknowledges what he doesn't know. . . . But a full generation has come of age with no memory of that day. It needs to hear anew what happened, and maybe learn that time, in fact, does not heal all wounds."

—*New York Times*

"[Zuckoff] has masterfully transformed this enormous record and his own interviews into a powerful narrative of that day."

—*Newsday*

"The horror and heroism of 9/11 are brought to life in this panoramic history. . . . The result is a superb, harrowing retelling of this most dramatic of stories."

—*Publishers Weekly* (starred review)

"A meticulously delineated, detailed, graphic history of the events of 9/11 in New York City, at the Pentagon, and in Pennsylvania. . . . Despite the story's sprawling cast, which could have sabotaged a book by a less-skilled author, Zuckoff ably handles all of the complexities. . . . [A]s contemporary history, *Fall and Rise* is a clear and moving success."

—*Kirkus Reviews* (starred review)

"Better and more comprehensive than any prior account. . . . Those of us who lived through those days will find the book cathartic; those rising generations who were too young to remember 9/11, or who weren't yet born, will find it revelatory."

—John Farmer, senior counsel to the 9/11 Commission and author of *The Ground Truth*

"With his rigorous research and moral clarity, Mitchell Zuckoff has provided us with an invaluable service. He has deepened our understanding of what happened on 9/11 and recorded the voices of the victims and the survivors. What's more, he has ensured that we never forget."

—David Grann, #1 *New York Times* bestselling author of *Killers of the Flower Moon*

"Compelling. . . . The victims, survivors, responders, and bereaved relatives central to his narrative are familiar to us, accessible and endearing because we can see ourselves reflected in the mirror of those lives that Zuckoff has restored through rigorous research, sensitive listening, and artful storytelling. Infused with empathy, *Fall and Rise* dignifies not only the web of individuals profiled in his book but all who were violated by the terrorist attacks of 9/11."

—Jan Ramirez, chief curator and executive vice president
of the National September 11 Memorial & Museum

"The 9/11 book we've been waiting for. A terrific storyteller and gifted researcher, Mitchell Zuckoff has rendered that world-changing day on a scale both intimate and monumental. This is narrative history at its very best."

—Cokie Roberts, *New York Times* bestselling author
and Emmy Award-winning journalist

"A triumph of great reporting and rigorous research, *Fall and Rise* by Mitchell Zuckoff recreates the tragic events of September 11, 2001, with harrowing precision and commendable empathy."

—Daniel James Brown, *New York Times*
bestselling author of *The Boys in the Boat*

"Both a deconstruction and a reconstruction, *Fall and Rise* is a definitive—perhaps *the* definitive—account of a day that will always live in infamy. In this forensic but also literary tour de force, Mitchell Zuckoff has not only created a moving and consequential work, he's performed a national service."

—Hampton Sides, bestselling author of
In the Kingdom of Ice and *On Desperate Ground*

"Mitchell Zuckoff's *Fall and Rise: The Story of 9/11* brings readers back to those harrowing days in 2001 that forever scarred the American psyche and left a wound a mile deep in the heart of our nation. Every page shivers in stark remembrance of New York; Washington, DC; and Shanksville, Pennsylvania, on those dark days. A masterpiece of historical scholarship!"

—Douglas Brinkley, author of *American Moonshot:
John F. Kennedy and the Great Space Race*

Fall and Rise

ALSO BY MITCHELL ZUCKOFF

*13 Hours: The Inside Account of What
Really Happened in Benghazi*

*Frozen in Time: An Epic Story of Survival and a
Modern Quest for Lost Heroes of World War II*

*Lost in Shangri-La: A True Story of
Survival, Adventure, and the Most Incredible
Rescue Mission of World War II*

Robert Altman: The Oral Biography

Ponzi's Scheme: The True Story of a Financial Legend

*Judgment Ridge: The True Story Behind the
Dartmouth Murders* (with Dick Lehr)

Choosing Naia: A Family's Journey

Fall
and Rise
The Story
of 9/11
Mitchell
Zuckoff

HARPER PERENNIAL

NEW YORK • LONDON • TORONTO • SYDNEY • NEW DELHI • AUCKLAND

HARPER PERENNIAL

A hardcover edition of this book was published in 2019 by HarprCollins Publishers.

HarperCollins books may be purchased for educational, business, or sales promotional use. For information, please email the Special Markets Department at SPsales@harpercollins.com.

FIRST HARPER PERENNIAL EDITION PUBLISHED 2020.

Designed by Leah Carlson-Stanisic

Photograph by © CATHERINE URSILLO/Getty Images

Map by Nick Springer, copyright © 2018 Springer Cartographics LLC.

Library of Congress Cataloging-in-Publication Data has been applied for.

ISBN 978-0-06-227565-3 (pbk.)

22 23 24 25 26 LBC 8 7 6 5 4

For my children—
and everyone else's

The ravages of many a forest fire of a bygone age may be read to-day in the scars left in the tree itself. The exact year that the fire occurred and some idea of its intensity are recorded in the wood, oftentimes grown over with living tissue and hid from the casual observer.

—FOREST PATHOLOGIST J. S. BOYCE, 1921

CONTENTS

PART III RISE FROM THE ASHES

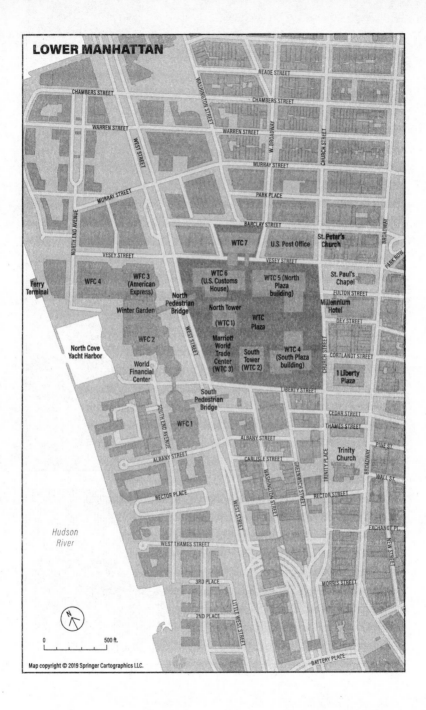

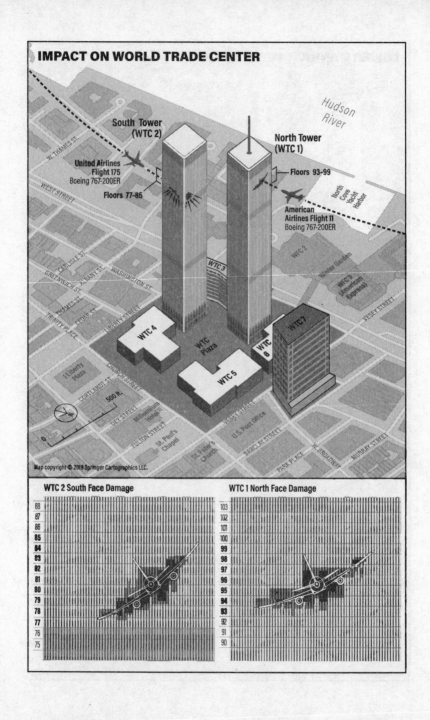

IMPACT ON WORLD TRADE CENTER

South Tower
(WTC 2)

North Tower
(WTC 1)

United Airlines
Flight 175
Boeing 767-200ER

Floors 93-99

Floors 77-85

American
Airlines Flight 11
Boeing 767-200ER

Hudson
River

North
Cove
Yacht
Harbor

WTC 3

WTC 2

WFC 2

Winter Garden

WFC 3
(American
Express)

WTC 4

WTC
Plaza

WTC 5

WTC
6

WTC 7

1 Liberty
Plaza

500 ft.

0

Millennium
Hotel

St. Paul's
Chapel

St. Peter's
Church

U.S. Post Office

W. THAMES ST.

WEST STREET

CARLISLE ST.

GREENWICH ST.

ALBANY ST.

WASHINGTON ST.

THAMES ST.

CEDAR ST.

TRINITY PLACE

LIBERTY STREET

CHURCH STREET

CORTLANDT ST.

DEY STREET

FULTON STREET

VESEY STREET

BARCLAY STREET

PARK PLACE

W. BROADWAY

MURRAY STREET

Map copyright © 2019 Springer Cartographics LLC.

WTC 2 South Face Damage

88
87
86
85
84
83
82
81
80
79
78
77
76
75

WTC 1 North Face Damage

103
102
101
100
99
98
97
96
95
94
93
92
91
90

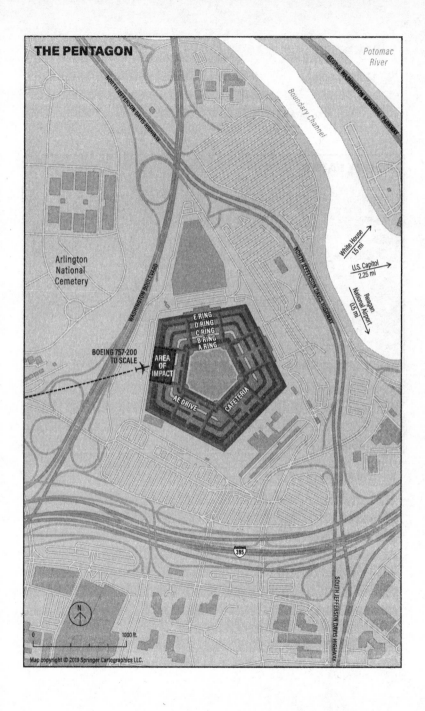

THE PENTAGON

Potomac River

GEORGE WASHINGTON MEMORIAL PARKWAY

NORTH JEFFERSON DAVIS HIGHWAY

Boundary Channel

Arlington National Cemetery

White House 1.5 mi

U.S. Capitol 2.25 mi

Reagan National Airport 0.5 mi

NORTH JEFFERSON DAVIS HIGHWAY

WASHINGTON BOULEVARD

E RING
D RING
C RING
B RING
A RING

BOEING 757-200 TO SCALE

AREA OF IMPACT

AE DRIVE

CAFETERIA

395

SOUTH JEFFERSON DAVIS HIGHWAY

N

0   1000 ft.

Map copyright © 2019 Springer Cartographics LLC.

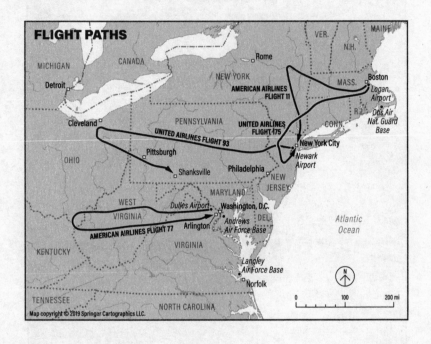

INTRODUCTION

"The Darkness of Ignorance"

ON OCTOBER 28, 1886, PRESIDENT GROVER CLEVELAND SAILED TO A teardrop-shaped island in New York Harbor to formally accept France's gift of the Statue of Liberty. Under leaden skies and a veil of mist, the president ended his speech with a tribute to the copper-clad lady's torch and her symbolic power: "A stream of light shall pierce the darkness of ignorance and men's oppression until Liberty shall enlighten the world."

Dignitaries pounded ceremonial last rivets as warship cannons boomed. Across the water in Lower Manhattan, revelers erupted in celebration. Cobblestone streets pulsed with braying horses, throbbing drums, and blooming flower carts. Brass bands marched like front-bound soldiers, and children scrambled up lampposts to avoid being trampled.

Out-of-towners drawn to the spectacle tilted their heads to gawk at the impossibly tall buildings that loomed over them. Amused by these sky-eyed rubes, an office boy in a high tower felt seized by a raffish idea. He opened a window and tossed out long ribbons of the narrow paper tape that normally recorded the drunkard's walk of stock prices. His pals followed suit.

"In a moment, the air was white with curling streamers," a reporter for the *New York Times* observed. "Hundreds caught in the meshes of electric wires and made a snowy canopy, and others floated downward and were caught by the crowd."

The fun was contagious. Serious men of finance became boys again, pressing against office windows to unspool paper onto the crowd. "There was seemingly no end to it," the *Times* reporter wrote. "Every window appeared to be a paper mill spouting out squirming lines of tape. Such was Wall Street's novel celebration."

With that, the ticker-tape parade was born.

During the next one hundred fifteen years, countless tons of celebratory confetti sailed from high-rise windows onto a stretch of Lower Broadway that became known as the Canyon of Heroes. Paper blizzards honored more than two hundred explorers and presidents, war heroes and athletes, astronauts and religious figures, luminaries from Einstein to Earhart, Churchill to Kennedy, Mandela to the Mets.

Then came September 11, 2001.

Torn open, aflame, weakening from within, the Twin Towers of the World Trade Center spewed paper like blood from an arterial wound. Legal documents and employee reviews. Pay stubs, birthday cards, takeout menus. Timesheets and blueprints, photographs and calendars, crayon drawings and love notes. Some in full, some in tatters, some in flames. A single scrap from the South Tower, tossed like a bottled message from a sinking ship, captured the day's horror. In a scrawled hand, next to a bloody fingerprint, the note read:

84th floor
west office
12 People trapped

After the paper came the people. After the people came the buildings. After the buildings came the wars. The ashes cooled, but not the anguish. For years, New Yorkers couldn't stomach a ticker-tape parade, especially so near the hallowed hole renamed Ground Zero. Yet with time, the unthinkable often becomes acceptable.

In February 2008, the underdog New York Giants won the Super Bowl. Tens of thousands of football fans gathered to celebrate, just blocks from where steel beams rose for a dazzling "Freedom Tower" at One World Trade Center, an audacious middle finger to America's enemies, taller and bolder than the boxy twins whose sanctified foot-prints the new building overlooked. As the victorious Giants rolled past, their joyful supporters danced in the streets as thirty-six tons of shredded paper fluttered down upon them.

Measured in ticker tape, the return to "normal" took less than seven years.

WITH TIME, NEWS becomes history. And history, it's been said, is what happened to other people. For anyone who lived through September 11, time might dull the anger and grief that followed the death and de-struction caused when terrorists turned four commercial passenger jets into guided missiles. But the memories won't die. The pain of the deadliest terrorist attacks in American history cut too deep, leaving knots of psychic scars that make each day an experience of before and after, of adapting to a world changed physically by every secu-rity checkpoint and psychologically by every mention of the "home-land," a word seldom used in the United States prior to the events now known as 9/11. (The month-and-day abbreviation became the univer-sal shorthand for the attacks largely because the digits corresponded to the nation's 9-1-1 emergency call system; there's no evidence the terrorists chose the date for that reason.)

Already an entire generation has no direct memory of 9/11, despite its daily effects on their lives. The historian Ian W. Toll described this progression in relation to another shocking enemy assault that also led to war: the raid on Pearl Harbor, sixty years earlier. "The pas-sage of time strips away the searing immediacy of the surprise attack and envelops it in layers of exposition and retrospective judgment," Toll wrote. "Hindsight furnishes us with perspective on the crisis, but it also undercuts our ability to empathize with the immediate concerns of those who suffered through it." He quoted John H. Mc-Goran, a sailor on the doomed battleship USS *California*: "If you didn't go through it, there are no words that can adequately describe it; if you were there, then no words are necessary."

Even if words might fail, they're the best hope to delay the descent of 9/11 into the well of history. That is the purpose of this book. The approach is to recount the chaotic day as a narrative in three parts: events in the air, on the ground, and in the aftermath, focusing on individuals whose actions and experiences range from heroic to heartbreaking to homicidal. For every account included here, a thousand others are equally important. I've tried to choose stories that reveal the depth and breadth of the day without turning this into an encyclopedia. The goal is to provide a fresh perspective among readers for whom the attacks remain "news," and to create something like memories for everyone else.

Another hope is more intimate: to attach names to some of the people directly affected by these events. Of the nearly three thousand men, women, and children killed on 9/11, arguably none can be considered a household name. The best "known" victim might be the so-called Falling Man, photographed plummeting from the North Tower of the World Trade Center. Yet even he remains nameless to most people, an anonymous icon.

THIS BOOK HAS its roots in the day itself. On September 11, 2001, as a reporter for the *Boston Globe*, I wrote the lead news story about the attacks, with contributions from several dozen colleagues. The work was at once historic and local: both hijacked planes that struck the Twin Towers took flight from Boston's Logan International Airport. Five days later, with help from four reporters, I published a narrative called "Six Lives" that became the scale model for this book. It wove the stories of six people affected by, responsible for, or otherwise connected to the hijacking of American Airlines Flight 11 and the North Tower calamity. As we explained at the time, the story was designed to reveal "a nation's shared experience, as told through their memories and the memories of their loved ones. It also creates a memorial to all those who were killed, and provides a record for all who lived."

Several years ago, I discussed "Six Lives" at Boston University, where I teach journalism and where at least twenty-eight 9/11 victims earned degrees. Talking afterward with my friend and agent Richard Abate, we feared that many of my students, as well as several of our own children, felt little or no personal connection to 9/11. To some it seemed as distant as World War I. That realization triggered an idea: I could expand "Six Lives" to cover not only the first flight and the first tower, but all four flights and their unscheduled destinations, along with the ripples of physical and emotional effects. Time would serve not as an eraser but as an ally, yielding information and perspective collected in the years since 9/11 to deepen the account while keeping it accessible and truthful.

Speaking of truth, this book follows strict rules of narrative nonfiction. It takes no license with facts, quotes, characters, or chronologies. Descriptions of events and individuals rely on firsthand or authoritative accounts, checked for accuracy and cited in the endnotes where appropriate. All references to thoughts and emotions come from the individual in whose mind they arose, either from interviews, first-person reports, or other primary sources.

The attacks of 9/11 are among the most heavily covered events in history. It shouldn't be surprising, then, that some individuals featured in this book have had their stories told elsewhere. Several are subjects of entire books, among them Rick Rescorla, Welles Crowther, Father Mychal Judge, former FBI counterterrorism chief John O'Neill, and several heroes of United Flight 93. Some accounts included here rely on testimony from the 2006 trial of al-Qaeda member Zacarias Moussaoui, who pleaded guilty to being involved in the 9/11 plot. Overall, I mined information from government documents, law enforcement reports, trial transcripts, books, periodicals, documentaries, and broadcast and online works from reputable sources, credited where appropriate. Mainly, I relied on my own interviews with survivors, family and friends of the lost, witnesses, emergency responders, government officials, scholars, and military men and women.

Despite my efforts, and despite years of investigations, unanswered questions remain. Certain details and timeline elements are vague or in dispute. I have pointed out some of those gaps and disagreements in the text or in the notes. I have not included unfounded allegations or pseudoscience from the cottage industry of 9/11 conspiracy theorists. Facts are stubborn and powerful: this is a true story.

THE ESSENTIAL JOB of journalism, from daily reporting to narrative history, is to answer six fundamental questions: who, what, where, when, why, and how. Motivation being the great mystery of human existence, the most elusive is usually "why." As in: "Why did terrorists who claimed to be acting on behalf of Islam hijack commercial airliners to crash them into U.S. civilian and government targets on 9/11?"

By focusing primarily on the day itself, I've left deep exploration of that question to others. Readers inclined toward further pursuit of "why" should seek out additional works. Three worth reading are Steve Coll's excellent *Ghost Wars: The Secret History of the CIA, Afghanistan, and Bin Laden, from the Soviet Invasion to September 10, 2001*; Terry McDermott's *Perfect Soldiers: The 9/11 Hijackers: Who They Were, Why They Did It*; and Lawrence Wright's Pulitzer Prize–winning book *The Looming Tower: Al-Qaeda and the Road to 9/11*.

Wright traced the forces, philosophers, and practitioners of the 9/11 brand of *jihad*, an Arabic word that translates as "struggle." His accomplishment cannot be reduced to a few lines, but he masterfully examined the mindset of those responsible for the attacks:

Christianity—especially the evangelizing American variety—and Islam were obviously competitive faiths. Viewed through the eyes of men who were spiritually anchored in the seventh century, Christianity was not just a rival, it was the archenemy. To them the Crusades were a continual historical process that would never be resolved until the final victory of Islam.

Wright also provided insight into the men who carried out the hijackings:

Radicalism usually prospers in the gap between rising expectations and declining opportunities. . . . Anger, resentment, and humiliation spurred young Arabs to search for dramatic remedies. Martyrdom promised such young men an ideal alternative to a life that was so sparing in its rewards. A glorious death beckoned to the sinner, who would be forgiven, it is said, with the first spurt of blood, and he would behold his place in Paradise even before his death.

Of the other exceptional books about 9/11, including those cited in the Select Bibliography, several deserve acknowledgment: *The Ground Truth: The Untold Story of America Under Attack on 9/11* by John Farmer, senior counsel to the 9/11 Commission, distills how government and military officials served (and misled) the public; *The Eleventh Day: The Full Story of 9/11,* by Anthony Summers and Robbyn Swan, is an impressive synthesis of information about these events; and *102 Minutes,* by Jim Dwyer and Kevin Flynn of the *New York Times,* lives up to its subtitle: *The Unforgettable Story of the Fight to Survive inside the Twin Towers.* The 9/11 Commission's final report is an essential resource, as are the commission's voluminous staff statements, hearing transcripts, and monographs. I benefited greatly from the work of former 9/11 Commission investigator Miles Kara, who maintains the insightful website "9-11 Revisited," at www.oredigger61.org.

In the pages ahead, my goal is to fulfill the promise I made in 2001 with "Six Lives": to create a memorial to all those who were killed and to provide a record for all who survived. Plus one more: to build understanding among those who follow.

—Mitchell Zuckoff, Boston

Fall and Rise

PROLOGUE

"A Clear Declaration of War"

THIS BOOK COULD BEGIN NEARLY FOUR DECADES BEFORE 9/11, IN 1966, with Egypt's execution of a fanatically anti-Western author named Sayyid Qutb, whose writings inspired two generations of Islamist terror groups. Or further back in time, to 1918, with the defeat of the last great Muslim empire, the Ottoman sultanate. Or even further, to 1798, the year Napoleon Bonaparte occupied Egypt. Or seven hundred years before that, with the start of the Crusades. Or five hundred years before that, when Muslims believe the first verses of the Quran were revealed to the Prophet Muhammad. Or more than two thousand years earlier, with the birth of Abraham.

When it comes to historical storytelling, it's impossible for one volume to capture everything that came before. Yet a story must start somewhere. In this case, consider a relatively recent date: February 23, 1998. On that day, a shadowy forty-year-old Islamic militant named Osama bin Laden issued a *fatwa*, a furious religious decree. His edict declared war on the United States and all its citizens, wherever they or their interests could be found.

Faxed to an Arabic newspaper in London, the fatwa was signed by bin Laden, a Saudi heir to a construction fortune who was living in Afghanistan, and three other belligerent Islamic leaders, from Egypt, Pakistan, and Bangladesh. Their declaration invoked a militant interpretation of jihad that they said obligated every Muslim to violently defend holy lands against enemies. Two years earlier, bin Laden had issued a narrower fatwa, aimed at military targets, that called for the removal of American troops from Saudi Arabia: "[E]xpel the enemy, humiliated and defeated, out of the sanctities of Islam." The new fatwa went much further.

In florid language, the February 1998 fatwa asserted that three primary offenses justified a declaration of global war: (1) the presence of American military forces on the holiest lands of Islam, the Arabian Peninsula; (2) the U.S.-led war in Iraq; and (3) the United States' support of Israel, in particular its control of Jerusalem. "All of these crimes and sins committed by the Americans," the statement said, "are a clear declaration of war on Allah, his messenger, and Muslims." In response, bin Laden and his cohort issued a command: "The ruling to kill the Americans and their allies—civilians and military—is an individual duty for every Muslim who can do it in any country in which it is possible to do it. . . . We—with Allah's help—call on every Muslim who believes in Allah and wishes to be rewarded to comply with Allah's order to kill the Americans and plunder their money wherever and whenever they find it."

By the time he released his more strident fatwa, the bearded, lanky bin Laden was no stranger to American intelligence agencies. Between 1996 and 1997, U.S. officials learned that he headed his own terrorist group and was involved in a 1992 attack on a hotel in Yemen that housed U.S. military personnel. They also discovered that bin Laden had played a role in the "Black Hawk Down" shootdown of U.S. Army helicopters in Somalia in 1993 and had possibly orchestrated a 1995 car bombing in Riyadh, Saudi Arabia, that killed five Americans working with the Saudi National Guard. After the fatwa, bin Laden's threat profile rose dramatically among U.S. officials, especially when, six months later, sources blamed him for the nearly simultaneous bombings of American embassies in Nairobi, Kenya, and Dar es Salaam, in neighboring Tanzania, which killed more than two hundred people. In response to those bombings, President Bill Clinton authorized an attack using Tomahawk missiles aimed at six sites in Afghanistan. American officials believed that bin Laden would be at one of the target locations, but he had left hours earlier, apparently tipped off by Pakistani officials.

Bin Laden remained a focus of kill or capture discussions, even as a federal grand jury in New York indicted him in absentia in 1998 for conspiracy to attack U.S. defense installations. The U.S. intelligence community formally described his terror group, called al-Qaeda, or "the Base," in 1999, fully eleven years after its formation. The attention only emboldened him. Bin Laden struck again in October 2000, when a small boat loaded with explosives tore a hole in a U.S. Navy destroyer, the USS *Cole*, as it refueled off the coast of Yemen. The blast killed seventeen crew members and injured dozens more.

Yet even as they tried to keep tabs on bin Laden, even as warning signals became sirens, American political and intelligence leaders never fully grasped how determined he was to execute his fatwa with mass murder inside the United States. Despite solid clues—which intensified during the summer of 2001—and sincere investigative efforts by a small number of individuals, overall the U.S. government response to bin Laden was characterized by missed connections, squandered opportunities, and overlooked signs of impending disaster. An intelligence-gathering structure built to monitor Russian men with bad suits and nuclear warheads didn't know what to make of a fanatical Saudi in flowing robes issuing fatwas by fax machine.

Even discounting for hindsight, overwhelming evidence shows that the U.S. government's failure to anticipate the attacks of 9/11 was as widespread as it was ultimately devastating. Scores of examples prove that point, but consider one. Several months before 9/11, the head of analysis for the U.S. government's Counterterrorism Center wrote: "It would be a mistake to redefine counterterrorism as a task of dealing with 'catastrophic,' 'grand,' or 'super' terrorism, when in fact most of these labels do not represent most of the terrorism that the United States is likely to face or most of the costs that terrorism imposes on U.S. interests." Those very labels—"catastrophic," "grand," "super-terrorism"—were in fact the perfect descriptions of what was about to happen.

WHILE GOVERNMENT AND intelligence officials tried to get a handle on bin Laden before and after his February 1998 fatwa, average Americans remained largely ignorant of him and his followers. For one thing, there was bin Laden's country of residence. Among journalists, Afghanistan had long been shorthand for any subject too far away for many Americans to care about.

When bin Laden's name did appear in the American media, journalists focused mainly on his wealth. Usually he'd be described something like this: "[A] multimillionaire Saudi dissident whom the State Department has labeled 'one of the most significant financial sponsors of Islamic extremist activities in the world today.'" Rarely did news accounts suggest that he might pose a direct threat to the United States as a terrorist leader, although a 1997 article in the *New York Times* tiptoed in that direction, noting that "recent reports" indicated that bin Laden had paid for a house in Pakistan that sheltered the mastermind of a 1993 World Trade Center truck bombing that killed six people and injured more than a thousand. But in general, at the time of the fatwa it would have been easy for a well-read American to claim little knowledge and less concern about bin Laden. Before he issued his declaration of war, his name had appeared in a grand total of fifteen articles in the *New York Times*, sometimes only in passing. Most other American news organizations mentioned him less, if at all.

Even bin Laden's February 1998 fatwa against Americans passed unnoticed by most U.S. news organizations. The first clear reference in the *Times* came nearly six months later, as an offhand line in a story about a search for suspects in the bombings of the American embassies in Kenya and Tanzania: "Earlier this year, Mr. bin Laden and a group of extremist Muslim clerics called on their followers to kill Americans." The story quickly moved on, mentioning only that bin Laden was the prime suspect in the 1996 bombing of Khobar Towers, an apartment complex in Saudi Arabia, that killed nineteen American airmen. However, a *New York Times* article in 1999 about the embassy bombings reversed course, sharply downplaying the apparent threat he posed. The story read, in part:

In their war against Mr. bin Laden, American officials portray him as the world's most dangerous terrorist. But reporters for The New York Times and the PBS program "Frontline," working in cooperation, have found him to be less a commander of terrorists than an inspiration for them. Enemies and supporters, from members of the Saudi opposition to present and former American intelligence officials, say he may not be as globally powerful as some American officials have asserted.

Yet in the years before 9/11, a few journalists offered darker perspectives about bin Laden's potential ability to violently carry out his fatwa. The *Washington Post* reporter Walter Pincus wrote a pointed story two days after bin Laden's 1998 declaration of war that cited a CIA memo that said U.S. intelligence officials took the threat seriously. Another prescient outlier was ABC's John Miller, who interviewed bin Laden in May 1998 at a training camp in Afghanistan. In the interview, bin Laden repeated his fatwa and said he would not distinguish between civilian and military targets. Writing about it later, Miller ruefully acknowledged that his interview barely registered with the public: "[W]e had our little story, and a few weeks later, in a few minutes of footage, Osama bin Laden would say 'hi' to America. Not many people would pay attention. Just another Arab terrorist."

One scholar who took serious note of the fatwa was Bernard Lewis, an eminent if controversial intellectual who studied relations between Islam and the West and coined the phrase "clash of civilizations." Writing in 1998 in *Foreign Affairs* magazine, Lewis concluded:

To most Americans, the declaration [by bin Laden] is a travesty, a gross distortion of the nature and purpose of the American presence in Arabia. They should also know that for many—perhaps most—Muslims, the declaration is an equally grotesque travesty of the nature of Islam and even of its doctrine of jihad. . . . At no point do the basic texts of Islam enjoin terrorism and murder. At no point do they even consider the random slaughter of uninvolved bystanders. Nevertheless, some Muslims are ready to approve, and a few of them to apply, the declaration's extreme interpretation of their religion. Terrorism requires only a few.

Lewis's warning went largely unheeded.

In the summer of 2001, not everyone in the United States felt confident in the state of the nation, but many relished, or took for granted, the privileges of life in the last superpower at the dawn of the twenty-first century. They had enjoyed the longest uninterrupted economic boom in the nation's history, and the spread of American culture, political ideas, and business interests to the world's farthest reaches seemed destined to continue indefinitely. Almost none lost sleep over threats emanating from a cave in Afghanistan. A Gallup poll taken on September 10, 2001, found that fewer than 1 percent of Americans considered terrorism to be the nation's No. 1 concern.

But they didn't know that a countdown had already begun. Nineteen bin Laden devotees, radicalized young Arab men living in the United States, awoke on September 11, 2001, determined to fulfill the fatwa. In twenty-four hours, the poll results would change, along with everything else.

PART I

FALL

From the Sky

CHAPTER 1

"QUIET'S A GOOD THING"

September 10, 2001

CAPTAIN JOHN OGONOWSKI
American Airlines Flight 11

"Dad, I need help with my math!"

John Ogonowski's eldest daughter, Laura, called out to her father the second he stepped inside his family's farmhouse in rural Dracut, Massachusetts.

"Laura!" yelled her mother, Margaret "Peg" Ogonowski, in response. "Let him walk in the door!"

Fifty years old, six feet tall and country-boy handsome, John gazed at his wife and sixteen-year-old daughter. His smile etched deep crinkles in the ruddy skin around his blue eyes. Dinner hour was near, and Peg suspected that John felt equal parts tired and happy to be home. As darkness fell on September 10, 2001, he'd just driven from Boston's Logan International Airport after piloting an American Airlines flight from Los Angeles. A day earlier, he'd flown west on American Flight 11, a daily nonstop from Boston to Los Angeles.

After twenty-three years as a commercial pilot, John's normal routine upon returning home was to go directly to the master bedroom and strip out of his navy-blue captain's uniform with the silver stripes on the sleeves. He'd pull on grease-stained jeans and a work shirt, then head to the enormous barn on the family's 130-acre farm, located thirty miles north of Boston, near the New Hampshire border. Quiet by nature, content working with his calloused hands, John inhaled the perfume of fresh hay bales and unwound by tackling one of the endless jobs that came with being a farmer who also flew jets.

But on this day, to Peg's surprise, John broke his routine. Changing clothes and doing chores would wait. Still in uniform, he sat at the kitchen counter with Laura and her geometry problems. "Let's remember," he often told his girls, "math is fun." They'd roll their eyes, but they liked to hear him say it.

Homework finished, the family enjoyed a dinner of chicken cutlets, capped by John's favorite dessert, ice cream. Also at dinner that night were Peg's parents, visiting from New York; his father's brother Al, who lived nearby; and their younger daughters Caroline, fourteen, and Mary, eleven.

At one point, Peg noticed something missing from John's uniform shirt. "Did you go to work without your epaulets?" she asked. "I had to stop for gas," John said. He'd removed the shoulder decorations so he wouldn't look showy, like one of those pilots who seemed to expect the world to salute them.

John's modesty and quiet confidence had attracted Peg nineteen years earlier, when she was a junior flight attendant for American. John had joined the airline as a flight engineer after serving in the Air Force during the Vietnam War, when he flew C-141 transport planes back and forth across the Pacific. Some of his return flights bore flag-draped coffins. In his early years at American, John was a rare bird: an unmarried pilot, easy on the eyes, respectful to all. On a flight out of Phoenix, a savvy senior flight attendant urged Peg to speak with him. When they landed in Boston he got her number.

They were married in less than a year. By the end of the decade John had been promoted to captain, Peg had risen in seniority, and they had three daughters. All that, plus their White Gate Farm, growing hay and picking fruit from three hundred blueberry bushes and an orchard of a hundred fifty peach trees John planted himself. Every spring, they put in pumpkins and corn to sell at John's parents' farm a couple of miles down the road, where he'd learned to drive a tractor at the age of eight. Peg often joked that the classic John Deere in their barn was her pilot husband's *other* jet.

John and Peg continued to work for American Airlines throughout their marriage, with John flying a dozen days a month and Peg working about the same. They alternated flight schedules so one or the other could be with the girls. When that failed, their families pitched in. John had spent a chunk of his career flying international routes, but the overnight flights wore him down, and he'd recently been recertified on the Boeing 767, the wide-bodied pride of American's domestic fleet. Lately he'd been flying regularly on the Boston–Los Angeles route, often on Flight 11, which Peg had flown hundreds of times as well.

John was scheduled to fly again the next morning, another six-hour trip to California, but he decided he didn't want to leave home so soon after returning from the West Coast. Also, federal agriculture officials and a team from Tufts University were coming to the farm in the morning to discuss a program John felt passionate about. He and Peg had set aside a dozen acres to allow Cambodian immigrant farmers to grow bok choy, water spinach, pigweed, and other traditional Asian vegetables, to sell at markets and to feed their families. John plowed for the immigrants and rarely collected the two-hundred-dollar monthly rent. He built greenhouses for early spring planting, provided water from the farm's pond, and taught the new Americans about New England's unforgiving soil, crop-killing pests, and short planting season. Soon the Ogonowskis' White Gate Farm was designated the first "mentor farm" for immigrants. When a reporter stopped by, John heaped credit on the Cambodians: "These guys are putting more care and attention into their one acre than most Yankee farmers put into their entire hundred acres."

After dinner, John went to the desktop computer in the TV room. He logged in to the American Airlines scheduling system, hoping that another pilot wanted to pick up an extra trip. A match would turn John's onscreen schedule green, allowing him to stay on the farm on September 11. He tried several times, with the same result each time.

"I'm just getting red lights," he told Peg.

The farm tour would go on without him, while once again John would serve as captain of American Airlines Flight 11, nonstop from Boston to Los Angeles.

PETER, SUE KIM, AND CHRISTINE HANSON
United Airlines Flight 175

In 1989, a vibrant young woman slalomed through a house party, weaving through the crowd to avoid a determined young man with red dreadlocks, freckles, and a closet stuffed with tie-dyed T-shirts. Peter Hanson was cute, but Sue Kim wasn't interested in a latter-day hippie desperate to convince her that the music of the Grateful Dead was comparable to the work of Mozart.

This sort of thing happened often to Sue, a first-generation Korean American. It made sense that a curious, intense man like Peter would meet her at a party and be smitten by her intelligence and effervescence. Sue's easy laugh made people imagine that she'd lived a charmed life. But she hadn't.

When Sue was two, her overworked parents sent her from their Los Angeles home to live with her grandmother in Korea. She returned to the United States four years later and learned that she had two younger brothers, who hadn't been sent away from their parents. Her mother died when Sue was fifteen, and she helped to raise her brothers. Later her father committed suicide after being diagnosed with cancer. Beneath her placid surface, Sue craved the bonds of a secure family.

After the house party, Peter engineered ways to see Sue again while he pursued a master's degree in business administration. When

Peter thought that he'd gained romantic traction, he cut off his dread-locks, stuffed them in a bag, and gave them to his mother, Eunice. She understood: Peter wanted to show Sue he'd be good marriage material. It marked a sharp turn toward responsibility for the free-spirited twenty-three-year-old. His parents worried that perhaps he wasn't quite ready for marriage, but he couldn't wait.

"If I don't nab her now, she won't be there," Peter told his mother. Eunice accompanied him on a shopping trip for an engagement ring. Sue said yes, accepting not only Peter but also his devotion to the Grateful Dead. Their wedding bands were antiques, handed down from the parents of Peter's father, Lee.

Peter earned an MBA from Boston University and became vice president of sales for a Massachusetts computer software company. He stayed close with his parents, with whom he'd traveled the world as a boy and occasionally enjoyed his favorite band's contact-high concerts. Even as he accepted adult responsibilities, Peter remained a prankster. One day while answering phones at the local Conser-vation Commission office where she worked, Eunice heard a stern male voice demanding permission to build a structure next to a pond on his property. Eunice calmly explained the review process and the permits needed, but the caller raged about his rights as a landowner. As the rant wore on, Eunice realized it was Peter.

Meanwhile, Sue developed into an impressive academic scientist. She'd worked her way through a biology degree at the University of California, Berkeley, then moved to Boston for a master's degree in medical sciences. With Peter's encouragement, Sue pursued a PhD in immunology, working with specially bred mice to explore the role of certain molecules in asthma and AIDS. Sue was scheduled to defend her dissertation that fall, but approval was a foregone conclusion. Her doctoral adviser envisioned Sue joining the faculty at Boston University.

Peter and Sue juggled their professional lives with taking care of their daughter, Christine, who was born in February 1999. She looked like Sue in miniature, a hug magnet with Peter's love of music. Chris-tine's middle name was Lee, for her paternal grandfather. Quietly,

Sue stocked up on pregnancy tests, hoping to give Christine a little brother and Peter's parents a grandson.

Lee and Eunice visited often from their Connecticut home. When Eunice arrived one day with a broken foot, Christine yelled, "I help you, Namma! Wait here!" She ran upstairs and returned with a colorful Band-Aid she applied to Eunice's cast. Lee found joy in watching Christine work with Peter in the yard. The little girl promised the young trees that she and her daddy would help them grow big and strong. When they said grace before meals, Christine insisted on a song from a television show about Barney the purple dinosaur: "I love you, you love me, we're a happy family. With a great big hug, and a kiss from me to you, won't you say you love me too?" If her grandparents missed a word, Christine made them start over.

Early in September, Peter needed to fly to California on business, so they decided to turn the trip into a family vacation and a visit with Sue's grandmother and brothers. The weekend before the September 11 flight, Christine told Eunice of her excitement about the upcoming trip, which included plans for an outing to Disneyland. During one phone call, Christine reported to her grandmother that she was going to California to see Mickey Mouse and Pluto. Then Christine expressed an even stronger desire: "I want to go to your house, Namma!"

On the night of September 10, Christine slept in her new big-girl bed with her favorite stuffed animal, Peter Rabbit holding a carrot. Before she left home the next morning, she'd tuck Peter under the covers, to keep him safe until she returned.

BARBARA OLSON
American Airlines Flight 77

Under the hot lights of the C-SPAN television show *Washington Journal*, host Peter Slen flipped open a copy of *Washingtonian* magazine for September 2001. The camera zoomed in to a headline, THE 100 MOST POWERFUL WOMEN IN WASHINGTON. Then it swung across the set to find conservative firebrand Barbara Olson, her telegenic smile dialed to full blast, her gleaming blond hair draped down the back of her red blazer. Slen asked Barbara: "Why are you listed as an influential political insider?"

Barbara knew perfectly well, but she answered modestly: "I don't know. That's where they put me." She changed the subject to a recent lunch where the magazine's honorees discussed who might be the first female president. Overwhelmingly, the capital's most powerful women named Hillary Clinton. Virtually alone in dissent was Barbara, who had just completed her second book lacerating the U.S. senator from New York and former First Lady.

"What does it mean to have influence in this town?" Slen asked. "How do you get it? Is it power, is it position, is it money, is it marriage?"

The question carried a sexist dagger, missed by audience members who didn't know that Barbara's husband was among the most powerful lawyers in the country: U.S. Solicitor General Ted Olson, the top legal strategist for the White House. President George W. Bush had given him the job after Olson had argued successfully before the U.S. Supreme Court to end the recount of votes in Florida from the 2000 election, a decision that led to Bush becoming president.

Barbara ignored the jab, replying with a laugh that long work paved the road to influence. She'd grown used to questions about whether a glamorous woman who drove a Jaguar and had a weakness for stiletto heels deserved her place at the center of the political world. But at forty-five, having earned a partnership in a prominent law firm, Barbara drew confidence from the knowledge that before marrying Ted, she'd been a professional ballet dancer, worked her way through law school, and prosecuted drug cases in the U.S. attorney's office in Washington. She'd also served as chief investigative counsel for the U.S. House Committee on Oversight and Government Reform.

During her five-year marriage to Ted, his third and her second, Barbara had seen her stock rise further as half of a Washington power couple. They hosted enormous parties for the conservative intelligentsia at their home in Virginia. They shared a love for Shakespeare, poetry, the opera, modern art, and their Australian sheepdogs: Reagan, for the president, and Maggie, for British prime minister Margaret Thatcher.

When the C-SPAN show took calls from viewers, Barbara's partisan nature was on full display. After lavish praise from one caller who loved her bestselling book about Hillary Clinton, *Hell to Pay*, another caller laced into Barbara for criticizing the Clintons. Weeks earlier, Barbara had apologized in the *Washington Post* for describing the former president's late mother as a "barfly who gets used by men."

The caller scolded her: "Miss Olson, you have to learn how to be more human. You're a very evil person. . . . You're not going to survive too long. You got too much hate and the devil in you."

Barbara smiled through the attack, though not as widely as before. Her blue eyes dimmed momentarily as she blinked away the criticism and the ominous prediction. "Well, we do have a First Amendment," Olson replied. "Everybody has a right to their own opinion. I don't have hate in me."

After the show ended, Barbara rushed on with her life. She needed to pack for a flight to Los Angeles, for her next performance as a face of conservatism: she was booked to appear on *Politically Incorrect with Bill Maher*. Her flight was set for Monday, September 10.

Barbara decided the schedule didn't work for her. Though it would require a dash from the airport to Maher's studio, she decided to push back her flight until the next day. Ted Olson would turn sixty-one years old on Tuesday, September 11. Before flying to California, Barbara wanted to wake up beside him, to wish him a happy birthday.

CEECEE LYLES
United Airlines Flight 93

As midnight approached on Monday, September 10, CeeCee Lyles lay on a futon bed in a tiny apartment she shared with four other United Airlines flight attendants near Newark International Airport in New Jersey. She clutched a teddy bear she'd named Lorne and talked on her cellphone to the bear's namesake, her husband, Lorne Lyles, back home in Florida.

At thirty-three, five foot seven, CeeCee had flashing brown eyes and a love of fine clothes that complemented her athletic figure. Years earlier, Lorne noticed her when each of them was taking a son to baseball practice. He nearly fell out of his car when she walked past. "Man! She is beautiful," he thought.

CeeCee had traveled a winding road to happiness with Lorne, and the cellphone was a lifeline when her work took her away from him. They'd talk for hours, often five or six times a day, sometimes as many as ten to fifteen. The comfort of the other's voice mattered as much as the subjects: their sons, two each from previous relationships; her work in airports and airplanes; his, as a police officer on the overnight shift in Fort Myers, Florida. Beyond work and kids, they'd talk about bills and chores and missing each other. As Lorne would say, they'd talk and talk, about "everything and nothing."

CeeCee had become a United flight attendant less than a year earlier, at Lorne's urging, after he recognized the emotional toll of her previous jobs, as a corrections officer in Miami and then as a police detective on the streets of Fort Pierce, Florida. When they began dating, Lorne was a police dispatcher in Fort Pierce, so to some extent they'd fallen in love over the airwaves, enchanted by the sound of each other's voice.

During her six years on the police force, CeeCee had put her good looks to use when she went undercover to portray a prostitute, but she got more satisfaction from helping women and children victimized by crime and drugs. She'd often stop by the Bible Way Soul Saving Station, where her uncle was the pastor, and she became a role model at a Christian women's shelter founded by two of her aunts. Her kindness had limits, though, replaced by toughness when dealing with criminals. CeeCee excelled in an Advanced Officer Survival course that included hand-to-hand fighting and takedown moves. Before marrying Lorne in May 2000, CeeCee picked up extra shifts and worked second and third jobs to support her sons, Jerome and Jevon, around whom her life revolved. She kept them focused on school, taught them to play baseball, and expected them to fight for loose balls on the basketball court.

Becoming a flight attendant allowed CeeCee to fulfill her dreams of traveling, meeting new people, and trading hardened criminals for the occasional drunken businessman. As a perk of the job, she and her family took sightseeing trips on days off and filled available seats on flights to Indianapolis, where Lorne's two sons, Justin and Jordan, lived with their mother. They'd done just that the previous weekend, then returned home so CeeCee's sons could be in school on Monday.

As the summer of 2001 flew past, CeeCee poured out her heart in a letter to the woman who had raised her, Carrie Ross, who was both CeeCee's adoptive mother and her biological aunt. CeeCee mentioned rough patches of her past, then wrote that she was as happy as she'd ever been. She loved her new job as a flight attendant, and she credited Ross's love and support for leading her to this high point in life.

Before flying to Newark on September 10, CeeCee squared away piles of laundry and filled the refrigerator with home-cooked meals. She hated to be away from her family, but she and Lorne didn't want to uproot from Florida to her airport base in New Jersey. So CeeCee joined a group of her fellow flight attendants, each paying $150 in monthly rent for the Newark crash pad, and bided her time until she'd earn enough seniority to gain greater control over her schedule.

The morning of Monday, September 10, Lorne drove CeeCee to the Fort Myers airport, walked her to her gate, kissed her goodbye, and began a new day of serial phone calls. CeeCee didn't reach the Newark apartment until eleven that night, and she wouldn't get much rest. She'd been assigned an early flight out of Newark, an 8:00 a.m. departure to San Francisco. Even as her energy flagged, she didn't want to stop talking with Lorne.

Two hours into their last call of September 10, which blended into their first call of September 11, CeeCee fell asleep clutching her cellphone and her teddy bear Lorne. The real Lorne hung up, certain that they'd speak again soon.

MAJOR KEVIN NASYPANY
Northeast Air Defense Sector, Rome, N.Y.

At forty-three, solidly built and colorfully profane, Kevin Nasypany had a name that rhymed with the New Jersey town of Parsippany, a military pilot's unflappable confidence, and a caterpillar mustache on a Saint Bernard's face.

On September 10, Nasypany woke with a full plate. He and his wife, Dana, had five children, three girls and two boys aged five to nineteen, and Dana was seven months pregnant. They also had a sweet new chocolate lab puppy that Nasypany had judged to be dumber than dirt. Their rambling Victorian house in upstate Waterville, New York, needed paint, the oversized yard needed care, and a half-finished bathroom needed remodeling. Plus, someone needed to close their aboveground pool for the season, a chore that Nasypany loudly proclaimed to be a royal pain in the ass.

To top it off, he had to protect the lives of roughly one hundred million Americans.

Nasypany was a major in the Air National Guard, working as a mission control commander at the Northeast Air Defense Sector, or NEADS (pronounced *knee*-ads). NEADS was part of the North American Aerospace Defense Command, or NORAD, the military organization with the daunting task of safeguarding the skies over the United States and Canada.

Protection work suited Nasypany, who'd been a leading defenseman on his college hockey team. At NEADS, he and his team stood sentry against long-range enemy bombers and intercontinental ballistic missiles sneaking past U.S. air borders, along with a catalog of other airborne dangers such as hijackings. Nasypany had joined NEADS seven years earlier, after an active duty Air Force career during which he earned the radio call sign "Nasty" and spent months aloft in a radar plane over Iraq, Kuwait, and Saudi Arabia during the First Gulf War.

On workdays, Nasypany drove his Nissan Stanza twenty-five miles to NEADS headquarters, a squat aluminum bunker that resembled a

UFO from a 1950s sci-fi movie. It was the last operating facility in a military ghost town, on the property of the decommissioned Griffiss Air Force Base in Rome, New York. The obscure location was fitting: in the grand scheme of U.S. military priorities, defending domestic skies had become something of a backwater, staffed largely by part-time pilots and officers in the Air National Guard.

Working eight-hour shifts around the clock, three hundred sixty-five days a year, Nasypany and several hundred military officers, surveillance technicians, communications specialists, and weapons controllers huddled in the green glow of outdated radar and computer screens. Bulky tape recorders preserved their spoken words as they kept a lookout for potential national security threats over Washington, D.C., and twenty-seven states in the Northeast, Mid-Atlantic, and Midwest.

One of the many challenges for Nasypany was to keep his crews sharp amid the daily tedium of a peacetime vigil. Entire shifts would pass with no hint of trouble, which was good for the country but potentially numbing to NEADS crews. Then, perhaps a dozen times a month, an "unknown" would appear on a radar scope, and everyone needed to react smartly and immediately, knowing that a mistake or a delay of even a few minutes theoretically could mean the obliteration of an American city.

In most of those cases, the NEADS crew would quickly identify the mystery radar dots. But three or four times a month, when initial efforts failed, NEADS staffers would carry out the most exciting part of their job: ordering the launch of supersonic military fighter jets to determine who or what had entered American airspace.

Nationally, NORAD and its divisions could immediately call upon fourteen fighter jets, two each at seven bases around the country. Those fighters remained perpetually "on alert," armed and fueled, pilots ready. The military had many more fighter jets spread among U.S. bases, but time would be needed to round up pilots and load fuel and weapons, and time would be an unaffordable luxury if America came under attack.

During the decade since the fall of the Soviet Union, America's leaders had behaved as though the airborne threat had nearly disappeared. At fourteen, the number of on-alert fighters nationwide marked a sharp drop since the height of the Cold War, when twenty-two military sites, with scores of fighter jets, were always ready to defend against a ballistic missile attack or any other threat to North America. In fact, by the summer of 2001, the number of on-alert fighter jet sites throughout the United States had been ordered to be cut from fourteen to only four, to save money, though that order had yet to be carried out.

NEADS directly controlled four of the on-alert fighter jets: two F-15s at Otis Air National Guard Base in Buzzards Bay, Massachusetts, on Cape Cod, and two F-16s at Langley Air Force Base in Hampton, Virginia.

When the fighters launched, time and again the unknown aircraft or mystery radar dot would turn out to be benign: a fish-spotting plane from Canada with faulty electronics, or a passenger jet from Europe whose pilots failed to use proper codes on their cockpit transponder, a device that sends ground radar a wealth of identifying information plus speed and altitude. In NEADS parlance, the end result would be a "friendly" plane that didn't "squawk," or properly identify itself by transponder, as a result of human or mechanical error. When the potential threat passed, the NEADS sentinels resumed their watch.

To stay ready for surprise inspections and, above all, a genuine threat by an unknown with nefarious intent, Nasypany and other NEADS officers regularly put their crews through elaborate exercises. They had one planned for September 11, with the impressive name Vigilant Guardian. The drill focused on a simulated attack by Russian bombers, with elaborate secondary scenarios including a mock hijacking by militants determined to force a passenger jet to land on a Caribbean island. Nasypany and some colleagues wanted the exercise to include a plot by terrorists to fly a cargo plane into the United Nations building in New York City, but a military intelligence officer had nixed that idea as too far-fetched to be useful.

Nasypany spent his September 10 shift preparing for the next day's exercise, but he also had to carry out a more mundane family responsibility. NEADS allowed tours by civilians, so Boy Scout troops, local politicians, and civic groups regularly clomped around the Operations Room, looking at the radar scopes after classified systems were switched off. In this case, his wife's sister Becky was visiting the Nasypanys from Kansas, and she'd always been curious about Kevin's work. He got approval for Becky to witness what she had long imagined to be the exciting world of national security surveillance in action.

As Nasypany toured NEADS with his wife and sister-in-law, disappointment spread across Becky's face. The nerve center of U.S. air defense didn't seem much different from the office of the air conditioning manufacturer where she worked.

"It looks like you guys don't do much," Becky said. "It's really quiet in here."

Nasypany couldn't help but smile. A good shift for NEADS, and for the nation, was eight hours of hushed monotony. "Quiet's a good thing around here," Nasypany told her. "When it starts getting loud, and people start raising their voices, that's a bad thing."

MOHAMED ATTA
American Airlines Flight 11

ZIAD JARRAH
United Airlines Flight 93

Inside a third-floor room in a middling Boston hotel, an unremarkable man prepared to move on. He pulled on a polo shirt, black on one shoulder and white on the other, and packed a flimsy vinyl Travelpro suitcase that resembled the rolling luggage preferred by airline pilots.

If not for the glare of his dark eyes, Mohamed Atta would have been easy to overlook: thirty-three years old, slim, five feet seven, clean-shaven, with brushy black hair, a drooping left eyelid, and a hard-set mouth over a meaty chin. After one night in room 308 of the Milner Hotel, Atta gathered his belongings before a final move that represented the last steps of a years-long journey that he believed would elevate him from angry obscurity into eternal salvation.

The youngest of three children of a gruff, ambitious lawyer father and a doting stay-at-home mother, Atta spent his early childhood in a rural Egyptian community. Atta's father, also named Mohamed, complained that Atta's mother pampered their timid son, making him "soft" by raising him like a girl alongside his two older sisters. Devout but secular Muslims—as opposed to Islamists, who wanted religion to dominate Egypt's political, legal, and social spheres—the family moved to Cairo when Atta was ten. While his peers played or watched television, Atta studied and obeyed his elders, a dutiful son determined to satisfy his disciplinarian father and follow the path of his intelligent sisters, on their way to careers as a doctor and a professor.

Atta graduated in 1990 from Cairo University with a degree in architectural engineering and joined a trade group linked to the Muslim Brotherhood, a political group that advocated Islamic rule and demonized the West. But his career hopes were hamstrung because he didn't earn high enough grades to win a place in the university's prestigious graduate school. At his father's urging, Atta studied English and German, and a connection through a family friend steered him toward graduate studies in Germany.

In 1992, at twenty-four, Atta enrolled at the Technical University of Hamburg-Harburg to pursue the German equivalent of a master's degree in urban planning. Some men in their early twenties from a traditional society might have viewed a cosmopolitan new home as an opportunity to expand their horizons, to explore their interests, or to rebel against a controlling father. Atta took another route, burrowing into his religion and trading his docile ways for fundamentalist fervor aimed at the West.

He shunned the pulsing social and cultural life of Hamburg, a wealthy city where the sex trade prospered alongside a thriving commercial district. He grew a beard and became a fixture in the city's most radical mosque, called al-Quds, the Arabic name for the city of Jerusalem. Most of the seventy-five thousand Muslims in Hamburg were Turks with moderate beliefs, but al-Quds catered to the small minority of Arabs drawn to extreme interpretations of Islam. The mosque's location placed the spiritual literally above the worldly: the rooms of the mosque sat atop a body building parlor in a seedy part of the city. Preachers tried to outdo one another in expressions of hatred toward the United States and Israel. Congregants could buy recordings of sermons by popular imams, including one who risked arrest under German antihate laws by declaring that "Christians and Jews should have their throats slit."

By 1998, nearly finished with his studies, Atta had surrounded himself with like-minded men who came to Germany for higher education but retreated into a radically distorted understanding of their religion.

One close confidant with whom he could engage in endless anti-American rants about the oppression of Muslims was named Marwan al-Shehhi, a native of the United Arab Emirates with an encyclopedic knowledge of Islamic scriptures. Ten years younger than Atta, Shehhi struggled in school but flourished as a fundamentalist.

Another member of Atta's inner circle was Ziad Jarrah, the only son of a prosperous family from Lebanon. Jarrah seemed an unlikely Islamic firebrand: he attended private Christian schools as a boy and later became a sociable, beer-drinking regular at Beirut discos. Jarrah found a girlfriend after he arrived in Germany, but later fell harder for the ferocious ideas he heard at al-Quds.

Along with at least one other member of their circle, the trio of Atta, Shehhi, and Jarrah decided to put their beliefs into action by waging violent jihad among Muslim separatists fighting Russians in Chechnya. While still in Germany, they connected with a recruiter for Osama bin Laden's terror group, al-Qaeda, who urged them to go first to Afghanistan, where they could receive training at jihadist camps. They reached Afghanistan in late 1999, where they pledged *bayat*, or allegiance, to bin Laden. The three well-educated men quickly drew attention from al-Qaeda's top leaders, including bin Laden himself. He'd been searching for men exactly like Mohamed Atta, Marwan al-Shehhi, and Ziad Jarrah.

In the months before the Hamburg group's arrival in Afghanistan, bin Laden had embraced the idea of a simultaneous suicide hijacking plot against the United States, and he needed certain recruits to serve as its key participants: men who possessed English language skills, knowledge of life in the West, and the ability to obtain travel visas to the United States. Known to al-Qaeda as the Planes Operation, the plot was the brainchild of a longtime terrorist named Khalid Sheikh Mohammed, who'd met bin Laden in the 1980s. Mohammed admired the murderous ambitions of his nephew Ramzi Yousef, who carried out the 1993 World Trade Center bombing. After Yousef's 1995 arrest in Pakistan, as the terrorist was flown by helicopter over Manhattan, a senior FBI agent lifted Yousef's blindfold and pointed out the World Trade Center's Twin Towers, aglow in the dark. The agent taunted his prisoner: "Look down there. They're still standing." Yousef replied: "They wouldn't be if I had enough money and explosives."

Al-Qaeda's Planes Operation sought to pick up where Yousef left off and to go much further. The plot had several iterations during its years of planning, but as envisioned by Khalid Sheikh Mohammed at least as far back as 1996, jihadists would hijack ten planes and use them to attack targets on the East and West Coasts of the United States. Bin Laden eventually rejected the idea as too complex and unwieldy. He wanted a combination of high impact and high likelihood of success. In a scaled-down version, approved by bin Laden in mid-1999, the plot intended to fulfill the threat of his 1998 fatwa against the United States and its people, and to inspire others to similar action, by striking key symbols of American political, military, and financial might.

Soon after meeting Atta, bin Laden personally chose him as the mission's tactical commander and provided him with a preliminary list of approved targets. Bin Laden sent the group back to Hamburg with instructions about what to do next. To avoid attracting attention and to appear less radical, Atta shaved his beard, wore Western clothing, and avoided extremist mosques. Next, in March 2000, he emailed thirty-one flight schools in the United States to ask about the costs of training and living accommodations, all of which would secretly be covered by wire transfers from al-Qaeda. Before they applied for visas to the United States, Atta, Shehhi, and Jarrah each claimed that he had lost his passport; their replacements eliminated evidence of potentially suspicious trips to Pakistan and Afghanistan. By late May 2000, all three men had new passports and tourist visas. By late summer they were studying in Florida to be pilots, with Atta and Shehhi at one flight school and Jarrah at another.

Meanwhile, both before and after the Hamburg group began flight school, sixteen other men who'd also pledged their lives to bin Laden and al-Qaeda entered the United States to play roles chosen for them in the Planes Operation. One, a twenty-nine-year-old Saudi named Hani Hanjour, had studied in the United States on and off for nearly a decade and had obtained a commercial pilot certificate in April 1999. While in Arizona, Hanjour fell in with a group of extremists, and by 2000 he was an al-Qaeda recruit in Afghanistan, where his flying and language skills, plus his firsthand knowledge of the United States, made him an ideal candidate in bin Laden's eyes to join the Planes Operation as a fourth pilot.

Thirteen of the others were between twenty and twenty-eight years old, all from Saudi Arabia except for one, who hailed from the United Arab Emirates. A few had spent time in college, but most lacked higher education, jobs, or prospects. All but one were unmarried. Like the Hamburg group, they'd joined al-Qaeda originally intending to fight in Chechnya. Bin Laden handpicked them for the plot and asked them to swear loyalty for a suicide operation. Although they weren't especially imposing, most no taller than five foot seven, he wanted them to serve as "muscle" for the men who were training to be pilots. Most returned home to Saudi Arabia to obtain U.S. visas, then returned to Afghanistan for training in close-quarters combat and knife killing skills. They began to arrive in the United States in April 2001, keeping to themselves and generally avoiding trouble.

The other two "muscle" group members originally were supposed to participate in Khalid Sheikh Mohammed's ten-plane plot. Experienced jihadists who'd fought together in Bosnia, Nawaf al-Hazmi and Khalid al-Mihdhar arrived in California on six-month tourist visas in January 2000, even before the Hamburg group began pilot training.

The U.S. intelligence community identified Mihdhar as a member of al-Qaeda before he landed in the United States, and Hazmi had been described as a bin Laden associate. Yet neither was on a terrorist watchlist available to border agents. By contrast, other countries had both Mihdhar and Hazmi on watchlists. Once the two men reached the United States, the CIA withheld from the FBI crucial information about them and their movements. Compounded by what a later investigation would call "individual and systemic failings" by the FBI, the result was a series of missed opportunities.

Once in the United States, the two natives of Mecca, Saudi Arabia, insinuated themselves into the Muslim community of San Diego and received help from fellow Saudis. Originally viewed by bin Laden as potential pilots, neither Mihdhar nor Hazmi had the necessary English language skills. Aptitude and intelligence might have been lacking, too—their flight training stalled after they told an instructor they wanted to learn how to fly a plane but showed no interest in takeoffs or landings.

During the spring and summer of 2001, as part of their final preparations, Atta, Jarrah, and Shehhi took cross-country flights to observe the workings of crews and to determine whether they might smuggle weapons on board. Atta flew to Spain to brief an al-Qaeda planner about the plot, then returned to the United States. Jarrah and Hanjour sought training on how to fly a low-altitude pathway along the Hudson River that passed New York landmarks including the World Trade Center, and they rented small planes for practice flights. "Muscle" group members busied themselves training at gyms.

As months passed, bin Laden became frustrated, pressuring Khalid Sheikh Mohammed to put the Planes Operation into motion. Bin Laden wanted it to be executed in May 2001, marking seven months since the bombing of the USS *Cole*, and then in June or July, when Israeli opposition party leader Ariel Sharon visited the White House. Each date passed as Atta hesitated to commit on timing until he felt absolutely ready.

Finally, in late August, Atta picked a day just weeks away: the second Tuesday in September. It's a mystery whether he made a simple logistical choice, based on his expectation that it would be a light travel day, which meant fewer passengers to deal with; whether he saw propaganda value in a date that matched America's 9-1-1 emergency telephone system; or whether he sought historical revenge by choosing the month and day of the start of the 1683 Battle of Vienna, a humiliating defeat for the Ottoman Empire against Christian forces that began a centuries-long decline of Islamic influence.

Whatever the trigger, Atta and his eighteen associates started buying flight tickets, some by using computers in public libraries. They kept enough money for expenses, then returned much of the rest to al-Qaeda operatives in the United Arab Emirates. All told, the entire plan cost less than half a million dollars.

The members of the Planes Operation broke into three groups of five and one group of four, each led by one of the four men who'd trained as a pilot: Atta, Shehhi, Hanjour, and Jarrah. By the second week of September, all had rented rooms at hotels or motels in or near Boston, Newark, and Washington, D.C.

Atta and the other pilots worked on final details, while some of the others focused on earthly desires. In Boston, "muscle" members Abdulaziz al-Omari and Satam al-Suqami paid for the company of two women from the Sweet Temptations escort service. One spent a hundred dollars on a prostitute two more times in a single day. In New Jersey, another paid twenty dollars for a private dance in the VIP room of a go-go bar, while another contented himself with a pornographic video.

On September 10, when everything and everyone was almost in place, Ziad Jarrah stepped outside a Days Inn in Newark, New Jersey, where he and three "muscle" men had checked in the previous day.

Jarrah's thoughts wandered to his girlfriend in Germany, a medical student of Turkish heritage named Aysel Sengün. They'd dated for five years, they emailed or spoke by phone almost daily, and she'd visited him in Florida eight months earlier. Jarrah showed off his new skills as a pilot, flying her in a single-engine plane to Key West. They'd discussed a future together, but Sengün's parents insisted that she marry a fellow Turk. When Jarrah asked for her father's blessing, the elder

Sengün threw Jarrah out of his house. They continued their relationship in secret, and weeks earlier, Jarrah had flown to Germany to see her. Over their years together, she'd watched as the happy-go-lucky man she met grew a beard and criticized her for being insufficiently devout, but more recently Sengün had been seeing what she thought was a return to his easygoing ways.

Jarrah left the Days Inn in a rented car and drove three miles to Elizabeth, New Jersey, to mail a letter he wrote that day to Sengün. He placed it in a package along with his private pilot's license, his pilot logbook, and a postcard showing a photo of a beach.

In a mix of German and Arabic, the letter began with expressions of love and devotion to *chabibi*, or "darling." Before signing it "Your man forever," Jarrah wrote:

> *I will wait for you until you come to me. There comes a time for everyone to make a move. . . . You should be very proud of me. It's an honor, and you will see the results, and everybody will be happy. . . .*

While Jarrah mailed his package, Atta prepared to leave his room at the Boston hotel. Some items in his Travelpro luggage made sense for a devout Muslim who'd received a commercial pilot's license nine months earlier: alongside a Koran and a prayer schedule, he packed videotaped lessons on how to fly two types of Boeing jets; a device for determining the effect of a plane's weight on its range; an electronic flight computer; a procedure manual for flight simulators; and flight planning sheets. Anyone who knew what he had planned would also have noted that he packed a folding knife and a canister of "First Defense" pepper spray. Finally, tucked into the black suitcase was a four-page letter, handwritten in Arabic, that charted Atta's physical and spiritual intentions.

Divided into three sections, the letter provided detailed instructions and exhortations on the subjects of martyrdom and mass murder. It covered demeanor and grooming, battle tactics, and the promise of eternal life in the company of "nymphs." After formal invocations, "In the name of Allah, the most merciful, the most compassionate," the first section addressed Atta's situation at that very moment. Titled "The Last Night," it began:

1. Embrace the will to die and renew allegiance.
   -*Shave the extra body hair and wear cologne.*
   -*Pray.*
2. Familiarize yourself with the plan well from every aspect, and anticipate the reaction and resistance from the enemy.
3. Read the Al-Tawbah [Repentance], the Anfal chapters [in the Quran], and reflect on their meaning and what Allah has prepared for the believers and the martyrs in Paradise.

Near the end of the first section, it offered this direction:

13. Examine your weapon before departure, and it was said before the departure, "Each of you must sharpen his blade and go out and wound his sacrifice."

Among the nineteen men with plans to wreak havoc in the next twenty-four hours, at least two others, one in New Jersey and the other outside Washington, had copies of the same letter.

Atta left the Milner Hotel and drove a rented blue Nissan Altima to a cheap hotel in the Boston suburb of Newton. There he picked up the man believed to have written, or at least copied, the instruction letter: Abdulaziz al-Omari, the same young Saudi who days earlier had ordered prostitutes like delivery pizza.

Atta and Omari headed toward Interstate 95 for a two-hour drive to Portland, Maine, a trip that could best be described as the first arc of a circular route. They held tickets for a commuter flight that would bring them back to Boston. The flight was scheduled to leave Portland at six o'clock the following morning, September 11.

CHAPTER 2

"HE'S NORDO"

American Airlines Flight 11

September 11, 2001

AMERICAN AIRLINES PILOT JOHN OGONOWSKI ROUSED HIMSELF BE-fore dawn on September 11, 2001, moving quietly in the dark to avoid waking his wife, Peg, or their three daughters. He slipped on his uniform and kissed Peg goodbye as she slept.

As the sun began its rise on that perfect late-summer morning, John stepped out the back door. Coffee would wait until he reached Boston's Logan International Airport, forty-five minutes away. He climbed into his dirt-caked green Chevy pickup, with hay on the floor and a bumper sticker that read THERE'S NO FARMING WITHOUT FARMERS.

John drove a meandering route as he left the land he loved. He could see the plots he'd set aside for the Cambodian immigrants, plus five acres of ripening pumpkins and ten acres of fodder corn whose stalks would be sold as decorations for Halloween and Thanksgiving. John steered down the long dirt driveway, through the white wooden gate that gave the farm its name. He passed the home of his uncle Al and tooted his horn in a ritual family greeting. It was nearly six o'clock.

Under sparkling blue skies, John drove southeast toward the airport, ready to take his seat in the cockpit and his place in a vast national air transport system that flew some 1.8 million passengers daily, aboard more than twenty-five thousand flights, to and from more than 563 U.S. airports.

He expected to be home before the weekend, for a family picnic.

AS JOHN OGONOWSKI neared the airport, Michael Woodward left his sleeping boyfriend at his apartment in Boston's fashionable Back Bay neighborhood and caught an early train for the twenty-minute ride to Logan. More than six feet tall and 200-plus pounds, Michael had a gentle face and a razor wit. Thirty years old, bright and ambitious, he'd risen from ticket agent to flight service manager for American Airlines, a job in which he ensured that planes were properly catered, serviced, and equipped with a full complement of flight attendants.

A salty breeze from Boston Harbor greeted Michael when he exited the train at the airport station, but that was the last he expected to see of the outdoors until the end of a long day. At 6:45 a.m., dressed in a gray suit and a burgundy tie, Michael walked to his office in the bowels of the airport's Terminal B, one level below the passenger gates. He wore a serious expression that revealed his discomfort.

Michael remained friends with many of the more than two hundred flight attendants he supervised, and now he had to scold one to get to work on time, keep her uniform blouse properly buttoned, and generally clean up her act or risk being fired. He called her into his office, took a deep breath, and delivered the reprimand. She accepted the criticism and Michael relaxed, confident that he had completed the worst part of his September 11 workday.

Outside Michael's office, flight attendants milled around a no-frills lounge where airline employees grabbed coffee and signed in by computer before flights. Michael brightened when he saw Betty Ong, a fourteen-year veteran of American Airlines whose friends called her Bee, sitting at a desk in the lounge, enjoying a few minutes of quiet before work.

Tall and willowy, forty-five years old, with shoulder-length black hair, Betty had grown up the youngest of four children in San Francisco's Chinatown, where her parents ran a deli. Betty loved Chinese

opera, carousel horses, Nat King Cole music, and collecting Beanie
Babies; she also excelled at sports. Betty walked with a lively hop in
her step and had a high-pitched laugh that brought joy to her friends.
She ended calls: "I love you lots!" After countless flights together,
she'd grown friendly with pilot John Ogonowski and his flight attendant wife, Peg, who often drove Betty home from Logan Airport to
her townhouse in suburban Andover, Massachusetts, not far from
the Ogonowskis' farm. Single after a breakup, between flights Betty
acted like a big sister to children who lived in her neighborhood and
took elderly friends to doctors' appointments. Betty had returned to
Boston the previous day on a flight from San Jose, California. Now
she was back at work, piling up extra trips before a Hawaiian vacation
later in the week with her older sister, Cathie.

Michael scanned the room and saw Kathleen "Kathy" Nicosia, a
green-eyed, no-nonsense senior flight attendant whom he'd taken
to dinner recently in San Francisco. Kathy had spent thirty-two of
her fifty-four years working the skies, and she'd developed a healthy
skepticism about managers, a skepticism that somehow didn't include
Michael. He walked over and she gave him a hug. A whiff of her perfume lingered after Kathy and Betty headed upstairs to the passenger
gates.

AROUND 7:15 A.M., on the tarmac outside the terminal at Gate 32,
Logan Airport ground crew member Shawn Trotman raised his fuel
nozzle and inserted it into an adapter underneath a wing of a wide-
bodied Boeing 767. The silver plane had rolled into place a little more
than an hour earlier, after an overnight flight from San Francisco. It
stretched 180 feet long, with red, white, and blue stripes from nose to
tail. The word "American" spanned the top of the first-class windows.
Bold red and blue A's, separated by a stylized blue eagle, adorned a
flaglike vertical stabilizer on the tail.

The work done, Trotman snapped shut the fueling panel. The
plane's two enormous wing tanks sloshed with highly combustible
Jet A fuel—essentially kerosene refined to burn more efficiently—for

the six-hour flight across the country. Trotman had filled the wings with fuel weighing 76,400 pounds, about the same weight as a forty-foot fire truck.

As Trotman moved on to another plane, the ground crew finished loading luggage and delivering catering supplies. While they worked, John Ogonowski walked under the plane to inspect the landing gear, part of a pilot's routine preflight check.

Meanwhile, inside the 767, flight attendant Madeline "Amy" Sweeney was upset. Blond and blue-eyed, thirty-five years old, Amy had recently returned to work after spending the summer at home with her two young children. This would be the first day she wouldn't be on hand to guide her five-year-old daughter Anna onto the bus to kindergarten. Amy used her cellphone to call her husband, Mike, who comforted her by saying she'd have plenty of days ahead to see their kids off to school.

AMY SWEENEY, BETTY Ong, and Kathy Nicosia were three of the nine flight attendants, eight women and one man, who'd be working with Captain John Ogonowski and First Officer Thomas McGuinness Jr., a former Navy fighter pilot. They were the eleven crew members of American Airlines Flight 11, a daily nonstop flight to Los Angeles with a scheduled 7:45 a.m. departure.

Boarding moved smoothly, a process made easier by the wide-bodied plane's two aisles between seats and a light load of passengers. The youngest passenger through the cabin door was twenty-year-old Candace Lee Williams of Danbury, Connecticut, a Northeastern University student and aspiring stockbroker en route to visit her college roommate in California. The oldest was eighty-five-year-old Robert Norton, a retiree from Lubec, Maine, heading west with his wife, Jacqueline, to attend her son's wedding.

Daniel Lee from Van Nuys, California, a roadie for the Backstreet Boys, had slipped away from the pop group's tour and bought a ticket on Flight 11 so he could be home for the birth of his second child. Cora Hidalgo Holland of Sudbury, Massachusetts, needed to interview health aides for her elderly mother in San Bernardino, California. Actress and photographer Berry Berenson, widow of the actor Anthony Perkins, was headed home to Los Angeles after a vacation on Cape Cod.

Also on board was seventy-year-old electronics consultant Alexander Filipov of Concord, Massachusetts, a gregarious, insatiably curious father of three. He knew how to say "Do you like Chinese food?" in more than a dozen languages, which enabled Filipov to strike up conversations with foreigners on business trips like this one. Nearby, missing his wife, Prasanna, after a three-week work trip, Los Angeles computer technician Pendyala "Vamsi" Vamsikrishna called home and left a message saying he'd be there for lunch.

As boarding continued, thirty-year-old Tara Creamer walked down the aisle and slid into window seat 33J. Tara was a woman who didn't rattle easily. Years earlier, on a first date, on Valentine's Day no less, a fellow college student took her to dinner at a red sauce Italian dive called Spaghetti Freddy's, then to the cannibal-versus-serial-killer movie *The Silence of the Lambs*. She married him.

Tara had met John Creamer at the University of Massachusetts Amherst when she was a vivacious, curly-haired brunette sophomore. John was a shy, blue-eyed football lineman for the UMass Minutemen, named for the Revolutionary War patriots. Tara lived on a dorm floor near some of John's friends, so he hung around her hallway long enough for her to notice him. After that first date of pasta and fava beans, they were a couple.

Both twenty-three when they wed, Tara and John had scraped together a down payment for a sweet yellow Cape Cod–style house with a screened porch in Worcester, Massachusetts, not far from where John's parents lived. They renovated it themselves after long days of work. After one seemingly endless paint-and-wallpaper binge, John lost patience. While he stewed, Tara, several months pregnant, calmly went to their unfinished basement, carrying a wide brush dripping white paint. On a rough gray wall, she painted "Tara ♥s John." The tension passed, but the sign and the sentiment endured. At night, they slept under a maroon, pink, and white quilt made by Tara's aunt, with a design of interlocking circles that symbolized their wedding rings. Written on the soft cloth were the words "Made with Love for Tara and John," and their wedding date, August 13, 1994.

Their son, Colin, arrived in 1997, followed three years later by a daughter, Nora. A fashion merchandising major in college, Tara became a planning manager at TJX Companies in Framingham, Massachusetts, parent firm of the big-box retailer T.J.Maxx. She sang to Colin and Nora on the way to work and spent lunch hours with them in an onsite company daycare center. Tara meticulously updated their milestone books, recording first crawls and first steps, first teeth and first words. When a company supervisor urged Tara to get into the habit of working late, Tara declined. Time with her children came first.

Tara and John never traveled together by air in the years after Colin's birth. She worried about leaving him, and then both Colin and Nora, in the event of a crash. Tara's mother died of cancer in 1995, and the loss still ached. But in May 2001, one of John's closest friends invited them to his wedding in Florida. Before the trip, Tara went on a planning spree, arranging insurance, guardians, and family finances, just in case. She also took the opportunity to explain the concept of death to Colin. He had only one grandmother, Tara told him, because the other one, her mother, was an angel in heaven, looking after him. Colin seemed to understand, but Tara couldn't be sure.

By late summer 2001, Nora had celebrated her first birthday, and Tara was ready to resume traveling for work. On this trip, Tara had the option of staying in California through the weekend to see a close friend, but she scheduled a red-eye return so she'd be home Friday morning. Packed and ready the night before Flight 11, Tara completed one last task before bed. Fulfilling her self-appointed role as family planning manager, she typed a detailed memo for John. Titled "Normal Daily Schedule," it was a mother's guide to caring for their children. It began: "Wake Colin up around 7–7:15. Let him watch a little cartoons (Channel 52). Nora—if she is not up by 7:30—wake her up. Just change her and give her milk in a sippy cup!"

In the seat next to Tara was auburn-haired Neilie Anne Heffernan

Casey of Wellesley, Massachusetts, also a TJX Companies planning manager. Two days earlier, Neilie and her husband, Mike, had run a 5K race to raise money for breast cancer research. They ran pushing a stroller with their six-month-old daughter, Riley. In nearby rows were five of their TJX colleagues, also headed to California on business: Christine Barbuto, Linda George, Lisa Fenn Gordenstein, Robin Kaplan, and Susan MacKay.

Susan's husband, Doug, was an FAA air traffic controller. He'd switched his schedule to an early shift that day so he could attend a nighttime school event for their eight-year-old daughter and make dinner for their thirteen-year-old son. When he got to work, Doug planned to radio American Flight 11's cockpit to ask Captain John Ogonowski to surprise Susan by saying hi for him.

IN A WIDE leather seat in the first row of first class sat financier David Retik of Needham, Massachusetts, a practical-joking, fly-fishing family man whose wife, Susan, was seven months pregnant with their third child. His colleagues considered David a rare bird: a venture capitalist whom everyone liked. On his drive to the airport, David had spotted a familiar car on the Massachusetts Turnpike. He sped up, pulled alongside, and waved to his surprised father, a doctor on his way to work.

Next to David sat travel industry consultant Richard Ross, whose family in Newton, Massachusetts, counted on him to spontaneously break into Sinatra songs; to raise money for brain cancer research; and to be chronically late. He held true to form this morning, as the last passenger to arrive for Flight 11. A flustered Richard told a gate agent that terrible traffic had made this the worst day of his life. Another agent took pity and upgraded him from business to first class.

One row behind David and Richard sat retired ballet dancer and philanthropist Sonia Puopolo of Dover, Massachusetts, looking elegant with a camel-colored pashmina scarf draped over her blazer. Her luggage bulged with baby pictures and childhood mementos for a visit to her Los Angeles-based son Mark Anthony, whom she called

Mookie. A wealthy patron of the arts and Democratic Party politicians, Sonia wore a distinctive wedding band with diamonds embedded in golden columns. The bejeweled shafts looked like the support pillars of a landmark building in miniature.

Nineteen passengers settled into business class, including Paige Farley-Hackel, of Newton, Massachusetts, in window seat 7A. A glamorous spiritual adviser and budding radio host, every night Paige left a five-item "gratitude list" for her husband, Allan, with items that ranged from "justice" to "skinny dipping" to "our happy marriage" to "airplanes." Paige's appreciation lists also included the names Ruth and Juliana: her closest friend, Ruth Clifford McCourt, and Ruth's four-year-old daughter, Juliana, who was Paige's goddaughter. Paige and Ruth had met years earlier, at a day spa Ruth owned before her marriage, and they considered each other kindred spirits. That morning, a driver delivered all three to Logan Airport after a night in Paige and Allan's home. Paige, Ruth, and Juliana had planned the trip to California together, but Ruth had mileage points for free tickets on United Airlines. She and Juliana booked a separate flight on United that left Boston at nearly the same time as American Flight 11. When both planes landed in Los Angeles, they planned to drive together to La Jolla for several days at the Center for Well Being, run by Deepak Chopra. Then they intended to reward Juliana with a trip to Disneyland before flying back to Boston.

Behind Paige, bound for home in Pasadena, California, sat humanitarian Lynn Angell and her husband of thirty years, David Angell. David was an award-winning television creator and executive producer of the sitcom *Frasier* who'd won two Emmys as a writer for *Cheers*. (By coincidence, in an episode David cowrote for *Frasier*, a stranger left a telephone message for the title character saying that she'd soon arrive on "American Flight 11.")

Behind the Angells, in seat 9B, sat a young man with thinning hair in Nike sneakers, jeans, and a green T-shirt who was a star of the new computer age. Daniel Lewin of Cambridge, Massachusetts, had built a business and a fortune before his thirtieth birthday by coinventing a way for the Internet to handle enormous spikes in traffic. But at the moment, Daniel was mired in a rough patch. He was flying west to a computer conference and to sign a $400 million deal he hoped would save his company, Akamai Technologies. Daniel had already seen his formerly billion-dollar fortune plummet as Akamai's stock fell to about three dollars a share, down from a hundred times that price two years earlier. His brilliant math mind notwithstanding, Daniel defied computer nerd stereotypes: the broad-shouldered, motorcycle-riding Internet visionary had won the weightlifting title Mr. Teenage Israel and spent four years as a commando in the Israeli military.

FLIGHT 11 ALSO carried five passengers from the Middle East with no plans to reach Los Angeles. Two were Egypt-born Mohamed Atta and Saudi Arabian native Abdulaziz al-Omari, who had taken that curiously roundabout route to Flight 11. On the evening of September 10 they had driven a rented car from Boston to a Comfort Inn motel in Portland, Maine. There they visited a Pizza Hut, a Walmart, a gas station, and two ATM machines. Before dawn on September 11, they drove to the Portland International Jetport for a US Airways commuter flight back to Boston.

It was something of a mystery why Atta and Omari went to Portland only to fly back to Boston the morning of September 11. One possibility was that they thought they'd seem less suspicious that way than if they drove to Logan Airport and arrived there at the same time as a large group of other Middle Eastern men. It's also possible that they expected to be subject to less stringent security screening at a smaller airport in Maine.

Once at the Portland airport, Atta checked two suitcases: his black rolling Travelpro and a green rolling bag that apparently belonged to Omari. The green suitcase contained innocuous items including Omari's Saudi passport and his checkbook, an Arabic-to-English dictionary, three English grammar books, a handkerchief, a twenty-dollar bill, Brylcreem antidandruff hair treatment, and a bottle of perfume.

At the Portland ticket counter, Atta asked an agent for his boarding pass for their next flight, departing Boston: American Flight 11. The agent told Atta he'd have to check in a second time when he reached Logan. Atta clenched his jaw and appeared on the verge of anger. He told the agent that he'd been assured he'd have "one-step check-in." The agent didn't budge or rise to Atta's hostility. He simply told Atta that he'd better hurry if he didn't want to miss the flight. Although Atta still looked cross, he and Omari left the ticket counter for the Portland airport's security checkpoint.

At 5:45 a.m., Atta and Omari walked without incident through the metal detector, which was calibrated to detect the amount of metal in a gun or a large knife. Their black carry-on shoulder bags traveled down the moving belt and passed cleanly through the X-ray machine. Omari also carried a smaller black case that looked like a camera bag, which also didn't raise alarms. Atta wore a stern expression and clothes that resembled a pilot's uniform: dark blue collared shirt and dark pants. Omari wore a cream-colored shirt and khaki pants. After clearing security, Atta and Omari sat in the last row of the small commuter jet for the short flight to Boston.

Meanwhile, Atta's checked bags were selected for added security screening, mainly to ensure that they didn't contain explosives. The selection was made by a program implemented in 1997 called the Computer Assisted Passenger Prescreening System, or CAPPS, which used an algorithm of classified factors, weighted by a computerized formula. The system also selected some passengers on each flight at random, to minimize complaints about discrimination based on race, ethnicity, or national origin and to prevent terrorists from learning ways to avoid being chosen. The very design of the system, which targeted only passengers who checked bags, reflected the Federal Aviation Administration's woefully mistaken belief in the summer of 2001 that hijackings were a thing of the past and that sabotage, in the form of a bomb sneaked onto a plane in the luggage of a passenger who didn't board, represented the greatest threat to air travel. It's unclear what led the FAA to that conclusion, especially because sixty-four hijackings had occurred worldwide between 1996 and 2001 versus only three cases of sabotage.

The Portland jetport didn't have explosive detection equipment, and the bags weren't opened and searched. Under FAA security rules, the only requirement was that the bags be held off the plane until the person who had checked them boarded. After Atta boarded, the ground crew tossed his checked suitcases into the small plane's luggage compartment.

Under previous, stricter airport security rules, abandoned by the FAA several years earlier, passengers whose bags were selected for explosive screening would also undergo a body pat-down and a thorough search of their carry-on bags. But those measures took time, and the FAA had come under harsh criticism for long airport lines, which led to costly and frustrating delays and declines in on-time arrivals. As a result, those pat-down and bag search rules were eliminated. No one patted down Atta or Omari or searched their carry-on bags.

It's unclear what effect those added security measures might have had on Atta and Omari's plans. During the two previous months, Atta had purchased two Swiss Army knives and a Leatherman multitool with a short knife. During his trip to Spain earlier in the summer, Atta reported to his al-Qaeda contact that he and two fellow pilot trainees, Marwan al-Shehhi and Ziad Jarrah, had been able to carry box cutters onto planes during their test flights. It's unknown whether Atta or Omari carried those or other weapons through security in Portland or Boston on September 11, but even if they had, it might not have mattered to ground screeners. Federal rules in place in the summer of 2001 allowed airline passengers to carry knives with blades shorter than four inches. Although security screeners had discretion to confiscate short-bladed knives using "common sense," government studies showed persistent gaps in the performance of low-wage human screeners. They worked for the airlines and consequently were encouraged to keep security lines short and fast-moving. Screeners were supposed to conduct "random and continuous" checks of carry-on bags, but in practice that rarely happened.

Gaps in airport security went deeper than screening rules and personnel. For instance, once a would-be hijacker passed the security checkpoint, he had every reason to think he was in the clear, with no worries about being confronted on a domestic flight by an armed air marshal. In 2001, the FAA employed only thirty-three such marshals, a sharp drop from the 1970s, and they were assigned exclusively to international flights considered to be high risk. That was the case despite a statement published by the FAA just eight weeks before September 11 in the *Federal Register*: "Terrorism can occur anytime, anywhere in the United States. Members of foreign terrorist groups, representatives from state sponsors of terrorism and radical fundamentalist elements from many nations are present in the United States. Thus, an increasing threat to civil aviation from both foreign sources and potential domestic ones exists and needs to be prevented and countered."

THE US AIRWAYS flight from Portland landed in Boston at 6:45 a.m., leaving Atta and Omari plenty of time to catch American Flight 11 to Los Angeles, departing from Terminal B. They passed through security screening at Logan Airport, again without incident. Within minutes of arriving in Boston, Atta received a call on his cellphone from a pay phone in Logan's nearby Terminal C, which served planes from United Airlines, among others. The caller was a key collaborator in Atta's deadly plan.

The two men waited at Gate 32 with other passengers, but before boarding Flight 11, Atta had a strange interaction. First Officer Lynn Howland had just arrived in Boston after copiloting the red-eye flight from San Francisco on the plane that would be redesignated American Flight 11. As she walked off the 767 and entered the passenger lounge, a man she didn't know approached her and asked if she'd be flying the plane back across the country. Based on his clothing, he looked like a pilot hoping to catch a ride to Los Angeles on an available jump seat.

"No, I just brought the aircraft in," Howland told him.

The man abruptly turned his back and walked away. Later she identified him as Mohamed Atta.

As he boarded Flight 11, Atta asked a gate agent whether the two bags he'd checked earlier in Portland had been loaded onto the plane.

Atta had reason for concern about his suitcase, especially if someone familiar with Arabic decided to search it prior to Flight 11's takeoff. Inside his black rolling Travelpro bag was the handwritten instruction letter about how to prepare spiritually and logistically to hijack a plane. Even without knowledge of Arabic, a sharp screener might have grown suspicious when he or she noticed the videotaped lessons on flying Boeing jets, the other pilot gear, the folding knife, and the canister of "First Defense" pepper spray. To track down Atta's bags, the gate agent called Flight 11's ground crew chief, Donald Bennett. He reported that the two bags had arrived, but too late. His crew had already loaded and locked the big jet's luggage compartment, and the airline's desire for on-time departures prohibited reopening it so close to takeoff. Because the bags had previously passed through security, no one had reason to inspect them just because they arrived late. Atta and Omari's bags got new tags, for a later flight to Los Angeles.

At 7:39 a.m., Atta and Omari stepped aboard and found seats 8D and 8G, the middle pair of Flight 11's two-two-two business cabin seat configuration.

Already seated were Saudi Arabian brothers Wail and Waleed al-Shehri, in seats 2A and 2B, in the first row of first class. The plane didn't have a Row 1, so those seats placed them directly behind the cockpit. Their checked bags were selected for explosives screening at Logan, just as Atta's and Omari's had been in Portland. No explosives were found, and the bags were loaded aboard Flight 11. Under the new FAA rules, just like Atta and Omari, neither Shehri brother had to undergo an added personal screening such as a pat-down or carry-on search for weapons or contraband.

The fifth member of their group, Saudi native Satam al-Suqami, didn't undergo added screening, either. Shortly after Atta and Omari boarded, Suqami made his way to seat 10B, on an aisle in business class.

The five men had chosen their seats aboard Flight 11 in a way that gave them access to the aisles and placed all of them close to the cockpit. By chance, Suqami's seat in business class put him directly behind tech entrepreneur and former Israeli commando Daniel Lewin.

FLIGHT 11 HAD a capacity of 158 passengers, but as the crew prepared for takeoff, only 81 seats were filled: 9 passengers in first class, 19 in business class, and 53 in coach.

Shortly before takeoff, American Airlines flight service manager Michael Woodward walked aboard for a final check.

In first class he found Karen Martin, the Number One, or head flight attendant, who was known for running an especially tight ship. Tall and blond, forty years old and fiercely competitive, Karen was described by friends as "Type A-plus." Nearby stood thirty-eight-year-old Barbara "Bobbi" Arestegui, the Number Five attendant, petite and patient, known for her ability to calm even the most difficult passenger.

Michael asked if they were ready to go.

"Yep, everything's fine," Karen Martin said. Michael spotted his friend Kathy Nicosia, the Number Two attendant, and waved.

Before he left, Michael scanned down the aisle, almost out of habit, to see if the attendants had closed all the overhead bins. As he looked through the business section, Michael locked eyes with the passenger in seat 8D. A chill passed through him, a queasy gut feeling he couldn't quite place and couldn't shake. Something about Mohamed Atta's brooding look seemed wrong. But the flight was already behind schedule, and Michael wouldn't challenge a passenger simply for glaring at him. He turned and stepped off Flight 11, and a gate agent closed the door behind him.

Buttoned up and ready to go, crew and passengers aboard Flight 11 began the usual drill: seats upright, belts fastened, tray tables secured into place, cellphones switched off. Flight attendants buckled into jump seats. Its wings loaded with fuel, the Boeing 767 rolled back from Gate 32. Inside their locked cockpit, Captain John Ogonowski and First Officer Thomas McGuinness Jr. taxied the silver jet away from the terminal.

Cleared for takeoff, they turned onto Logan's Runway 4R and checked the wind speed and air traffic. They took flight at 7:59 a.m., becoming one of the roughly forty-five hundred passenger and general aviation planes that would be airborne all across the United States by late morning.

Moments after takeoff, the pilots made a U-turn over Boston Harbor and pointed the plane west, flying through clear skies several miles above the wide asphalt ribbon of the Massachusetts Turnpike, headed toward the New York border.

DURING THE FIRST fourteen minutes of Flight 11, pilots John Ogonowski and Tom McGuinness followed instructions from an FAA air traffic controller on the ground and eased the 767 up to 26,000 feet, just under its initial cruising altitude of 29,000 feet. They spoke nineteen times with air traffic control during the first minutes of the flight, all brief, routine exchanges, automatically recorded on the ground, mostly polite hellos and instructions about headings and altitude.

The smell of fresh-brewed coffee wafted through the cabin as flight attendants waited for the pilots to switch off the Fasten Seatbelt signs. First-class passengers would soon enjoy "silver service," provided by Karen Martin and Bobbi Arestegui, with white tablecloths for continental breakfast. Business passengers would receive similar but less fancy options from attendants Sara Low and Jean Rogér, with help from Dianne Snyder. Muffins, juice, and coffee would have to sustain passengers in coach, served by Betty Ong and Amy Sweeney, with Kathy Nicosia working in the rear galley. The lone male attendant, Jeffrey Collman, would help in coach or first class as needed. Regardless of seating class, everyone on board would be invited to watch comedian Eddie Murphy talk to animals during the in-flight movie, *Dr. Dolittle 2*.

Fifteen minutes into the flight, shortly before 8:14 a.m., the pilots verbally confirmed a radioed request from an air traffic controller named Peter Zalewski to make a 20-degree right turn. The plane turned. Sixteen seconds later, Zalewski instructed Flight 11 to climb to a cruising altitude of 35,000 feet. The plane climbed, but only to 29,000 feet. No one in the cockpit replied to Zalewski's order. Ten seconds passed.

Zalewski tried again. Soft-spoken, forty-three years old, after nineteen years with the FAA, Zalewski had grown accustomed to the relentless pressure of air traffic control. He spent his days in a darkened, windowless room at an FAA facility in Nashua, New Hampshire, known as Boston Center, one of twenty-two regional air traffic control centers nationwide. A simplified way to describe the job done by Zalewski and the 260 other controllers at Boston Center would be to call it flight separation, or doing everything necessary to keep airplanes a safe distance from one another. Zalewski's assignment called for him to keep watch on his radar screen, or "scope," for planes flying above 20,000 feet in a defined area west of Boston. When they left his geographic sector, they became another FAA controller's responsibility.

When Zalewski received no reply from the pilots of Flight 11, he wondered if John Ogonowski and Tom McGuinness weren't paying attention, or perhaps had a problem with the radio frequency. But he didn't have much time to let the problem sort itself out. He began to grow concerned that, at its current altitude and position, Flight 11 might be on a collision course with planes flying inbound toward Logan. Zalewski checked his equipment, tried the radio frequency Flight 11 used when it first took off, then used an emergency frequency to hail the plane. Still he heard nothing in response.

"He's NORDO," Zalewski told a colleague, using controller lingo for "no radio." That could mean trouble, but this sort of thing happened often enough that it didn't immediately merit emergency action. Usually it resulted from distracted pilots or technical problems that could be handled with a variety of remedies. Still, silent planes represented potential problems for controllers trying to maintain separation. As one of Zalewski's colleagues tracked Flight 11 on radar, moving other planes out of the way, Zalewski tried repeatedly to reach the Flight 11 pilots.

8:14:08 a.m.: "American Eleven, Boston."

Fifteen seconds later, he called out the same message.

Ten seconds later: "American one-one . . . how do you hear me?"

Four more tries in the next two minutes. Nothing.

8:17:05 a.m.: "American Eleven, American one-one, Boston."

At one second before 8:18 a.m., flight controllers at Boston Center heard a brief, unknown sound on the radio frequency used by Flight 11 and other nearby flights. They didn't know where it came from, and they couldn't be certain, but it sounded like a scream.

ZALEWSKI TRIED AGAIN. And again. And again. Still NORDO.

Another Boston Center controller asked a different American Airlines pilot, on a plane inbound to Boston from Seattle, to try to hail Flight 11, but that didn't work, either. That pilot reported Flight 11's failure to respond to an American Airlines dispatcher who oversaw transatlantic flights at the airline's operations center in Fort Worth, Texas.

Then things literally took a turn for the worse.

Watching on radar, Zalewski saw Flight 11 turn abruptly to the northwest, deviating from its assigned route, heading toward Albany, New York. Again, Boston Center controllers moved away planes in its path, all the way from the ground up to 35,000 feet, just in case. This was strange and troubling, but sometimes technology failed, and still neither Zalewski nor anyone else at Boston Center considered it a reason to declare an emergency.

Then, at 8:21 a.m., twenty-two minutes after takeoff, someone in the cockpit switched off Flight 11's transponder. Transponders were required for all planes that fly above 10,000 feet, and it would be hard to imagine any reason a pilot of Flight 11 would purposely turn it off.

Without a working transponder, controllers could still see Flight 11 as a dot on their primary radar scopes, but they could only guess at its speed. They also had no idea of its altitude, and it would be easy to "lose" the plane amid the constant ebb and flow of air traffic. Seven minutes had passed since the pilots' last radio transmission, after which they failed to answer multiple calls from Zalewski in air traffic control and from other planes. The 767 had veered off course and failed to climb to its assigned altitude. Now it had no working transponder. All signs pointed to a crisis of electrical, mechanical, or human origin, but Zalewski still couldn't be sure.

Zalewski turned to a Boston Center supervisor and said quietly:

"Would you please come over here? I think something is seriously wrong with this plane."

But he refused to think the worst without more evidence. When the supervisor asked if he thought the plane had been hijacked, Zalewski replied: "Absolutely not. No way." Perhaps it was wishful thinking, but it remained possible in Zalewski's mind that an extraordinarily rare combination of mechanical and technical problems had unleashed havoc aboard Flight 11.

Zalewski's mindset had roots in his training. FAA controllers were taught to anticipate a specific sign or communication from a plane before declaring a hijacking in progress. A pilot might surreptitiously key in the transponder code "7500," a universal distress signal, which would automatically flash the word HIJACK on the flight controller's green-tinted radar screen. If the problem was mechanical, a pilot could key in "7600" for a malfunctioning transponder, or "7700" for an emergency. Otherwise, a pilot under duress could speak the seemingly innocuous word "trip" during a radio call when describing a flight's course. An air traffic controller would instantly understand from that code word that a hijacker was on board. Boston Center had heard or seen no verbal or electronic tipoffs of a hostile takeover of Flight 11.

But all that training revolved around certain narrow expectations about how hijackings transpired, based on decades of hard-earned experience. Above all, those expectations relied on an assumption that one or both of the pilots, John Ogonowski and Tom McGuinness, would remain at the controls.

The idea that hijackers might incapacitate or eliminate the pilots and fly a Boeing 767 themselves didn't register in the minds of Boston Center controllers. To them, the old rules still applied. Zalewski kept trying to hail the plane.

CHAPTER 3

"A BEAUTIFUL DAY TO FLY"

United Airlines Flight 175

LEE AND EUNICE HANSON SAT IN THE KITCHEN OF THEIR BARN-RED home in Easton, Connecticut, nestled on a winding country road past fruit farms and signs offering fresh eggs and fresh manure. As they ate breakfast, the Hansons talked about their bubbly two-and-a-half-year-old granddaughter, Christine, who that morning was taking her first airplane flight. Christine would be flying from Boston to Los Angeles with her parents, Peter and Sue Kim, Lee and Eunice's son and daughter-in-law, aboard United Flight 175. Lee and Eunice spent the morning watching the clock and imagining each step along the young family's journey, turning routine stages of the trip into exciting milestones, as only loving grandparents could.

"Boy, have they got a beautiful day to fly!"

"They're probably in the tunnel on the way to the airport!"

"I bet they're boarding!"

Peter, Sue, and Christine were due home in five days, after which they planned to visit Eunice and Lee for a friend's wedding. As soon as the trio walked through the door, Lee and Eunice intended to quiz them for a minute-by-minute account of their California adventure.

The Hansons didn't know it, but that morning's flight path for United Flight 175 crossed the sky directly northwest of their property. If they had stepped away from breakfast, walked outside to their wooden back deck, and looked above the trees on their three sylvan acres, they might have spotted a tiny dot in the morning sky that was their family's plane. Lee and Eunice could have waved goodbye.

UNITED FLIGHT 175 was the fraternal twin of American Flight 11: a wide-bodied Boeing 767, bound for Los Angeles, loaded with fuel and partly filled with passengers. The two planes left the ground fifteen minutes apart.

Minutes after its 8:14 a.m. takeoff, United Flight 175 crossed the Massachusetts border and cruised smoothly in the thin air nearly six miles above northwestern Connecticut. The blue skies ahead were "severe clear," with unlimited visibility, as Captain Victor Saracini gazed through the cockpit window.

Saracini, a former Navy pilot, had earned a reputation as the "Forrest Gump Captain" for entertaining delayed passengers with long passages of memorized movie dialogue. Alongside him sat First Officer Michael Horrocks, a former Marine Corps pilot who called home before the flight to urge his nine-year-old daughter to get up for school. "I love you up to the moon and back," he told her. With that, she rose from bed.

The calm in the cockpit was broken when, more than twenty minutes after takeoff, a Boston Center air traffic controller working alongside Peter Zalewski asked the Flight 175 pilots to scour the skies for the unresponsive American Flight 11. Saracini and Horrocks's initial hunt for the American plane failed, but after a second request from air traffic control, at 8:38 a.m., the United pilots spotted the silver 767 five to ten miles away, two to three thousand feet below them.

They reported their discovery, then followed the controller's instructions to ease their jet 30 degrees to the right, to keep away from the American Airlines plane. Whatever was happening aboard Flight 11, Boston Center controllers continued to want other planes to give it wide berth.

Saracini and Horrocks acknowledged the controller's orders and turned Flight 175 to veer away from Flight 11. As they did, something nagged at the United pilots. They didn't mention it to their Boston Center air traffic controllers, but shortly after taking flight, Saracini and Horrocks had heard a strange and troubling transmission on a radio frequency they shared with nearby planes, including Flight 11.

LIKE AMERICAN FLIGHT 11, United Flight 175 had lots of empty seats, flying at about one-third capacity. The fifty-six passengers were in the hands of nine crew members: the two pilots plus seven flight attendants. Perhaps the biggest difference between the United and American flights to Los Angeles on the morning of September 11 was the sounds inside the cabins: only adults boarded Flight 11, while the high-pitched voices of young children rang through United 175.

Heading home to California from a vacation on Cape Cod, three-year-old David Gamboa Brandhorst, who had a cleft chin and a deep love of Legos, sat in business class Row 8 between his fathers. To his left sat his serious "Papa," Daniel Brandhorst, a lawyer and accountant. To his right sat his happy-go-lucky "Daddy," Ronald Gamboa, manager of a Gap store in Santa Monica.

Nestled in 26A and 26B of coach were Ruth Clifford McCourt and her four-year-old daughter, Juliana. They'd driven to the airport and spent the previous night with Ruth's best friend and Juliana's godmother, Paige Farley-Hackel, who was heading to Los Angeles aboard American Flight 11. Blond and big-eyed, with porcelain skin, Juliana loved creatures large and small. She'd tell anyone who'd listen that she had recently learned to ride a pony. That day she'd smuggled aboard an unticketed passenger: a green praying mantis she'd found in the garden of her family's Connecticut home. It resided in an ornate little cage on Juliana's lap, her companion for when they reached California. Her mother, Ruth, a striking woman who spoke with a trace of her native Ireland, carried a special item, too: a papal coin from her wedding at the Vatican, tucked safely into her Hermès wallet.

The third little passenger of Flight 175 sat in Row 19: Lee and

Eunice Hanson's granddaughter, Christine Lee Hanson, flanked by Peter and Sue. Surrounding the Hansons and the other families on United Flight 175 were a mix of business and pleasure travelers.

The Reverend Francis Grogan, heading west to visit his sister, occupied a first-class seat with a ticket given to him by a friend. After serving as a sonar expert on a Navy destroyer during World War II, Father Grogan had traded conflict for conciliation. He spent his life as a teacher, a chaplain, and a parish priest.

In business class sat former pro hockey player Garnet "Ace" Bailey, a fierce competitor who spent ten seasons in the National Hockey League and won two Stanley Cups with the Boston Bruins in the 1970s. Hardly a delicate ice dancer, Bailey served a total of eleven hours in the penalty box during his bruising NHL career. At fifty-three, still tough, Ace Bailey had become director of scouting for the Los Angeles Kings of the NHL. He'd also cemented a relationship as friend and mentor to hockey legend Wayne Gretzky, in part by helping Gretzky to overcome his fear of flying. When Ace wasn't on the road, he treated his wife, Kathy, and son, Todd, to a dish he called "Bailey-baisse," a medley of sautéed meats baked with onions and tomatoes. A few rows back, in coach, sat the Kings' amateur scout, Mark Bavis, a former hockey standout at Boston University. Training camp would soon begin, and Bailey and Bavis were needed on the ice in Los Angeles.

Retired nurse Touri Bolourchi expected to be home in Beverly Hills by the afternoon, after visiting her daughter and grandsons in Boston. An Iranian-born Muslim, she'd fled to the United States two decades earlier when the Ayatollah Ruhollah Khomeini closed the schools. Touri added to Flight 175's international mix: also aboard were three German businessmen, an Israeli woman, and a British man.

Among the crew members were two flight attendants in love: Michael Tarrou, a part-time musician, and Amy King, a onetime homecoming queen. They'd recently moved in together, and they had arranged to work the same flight so they could spend time together during a layover. All signs pointed to marriage.

Two other flight attendants had recently switched careers: Alfred

Marchand had turned in his badge and gun as a police officer to become a flight attendant a year earlier; and Robert Fangman had begun flying for United just eight months earlier. He gave up half his income and a job he hated, as a cellphone salesman, to follow his dream of international travel.

Airplanes bring together people from different worlds and worldviews. On Flight 175, that held true for two strangers, Robert LeBlanc and Brian "Moose" Sweeney, one old, one young, one a pacifist, the other a U.S. Navy veteran of the Iraq War who considered himself a warrior and imagined himself to be the descendant of Vikings. The two men sat in window seats on opposite sides of the coach cabin.

THE PREVIOUS DAY, with many miles ahead on a seven-hour drive, Robert LeBlanc gripped the steering wheel of his Audi sedan and prepared to pop the question. He was seventy years old, spry and fit, a snowy beard and tanned, craggy face giving him the look of an arctic explorer.

After a weekend visiting Rochester, New York, Bob and his driving companion were headed home to the little town of Lee, New Hampshire. A retired professor, Bob would be leaving before dawn the next day, September 11, for a geography conference on the West Coast. Now he'd reached a decision: he knew how he wanted to spend his remaining seasons, and with whom. He turned toward the woman he loved.

"I have a ten-year plan," Bob said. "I know you might not be ready, but I want you with me."

Sitting in the passenger seat, Andrea LeBlanc understood what Bob was asking. After all, she'd been married to him for twenty-eight years. Bob hoped Andrea would dramatically scale back her busy veterinary practice so they could travel the world together. Bob's ten-year plan involved "hard" trips to developing countries at the farthest corners of the globe, after which they'd ideally spend a decade visiting the "easy places." That is, if Andrea would agree.

Bob was asking a lot, and he knew it. Along with raising their children—two from her first marriage, three from Bob's—Andrea's Oyster River Veterinary Hospital had been her life's work. Nearly fourteen years his junior, Andrea would have to choose between spending the bulk of her time with her four-legged patients or with her best friend. As they drove, Bob's question hung in the air long enough for Andrea to consider the man she loved and the life they shared.

Born in 1930, Bob grew up in the French Canadian neighborhood of Nashua, New Hampshire, at the time a spent mill city. A restless boy, he often rode his bicycle downtown to see trains pull in from Montreal. Bob grew fascinated by why people lived where they lived, and how their physical world shaped their culture, from language to music, religion to livelihood, relationships to diet.

After high school, Bob enlisted in the Air Force. After a flirtation with geology at the University of New Hampshire, he earned a doctorate in cultural geography from the University of Minnesota. Then Bob returned to UNH as a professor and remained there for thirty-five years, until he retired in 1999. Along the way, he developed a reputation as a gifted teacher, frugal toward himself and generous toward others; a master cook who loved candlelit dinners; and a passionate traveler whose been-almost-everywhere map included Nepal, Bhutan, China, Morocco, Peru, South Africa, Botswana, Namibia, and Burma.

On a 1999 trip to Java, Bob led Andrea to the world's largest Buddhist temple, called Borobudur. While he explored, Andrea set off in search of a rare bas-relief panel that depicted the Buddha among animals. Ten Muslim teenagers followed her, inching closer in the hope of practicing their English. Eager to return to quiet contemplation and her search for the sculpture, she made a suggestion.

"Go look for a man with a white beard," Andrea told them. "That's my husband. He'd *love* to talk with you."

Forty minutes later, as she neared an exit, she heard gales of laughter: Bob was leading an impromptu class, asking questions, drawing his new friends into his sphere. Andrea snapped a photograph of the Muslim teens squeezed against a smiling Bob.

Two years earlier, in Chiapas, Mexico, they had watched as leftist Zapatista revolutionaries marched through the streets. Andrea asked Bob what drove the young men to take up arms. "When people aren't heard long enough," he said, "they'll resort to violence."

During their just-completed weekend in Rochester, Andrea and her daughter Nissa had wandered around a craft fair. Andrea said: "It's just so strange. I cannot even imagine being any happier."

"That is *so* weird," Nissa said, stopping in her tracks. "Dad said the same thing to me last night."

In the car, drawing closer to home, Bob waited as Andrea weighed his question about their future travels together. Reflecting on the man behind the wheel, Andrea felt that he had given her so much, asking relatively little in return. She turned to Bob with her answer: "Okay, I'll do it."

Bob woke before dawn to catch United Flight 175. As he left their bedroom, he promised Andrea that he'd call her that night. He put a copy of his California itinerary on the refrigerator, alongside newspaper clippings of recipes he intended to try. As he moved through the house, Bob looked sharp, his thick white hair freshly cut by Andrea the night before. On his desk were plans for no fewer than five trips, starting in ten days with leading a group of older travelers to Argentina, followed by jaunts to India and Norway.

Waiting outside at 5 a.m. to drive him to Logan Airport was Bob's daughter Carolyn. On the way, they became so lost in conversation they almost missed the exit. Bob loved airports the way some children love construction sites. Happy, he bounded into Terminal C, holding a boarding pass for seat 16G.

ON THE OTHER side of the aisle, the self-described warrior in 15A was named Brian David "Moose" Sweeney (no relation to Flight 11 flight attendant Amy Sweeney).

Brian grew up in the little Massachusetts town of Spencer, where nothing much had happened since Elias Howe perfected the sewing machine there in 1846. He earned a football scholarship to Boston

University, where opposing players noticed his bright blue eyes just before they saw stars. Known as Sweenz to his friends, Brian and a fellow lineman shared another nickname: the Twin Towers.

After college, Brian searched fruitlessly for a challenge, until he saw an air show display by F-14 fighter jets. He enlisted in the Navy and graduated at the top of his class to become a naval aviator. Brian served in the Persian Gulf War, enforcing the "no-fly zones" in Iraq, then taught at the Navy Fighter Weapons School, better known by its movie title name, Top Gun. He convinced himself that generations earlier, Norse warrior blood had mixed with his Irish heritage, so he fashioned a two-bladed battle-ax and a Viking helmet, complete with horns. He wore it on Halloween and whenever the mood struck.

While teaching at the Top Gun school, Brian twisted his neck during a flight maneuver and shattered two cervical disks, leaving him partially paralyzed while in midair. The military crash-and-burn team rushed out, but it left empty-handed when Brian somehow landed safely. Brian loved the Navy, but after surgery he faced an agonizing choice between a desk job and an honorable discharge. His commanding officer told him: "You have the heart of a warrior and the soul of a poet. You've proven your mettle as a warrior, now go find your spirit." Brian stayed close to military service by working as an aeronautical systems consultant for defense contracting companies.

In 1998, Brian strolled into a snooty Philadelphia bar crawling with Wall Street types in custom suits. At six feet three and a rugged 225 pounds, wearing jeans, a denim shirt, hiking boots, and a baseball cap, Brian stood out like a linebacker among jockeys. A fit, pretty young woman named Julie spotted him from across the bar. She told her friend: "That's the kind of guy that I can marry and sit in front of a fireplace in the Poconos with, and be happy." The attraction was mutual.

Brian handed her a business card that read LT. BRIAN "MOOSE" SWEENEY—INSTRUCTOR, TOP GUN FIGHTER WEAPONS SCHOOL, MIRAMAR, CALIF. Julie thought it was a gag he used to impress women in bars. It was real, if somewhat dated. Julie was, in fact, impressed, and seven months later she became Mrs. Brian Sweeney.

By the summer of 2001, Brian and Julie had bought a house in Barnstable on Cape Cod, where she'd been hired as a high school health teacher. They had two dogs, and their talks about parenthood had grown more frequent. More than two years into their marriage, the twenty-nine-year-old Julie remained awestruck by her thirty-eight-year-old husband. She admired his self-confidence; she loved how this large and powerful man had a gentle voice that calmed her; she treasured the way he made her feel safe; she marveled at the practical intelligence that enabled him to build a house, while his spiritual side gave him peaceful assurance about an afterlife.

During the weeks before September 11, they'd talked about death. Brian told Julie that if he died, she should throw a party. "You celebrate life," he said. "You invite all my friends and you drink Captain Morgan and you live. And if you find somebody, you remarry. I won't be angry or jealous or whatever."

Julie looked straight back at him and said: "Well, listen, if I ever die, you are not to do any of that. You are *not* to find anybody else."

Brian laughed. "Someday you'll figure that out."

Brian traveled to California for work one week per month, regularly aboard United Flight 175. He'd normally be gone Monday through Friday, but he'd decided to extend his summer weekend and instead leave on Tuesday, September 11.

The night before the flight, they ate Chinese food, then Brian gave himself a haircut before starting to pack. Several weeks earlier, Brian found a photograph of Julie when she was five years old, with wet hair and a goofy smile. "This is the sweetest picture I've ever seen of you," Brian told her when he discovered it. While Brian packed, Julie sneaked the photo into his suitcase, so he'd find it again when he reached California.

The morning of September 11, Julie drove Brian to the Cape Cod airport in Hyannis for a connecting flight to Boston. He was dressed in the same "Sweeney uniform" of jeans, denim shirt, work boots, and baseball cap he'd worn when they met. Brian kissed Julie, then surprised her with news that he'd be back a day early, so they could spend the last summer weekend together.

AMID THE FAMILIES, business travelers, and tourists were five Middle Eastern men who fit none of those categories. They selected seats almost exactly in the pattern Mohamed Atta and his four collaborators used aboard American Flight 11. Once again, the tactical arrangement placed members of their group close to the cockpit, while others could cover both aisles if anyone came forward to challenge them from the rear of the plane.

The first two to board United Flight 175 were Fayez Banihammad, of the United Arab Emirates, and Mohand al-Shehri, from Saudi Arabia, who sat in first-class seats 2A and 2B. Four weeks earlier, Banihammad had bought a multitool with a short blade, called a Stanley Two-Piece Snap-Knife Set.

Next came Marwan al-Shehhi, the native of the United Arab Emirates who'd met Atta and Jarrah in Hamburg and traveled with them to the al-Qaeda camps in Afghanistan and then to Florida for flight training. Nine months earlier, Shehhi had received his FAA commercial pilot certificate at the same flight school, on the same day, as Atta, who sometimes referred to Shehhi as his "cousin." On the same day as Banihammad's knife purchase, in the same city, Shehhi had bought two short-bladed knives, one called a Cliphanger Viper and the other called an Imperial Tradesman Dual Edge.

Shehhi seemed the most likely person to have made the 6:52 a.m. call to Atta's cellphone. The call was made from Logan's Terminal C, from a pay phone located between the security screening checkpoint and the departure gate for Flight 175. Based on location and timing, the three-minute call to Atta might have been a final confirmation that they were ready to move forward with their plan.

When he reached the plane, Shehhi sat in the middle of business class, in seat 6C, just as Atta had chosen a seat in the middle of business class on Flight 11.

The last two to board Flight 175, Ahmed al-Ghamdi and Hamza al-Ghamdi, possibly cousins, came from the same small town in Saudi Arabia. Hamza al-Ghamdi had purchased a Leatherman Wave multitool the same day and in the same city where Banihammad and Shehhi had purchased their knives. Whether they carried those particular knives aboard Flight 175 isn't known.

Hamza al-Ghamdi apparently took to heart the instruction in the handwritten Arabic letter to "wear cologne." Earlier that morning, his overbearing fragrance had made a lasting, unpleasant impression on the desk clerk when he checked out of the off-brand Days Hotel, a few miles from the airport. He made no better impression on the cabdriver who drove them to the airport when he left a fifteen-cent tip.

Upon their arrival at Logan's United Airlines ticket counter shortly before 7 a.m., the Ghamdis had seemed confused. One told a customer service agent that he thought he needed to buy a ticket for the flight, not realizing that he already had one. Both had limited English skills, so they had difficulty answering standard security questions about unattended bags and dangerous items. The customer service agent repeated the questions slowly, and the Ghamdis eventually gave acceptable answers. Aboard Flight 175, the Ghamdis sat together in the last row of business class, in the center two seats, 9C and 9D.

None of the five men or their luggage was chosen by the computerized system or by airport workers for additional security screenings.

CHAPTER 4

"I THINK WE'RE BEING HIJACKED"

American Airlines Flight 11

WHEN AMERICAN FLIGHT 11 TOOK OFF, FLIGHT ATTENDANT BETTY "Bee" Ong sat buckled into a jump seat in the tail section, on the left side of the plane, ready to begin her onboard routine. From that vantage point, she had a direct view up an aisle through coach and business into first class.

Less than twenty minutes after takeoff, just as she normally would have begun serving passengers breakfast, Betty witnessed the reason why Flight 11 changed direction without authorization, why someone switched off the transponder, why the cockpit stopped communicating with air traffic controller Peter Zalewski at the FAA's Boston Center, and why it didn't answer calls from other planes.

At 8:19 a.m., six minutes after Flight 11 pilots John Ogonowski and Tom McGuinness stopped responding to Zalewski's calls, Betty grabbed an AT&T telephone called an Airfone, built into the 767. Airfones were common on cross-country flights in 2001, and many planes had an Airfone, for use by passengers with credit cards, on the back of every middle seat in coach. Betty dialed a toll-free reservations number for American Airlines, a number she often used to help passengers make connecting flights. The call went through to the airline's Southeastern Reservations Office in central North Carolina, where a reservations agent named Vanessa Minter answered.

"I think we're being hijacked," Betty said, her voice calm but fearful.

Vanessa Minter asked Betty to hold. She searched for an emergency button on her phone but couldn't find one. Instead, she speed-dialed the American Airlines international resolution desk on the other side of her office and told agent Winston Sadler what Betty had said. Sadler jumped onto the call and pressed an emergency button on his phone. That allowed the airline's call system to record about four minutes of what would be a more than twenty-five-minute call from Betty that would provide crucial information about what occurred and who was responsible. Sadler also sent an alarm that notified Nydia Gonzalez, the reservations office supervisor, who also joined the call.

"Um, the cockpit's not answering," Betty said. "Somebody's stabbed in business class, and, um, I think there is Mace—that we can't breathe. I don't know, I think we're getting hijacked."

For employees of a call center who normally helped stranded travelers find new flights, Betty's call was beyond shocking. After some confusion about who Betty was and what flight she was on, during which the airline employees asked Betty to repeat herself several times, eventually they understood that Betty was the Number Three flight attendant on American Airlines Flight 11. Once that was established, Betty stammered at times as she did her best to describe a bloody, chaotic scene.

"Our, our Number One got stabbed. Our purser is stabbed. Ah, nobody knows who stabbed who and we can't even get up to business class right now because nobody can breathe. Our Number One is, is stabbed right now. And our Number Five. Our first-class passenger that, ah, first, ah, class galley flight attendant and our purser has been stabbed and we can't get to the cockpit, the door won't open. Hello?"

She remained polite and self-possessed, even as her throat tightened with fear. Betty repeated herself several more times in response to the questions of reservation office employees.

Supervisor Nydia Gonzalez asked if Betty heard any announcements from the cockpit, and Betty said there had been none.

Two minutes into Betty's call, at 8:21 a.m., Gonzalez called Craig
Marquis, the manager on duty at American Airlines' operations con-
trol headquarters in Fort Worth, Texas, to report an emergency aboard
Flight 11, with stabbings and an unresponsive cockpit.

Meanwhile, Betty turned to other flight attendants clustered around
her at the back of the plane: "Can anybody get up to the cockpit? Can
anybody get up to the cockpit?" Then she returned to the call: "We
can't even get into the cockpit. We don't know who's up there."

At that point, reservations agent Winston Sadler displayed the widely
held but tragically mistaken belief that only the airline's pilots could
fly a Boeing 767. "Well," Sadler told Betty, "if they were shrewd"—
meaning the original crew—"they would keep the door closed and . . ."

Betty: "I'm sorry?"

Sadler: "Well, would they not maintain a sterile cockpit?"

Betty: "I think the guys [hijackers] are up there. They might have
gone there, jammed their way up there, or something. Nobody can
call the cockpit. We can't even get inside."

Sadler went silent.

Betty: "Is anybody still there?"

Sadler: "Yes, we're still here."

Betty: "Okay. I'm staying on the line as well."

Sadler: "Okay."

Nydia Gonzalez returned to the call. After asking Betty to repeat
herself several times, Gonzalez asked: "Have you guys called anyone
else?"

"No," Betty answered. "Somebody's calling medical and we can't
get a doc—"

The tape ended, but the call continued for more than twenty minutes as Nydia Gonzalez and Vanessa Minter took notes and relayed
information from Betty to Craig Marquis at the airline's control headquarters in Fort Worth. Throughout, Gonzalez reassured Betty, urging her to stay calm and telling her she was doing a wonderful job.

"Betty, how are you holding up, honey?" Gonzalez asked. "Okay. You're gonna be fine. . . . Relax, honey. Betty, Betty."

Several times Betty reported that the plane was flying erratically, almost turning sideways.

"Please pray for us," Betty asked. "Oh God . . . oh God."

EVEN AS HIS anxiety rose about American Flight 11, Boston Center air traffic controller Peter Zalewski knew nothing about Betty Ong's anguished, ongoing call. No one from American Airlines' Fort Worth operations control headquarters relayed information to the FAA's Command Center in Herndon, Virginia, to FAA headquarters in Washington, or to anyone else. As minutes passed and commandeered Flight 11 flew west across Massachusetts and over New York, communications among the airline, the FAA, and U.S. military officials were sporadic at best, incomplete or nonexistent at worst.

Adding to the stress, Zalewski couldn't devote his entire attention to the troubled American Airlines flight. Other planes continued to take off from Logan Airport and enter Zalewski's assigned geographic sector. One of those flights was United Airlines Flight 175. For eleven minutes, an unusually long time, Zalewski had no contact with Flight 11.

Then, at 8:24 a.m., five minutes after the start of Betty Ong's ongoing call to American Airlines' reservations center, Zalewski heard three strange clicks on the radio frequency assigned to Flight 11 and numerous other flights in his sector.

"Is that American Eleven, trying to call?" Zalewski asked.

Five seconds passed. Then Zalewski heard an unknown male voice with a vaguely Middle Eastern accent. Zalewski handled a great deal of international air traffic, so an Arab pilot's voice wasn't entirely unexpected. The unknown man's radio message wasn't clear, and Zalewski didn't comprehend it.

Unknown at that point to anyone at Boston Center, the foreign-sounding man, almost spitting his words directly into the microphone, had said: "We have some planes. Just stay quiet, and we'll be okay. We are returning to the airport."

The comment apparently wasn't intended for Zalewski or other FAA ground controllers. Rather, it sounded like a message from the cockpit intended to pacify Flight 11's passengers and crew, none of whom heard it. The person in the pilot's seat—almost certainly Mohamed Atta—keyed the mic in a way that transmitted the message to air traffic control on the ground, as well as to other planes using the same radio frequency, and not to passengers and crew in the cabin behind him. To have been heard inside the plane, the hijacker-pilot would have needed to flip a switch on the cockpit radio panel.

At a time when every piece of information counted, and every minute was crucial, the fact that Zalewski didn't comprehend that chilling message marked a major misfortune on a day filled with them. The first sentence of the hijackers' first cockpit transmission at 8:24:38 a.m. not only announced the terror aboard American Flight 11, it included a seemingly unintentional warning about an unknown number of similar, related plots already in motion, but not yet activated, on other early-morning transcontinental flights. Whoever was flying Flight 11 didn't simply say that he and his fellow hijackers had seized control of *that* plane. He said: "We have *some planes.*"

If the message had been understood immediately, the plural use of "planes" conceivably might have prompted Zalewski and other air traffic controllers to warn other pilots to enforce heightened cockpit security. Those pilots, in turn, might have told flight attendants to be on guard for trouble. But that's a best-case scenario. It's also possible that the comment would have been overlooked or dismissed as an empty boast or downplayed as a misstatement by a hijacker with limited English skills. There was no way to know, because Zalewski didn't catch it.

Zalewski answered: "And, uh, who's trying to call me here? . . . American Eleven, are you trying to call?"

Seconds later, Zalewski heard another communication from the cockpit, also apparently intended for the passengers and crew of Flight 11:

"Nobody move. Everything will be okay. If you try to make any moves, you will injure yourselves and the airplane. Just stay quiet."

Zalewski heard that message loud and clear. He screamed for his supervisor, Jon Schippani: "Jon, get over here right now!"

Zalewski announced to the room of flight controllers that Flight 11 had been hijacked. Feeling ignored, as though not everyone at Boston Center appreciated the urgency, Zalewski flipped a switch to allow all the air traffic controllers around him to hear all radio communications with Flight 11. He handed off his other flights to fellow controllers. All the while, Zalewski wondered what essential information he might have missed in the first radio transmission. On the verge of panic, Zalewski turned to another Boston Center employee, a quality assurance supervisor named Bob Jones.

"Someone has to pull these fucking tapes—right now!" Zalewski told Jones.

Jones rushed to the basement to find the recording on the center's old-fashioned reel-to-reel recording machines so he could decipher the hijacker's first message.

Zalewski's first thought was that the hijackers of Flight 11 might make a U-turn and return to Logan Airport, putting the plane dangerously in the path of departing westbound flights. But the radicals in the cockpit had another destination in mind.

The Boeing 767 turned sharply south over Albany, New York. Its flight path followed the Hudson River Valley in the general direction of New York City at a speed of perhaps 600 miles per hour. Even if the plane slowed somewhat, it could fly from Albany to Manhattan in as little as twenty minutes.

Between 8:25 and 8:32 a.m., Boston Center managers alerted their superiors within the FAA that American Flight 11 had been hijacked and was heading toward New York City. Zalewski felt what he could only describe as terror.

Yet just as American Airlines employees failed to immediately pass along information from Betty Ong's call, more than twelve minutes passed before anyone at Boston Center or the FAA called the U.S. military for help.

One explanation for the delay was a hardwired belief among airline, government, and many military officials that hijackings followed a set pattern, in which military reaction time wasn't the most important factor. The established playbook for hijackings went something like this: Driven by financial or political motives, such as seeking asylum, ransom, or the release of prisoners, hijackers took control of a passenger plane. Once in command, they used the radio to announce their intentions to government officials or media on the ground. They ordered the airline's pilots to fly toward a new destination, using threats to passengers and crews as leverage. Eventually the hijackers ordered the pilots to land so they could refuel, escape, arrange for their demands to be met, or some combination. Under those circumstances, the appropriate, measured response from ground-based authorities was to clear other planes from the hijacked plane's path and to seek a peaceful resolution that would protect innocent victims.

If the takeover of Flight 11 followed that "traditional" hijacking approach, a delay of a few minutes when sharing information shouldn't have been a significant problem. There would have been plenty of time to seek military help or assistance from the FAA once the hijackers issued demands and announced a destination. But this hijacking didn't follow "normal" rules. No demands were forthcoming, and no one in contact with Flight 11 anticipated that hijackers might kill or incapacitate the pilots and fly the plane.

Meanwhile, American Airlines employees at the airline's control center in Texas tried multiple times, including at 8:23 a.m. and 8:25 a.m., to reach the original Flight 11 pilots. They used a dedicated messaging system that linked the ground and the cockpit, known as the Aircraft Communications Addressing and Reporting System, or ACARS.

"Plz contact Boston Center ASAP," one ACARS message read. "They have lost radio contact and your transponder signal."

Flight 11 didn't reply.

AS FLIGHT 11 flew erratically through the sky, flight attendant Amy Sweeney sat in a rear jump seat next to Betty Ong. Amy had called her husband an hour earlier, upset about missing their daughter's sendoff to kindergarten. Now she tried to call the American Airlines flight services office in Boston with horrific news.

After two failed tries, Amy sought help from fellow flight attendant Sara Low, a high-spirited, athletic young woman with a pixie haircut who'd left a job at her father's Arkansas mining company to satisfy her desire for adventure. Sara gave Amy a calling card number that allowed her to charge the call to Sara's parents.

On her third try, at 8:25 a.m., Amy got through to Boston and reported that someone was hurt on what she mistakenly called Flight 12, an error that Betty also made early in her call.

A manager on duty, Evelyn "Evy" Nunez, asked for more details. "What, what, what? . . . Who's hurt? . . . What?" She got some information, but the call was cut off. Overhearing the loud conversation, flight services manager Michael Woodward asked what was happening. Nunez said she'd received a strange call about a stabbing on Flight 12.

The report was confusing, so Michael and another Boston-based American Airlines employee ran upstairs to Logan's Terminal B gates to see if there was maybe a case of "air rage" on a parked plane, or a violent person wandering drunk in the terminal. But all was quiet, and all morning flights had already left. Then it dawned on him.

"Wait a minute," Michael told his colleague. "Flight 12 comes in at night. It hasn't even left Los Angeles yet."

They rushed back to the office, where Michael learned that another emergency call had come in. This time they quickly understood that the caller was flight attendant Amy Sweeney, whom Michael had known for a decade. He'd seen off Flight 11 less than a half hour earlier, after that disturbing moment when he locked eyes with Mohamed Atta.

Michael took over the call.

"Amy, sweetie, what's going on?" he asked.

In a tightly controlled voice, Amy answered: "Listen to me very, very carefully."

Michael grabbed a pad of paper to take notes.

AT 8:29 A.M., a half hour after takeoff, American Flight 11 turned south-southeast, putting it more directly on a route to Manhattan. The 767 climbed to 30,400 feet. Two minutes after adjusting course, it descended to 29,000 feet.

One second before 8:34 a.m., air traffic controllers at Boston Center heard a third disturbing transmission from the cockpit, a lie apparently intended for the passengers and crew but never heard by them: "Nobody move, please. We are going back to the airport. Don't try to make any stupid moves."

Controllers at Boston Center fell silent. Then they decided to do something: FAA air traffic control managers called in the military.

Normally, if the system had worked as designed, top officials at the FAA in Washington would contact the Pentagon's National Military Command Center, which in turn would call the North American Aerospace Defense Command, or NORAD, the military organization responsible for protecting the skies over the United States and Canada. NORAD, in turn, would ask approval from the Secretary of Defense to use military jets to intervene in the hijacking of a commercial passenger jet. None of that was necessarily a smooth or rapid process.

Boston Center controllers concluded that it would take too long to bob and weave through the FAA bureaucracy, to get approval from someone in the Defense Department, to scramble fighter planes to chase Flight 11. They knew it wasn't correct protocol, but they took matters into their own hands. First, they called their colleagues at an air traffic control facility on Cape Cod and asked them to place a direct call seeking help from fighter jets stationed at Otis Air National Guard Base. Then they concluded that even that wasn't enough. As Flight 11 streaked toward Manhattan, Boston Center air traffic controllers urgently wanted to get the military involved. At the very least, the military might have better luck tracking the hijacked plane; some Boston Center controllers knew that the military had radar that could reveal a plane's altitude even with its transponder turned off.

They tried to call a NORAD military alert site in Atlantic City, unaware that it had been shut down as part of the post–Cold War cuts in rapid air defense. Then, at 8:37 a.m., three minutes after first seeking help through controllers on Cape Cod, a supervisor at Boston Center named Dan Bueno called the Otis Air National Guard base directly. At roughly the same time, a Boston Center air traffic controller named Joseph Cooper called NORAD's Northeast Air Defense Sector, or NEADS, in Rome, New York. That's where Major Kevin Nasypany had arrived earlier that morning expecting to put his team through the training exercise called Vigilant Guardian.

A few seconds before 8:38 a.m., Cooper made the first direct notification of a crisis on board American Flight 11 to the U.S. military: "[W]e have a problem here," Cooper said. "We have a hijacked aircraft headed towards New York, and we need you guys to, we need someone to scramble some F-16s or something up there. Help us out."

"Is this real-world, or exercise?" asked NEADS Technical Sergeant Jeremy Powell. Powell's question reflected the fact that he knew Vigilant Guardian was planned for later in the day, and he wondered if it had begun early.

"No," Cooper answered, "this is not an exercise, not a test."

The air traffic controllers' calls to the military sent the nation's air defense system into high gear. But it did so outside of normal operating procedures, with delayed or at times nonexistent communications with the FAA, and without anything remotely resembling a well-defined response plan. Nasypany and his team at NEADS would have to rely on training and instinct, reacting moment by moment and making it up as they went along.

WHEN JOSEPH COOPER of the FAA's Boston Center called for military help with Flight 11, Nasypany wasn't on the NEADS Operations Floor among the radar scopes. When no one could find him, a voice boomed over the loudspeaker: "Major Nasypany, you're needed in Ops, pronto!"

The hijacking of Flight 11 surprised America's airline, air traffic, airport security, political, intelligence, and military communities. But it literally caught Nasypany with his pants down. Roused by the public address announcement, he zippered his flight suit and rushed from the men's room to the war room: the NEADS Operations Floor, or Ops, a dimly lit hall with four rows of radar and communications work stations that faced several fifteen-foot wall-mounted screens. A glassed-in command area called the Battle Cab watched over the men and women scanning the electronic sky for danger.

When Nasypany reached the Ops floor, he felt annoyed by his team's talk of a hijacking. Nasypany thought that someone had prematurely triggered the Vigilant Guardian exercise. He growled at no one in particular, "The hijack's not supposed to be for another hour!"

Nasypany quickly discovered that the hijacking of American Airlines Flight 11 was "real-world," and that Cooper had skipped protocol and called NEADS directly for help. Hijackings were on Nasypany's list of potential threats, but they weren't a top priority in his normal routine. Later, when one of his subordinates seemed on the verge of falling apart under the stress of the day's events, Nasypany tried to lighten the mood by publicly admitting that he'd been "on the shitter" when summoned by the loudspeaker. At a more reflective moment, Nasypany confessed that he'd remember that announcement for the rest of his life.

IMMEDIATELY AFTER COOPER'S call for help, a young NEADS identification technician named Shelley Watson spoke with Boston Center's military liaison, Colin Scoggins. Watson quizzed Scoggins for whatever information he possessed about Flight 11, a rushed conversation that revealed how little the air traffic controllers on the ground knew about what was happening in the air.

Watson: "Type of aircraft?"

Scoggins: "It's a—American Eleven."

Watson: "American Eleven?"

Scoggins: "Type aircraft is a 767 . . ."

Watson: "And tail number, do you know that?"

Scoggins: "I, I don't know—hold on."

Scoggins turned to air traffic supervisor Dan Bueno to ask for more information, including the number of "souls on board." But Bueno didn't know either.

Scoggins: "No, we—we don't have any of that information."

Watson: "You don't have any of that?"

Scoggins explained that someone had turned off the cockpit transponder, so they didn't have the usual tracking information. Boston Center could see Flight 11 only on what was called primary radar, which made it difficult to keep track of it among the constellations of radar dots representing the many planes in the sky.

Watson: "And you don't know where he's coming from or destination?"

Scoggins: "No idea. He took off out of Boston originally, heading for, ah, Los Angeles."

NASYPANY QUICKLY RECEIVED authorization from his boss, Colonel Robert Marr, to prepare to launch the two F-15 fighter jets on alert at Otis on Cape Cod, roughly one hundred fifty miles from New York City.

The call from NEADS triggered a piercing klaxon alarm at Otis, and a voice blared on the public address system: "Alpha kilo one and two, battle stations." The on-alert fighter pilots, Lieutenant Colonel Timothy Duffy and Major Daniel Nash, ran into a locker room to pull on their G-suits and grab their helmets. Then they sprinted to a Ford pickup and raced a half mile to the hangar housing their F-15s, strapped in, and waited for further orders.

At a top speed of more than twice the speed of sound, an F-15 Eagle fighter jet could reach New York from Cape Cod in ten minutes. But the F-15s from Otis were fourteen years old and loaded with extra fuel tanks, so it would take them perhaps twice as long. And they weren't going anywhere until Duffy and Nash received orders to scramble, or launch.

If orders did come, based on expectations of a "traditional" hijacking, the fighter pilots would try to quickly locate the commandeered plane. Then they would act only as military escorts, with orders to "follow the flight, report anything unusual, and aid search and rescue in the event of an emergency." While following the flight, Duffy and Nash would be expected to position their F-15s five miles directly behind the hijacked plane, to monitor its flight path until, presumably, the hijackers ordered the pilots to land. Under the most extreme circumstances envisioned, the fighter pilots might be ordered to fly close alongside, to force a hijacked plane to descend safely to the ground.

But with Flight 11's transponder turned off, the F-15 pilots would have problems doing any of that. No one knew exactly where to send the fighter jets. Although military radar could track a plane with its transponder off, military air controllers needed to locate it first and mark its coordinates. As minutes ticked by, controllers working for Major Nasypany at NEADS searched their radar screens in a frustrating attempt to find the hijacked passenger jet.

Complicating matters, the FAA and NEADS used different radar setups to track planes. In key respects, they spoke what amounted to different controller languages. At one point during the search, a civilian air traffic controller from New York Center told a NEADS weapons controller that his radar showed Flight 11 "tracking coast." To an FAA controller, the phrase described a computer projection of a flight path for a plane that didn't appear on radar. But that wasn't a term used by military controllers. NEADS controllers thought "tracking coast" meant that Flight 11 was flying along the East Coast.

"I don't know where I'm scrambling these guys to," complained Major James Fox, a NEADS weapons officer whose job was to direct the Otis fighters from the ground. "I need a direction, a destination."

Nasypany gave Fox a general location, just north of New York City. That way, until someone located the hijacked plane, the fighter jets would be in the general vicinity of Flight 11.

Meanwhile, Colonel Marr called NORAD's command center in Florida to speak with Major General Larry Arnold, the commanding general of the First Air Force. Marr asked permission to scramble the fighters without going through the usual complex Defense Department channels and without clear orders about how to engage the plane.

Arnold made a series of quick calculations. Hijackers had seized control of a passenger jet headed toward New York. They'd made the plane almost invisible to radar by turning off its transponder. They wouldn't answer radio calls and showed no sign of landing safely or making demands. This didn't seem like a traditional hijacking, though he couldn't be sure. He wouldn't wait to find out—they'd worry about getting permission later.

Arnold told Marr: "[G]o ahead and scramble the airplanes."

UNDER THE HIJACKERS' command, American Flight 11 adjusted its route again, turning more directly to the south. The plane slowed and began a sharp but controlled descent, dropping at a rate of 3,200 feet per minute.

With its transponder switched off, Flight 11 remained largely a mystery to air traffic controllers. If they could see it at all, it appeared as little more than a green blip on their screens. Trying to determine its speed and altitude, they sought help from other pilots. When Flight 11 turned to the south and began to descend, a controller from Boston Center named John Hartling called a nearby plane that had taken off from Boston minutes after Flight 11 and was headed in the same general direction. That plane was United Airlines Flight 175.

Hartling asked the United 175 pilots if they could see American Flight 11 through their cockpit windshield.

At first the sky looked empty, so Hartling asked again.

"Okay, United 175, do you have him at your twelve o'clock now, and five, ten miles [ahead]?"

"Affirmative," answered Captain Victor Saracini. "We have him, uh, he looks, ah, about twenty, uh, yeah, about twenty-nine, twenty-eight thousand [feet]."

Hartling instructed United 175 to turn right: "I want to keep you away from this traffic."

Saracini and his first officer, Michael Horrocks, banked the plane to the right, as told. They didn't ask why, and Hartling didn't tell them.

As far as anyone knew, the only action needed to keep United

Flight 175 and every other plane safe would be separation—that is, steering them away from hijacked American Flight 11. No one had yet deciphered the first sentence of the hijackers' first radio transmission from Flight 11: "We have some planes." No one imagined that more than one flight might soon be in danger.

The concept of more than one hijacking simultaneously and in coordination wasn't on anyone's radar, literally or figuratively. Years had passed since the last hijacking of a U.S. air carrier, and coordinated multiple hijackings had never happened in the United States. Almost no one in the FAA, the airlines, or the military had dealt with such a scenario or considered it a likely threat. The last organized multiple hijackings anywhere in the world had occurred more than three decades earlier, in September 1970, when Palestinian militants demanding the release of prisoners in Israel seized five passenger jets bound from European cities to New York and London. They diverted three planes to a Jordanian desert and one to Cairo. The crew and passengers of the fifth jet, from the Israeli airline El Al, subdued the hijackers, killing one, and regained control of their plane. No passengers or crew members aboard those hijacked planes died.

AS FLIGHT 11 bore down on New York City, flight attendants Betty Ong and Amy Sweeney sat side by side in the back of the plane. During separate, overlapping calls, they provided a chilling account of the crisis around them. The calls alone were acts of extraordinary bravery. Flight attendants were trained to anticipate that hijackers might have "sleeper" comrades embedded among the passengers, waiting to attack anyone who posed a threat or disobeyed commands.

Just as the military operated under certain expectations about how hijackings played out, aircrews were taught that they should refrain from trying to negotiate with or overpower hijackers, to avoid making things worse. Under a program known as the Common Strategy, crews were told to focus on attempts to "resolve hijackings peacefully" and to get the plane and its passengers on the ground safely. The counterstrategy to a hijacking also called for delays, and if that didn't work, cooperation and accommodation when necessary. In these scenarios, "suicide wasn't in the game plan," as one study phrased it. Neither was a hijacker piloting the plane. Further, no one considered the possibility that a hijacked airplane would attempt to disappear from radar by someone in the cockpit turning off a transponder.

Still operating under the old set of beliefs, some air traffic controllers predicted that Flight 11 would land at John F. Kennedy International Airport in Queens, New York, although at least one put his money on the plane making a run for Cuba. But the old playbook on hijackings had become dangerously obsolete. That was clear as soon as Betty Ong and Amy Sweeney revealed details of a precise, multilayered plot aboard Flight 11.

AFTER THE LAST routine contact between pilots John Ogonowski and Tom McGuinness and Boston Center's Peter Zalewski at 8:13 a.m., Atta and his men pounced. Based on the timing, about fifteen minutes into the flight, they might have used a predetermined signal: when the pilots turned off the Fasten Seatbelt signs.

One or more of the hijackers, possibly the brothers Wail and Waleed al-Shehri, who were sitting in the first row of first class, sprayed Mace or pepper spray to create confusion and force passengers and flight attendants away from the cockpit door.

Using weapons smuggled aboard, perhaps the short-bladed knives bought in the months before September 11, they stabbed or slashed first-class flight attendants Karen Martin and Bobbi Arestegui. Amy Sweeney told Michael Woodward in Boston that Karen was critically injured and being given oxygen. Betty Ong told Nydia Gonzalez in Fort Worth that Karen was down on the floor in bad shape. Both said Bobbi Arestegui was hurt but not as seriously as Karen. They didn't say what kind of knives or weapons the hijackers used. Neither indicated that the terrorists had guns or other weapons.

All nine flight attendants had keys to the cockpit, but it's not clear how the hijackers gained entry—Atta and his crew might have attacked Karen and Bobbi to steal their keys, or the hijackers might have gotten into the cockpit another way. When the plane dipped and pitched erratically, Betty suspected that a hijacker was already in control. She said she thought they had "jammed the way up there." In fact, the cockpit doors were relatively flimsy and weren't strong enough to prevent forced entry. Another possibility was that the hijackers stabbed the first-class flight attendants to induce the pilots to open the cockpit door. Or maybe it was even simpler: In September 2001, one key opened the cockpit doors of all Boeing planes. Maybe the hijackers brought a key on board with them.

Whichever way Atta and the others gained access, no one who knew John Ogonowski, the Vietnam veteran who farmed when he wasn't flying, or Tom McGuinness, the former Navy pilot, would believe that either went down quietly. Possible evidence of that came during Betty Ong's call. She said she heard loud arguing after the hijackers entered the cockpit. If Ogonowski and McGuinness were in their seats, low and strapped in, they would have been at a distinct disadvantage against knife-wielding attackers coming at them from behind with the element of surprise.

As the only hijacker with pilot training, Mohamed Atta almost certainly took control of the plane after the hijackers killed or disabled the pilots. He spoke English fluently, so he likely made the radio transmissions heard by Peter Zalewski in Boston Center. It's also possible that Atta's seatmate, Abdulaziz al-Omari, accompanied him into the cockpit.

Betty Ong reported to Nydia Gonzalez that a passenger's throat had been slashed and that the man appeared to be dead. On her call to Michael Woodward, Amy Sweeney said the passenger was in first class. Betty told Nydia Gonzalez the passenger's name was "Levin or Lewis" and that he'd been seated in business class seat 9B. In her first, brief call, Amy identified the attacker as the passenger who'd been seated in 10B.

Nydia Gonzalez tried to confirm the killer's identity. She asked Betty: "Okay, you said Tom Sukani?"—a name phonetically similar to that of "muscle" hijacker Satam al-Suqami. "Okay—okay, and he was in 10B. Okay, okay, so he's one of the persons that are in the cockpit. And as far as weapons, all they have are just knives?"

Based on Betty and Amy's calls, it's possible that the brilliant computer entrepreneur and former Israeli commando Daniel Lewin saw the hijackers attack the flight attendants and heroically leapt into action. But unknown to Lewin, seated directly behind him was the fifth hijacker, Satam al-Suqami, whose name Nydia Gonzalez heard as "Tom Sukani." In that scenario, when Lewin rose to fight back, Suqami slit his throat, making Lewin perhaps the first casualty of 9/11. Another possibility was that the hijackers had planned all along to begin the takeover by attacking crew members and at least one passenger, to frighten the rest into compliance. In that scenario, Lewin would have been an unwitting victim who happened to be sitting in the targeted seat.

Amy told Michael Woodward the three hijackers in business class were Middle Eastern and gave him their seat numbers, key pieces of evidence to identify the terrorists. Betty identified the seat numbers of the two hijackers in first class, the Shehri brothers.

Amy told Michael she saw one of the hijackers with a device with red and yellow wires that appeared to be a bomb. He wrote "#cockpit bomb" on his notepad. Betty didn't mention a bomb, and no one knew if whatever Amy saw was real or a decoy.

On their separate calls, the two flight attendants said they didn't know whether coach passengers fully understood the peril. Amy told Michael that she believed the coach passengers thought the problem was a routine medical emergency in the front section of the plane. First-class passengers were herded into coach, but in the uproar, it wasn't clear whether former ballet dancer Sonia Puopolo, business consultant Richard Ross, venture capitalist David Retik, or anyone else who'd been up front mentioned the violence they'd seen.

Amy told Michael that in addition to Betty, the flight attendants who weren't injured—Kathy Nicosia, Sara Low, Dianne Snyder, Jeffrey Collman, and Jean Rogér—kept working throughout the crisis, helping passengers and finding medical supplies.

Betty and Amy relayed all the information they could, as quickly and completely as they could, for as long as they could. At 8:43 a.m., roughly a half hour after the hijacking began, Flight 11 changed course again, to the south-southwest. The move put the Boeing 767, still heavy with fuel, on a direct course for Lower Manhattan, the heart of America's financial community.

At the American Airlines center in Fort Worth, Nydia Gonzalez begged for information: "What's going on, Betty? . . . Betty, talk to me. . . . Betty, are you there? . . . Betty?"

Betty didn't answer.

Nydia turned to her colleagues: "Do you think we lost her? Okay, so we'll like—we'll stay open."

Then Nydia Gonzalez added an unintentionally haunting coda to Betty Ong's bravery: "We—I think we might have lost her."

AROUND THE SAME time, Amy Sweeney told Michael Woodward: "Something is wrong. We're in a rapid descent. . . . We are all over the place." Another American Airlines employee who overheard the call said she heard Amy scream.

Michael tried his best to calm Amy. He told her to look out the window and tell him what she saw. "We are flying low," she said. Amy told Michael she saw water and buildings. "We are flying very, very low. We are flying way too low!"

Amy paused. Powerless on the other end of the phone, Amy's colleague and friend Michael Woodward waited, every second stretching into a lifetime. Less than an hour earlier, he'd stood inside the plane, locked eyes with Mohamed Atta, and waved goodbye to his friends.

Michael heard Amy's last words, before the call dissolved into static: "Oh my God!—We are way too low!"

UNDER THE COMMAND and control of fanatics bent on murder and determined to commit suicide, American Airlines Flight 11 had been transformed from a passenger jet into a guided missile. Atta's radio transmissions about returning to the airport and everything being okay were elements of a cruel ruse to pacify passengers and to prevent an uprising against his outnumbered men. He had played on old beliefs about how hijackings occurred and were usually resolved without violence. Even though his lies weren't heard by Flight 11's passengers, the radio calls and the hijackers' advance training and in-flight actions revealed a carefully calibrated plan built on surprise, violence, trickery, and a studied understanding of their targets, all to achieve a barbaric goal.

The Boeing 767 that was American Airlines Flight 11 completed an unapproved, L-shaped path through bright blue skies that covered roughly three hundred miles from Boston, west to Albany, then south over the streets of Manhattan. At the last millisecond of its trip, at a speed estimated at 440 miles per hour, the silver plane's nose touched the glass and steel of the north face of the 96th floor of the North Tower of the World Trade Center.

TJX planning manager Tara Creamer's instructions to her husband, John, on how to care for their children would need to last a lifetime.

Cambodian farmers who relied on John Ogonowski would have to find a new teacher and patron. His wife and daughters would be set adrift without their anchor.

Amy Sweeney's children would have to get to school, and through life, without her.

Betty Ong's elderly friends would need new rides to doctors' appointments. Her sister Cathie would never again hear her say "I love you lots."

Robert Norton's stepson would have to get married without him.

Daniel Lee's soon-to-be-born daughter would spend her entire life without him.

Daniel Lewin's family and his company would have to forge new paths without his genius or his guidance.

Someone else would have to find a health aide for Cora Hidalgo Holland's mother.

Susan MacKay's air traffic controller husband, Doug, who planned to have John Ogonowski say hi for him, would have to live with the knowledge that Susan passed through airspace over the Hudson River Valley that he normally controlled.

Dozens of children would grow up without a mother or a father, an aunt or uncle, a grandmother or grandfather. Parents would grow old without a daughter or a son; husbands, wives, and partners would be forced to carry on alone. Grief would grip untold families, friends, colleagues, and strangers, wounded by the deaths of seventy-six passengers and eleven crew members, all murdered by five al-Qaeda hijackers aboard American Airlines Flight 11.

The time of their deaths was 8:46:25 a.m.

More than six minutes later, the two F-15 fighter jets from Otis took flight. They soared south toward New York in pursuit of a passenger jet that no longer existed.

But something terrible was happening on another passenger plane also bound from Boston to Los Angeles.

CHAPTER 5

"DON'T WORRY, DAD"

United Airlines Flight 175

TWENTY-FIVE MINUTES AFTER TAKEOFF FROM LOGAN, WHILE AMERican Flight 11 was still airborne, United Flight 175 reached the crystal-blue skies over upstate New York.

With the passage of time and distance came a transfer of ground control from Boston Center to a similar FAA facility called New York Center, located in a sleepy suburb on Long Island. New York Center's radar scopes covered some of the world's busiest skies: air traffic over the New York metro area and parts of Pennsylvania.

With the handoff of the flight from Peter Zalewski in Boston Center to his colleagues in New York Center came a new radio frequency for the pilots of Flight 175. At 8:41 a.m., several minutes after being instructed to veer away from Flight 11, they used that new frequency to report the disturbing radio communication they'd heard earlier.

"Yeah. We figured we'd wait to go to your center," said Captain Victor Saracini. "We heard a suspicious transmission on our departure out of Boston. Ah, with someone, ah, it sounded like someone keyed the mic and said, ah, 'Everyone, ah, stay in your seats.'"

Saracini didn't say exactly when he heard the message, and he didn't know its source among the multiple planes that used the same frequency. The words he used didn't precisely fit any of the three accidental transmissions believed to be from Mohamed Atta from Flight 11's cockpit.

Saracini might have been referring to Atta's second message: "Nobody move. Everything will be okay. If you try to make any moves, you will injure yourselves and the airplane. Just stay quiet." But the fact that Saracini heard only one message made it more likely to have been the hijacker's third and last transmission. Otherwise, Saracini presumably would have mentioned one or both of the earlier threatening calls. The third call began at one second before 8:34 a.m.: "Nobody move, please. We are going back to the airport. Don't try to make any stupid moves."

Saracini didn't say why he waited to report the message, but one logical explanation would be that he assumed that the controllers at Boston Center already knew about it because they used the same frequency. Once Flight 175 passed to New York Center and changed radio frequencies, Saracini presumably wanted to be certain that his new ground controller knew about it, too. If so, Saracini showed good foresight: Flight 175's air traffic controller in New York Center was Dave Bottiglia, who hadn't yet heard anything about the chaos unfolding aboard American Flight 11 as it raced toward New York City.

"Oh, okay," Bottiglia answered. "I'll pass that along over here."

Under normal circumstances, an official at the FAA would share a report of a suspicious cockpit communication with airline officials, in this case the United Airlines System Operations Control center, just outside Chicago. But no one from Boston Center or New York Center did so. Bottiglia could have remedied that, but other worries suddenly demanded his full attention.

Moments after Bottiglia heard from Saracini, another New York Center flight controller walked over and showed Bottiglia a point, or "target," on his radar screen.

"You see this target here?" the controller asked. "This is American Eleven. Boston Center thinks it's a hijack."

Now Dave Bottiglia had responsibility for both American Airlines

Flight 11 and United Airlines Flight 175. Having been made aware that Flight 11 was racing toward Manhattan, descending rapidly with its transponder turned off and someone in the cockpit making threats, Bottiglia did his best to keep track of the American Airlines plane.

THE FIRST OUTWARD sign of trouble aboard United Airlines Flight 175 came at 8:47 a.m., about one minute after Flight 11 hit the World Trade Center's North Tower.

Someone in the United plane's cockpit changed the plane's transponder code twice within a minute. Bottiglia didn't notice because he was furiously searching for American Flight 11, which by then no longer existed.

Three minutes later, at 8:50 a.m., the pilot of a Delta Air Lines flight radioed Bottiglia to let him know about "a lot of smoke in Lower Manhattan." The pilot said it looked as if the World Trade Center was on fire. Bottiglia acknowledged the call as he continued to search his screen for missing Flight 11.

Meanwhile, United Flight 175 remained on a southwesterly route, crossing over New Jersey and then over Pennsylvania. At 8:51 a.m., roughly four minutes after someone switched the assigned transponder code, Bottiglia noticed the change. He radioed the plane to order the pilots to return to their proper code but received no reply.

Whoever sat at the controls of Flight 175 did something else: the pilot changed altitude without approval, climbing several thousand feet, then plunging into a steep descent.

For New York Center controllers watching on radar screens in darkened rooms many miles away, pieces of a horrifying puzzle began to fall into place. First, American Flight 11 vanished as it descended toward Manhattan, and soon after, a Delta pilot reported a fire at the World Trade Center. Now United Flight 175 had changed its altitude and transponder code. The cockpit failed to respond to radio calls shortly after Captain Victor Saracini radioed his report about the suspicious message.

Flight controllers are trained to use logic and reason and not to jump to conclusions. Dave Bottiglia felt in his gut that these abnormal events were related, though he wouldn't yet call them elements of a coordinated, nearly simultaneous hijacking plot.

Bottiglia tried to focus his attention on United Flight 175 and, if he could find it again, American Flight 11. He began to move all other flights in his sector out of the way of Flight 175's self-assigned path, and he continued his attempts to reach the United pilots. Five times he called them, with no reply.

His voice starting to shake, Bottiglia turned to a flight control colleague in New York Center. Bottiglia asked him to watch all the other planes in his sector: "Please just take everything and don't ask any questions."

AT 8:52 A.M., the phone rang in Lee and Eunice Hanson's cozy kitchen in Easton, Connecticut. Lee answered and heard the voice of their son Peter, talking quietly, his tone somber.

"Dad, we are on the airplane. It's being hijacked," Peter said.

Lee couldn't process what his son was telling him. He hoped it was a joke, like the prank call Peter made years earlier to Eunice at the conservation commission.

"What are you talking about?" Lee answered. "C'mon, don't scare everybody."

"No, it's true," Peter said. He spoke in a soft, clear voice. The longer Peter spoke, the more his father heard tremors of nerves, the clearer it became that Peter wasn't joking.

"I think they've taken over the cockpit. . . . An attendant has been stabbed . . . and someone else up front may have been killed. The plane is making strange moves. Call United Airlines. . . . Tell them it's Flight 175, Boston to L.A."

Lee asked about Christine and Sue, and Peter told him they were okay. All the passengers had crowded together in the rear of the plane. "It's very tight here, Dad."

"I'm going to hang up," Peter said. "Call United Airlines."

Lee repeatedly tried the airline, but the line was busy. He called the Easton Police Department, told an officer what Peter said, and asked the officer to call United and to contact the town's police chief. Shortly after, Lee called back to make sure the police had reached the airline. This time an officer told him: "Gee, Mr. Hanson, a plane has hit the World Trade Tower. You should turn the television on."

Lee and Eunice turned to CNN.

His son's voice echoed in Lee's mind. He shared what he knew with Eunice, who could barely breathe. Lee felt an urge to call Peter on his cellphone, but he stopped himself, fearing that a ringing phone might endanger everyone on board.

Distraught, in shock and disbelief, Lee and Eunice stared at the horrendous scenes on television.

CNN ANCHOR CAROL LIN broke into the cable network's morning news report shortly before 8:49 a.m., several minutes before the Hansons tuned in. The screen filled with horrifying images of the North Tower, its top floors engulfed in fire and smoke.

"This just in," Lin said. "You are looking at obviously a very disturbing live shot there. That is the World Trade Center, and we have unconfirmed reports this morning that a plane has crashed into one of the towers of the World Trade Center. CNN Center right now is just beginning to work on this story, obviously calling our sources and trying to figure out exactly what happened. But clearly, something relatively devastating is happening this morning there on the south end of the island of Manhattan."

From that moment, the global audience expanded exponentially, as seemingly everyone watching rushed to a phone to tell someone else: "Turn on CNN."

Live on the air, Lin and her viewers heard a first-person account via telephone from CNN's vice president of finance, Sean Murtagh, who'd been in a meeting on the 21st floor of a building facing the World Trade Center. Murtagh was conscripted by circumstance into working as a reporter: "I just witnessed a plane that appeared to be cruising at slightly lower than normal altitude over New York City, and it appears to have crashed into—I don't know which tower it is—but it hit directly in the middle of one of the World Trade Center towers."

Lin: "Sean, what kind of plane was it? Was it a small plane, a jet?"

Murtagh: "It was a jet. It looked like a two-engine jet, maybe a 737."

Lin: "You are talking about a large passenger commercial jet."

Murtagh: "A large passenger commercial jet."

They discussed Murtagh's location and other details. Then Lin asked a question that suggested she suspected that the crash was caused by a mechanical failure: "Did you see any smoke, any flames coming out of engines of that plane?"

"No, I did not," Murtagh answered. "The plane just was coming in low, and the wingtips tilted back and forth, and it flattened out. It looks like it hit at a slight angle into the World Trade Center. I can see flames coming out of the side of the building, and smoke continues to billow."

Other than the exact model of the plane, CNN immediately got the basics right about Flight 11's crash, although they didn't yet know the flight number or much else. But several other early media reports suggested that it might have been a small commuter plane. As every broadcast and print newsroom leapt onto the story, early speculation raged that the crash was an accident, caused by a lost or inexperienced pilot. Americans old enough to remember perhaps flashed back to July 28, 1945, when a B-25 bomber lost in morning fog crashed into the Empire State Building, killing three crew members and eleven others.

Confusion reigned in the government as well, including at the FAA and the FBI, as officials struggled to confirm that a plane had in fact hit the North Tower. Others questioned whether it wasn't a plane at all, but a bomb more powerful than the one driven by truck into an underground World Trade Center garage in February 1993.

One example of the confusion: Around 8:55 a.m., the flight control manager at New York Center tried to notify regional FAA officials that United Flight 175 had apparently been hijacked. But the regional FAA officials refused to be disturbed. They were too busy discussing the hijacking of American Flight 11, which they didn't realize had scythed into the North Tower almost ten minutes earlier.

During the first frenzied minutes before and after 9 a.m. on September 11, only a few people recognized the enormity of the unfolding catastrophe. Tragically, some of those who best understood key pieces of the crisis were inside American Flight 11 before it crashed and United Flight 175 as it sped toward New York City.

AT NEARLY THE same time as Peter Hanson called his parents' Connecticut home, a telephone rang at a United Airlines facility in San Francisco where in-flight crews called to report minor maintenance problems. Flight attendants knew they could dial "f-i-x," using the corresponding numbers on the keypad, 3-4-9, and automatically be connected to the airline's maintenance center.

From an Airfone near the rear of Flight 175, a male flight attendant, believed to be former cellphone salesman Robert Fangman, reported details of the hijacking to a maintenance worker. The information dovetailed with the report Peter Hanson gave his father. The flight attendant said that both pilots of United Flight 175 had been killed, a flight attendant had been stabbed, and hijackers were probably flying the plane.

The unrecorded call cut off after about two minutes. The United maintenance worker and a colleague tried to recontact the flight using the ACARS digital message system linked to the cockpit: "I heard of a reported incident aboard your [aircraft]," they wrote. "Plz verify all is normal."

They received no reply. Minutes passed before someone from the San Francisco maintenance center reported the call to United headquarters in Chicago.

Not every attempt to sound the alarm or to reach loved ones from Flight 175 proved successful. Between 8:52 a.m. and 8:59 a.m., former pro hockey player Garnet "Ace" Bailey tried four times to call his wife, Kathy, on her business and home phones. The calls dropped or didn't connect. He never got through.

FLIGHT 175 COMPLETED a fishhook turn over Allentown, Pennsylvania, banking to the left and descending as it crossed back over New Jersey, and headed toward New York City. The pilot almost certainly was Marwan al-Shehhi, the companion of Mohamed Atta and the only one of the five Flight 175 hijackers trained to fly a passenger jet.

Nothing was safe in their path. A New York Center controller watched as the United plane turned toward a Delta 737 flying southwest at 28,000 feet.

"Traffic two o'clock! Ten miles," the controller warned the Delta pilots. "I think he's been hijacked. I don't know his intentions. Take any evasive action necessary."

The Delta flight ducked away from United 175, but soon after, the hijacked plane put itself on a collision course with a US Airways flight. An alarm sounded in the US Airways cockpit, and the pilots dived to avoid a midair crash.

AFTER DROPPING HER husband, Brian "Moose" Sweeney, at the Hyannis airport, Julie Sweeney got ready for her fifth day of work as a high school health teacher on Cape Cod. She'd already left for work when the phone rang in the home she shared with the former Navy F-14 pilot, Top Gun instructor, "Twin Tower" college football player, and costume-party Viking.

Brian's call, at shortly before 8:59 a.m., went to their answering machine.

He spoke in a calm, serious tone, and his message echoed what he'd told Julie weeks earlier about how he wanted her to "celebrate life" if anything happened to him:

"Jules, this is Brian. Listen, I'm on an airplane that's been hijacked. If things don't go well, and it's not looking good, I just want you to know I absolutely love you. I want you to do good, go have a good time. Same to my parents and everybody. And I just totally love you, and [anticipating heaven or an afterlife] I'll see you when you get there. 'Bye, babe. Hope I'll call you."

Brian hung up. Then he punched in the numbers for another telephone call.

AS LOUISE SWEENEY prepared to leave home to run errands, the phone rang. She picked up to hear her son's voice: "Mom, It's Brian. I'm on a hijacked plane and it doesn't look good. I called to say I love you and I love my family."

Brian told his mother that he didn't know who the hijackers were. Calling from the rear of the plane, he said he thought they might come back, so he might have to hang up quickly. He told her he believed that they might be flying somewhere over Ohio. Brian said he and other passengers might storm the cockpit.

Louise knew her son, and she recognized how he sounded when he became "pissed off." He had that tone.

Before ending the call, Brian told her, "Remember *Crossing Over*. Don't forget *Crossing Over*."

Louise remembered, and she wouldn't forget. As Flight 175 streamed toward New York City, descending by the second, flying erratically, the original pilots apparently dead and a flight attendant wounded, Brian tried to comfort his mother. *Crossing Over* was a book and television program featuring a self-professed psychic named John Edward, who claimed to communicate with the dead. Brian wanted his mother to know that somehow, someday, he would see her again.

Now Brian had to go.

"They are coming back," he said. Brian told his mother goodbye as he hung up.

Louise Sweeney turned on the television.

AT NINE O'CLOCK, Peter Hanson called his parents' home a second time from Flight 175.

"It's getting bad, Dad," Peter told his father, Lee. "A stewardess was stabbed. They seem to have knives and Mace."

"They said they have a bomb. It's getting very bad on the plane. Passengers are throwing up and getting sick."

"The plane is making jerky movements. I don't think the pilot is flying the plane. I think we are going down."

"I think they intend to go to Chicago or someplace and fly into a building."

Peter's description of the hijackers' weapons, claims, and tactics echoed the calls only minutes earlier from Betty Ong and Amy Sweeney on Flight 11.

Then, just as Brian Sweeney had tried to reassure his wife and mother, Peter Hanson sought to comfort his father: "Don't worry, Dad. If it happens, it'll be very fast."

AT 9:01 A.M., as United Flight 175 rapidly descended, Peter Mulligan, a flight control manager at New York Center, told the FAA Command Center in Virginia: "We have several situations going on here. It's escalating big, big time. We need to get the military involved with us."

As it barreled toward the World Trade Center, already the scene of disaster from Flight 11's crash, United Flight 175 looked as though it might hit the Statue of Liberty.

Air traffic controllers stared rapt at their screens, even as they continued to warn other nearby planes. Everything in their long years of training and experience channeled their minds toward the hope and expectation that hijacked planes would land and passengers would be held hostage until demands were met, or until the military either forced the hijackers to surrender or killed them. To the very last moments, some controllers held tight to the notion that the pilots were racing toward the nearest airport, beset by a routine mechanical or electrical emergency.

Reality overtook that fantasy.

United Airlines Flight 175 flew low and fast, banking toward the southern twin of the burning North Tower of the World Trade Center. Flight controllers, airline officials, government and military experts, and everyone else would need to accept a new script for hijackings, one that featured a multipronged murder-suicide plot designed to maximize civilian casualties and terrorize survivors through the destruction of physical and symbolic pillars of America's power.

The evidence flashed on the air traffic controllers' radar screens.

"No!" a New York controller shouted. "He's not going to land. He's going in!"

FROM THE BACK of the plane, with his wife and daughter pressed against him, Peter Hanson spoke his final words to his father: "Oh my God. . . . Oh my God, oh my God."

Lee Hanson heard a woman shriek.

AT THAT MOMENT, a battalion of television and still cameras on the ground and in helicopters trained their lenses on Lower Manhattan. Every network joined CNN live on the air, yet still almost no one knew what was happening or what kind of planes were involved.

On ABC's *Good Morning America*, the smoke-obscured North Tower filled the screen while hosts Diane Sawyer and Charles Gibson interviewed reporter Don Dahler on the scene. As Dahler described scores of fire crews and other first responders rushing toward the World Trade Center, a Boeing 767 zoomed into view on the right side of the screen.

At 9:03:11 a.m., Lee and Eunice Hanson, Louise Sweeney, and millions of others became witnesses to murder. They watched live on television as United Flight 175, traveling between 540 and 587 miles per hour, slammed on an angle into the 77th through 85th floors of the South Tower of the World Trade Center. A bright orange fireball exploded. The building rocked and belched smoke, glass, steel, and debris. The plane and everyone inside it disappeared forever.

In her kitchen, Eunice Hanson screamed.

In her television studio in New York, Diane Sawyer gasped, "Oh my God."

"That looks like a second plane," her colleague Charles Gibson said.

"That just exploded!" said reporter Don Dahler, still on the phone to the studio, his location preventing him from seeing the crash.

Gibson composed himself. On some level, every professional broadcaster feared becoming known for a histrionic narration of a terrible event, like the radio reporter who nearly fell apart while witnessing the crash of the German airship *Hindenburg* in 1937.

"We just saw another plane coming in from the side," Gibson said soberly. "So this looks like it is some sort of a concerted effort to attack the World Trade Center that is under way."

After replaying the video to be certain about what they'd seen, Gibson's voice went slack.

"Oh, this is terrifying. . . . Awful."

Sawyer spoke for Eunice and Lee Hanson, Louise Sweeney, and countless others who saw United Flight 175's final seconds. "To watch powerless," she said, "is a horror."

THE TOLL WAS incalculable, just as it had been less than seventeen minutes earlier from the crash of American Flight 11. The immediate victims of United Flight 175 were two pilots, seven crew members, and fifty-one passengers, including three small children. All of them slaughtered in public view, preserved on film, by five al-Qaeda terrorists.

Two-year-old Christine Hanson and four-year-old Juliana McCourt would never visit Disneyland. Neither they nor David Gamboa-Brandhorst would know first days of school, first loves, or any other milestone, from triumph to heartbreak, of a full life. Andrea LeBlanc would never again travel the world with her gregarious, pacifist husband, Bob. Julie Sweeney wouldn't bear children, grow old, and feel safe with her confident warrior husband, Brian.

Delayed passengers wouldn't hear recitals of *Forrest Gump* dialogue from Captain Victor Saracini. First Officer Michael Horrocks's daughter wouldn't rise from bed with the promise that her daddy loved her to the moon. Ace Bailey and Mark Bavis would never again share their gifts with young hockey players or with their own families.

Retired nurse Touri Bolourchi, who'd fled Iran and the Ayatollah

Khomeini, wouldn't see her grandsons grow up as Americans. The Reverend Francis Grogan, who survived World War II on a Navy destroyer, would never again see his sister or comfort his flock. Flight attendants Alfred Marchand and Robert Fangman, who'd changed careers to fly, wouldn't see the world or their loving families. Flight attendants Michael Tarrou and Amy King would never marry.

Lee and Eunice Hanson would never see Peter and Sue Kim fulfill their professional promise or expand their loving family with more children. Christine would never again visit "Namma's house" or insist that her grandparents sing the correct words to Barney's "I love you" song.

And still the day had just begun.

AT ALMOST PRECISELY the same time as United Flight 175 hit the South Tower, a Boston Center flight control manager named Terry Biggio reported to a New England FAA official that his team had deciphered the hijacker's first accidental radio transmission from American Flight 11, spoken nearly forty minutes earlier.

Biggio said: "I'm gonna reconfirm with, with downstairs, but the, as far as the tape . . . [He] seemed to think the guy said that 'We have planes.' Now I don't know if it was because it was the accent, or if there's more than one, but I'm gonna . . . reconfirm that for you, and I'll get back to you real quick. Okay?"

To be certain the message came across loud and clear, Biggio repeated himself and emphasized: "Planes, as in plural."

Unknown to Biggio, during the previous ten minutes strange and suddenly familiar events had begun aboard a third transcontinental passenger jet.

CHAPTER 6

"THE START OF WORLD WAR III"

American Airlines Flight 77

AFTER A CELEBRATORY DINNER THE NIGHT BEFORE, BARBARA OL-son woke beside her husband, Ted, on his birthday, just as she'd planned. The lawyer, author, and conservative activist got ready for an early flight to Los Angeles, where she was to appear on that night's edition of *Politically Incorrect with Bill Maher*.

Before leaving her Virginia home for Dulles International Airport, before Flight 11 or Flight 175 met their fiery ends, Barbara placed a note on Ted's pillow: "I love you. When you read this, I will be thinking of you and I will be back on Friday."

AS THE MORNING progressed, the defenders of American airspace were forced to rely almost as much at times on television news up-dates as on their radar scopes and official reports. From their limited vantage point inside the NEADS bunker in upstate New York, Major Kevin Nasypany, Colonel Robert Marr, and their team struggled to make sense of confusing, conflicting, inaccurate, and occasionally devastating information about events in New York and whether more threats loomed.

When a NEADS technician saw the burning North Tower on television shortly before nine, those images marked the first notice anyone there received about what had happened. The technician gasped, "Oh God!" Her colleague answered, "God save New York."

A report soon reached them that the plane was a Boeing 737, perhaps as a result of the CNN broadcast that mentioned that model. Otherwise, the plane that struck the North Tower appeared to match the Boeing 767 passenger jet they'd been trying without luck to find: American Flight 11. The NEADS team still hadn't heard about United Flight 175 or any other hijacked planes. When they confirmed that the North Tower crash involved Flight 11, that presumably would mean the end of NEADS mission. NEADS staffers asked Nasypany what he wanted to do with the two F-15s they'd scrambled from the Otis base on Cape Cod.

Unsure whether the CNN report and other information they'd received was accurate, concerned that the plane they sought was a Boeing 767, not a 737, and lacking official confirmation, Nasypany continued to play defense. "Send 'em to New York City still," he ordered. "Continue! Go!"

A NEADS identification technician, Senior Airman Stacia Rountree, sought more information about the crashed plane from the FAA Boston Center's military liaison, Colin Scoggins. The call initially seemed to confirm the loss of Flight 11, but soon it did the opposite, increasing confusion about which plane had struck the tower.

Scoggins: "Yeah, he crashed into the World Trade Center."

Rountree: "That is the aircraft that crashed into the World Trade Center?"

Scoggins: "Yup. Disregard the tail number [for American Flight 11]."

Rountree: "Disregard the tail number? He did crash into the World Trade Center?"

Scoggins: "That, that's what we believe, yes."

Another NEADS technician interrupted, saying that the military hadn't received official confirmation that the North Tower crash involved American Flight 11. Media reports still mentioned a small Cessna that had supposedly gotten lost over Manhattan. To top it off, American Airlines officials had yet to confirm to anyone that Flight 11 had even been hijacked, much less that it had crashed. Rountree's supervisor, a no-nonsense master sergeant named Maureen "Mo" Dooley, took over the call.

Dooley: "We need to have—are you giving confirmation that American 11 was the one?"

Scoggins: "No, we're not gonna confirm that at this time. We just know an aircraft crashed in and—"

On the other hand, Scoggins acknowledged, that didn't mean they had any idea where to find American Flight 11. Dooley asked him: "[I]s anyone up there tracking primary [radar] on this guy still?"

Scoggins replied: "No. The last [radar sighting] we have was about fifteen miles east of JFK [Airport], or eight miles east of JFK was our last primary hit. He did slow down in speed. The primary that we had, it slowed down below, around to three hundred knots."

Dooley: "And then you lost 'em?"

Scoggins: "Yeah, and then we lost 'em."

With incomplete information, Nasypany couldn't rule out the possibility that American Flight 11, with a hijacker at the controls, remained airborne and hiding from radar with its transponder off, somewhere over one of the most heavily populated areas of the United States. Meanwhile, Nasypany and the NEADS team didn't learn about United Flight 175 until 9:03 a.m.

Rountree cried out: "They have a second possible hijack!"

But again, just as with Flight 11, the notification came far too late. At almost that exact moment, Flight 175 smashed into the South Tower. Colonel Marr and others at NEADS watched it live on CNN. The two F-15 fighter jets from Otis still hadn't reached New York.

America's air defense system couldn't stop those crashes, but Nasypany still wanted the F-15s in the sky over New York. The United States had just experienced its first simultaneous multiple hijackings, and no one could say whether the terrorists had more planned. As he prowled the room at NEADS, bottling his frustration while he pressured, calmed, and cajoled his team, Nasypany hadn't yet heard Mohamed Atta's ominous statement, "We have some planes." But he didn't need to.

"We've already had two," Nasypany thought. "Why not more?"

EARLIER THAT MORNING at Dulles International Airport outside Washington, D.C., before either Flight 11 or Flight 175 was hijacked, passengers walked calmly onto the sparsely filled American Airlines Flight 77. The plane was a Boeing 757, a single-aisle passenger jet smaller and slimmer than the wide-bodied 767, but nonetheless a large plane suited to transcontinental flights. Bound nonstop for Los Angeles, Flight 77's fuel weighed just under 50,000 pounds, more than a fully loaded city bus.

Two Flight 77 passengers, one in first class, the other in coach, represented the two distinct worlds of Washington, D.C. One, Barbara Olson, enjoyed great celebrity and clout as a member of the capital's ruling elite. The other embodied great possibility.

Bernard C. Brown II stepped aboard Flight 77 with a complete set of useful tools: looks, brains, charisma, an eye for sharp clothes, and a fair shot at fulfilling his dream of becoming either a professional basketball player or a scientist. But Bernard was still only eleven, which meant that his nimble mind sometimes wandered to subjects other than school.

Fifth grade had gone well, and Bernard's parents and teachers wanted him to remain on a high-achieving trajectory at the Leckie Elementary School in the southwest corner of Washington, D.C., near what was known as Bolling Air Force Base. Some students at Leckie lived in a homeless shelter, but Bernard was among the fortunate ones: he lived in military housing with his younger sister, his mother, Sinita, and his father, Bernard Brown Sr., a chief petty officer in the Navy who worked at the Pentagon. The two men of the family were known as Big Bernard and Little Bernard.

As the new school year began, Little Bernard's fifth-grade teacher successfully urged her best friend at Leckie, sixth-grade teacher Hilda Taylor, to pick Bernard to join her for a special treat: a four-day trip to study marine biology at a sanctuary off the California coast. A native of Sierra Leone, Hilda Taylor believed that American children needed to look beyond their borders to gain a deeper understanding of the wider world. With that goal in mind she'd become involved with the National Geographic Society, which sponsored the trip.

Two National Geographic staff members also found seats aboard Flight 77, along with two other pairs from Washington schools: teacher James Debeuneure and eleven-year-old Rodney Dickens, and teacher Sarah Clark and eleven-year-old Asia Cottom.

Bernard had been nervous about his first flight, but he felt reassured by Big Bernard, who coached his precocious son in basketball and life. For added confidence, and to stay true to his alternate career choice, Little Bernard marched down the aisle toward seat 20E wearing a new pair of Air Jordan sneakers.

BARBARA OLSON, BERNARD BROWN II, and the National Geographic group were among the fifty-eight passengers who filed through the door onto Flight 77, less than one-third the plane's capacity. They ranged across every age, stage, and station in life.

In the seat next to Bernard was Mari-Rae Sopper, who before boarding wrote an email to family and friends with the subject line "New Job New City New State New Life." Thirty-five years old, she'd quit working as a lawyer to head west for her dream job: women's gymnastics coach at the University of California, Santa Barbara. Five foot two, so determined that even her mother called her bullheaded, Mari-Rae had been an All-American gymnast at Iowa State University. She upended her life and accepted the coaching job even though she knew the school intended to phase out women's gymnastics after one year. Mari-Rae had a stubborn plan: she intended to persuade her new bosses to reverse the decision and continue the women's gymnastics program.

Scrambling into four seats of Row 23 were economist Leslie Whittington, her husband, Charles Falkenberg, and their daughters, Zoe and Dana, about to begin a two-month adventure in Australia. An associate dean and associate professor of public policy at Georgetown University, Leslie had accepted a visiting fellowship at Australian National University in Canberra. Along with teaching, the trip would allow her to test theories for a book she was writing about women, work, and families. A computer engineer and scientist, Charles took a leave from his work developing software that organized and managed scientific data. Earlier in his career, he developed a software system for researchers in Alaska trying to measure impacts of the Exxon Valdez oil spill. At eight, Zoe was a Girl Scout, a swim team member, a ballet student, an actress in school musicals, and a devoted reader of the Harry Potter books. At three, curly-haired, irrepressible Dana found comfort in her stuffed lamb and joy in stories about princesses. (She regularly dressed as one.)

A married couple occupied the other two seats in Row 23: quiet, retired chemist Yugang Zheng and his outgoing, retired pediatrician wife, Shuying Yang. They were on their way home to China after a nearly yearlong visit with their daughter, a medical student and cancer researcher at Johns Hopkins University in Baltimore. They'd just returned from a week of sightseeing, hiking, and swimming in Maine and had delayed their flight to spend one more day with their daughter and her husband. As a wedding gift, they'd given the young couple a statue of the goddess of compassion, Bodhisattva Guanyin, who hears the cries of the world and brings care to those in need.

In the row in front of them, Retired Rear Admiral Wilson "Bud" Flagg and Darlene "Dee" Flagg had plans for a family gathering in California. Both sixty-two, the high school sweethearts had recently celebrated their fortieth wedding anniversary and Bud's fortieth reunion at the U.S. Naval Academy. One story that made the rounds at the reunion explained how Bud had stopped his classmates from raiding his stash of Dee's cookies: he substituted a batch he'd baked with laxatives. (It was a lesson he didn't have to teach twice.) Bud served three tours as a fighter pilot in Southeast Asia during the Vietnam War. Later, he had a dual career as a pilot for American Airlines and an officer in the Naval Reserve. The Flaggs had two sons and four grandchildren, and together they ran a Virginia cattle farm.

In a window seat seven rows away sat Dr. Yeneneh Betru. He'd moved to the United States from Ethiopia as a teenager in order to fulfill a promise to his grandmother that he would become a doctor and cure whatever ailed her. Soft-spoken but determined, thirty-five years old, Yeneneh traveled throughout the United States training other doctors in the care of hospitalized patients, while spending his personal time and money assembling equipment to create the first public kidney dialysis center in Addis Ababa.

In 5B of business class sat a man known for his dapper clothes and mastery of dominoes and whist: Eddie Dillard. At fifty-four, Eddie had retired four years earlier from a career as a district manager for the tobacco company Philip Morris. Since then, he'd transformed into a savvy real estate investor. He was flying to California to work on a rental property he owned with his wife, Rosemary, an American Airlines base manager at Reagan National Airport in Washington.

In first class, newlyweds Zandra and Robert Riis Ploger III buckled into second-row seats on the first leg of a two-week honeymoon to Hawaii. Despite previous marriages and four grown children between them, Zandra and Robert acted like teenagers, holding hands and exchanging pet names: Pretty for her, Love for him. He was a systems architect at Lockheed Martin, she was a manager at IBM. Both were dedicated fans of *Star Trek*.

Predictably for a flight from Washington, spread throughout the cabin were passengers with connections to the government and the military. Bryan Jack was a PhD numbers cruncher for the Defense Department who'd won the department's Exceptional Service Medal twice in the past three years. William Caswell was a physicist with a PhD from Princeton who served in the Army during Vietnam and now worked for the Navy as a civilian. Both men were on official business trips that took them away from their offices in the Pentagon.

Dr. Paul Ambrose was a fellow at the Department of Health and Human Services, on his way to California for a conference on how to prevent youth obesity; Charles Droz was a retired lieutenant commander in the Navy who'd built a career in computer technology; Dong Chul Lee spent eighteen years working for the U.S. Air Force and the National Security Agency before taking an engineering job with Boeing; consultant Richard Gabriel had lost a leg in battle during the Vietnam War; and John Yamnicky Sr. was a barrel-chested retired Navy captain who flew fighter jets in Korea and on three tours in Vietnam.

In the cockpit, Captain Charles "Chic" Burlingame formerly flew F-4 Phantom fighters as a medal-winning pilot and honors graduate of the Navy's "Top Gun" school. Married to an American Airlines flight attendant, Chic Burlingame was an Eagle Scout, an Annapolis graduate, a father, a grandfather, and a stepfather of two. He was one day shy of fifty-two. Tucked in his wallet was a laminated prayer card from his mother's funeral, ten months earlier, with part of a poem: "I am the soft stars that shine at night. Do not stand at my grave and cry; I am not there, I did not die." Joining him at the controls was First Officer David Charlebois, a young pilot dedicated to his partner. Together they enjoyed their row house in Washington, D.C., and the border collie he'd rescued when it was a puppy.

Strapped into a jump seat in the back of the plane was senior flight attendant Michele Heidenberger, wife of a US Airways pilot and mother of two, who'd been flying for thirty-one years. Before takeoff she called her husband, Thomas, to make sure their fourteen-year-old son was awake and had packed a lunch for school.

Serving first class, flight attendant Renée May was an artist who knitted blankets for her friends and had recently accepted her boyfriend's proposal. At thirty-nine, Renée had learned only a day earlier that she was seven weeks pregnant. After landing in Los Angeles, she planned to hop a quick flight to visit her parents in Las Vegas. She'd spoken with them twice in the past two days but had told them only that she had big news to share.

Also working the flight was a couple whose friends called them Kennifer. Married for eight years, flight attendants Ken and Jennifer Lewis normally flew separately, but they used their seniority to mesh their schedules so they could vacation when they reached Los

Angeles, their favorite city. A magnet on their refrigerator read HAP-
PINESS IS BEING MARRIED TO YOUR BEST FRIEND. When they were home,
in the foothills of Virginia's Blue Ridge Mountains, Ken and Jennifer
liked to drag lawn chairs to the end of their driveway, trailed by their
five cats. As night fell, they would gaze at the stars.

ALSO ON BOARD were five young Saudi Arabian zealots who'd pledged
their lives to al-Qaeda. Like their collaborators on American Flight 11
and United Flight 175, the men chose seats strategically, clustered toward the front of the plane.

Unlike their associates aboard the other two flights, three of the al-
Qaeda members on American Flight 77 nearly had their plans foiled
by airport security.

At 7:18 a.m., Majed Moqed and Khalid al-Mihdhar set off alarms
when they walked through a Dulles Airport metal detector. Security
workers sent them to a second metal detector. Mihdhar passed, but
Moqed failed again. A private security officer hired by a contractor for
United Airlines scanned Moqed with a metal detection wand and sent
him on his way. Neither was patted down.

Almost twenty minutes later, Nawaf al-Hazmi set off alarms at
both metal detectors at the same security checkpoint. Two weeks
earlier, he'd purchased Leatherman multitool knives, and a security
video showed that he had an unidentified item clipped onto his rear
pants pocket. A security officer hand-wanded Hazmi and swiped his
shoulder bag with an explosive trace detector. No one patted him
down, and he walked on toward Flight 77 with his brother, Salem
al-Hazmi.

All five were chosen for another security screening, three by the
CAPPS computer algorithm and two, the Hazmi brothers, because an
airline customer service representative judged them to be suspicious.
One, apparently Salem al-Hazmi, offered an identification card without a photograph and didn't seem to understand English. The airline
worker who checked them in thought he seemed anxious or excited.

In the end, the selection of all five men for a second layer of security screening proved meaningless. Just as with their collaborators, it only meant that their checked bags were held off the plane until after they boarded.

Hani Hanjour, who'd trained as a pilot, took seat 1B in first class. Four rows back in the same cabin, in seats 5E and 5F, sat the Hazmi brothers. They were the only two passengers on Flight 77 to request special meals: the Hindu option, with no pork.

On the opposite side of the plane, in coach seat 12A sat Majed Moqed. Next to him, in 12B, was Khalid al-Mihdhar, slim and dark-haired, a man who U.S. intelligence officials had known for several years was a member of al-Qaeda, yet who traveled under his real name.

AMERICAN AIRLINES FLIGHT 77 pushed back from Dulles Gate D-26 at 8:09 a.m. It was airborne eleven minutes later.

At that moment, United Flight 175 had been in flight for six minutes, with no signs of trouble. American Flight 11 had already stopped communicating with air traffic controllers, and soon after, flight attendant Betty Ong began her distress call to American Airlines.

THREE MINUTES INTO their flight from Cape Cod to New York in pursuit of American Flight 11, Otis F-15 pilots Tim Duffy and Dan Nash learned that the World Trade Center had been struck by a plane, presumably the one they were supposed to find. They saw rising smoke from more than a hundred miles away. The clouds of smoke intensified minutes later with the strike on the South Tower.

As the fighter pilots approached a crime scene of almost unimaginable proportions, NEADS Major Kevin Nasypany ordered them to fly in a holding pattern in military-controlled airspace off Long Island. That way, they'd stay clear of scores of passenger planes that still flew nearby.

At 9:05 a.m., two minutes after the crash of United Flight 175, FAA controllers issued an order that barred all nonmilitary aircraft from taking off, landing, or flying through New York Center's airspace until further notice. Meanwhile, Boston Center had stopped all departures from its airports. Soon after, all departures were stopped nationwide for planes heading toward or through New York or Boston airspace.

Around the same time, fearing more hijackings, the operations manager at Boston Center told the controllers he supervised to warn airborne pilots by radio to heighten security, with the aim of preventing potential intruders from gaining access to cockpits. He urged the national FAA operations center in Virginia to issue a similar cockpit safety notice nationwide, but there's no evidence that that happened.

As the NEADS team absorbed news of the second crash into the World Trade Center, a technician uttered an offhand comment charged with insight: "This is a new type of war, that's what it is."

At first, almost no one could fathom the idea of terrorist hijackers who'd been trained as pilots at U.S. flight schools. Several technicians at NEADS held on to the idea that the original pilots had somehow remained at the controls, flying under duress from the terrorists and unable to use their transponders to issue an alert, or "squawk," using the universal hijacking code 7500.

"We have smart terrorists today," a NEADS surveillance officer said. "They're not giving them [the pilots] a chance to squawk."

Shortly before 9:08 a.m., five minutes after the South Tower explosion, Nasypany decided that he wanted the Otis fighter jets to be ready for whatever might come next from the terrorists. No simulations, exercises, or history had prepared any of them for this, and other than Boston Center's unapproved calls to NEADS, the FAA still had yet to make contact with the military. Nasypany improvised.

"We need to talk to FAA," Nasypany told his team. "We need to tell 'em if this stuff is gonna keep on going, we need to take those fighters, put 'em over Manhattan. That's the best thing, that's the best play right now. So, coordinate with the FAA. Tell 'em if there's more out there, which we don't know, let's get 'em over Manhattan. At least we got some kind of play."

Nasypany wanted to launch two more fighter jets, the pair of on-alert F-16s ready and waiting at Langley Air Force Base in Virginia.

The fighters were part of the North Dakota Air National Guard's 119th Fighter Wing, nicknamed the Happy Hooligans.

But Colonel Marr rejected that plan. He wanted the Langley fighters to remain on the ground, on runway alert. With only four ready-to-launch fighter jets in his arsenal, the colonel didn't want all of them to run out of fuel at the same time. Unaware that airborne fuel tankers would have been available, Marr thought that putting the Langley fighters in the air might leave the skies relatively unprotected if something else happened in the huge area of sky that NEADS was sworn to protect.

Nasypany's mind kept churning. Two suicide hijackers in fuel-laden jets had slammed into the North and South Towers of the World Trade Center, both of which burned on television screens all around him. They'd killed everyone on board and an unknown number of people in the buildings. Nasypany had positioned two F-15s in the sky over New York, and it remained anyone's guess if they'd soon be chasing other hijacked planes with similar deadly intentions.

"We need to do more than fuck with this," Nasypany declared.

Nasypany wondered aloud how he and his team would respond if the nation's military commanders, starting with the president of the United States, gave a shoot-down order for a plane filled with civilians. He asked members of his staff how they would react to such an order. As they scrambled to absorb the moral and practical implications, Nasypany focused on the weaponry they would use, if necessary.

"My recommendation," Nasypany told his team, "if we have to take anybody out, large aircraft, we use AIM-9s in the face."

The AIM-9 is a short-range, air-to-air missile known as the Sidewinder, with a twenty-pound warhead and an infrared guidance system that locks onto its target. Each fully armed F-16 fighter carried six of them, while each F-15 carried two Sidewinders and two larger missiles called Sparrows.

Nasypany made the comment with the professional air of a military airman who might receive a wrenching command. Then he paused a moment, as though unsettled, and added more obliquely, "If need be . . ."

The potential need to shoot down a commercial airliner filled with innocent men, women, and children remained unresolved. A short time later, a female NEADS staffer said to no one in particular: "Oh God, they better call the president."

Another staffer said: "Believe me, he knows."

In fact, President George W. Bush had learned about the World Trade Center crashes only minutes earlier, and no discussions had yet taken place about what action the military should take if more terrorists turned passenger jets into weapons of mass destruction.

Still, at NEADS they wanted to be ready. On Nasypany's orders, Otis F-15 fighter pilots Duffy and Nash left their holding pattern and established a CAP, or Combat Air Patrol, over Manhattan. A staffer from NEADS radioed Duffy to ask if he'd have a problem with an order to shoot down a hijacked passenger jet. Having seen the destruction already caused by suicide hijackers, Duffy answered flatly: "No."

Nash looked down from his cramped cockpit at the burning towers. Thick black smoke spiraled upward to space. Nash thought: "That was the start of World War III."

If Nash was correct, the next battle had already begun, and the battlefield would be Washington, D.C.

THE SKIES WERE blue, the air was smooth, and all was normal during the first half hour of American Flight 77's voyage west.

Shortly after the flight took off from Dulles, before the hijackings of Flight 11 or Flight 175 were known beyond a tiny circle of people, an FAA flight controller at the Washington Center, Danielle O'Brien, made a routine handoff of Flight 77 to a colleague at the FAA's Indianapolis Center. For reasons she couldn't explain and would never fully understand, O'Brien didn't use one of her normal sendoffs to the pilots: "Good day," or "Have a nice flight." Instead she told them, "Good luck."

Indianapolis Center air traffic controllers managed separation in the airspace over 73,000 square miles of the Midwest, including parts of Kentucky, Illinois, Indiana, Ohio, Tennessee, Virginia, and West

Virginia. When initially under the control of Indianapolis Center, Flight 77's pilots climbed, as instructed, to 35,000 feet and turned right 10 degrees. At just before 8:51 a.m., five minutes after the crash of Flight 11, the pilots acknowledged routine navigational instructions from Indianapolis controller Chuck Thomas, an eleven-year FAA veteran who tracked the flight along with fourteen other planes in his sector.

Focused on their work, neither Thomas nor the other controllers in Indianapolis Center had seen television reports about the crash of American Flight 11. They also had no knowledge of the then ongoing crisis aboard United Flight 175.

From the perspective of Chuck Thomas and other controllers in Indianapolis Center, all of them unaware of the emerging pattern of suicide hijackings, American Flight 77 began behaving in strange and unexpected ways starting at 8:54 a.m. First, the plane made an unauthorized turn to the southwest. Three minutes later, someone turned off its transponder and the plane disappeared from Thomas's radar screen.

Concerned, but with no reason to fear the worst, Thomas searched for Flight 77 on his screen along its projected flight path to the west and to the southwest, the direction in which he had seen it turn. Nothing.

"American Seventy-Seven, Indy," Thomas called over the radio. He tried five more times over the next two minutes, starting at 8:56 a.m. No reply.

Thomas called American Airlines for help in contacting Flight 77 pilots Chic Burlingame and David Charlebois in the cockpit, but the airline's dispatchers also couldn't reach them by radio. Airline officials sent a text message to the cockpit instructing the pilots to contact Indianapolis Center air traffic controllers. That went unanswered, too.

Thomas and other controllers spread word throughout Indianapolis Center that they had lost contact with Flight 77. As a precaution, they agreed to "sterilize the airspace," moving other planes out of the way of Flight 77's projected westerly path. But its failure to respond and the loss of its transponder signal made the controllers doubt the plane was still airborne. Still in the dark about what had happened in New York, they suspected that Flight 77 had experienced a catastrophic electrical or mechanical failure and that the plane had crashed. Controllers tried to call the flight for several more minutes but heard only silence. They made a final radio call to Flight 77 at shortly after 9:03 a.m.—coincidentally, at almost the exact moment United Flight 175 hit the South Tower.

Meanwhile, a conference call about the hijackings of American Flight 11 and United Flight 175 among air traffic controllers in the Boston, New York, and Cleveland centers didn't include Indianapolis Center. The reason was at once straightforward and tragically wrongheaded: no one thought a hijacked plane was headed that way, so the FAA didn't want to distract them from their work.

At 9:08 a.m., more than twenty minutes after American Flight 11 had hit the North Tower, Indianapolis controllers remained unable to find Flight 77 on radar or raise the cockpit by radio. They called the West Virginia State Police, Air Force Search and Rescue in Langley, Virginia, and the FAA's Great Lakes Regional Office to alert them to the possible crash of Flight 77. They considered a downed plane the most likely outcome; in an information vacuum about the other incidents, nothing of what the Indianapolis controllers saw fitted their expectations or training about a "traditional" hijacking.

But in fact, American Flight 77 was still airborne.

Someone in the cockpit had made a hairpin turn over Ohio. As a result, Indianapolis air traffic controllers were looking for the plane in exactly the wrong direction. While they searched to the west and southwest, because that's where the plane had been heading, Flight 77 now pointed east.

Its autopilot was set for a new course: to Reagan National Airport, in the heart of Washington, D.C.

CHAPTER 7

"BEWARE ANY COCKPIT INTRUSION"

United Airlines Flight 93

UNITED AIRLINES FLIGHT ATTENDANT CEECEE LYLES'S CELLPHONE rang before 5 a.m. as she slept on the futon in her crash pad apartment in Newark, New Jersey. Only a few hours had passed since she fell asleep midconversation with her husband, Lorne. Now he called to wake her up, so she wouldn't miss her flight. As soon as she opened her eyes, they resumed their seemingly endless, "everything and nothing" phone conversation.

As CeeCee got ready for work, she quizzed Lorne about his overnight shift as a Fort Myers, Florida, police officer. She dawdled in the bathroom, fixing her hair and perfecting her makeup before donning her navy-blue uniform. Three minutes before an airport bus made its 6:15 a.m. stop at the apartment building, a flight attendant she roomed with called out: "Girl, you're going to miss that shuttle!"

CeeCee grabbed her bags and bolted out the door. She and Lorne kept talking on her ride to Newark International Airport, reviewing what bills and chores he should handle while she traveled. They talked until her seven o'clock briefing with fellow flight attendants at the United Airlines operations center, located beneath baggage claim in Terminal A. They resumed talking at 7:20 a.m. and continued their conversation until CeeCee reached the security checkpoint. They talked again as she walked through the terminal to United Flight 93, a Boeing 757 parked at Gate 17A.

CeeCee told Lorne she expected an easy day. The nonstop flight to San Francisco had been assigned five flight attendants, she told him, despite a sparsely filled cabin. First class would have ten passengers and coach would have only twenty-seven, which meant that four out of every five seats on Flight 93 would be empty.

Work had begun and CeeCee said goodbye. Lorne told her he loved her and to call when she landed.

THROUGHOUT THE MORNING, as minutes ticked past and the terror swelled, only the hijackers and their al-Qaeda bosses knew how many planes they intended to seize. It could be two, ten, or more. But from the terrorists' perspective, the first hour of their attack went like clockwork: so far, they'd hijacked three planes, two of which had struck their targets in New York and the third was under their control, coursing toward Washington, D.C.

Those results were the fruits of a poisoned tree. After months of research and reconnaissance led by Mohamed Atta, the hijackers had guessed correctly about how their victims in the air and their enemies on the ground would and wouldn't react to a hostile airborne takeover. During the first three hijackings, fifteen terrorists had used planning, training, subterfuge, and deadly violence to exploit preconceived notions and gaping weaknesses they'd identified in U.S. airline security, all in service to Osama bin Laden's 1998 declaration of war against the United States and its people.

The hijackers on American Flight 11, United Flight 175, and American Flight 77 had boarded without incident, despite their apparent possession of short-bladed knives, not to mention previous travels and associations that should have been flaming red flags. They'd swiftly gained access to cockpits and replaced pilots with men who'd trained to fly jets expressly for the purpose of becoming martyrs. "Muscle" hijackers spread fear by attacking several crew members and passengers. They herded the rest to each plane's rear section to keep them out of the way. Claims about bombs, whether true or (more likely) false, confused and frightened passengers and crew members into obedience, perhaps with the exception of former Israeli commando Danny Lewin on Flight 11, whose throat was apparently slashed by the hijacker who sat behind him. Announcements from terrorist pilots in the cockpits, even if not all were heard by passengers and crew members, were lies designed to trick their hostages into believing that these were "ordinary" hijackings, with political or monetary goals, and that no one else would be hurt if the terrorists were allowed to fly to their chosen destination and if authorities on the ground satisfied their demands.

During the first three flights, the tightly choreographed strategy worked. And one of the most important elements was timing.

The plan to use the hijacked planes as weapons of mass destruction depended on the hijackers' ability to commandeer and maintain control of fuel-heavy transcontinental flights that took off within a few minutes of one another. That narrow window maximized the element of surprise, which the hijackers understood or hoped would lead to a chaotic response, too late to stop them from reaching their intended targets. Conversely, delays would increase the chance that they'd be stopped on the ground by a shutdown of air traffic, confronted in the air by fighter jets, or challenged on board by passengers and crew members who might discover that other hijackings hadn't ended with safe landings and the release of innocents.

Just as Atta intended, American Flight 11 and United Flight 175 took off from Boston's Logan Airport only fifteen minutes apart, at 7:59 a.m. and 8:14 a.m. respectively, each fourteen minutes after its scheduled departure time. American Flight 77 left Washington's Dulles Airport at 8:20 a.m., ten minutes after its scheduled departure. In fact, all three planes could be described as being on schedule. Departure times typically specified when a plane was supposed to leave the gate, before taxiing and takeoff. Considering the long delays that often dogged air travel, time had been on the hijackers' side. So far.

A fourth transcontinental flight, scheduled to depart at 8:00 a.m. from another airport in the Northeast, didn't get off the ground as quickly. And that made all the difference.

THE PASSENGERS OF that flight, United Flight 93, swiftly boarded the lightly booked plane.

Mark "Mickey" Rothenberg always flew first class, thanks to his bulging frequent flier account from far-flung business trips. Trim, fifty-two years old, a husband and father of two, Rothenberg was a devotee of black cashmere sweaters, a pack-a-day smoker, and a math whiz. He settled into seat 5B for the first leg of a business trip to Taiwan for his import business.

Around him was a collection of strangers with a great many similarities, young and young-in-spirit men and women, many of whom had been shaped by sports in their youths and who channeled their competitive fires into successful careers.

Directly in front of Mickey sat thirty-eight-year-old Thomas E. Burnett Jr., tall and square-jawed, who'd parlayed a sharp mind and a knack for sales into a job as chief operating officer of a company that manufactured heart pumps for patients awaiting transplants. Analytical and ambitious, a former high school quarterback, Tom originally booked a later flight, but he'd switched onto Flight 93 to get home sooner to his wife, Deena, a former flight attendant, and their three young daughters.

Across the aisle in 4D was Mark Bingham, a goateed, thirty-one-year-old public relations executive. Six foot four and more than 200 pounds, Mark ran with the bulls in Pamplona and dressed as what he described as a "transvestite lumberjack" for Halloween. During college, he played on national championship rugby teams at the University of California, Berkeley. He still loved the bone-crushing game: he cofounded a gay-inclusive team called the San Francisco Fog. Mark's toughness extended beyond the field. Six years earlier, two muggers, one with a gun, demanded cash and watches from Mark and his then partner. Mark jumped the armed mugger, who smashed him on the head with the gun, drawing blood. Mark knocked away the gun and the muggers fled. United flights felt like homecomings for Mark: his mother and his aunt were United flight attendants. Headed to California for the wedding of a fraternity brother who happened to be a Muslim, Mark overslept and nearly missed Flight 93—a kindly gate agent had opened the jetway door and let him board.

In a small-world coincidence, six rows back sat Todd Beamer, who graduated one year ahead of Mark Bingham from the same high school in Los Gatos, California. Although both were school-boy athletes, Todd spent only his senior year there, and it's unknown whether he and Mark knew each other at school or recognized each other on the plane. Todd had a boyish face, a warm smile, and a drive for success that made him an ace salesman for computer software maker Oracle Corp. When he wasn't working, Todd devoted himself to teaching Sunday school, playing in a church softball league, and above all, spending time with his pregnant wife, Lisa, and their two young sons. At his church men's group, Todd was studying a book called *A Life of Integrity*.

In a window seat one row back sat an affable thirty-one-year-old man with curly hair, sympathetic eyes, and the thickly muscled shoulders of a powerful athlete. Jeremy Glick worked as a sales rep for a web management company, but he looked as though he'd be more comfortable in a weight room. Jeremy carried 220 pounds on his six-foot frame and held a black belt in judo. In college, he showed up alone, without a coach or a team, to a national collegiate judo championship—and he won. Jeremy and his wife, Lyz, were high school sweethearts; she had given birth three months earlier to a daughter they named Emerson, after Jeremy's favorite poet, Ralph Waldo Emerson. They called her Emmy. Jeremy reluctantly tore himself away from home for a business trip to California. A fire on September 10 at the Newark airport forced him to switch his plans to Flight 93.

In the next row, Louis "Joey" Nacke II packed almost 200 solid pounds onto his five-foot-nine frame. Joey had a taste for wine and cigars and sported a Superman logo tattoo on his left shoulder. At forty-two, with a new wife and two teenage sons from a previous marriage, Joey ran a distribution center for K-B Toys.

A few rows back sat Toshiya Kuge, an angular twenty-year-old who played linebacker for his college football team in his native Japan. Returning home after his second visit to the United States, Toshiya had spent two weeks sightseeing and sharpening his English language skills, part of his plan to earn a master's degree in engineering from an American university.

Not as young as the others, William Cashman was as tough as almost any of them: at sixty, wiry and strong, he was an ironworker who'd helped to build New York's World Trade Center. He studied martial arts and, in his youth, served as an Army paratrooper in the 101st Airborne Division. His friend Patrick "Joe" Driscoll, a retired software executive seated beside him, had spent four years on a Navy destroyer during the Korean War. Together they planned to test themselves hiking in Yosemite National Park.

Others aboard Flight 93 represented a cross-section of American life, ranging in age from twenty to seventy-nine. The oldest was Hilda Marcin, a retired bookkeeper and teacher's aide traveling to California to move in with her daughter. Flying home after visiting friends in New Jersey, the youngest passenger was Deora Bodley, a junior at Santa Clara University who dreamed of becoming a child psychologist. U.S. Census workers Marion Britton and Waleska Martinez were heading west for a conference. Computer engineer Edward Felt was rushing to San Francisco on a last-minute business trip. Attorney and engineer Linda Gronlund and her boyfriend, computer software designer Joseph DeLuca, were going to California's wine country to celebrate Linda's forty-seventh birthday.

Donald and Jean Peterson, the only married couple on the plane, were, like Cashman and Driscoll, headed to Yosemite National Park, for a vacation with Jean's parents and her brother. They originally held tickets for a later flight but arrived early at Newark and were given seats on Flight 93. A retired electric company executive, Don counseled men struggling with alcohol and drug dependency. Packed among his belongings was a Bible in which he'd tucked a handwritten list of the names of men he was praying for.

Donald Greene, an experienced pilot who worked as an executive in an aircraft instrument company, planned to join his brothers at Lake Tahoe for a hiking and biking trip. He'd packed his gear in a green duffel bag adorned with the words "Courageous Challenge." Honor Elizabeth "Lizz" Wainio was a district manager in the retail arm of the Discovery Channel Stores, heading west on business. Andrew "Sonny" Garcia—who had worked as an air traffic controller years earlier with the California National Guard—was going home after a meeting for his industrial supply business. Richard Guadagno, a biologist who'd studied close-quarters fighting as part of his training as a federal law enforcement officer, was returning to his job as manager of the Humboldt Bay National Wildlife Refuge in Eureka, California, after celebrating his grandmother's hundredth birthday.

At thirty-eight, expecting her first child, Lauren Grandcolas was an advertising executive and aspiring author, returning home to California from a memorial service for her grandmother. Retired bartender John Talignani was traveling west to support his family after the death of his stepson, who'd been killed in a car crash on his honeymoon. A cane and a mobility scooter hadn't stopped Colleen Fraser from becoming a fierce advocate for the disabled and helping to draft the Americans with Disabilities Act. When Congress debated the bill, Colleen commandeered a paratransit bus and drove fellow activists to Washington to lobby senators. She was on hand when President George H. W. Bush signed the bill into law.

The thirty-seven passengers on Flight 93 would be cared for by CeeCee Lyles and four others: chief flight attendant Deborah Welsh, who loved exotic places and donated extra airline meals to homeless people in her Manhattan neighborhood; Lorraine Bay, an easygoing veteran of thirty-seven years in the sky who mentored younger flight attendants; Sandra Bradshaw, who'd cut back on her schedule to spend more time with her two toddlers, her stepdaughter, and her husband, Phil, a pilot for US Airways; and Wanda Green, who served as a deacon in her church, was a single mother to her daughter and son, and nearly thirty years earlier had become one of United's first African American flight attendants.

Pilot Jason Dahl learned to fly at thirteen and rose swiftly at United to become a "standards" pilot who trained and tested his fellow pilots. When his son Matt's sixth-grade class went to Washington, D.C., Jason arranged to fly the plane, to make sure they arrived safely. Whenever he flew, Jason carried a small box of rocks, a treasured keepsake from Matt. Jason planned to be home in Colorado on Friday for his wedding anniversary. He had a cascade of surprises planned for his wife, Sandy, a United flight attendant, starting with a baby grand piano programmed with their wedding song. He'd also arranged a manicure, a pedicure, and massage; after that, he planned to prepare a gourmet dinner for Sandy and sixteen of their friends. Then they'd fly to London.

Jason had never flown with his copilot, LeRoy Homer Jr., but they were cut from the same cloth. LeRoy had filled his boyhood bedroom with model planes and started flying lessons at fifteen. He had graduated from the Air Force Academy, served in Operation Desert Shield and Operation Desert Storm against Iraq, and flown humanitarian missions in Somalia. Thirty-six, soft-spoken and charming, LeRoy had served as a major in the U.S. Air Force Reserves. He traveled regularly with his wife, Melodie, a nurse he'd met through mutual friends, but they'd scaled back their adventures since the birth of their daughter, Laurel, eleven months earlier. Inscribed inside his wedding band was part of a Bible verse on life's blessings: faith, hope, and love. The inscription was the next line, "And the greatest of these is love."

SEATED IN FIRST class, four men from the Middle East—three from Saudi Arabia and one from Lebanon—had murder and martyrdom in mind. All four had checked out of the Newark airport's Days Inn that morning and had passed through security without incident. The CAPPS security system selected one, Ahmed al-Haznawi, for additional screening. Following the same steps as the screeners at Logan and Dulles airports, Newark's security staff checked his suitcase for explosives, didn't find any, and held it off the flight until Haznawi boarded.

Ziad Jarrah, the onetime Lebanese disco habitué who became part of Atta's extremist Hamburg crew and trained as a pilot, sat in seat 1B, closest to the cockpit.

Before boarding the plane, Jarrah made five telephone calls to Lebanon, one to France, and one to his girlfriend, Aysel Sengün, in Germany, to whom he'd sent a farewell letter and a package of mementos a day earlier. She was in the hospital after having her tonsils removed. The connection was clear, the conversation banal. Sengün heard no noises in the background, and she claimed to detect nothing strange or suspicious about the call. He asked how she was doing, then told her, "I love you."

Sengün asked, "What's up?" Jarrah said "I love you" again, then hung up.

Ahmed al-Haznawi sat in the last row of first class, in seat 6B, directly behind glassware importer and consultant Mickey Rothenberg. Saeed al-Ghamdi and Ahmed al-Nami sat in 3D and 3C. At least one of the four men possessed the terrorist instruction sheet that began with "The Last Night," tucked into either his carry-on or his checked luggage. Among the commands for the last phase, once they boarded the plane, were the following:

> Pray that you and all your brothers will conquer, win, and hit the target without fear. Ask Allah to bless you with martyrdom, and welcome it with planning, patience, and care. . . .
>
> When the storming begins, strike like heroes who are determined not to return to this world. Glorify [Allah—that is, cry "Allah is Great"], because this cry will strike terror in the hearts of the infidels. He said, "Strike above the necks. Strike all mortals." And know that paradise has been adorned for you with the sweetest things. The nymphs, wearing their finest, are calling out to you, "Come hither, followers of Allah!"

If the group of terrorists on United Flight 93 tried to follow the pattern of their collaborators aboard Flights 11, 175, and 77, they were clearly one hijacker short. A Saudi man who authorities later suspected was supposed to have been the twentieth hijacker had landed a month earlier at Florida's Orlando International Airport, arriving on a flight from London. He landed with no return ticket or hotel reservations, carried $2,800 in cash and no credit cards, spoke no English, and claimed he didn't know his next destination after he intended to spend six days in the United States. He grew angry when questioned by an alert immigrations inspector named José E. Melendez-Perez, who suspected that the man was trying to immigrate illegally. Melendez-Perez thought the Saudi fit the profile of a "hit man." He consulted with supervisors, then forced the man onto a flight to Dubai, via London.

Waiting in vain that day at the Orlando airport was Mohamed Atta.

AT 8:00 A.M., Flight 93's scheduled departure time, the 757 pushed back from the Newark gate, but it didn't get far. It fell into a tarmac conga line with perhaps fifteen other planes, stopping and starting, slowly taxiing toward the runway. Passengers in first class drank juice, while those in coach went thirsty. Ten, twenty, forty minutes crawled past.

A few seconds before 8:42 a.m., pilots Jason Dahl and LeRoy Homer Jr. heard the command from the tower: "United Ninety-Three . . . cleared for takeoff."

Nearly a half hour had elapsed since the start of the hijacking of American Flight 11. Betty Ong and Amy Sweeney had already called American Airlines offices in North Carolina and Boston and had provided information about the hijackers' identities and tactics. Major Kevin Nasypany's team at NEADS had been notified about Flight 11 five minutes earlier. The F-15s at Otis Air Force Base had been ordered to battle stations less than a minute before. The pilots of United Flight 175 had just notified air traffic control about a strange radio transmission they had heard from Flight 11.

At the World Trade Center in New York, a businessman from New Jersey named Ron Clifford straightened his yellow tie and pushed through the revolving doors leading to the lobby of the North Tower. If United Flight 93's runway delay had lasted a little longer, the pilots, flight attendants, and passengers aboard that plane might have seen American Airlines Flight 11 zooming through cloudless blue skies toward the very tower, only fourteen miles to the northeast, where Ron awaited what he thought would be the most important meeting of his career.

It's also possible that if Flight 93 had been delayed a bit longer, it would have been caught in a "ground stop" and would never have taken off at all.

AS UNITED FLIGHT 93 took flight and headed west, the men and women on board were in a kind of suspended animation, unaware that the world had already changed.

At 8:52 a.m., ten minutes after Flight 93 became airborne, a male flight attendant aboard United Flight 175 called the airline's maintenance center to report the murder of both pilots, the stabbing of a flight attendant, and his belief that hijackers were flying the plane. Ten minutes passed, then a maintenance supervisor called United's operations center in Chicago to report the hijacking of Flight 175. After initial confusion about whether the report actually involved American Flight 11, United's managers spread word up their chain of command to United's chief operating officer, Andy Studdert, and the company's chief executive, James Goodwin. It took another thirty minutes to activate a crisis center at United's Chicago headquarters.

Beginning at 9:03 a.m., several United flight dispatchers used the cockpit email system called ACARS to inform pilots that planes had crashed into the World Trade Center. But those messages didn't include specifics about hijackings, warnings to enhance cockpit security, or suggestions about other precautions.

At 9:08 a.m., United Airlines flight dispatchers based at the company's operations headquarters in Chicago sent messages to transcontinental planes waiting to take off, informing crew that a ground stop had been placed on commercial flights at airports around New York.

Still no one sent word to Flight 93 or other vulnerable flights already in the air.

By 9:15 a.m., as the Twin Towers burned, Flight 93 had spent more than ten minutes at its cruising altitude of 35,000 feet. Flight attendants would have begun cabin service. Pilots Jason Dahl and LeRoy Homer Jr. engaged the 757's autopilot as they flew west over Pennsylvania. All seemed normal.

They remained oblivious to the hijackings and suicide-murder crashes of American Flight 11 and United Flight 175 by men from the Middle East who sat in first class and business, who killed passengers and crew members, who forced their way into cockpits and took control. No one told them that a hijacker on Flight 11 had said "planes," plural. They also hadn't been told about the disappearance of American Flight 77, which had occurred roughly twenty minutes earlier. During communications with ground controllers, the Flight 93 pilots' biggest worry seemed to be some light chop and a headwind that might hinder their plan to make up for the ground delay and land in San Francisco close to their scheduled arrival time of 11:14 a.m.

During fourteen routine communications from FAA ground controllers in the first minutes of Flight 93's journey, no one mentioned to Jason Dahl or LeRoy Homer Jr. the crisis affecting at least three other westbound transcontinental flights, the fighter jets patrolling the sky over New York City, or the possibility that other commercial flights might be victimized.

Then, almost simultaneously, worry struck two individuals on the ground who had personal connections to the pilots of Flight 93. Both tried to reach the men in the cockpit.

MELODIE HOMER HEARD her alarm early that morning, then fell back asleep. As always when he flew, her husband, LeRoy, had laid out his uniform the night before, with his epaulets and ID in his pockets, so he could dress silently in the bathroom without waking her. Before he left for the ninety-minute drive from their southern New Jersey home to the Newark airport, LeRoy whispered that he was leaving. He said he'd call when he landed and that he loved her.

Later that morning, after dropping off their infant daughter at a neighbor's house, Melodie returned home and turned on the television as she made breakfast. She watched, stunned, as a plane crashed into the World Trade Center's South Tower. As her mind reeled, through her shock Melodie vaguely heard a newscaster say something about a possible problem with air traffic control. She grabbed a sheet of paper from the refrigerator with LeRoy's flight information and called the United Airlines flight operations office at New York's John F. Kennedy International Airport, which handled all New York–area flights for the airline. She told a receptionist that LeRoy was the first officer on Flight 93, and that she worried whether he was all right. After a short hold, the receptionist returned and assured her, "I promise you, everything is okay."

Melodie sobbed with relief. The receptionist thoughtfully asked if she wanted to send LeRoy a message through the Aircraft Communications Addressing and Reporting System, or ACARS. Melodie took several deep breaths. Her voice cracking, she asked that the message to her husband read, "Just wanted to make sure you're okay."

As sent by Tara Campbell, a United flight operations service representative, the message read: "LeRoy, Melody [*sic*] wants to make sure you are O.K.! Send me back a message."

Melodie's message reached Flight 93 at 9:22 a.m., the same time as either Jason or LeRoy casually complained about the headwinds to an air traffic controller.

ACARS messages generally arrive in the cockpit in one of two ways: either an indicator light flashes MSG, to alert pilots to a digital message on their screens, or a hard copy automatically prints out at a console between the pilots' seats. Airline dispatchers can also alert pilots with a bell that chimes when an electronic ACARS message arrives. Campbell had the ability only to send Melodie's message to the Flight 93 cockpit printer.

Personal messages were unusual on the ACARS system, yet despite the request that he reply, LeRoy didn't do so. It's possible that neither he nor Jason noticed the message, as they carried out routine duties. When she didn't hear back, Tara Campbell sent Melodie's message to the cockpit printer a second time, and then a third. There was still no response.

There might have been a benign if multifaceted explanation why LeRoy didn't answer: not having been warned about multiple hijackings that had begun roughly an hour earlier; unaware of the World Trade Center crashes that had begun more than a half hour earlier; uninformed about the burning towers that Melodie had seen on television; not knowing that another transcontinental flight had disappeared from radar—without all this information, it's possible that LeRoy couldn't imagine why his wife was worried. With blue skies ahead and a job to do, perhaps he didn't see a reason to reply immediately.

While she waited, Melodie held tight to the receptionist's promise that "everything is okay."

AT NEARLY THE same time, without direction from airline officials, the FAA, or anyone else, one midlevel United Airlines employee felt stirred by the same cautious impulse that seized Melodie Homer.

At sixty-two, balding and ruddy-cheeked, a hobby sailor in his free time, Ed Ballinger had started working for United Airlines in 1958 as a teenage weather clerk. Forty-three years later, he'd risen to transcontinental dispatcher in the airline's Chicago operations headquarters. Ballinger wasn't scheduled to work September 11, but he owed his employer a day, so he arrived at eight o'clock Eastern time and began his shift.

Ballinger's job at United called for him to monitor the progress of flights assigned to him, to inform pilots of safety information, and to cancel or redirect flights that he and the pilots believed couldn't operate without undue risk. He based his decisions on a company-wide priority list called the Rule of Five: Safety, Service, Profitability, Integrity, and Responsibility to the Passenger.

When he arrived at work, Ballinger harked back to his first job at United and took note of the perfect weather across the United States for the sixteen flights he'd track. Two of those were United 175 from Boston and United 93 from Newark.

Unlike FAA air traffic controllers, Ballinger normally didn't use radar to track his flights; he followed their progress with a computer system that anticipated where a plane presumably would be along its route based on its flight plan. He focused much of his time on reviewing preflight plans such as fuel load and flight path before approving takeoffs, while keeping track of real and potential delays. Once flights were in the air, United pilots primarily communicated with FAA controllers. Ballinger and other dispatchers couldn't monitor radio calls between flights and the FAA, so to a large degree he remained in the dark, too.

Sometimes even Ballinger's fellow United Airlines employees weren't much help, either. When a flight attendant aboard Flight 175, believed to be Robert Fangman, reported the plane's hijacking to the United maintenance center in San Francisco, roughly ten minutes passed before that information reached Ballinger in Chicago. Immediately, Ballinger sent a carefully worded, purposely vague ACARS message to the United Flight 175 cockpit: "How is the ride. Any thing [*sic*] dispatch can do for you."

If Flight 175 pilots Victor Saracini and Michael Horrocks had been at the controls under duress from hijackers, they might have signaled trouble, perhaps by using the hijack code word "trip." But based on the telephone calls from United 175's passengers and crew, the pilots almost certainly were already dead. Either way, they would soon be. Ballinger sent that message at 9:03 a.m., at almost the precise moment that Flight 175 plowed into the South Tower.

Five minutes later, Ballinger learned about the ground stop around New York City, so he sent messages to a half dozen United planes at New York–area airports, telling them to stay put.

As information churned around United's headquarters, Ballinger pieced together what he knew: two planes had hit the World Trade Center; Flight 175 had been hijacked; and the FAA had ordered a ground stop. The first priority on United's Rule of Five rang clear in his mind: safety. He needed to spread the word, by alerting "his" pilots to the violent cockpit takeover tactics hijackers had used aboard Flight 175.

At 9:19 a.m., Ballinger hurriedly began to send ACARS messages to his flights, one after another, first to planes that hadn't yet taken off, and then in order of departure time: "Beware any cockpit introusion [*sic*]. Two aircraft in NY, hit Trade C[e]nter Builds." Ballinger sent the message in batches, to several flights at a time. One message went to Flight 175, which had crashed twenty minutes earlier. In the heat of the moment, Ballinger sent the message despite already knowing that Flight 175 had been hijacked; he didn't yet know that it was the plane that had hit the South Tower.

Ballinger's ACARS messages marked the first direct warnings of danger to planes by United Airlines or American Airlines, or from air traffic control, for that matter. To be certain that his warnings reached the pilots, Ballinger sent them as both digital messages, with a chime, and as printed-out text messages. He knew that every cockpit contained a fire ax, located behind the first officer's seat. Ballinger expected pilots who received his message to move the hammer-sized weapon to the floor near their feet, for easy access, to defend their planes, their lives, and the innocents on board.

Shortly before he sent the warning to Flight 93, Ballinger received a happy-go-lucky ACARS message from Captain Jason Dahl: "Good morning . . . Nice clb [climb] outta EWR [Newark Airport]." Jason commented about the sights from the cockpit and the weather, then signed off with his initial, *J*.

After Ballinger began notifying his flights to guard their cockpits, United's air traffic control coordinator sent his own message of warning to the airline's dispatchers: "There may be [additional] hijackings in progress. You may want to advise your [flights] to stay on alert and shut down all cockpit access [inflight]." Ballinger didn't notice the message; he was already too busy contacting his flights.

While Ballinger progressed through his list, Melodie Homer's ACARS message reached the Flight 93 cockpit first. One minute later, at 9:23 a.m., Ballinger sent Jason Dahl and LeRoy Homer Jr. his cautionary message to "beware."

Less than a minute later, Ballinger and other dispatchers received word from United's chief operating officer, Andy Studdert, that "Flt 175–11 [denoting the date] BOS/LAX has been involved in an accident at New York."

Either before they received Ballinger's warning or before they read it, Jason Dahl or LeRoy Homer Jr. checked in with a routine altitude and weather report to an air traffic controller at the FAA's Cleveland Center: "Morning Cleveland, United Ninety-Three with you at three-five-oh [thirty-five thousand feet], intermittent light chop." The controller didn't reply; he was busy rerouting planes affected by the ground stop. At 9:25 a.m., Flight 93 checked in again with Cleveland Center. This time the controller answered, but still he didn't warn them.

One minute later, at 9:26 a.m., Ed Ballinger's intrusion warning registered with the pilots of Flight 93. Jason Dahl's chatty messaging tone changed. He wrote a hasty, misspelled ACARS reply: "Ed cofirm latest mssg plz—Jason."

In a stressful atmosphere, it wouldn't have been hard to overlook the "plz" in Jason Dahl's reply and focus instead on the misspelled word "confirm." That was especially true for Ballinger, as he kept track of fifteen flights after having just learned from one of United's top officials that the sixteenth plane on his roster, United Flight 175, had crashed in New York. Without the word "plz," the response from Flight 93 pilot Jason Dahl could easily read as a simple acknowledgment of a message received—"Ed cofirm latest mssg"—as opposed to a worried request for more information.

Ballinger didn't immediately reply to Flight 93. In the meantime, at 9:27 a.m., the pilots responded to a routine radio call from a Cleveland air traffic controller, who told them to watch for another plane twelve miles away and two thousand feet above them.

"Negative contact," Jason Dahl replied. "We're looking."

Seconds later, at 9:28 a.m., every missed opportunity, every minute of delay in the spread of information and warning, every bit of bad luck and timing, coalesced in the cockpit of United Flight 93. The terrorists' element of surprise remained intact, and Melodie Homer's and Ed Ballinger's worst fears came true.

CHAPTER 8

"AMERICA IS UNDER ATTACK"

American Airlines Flight 77

EVEN AFTER TWO HIJACKED JETS STRUCK THE WORLD TRADE CENTER, even as concern mounted among Indianapolis Center controllers about strange behavior by American Airlines Flight 77, no one from the FAA informed the U.S. military that a plane that took off from Dulles Airport had stopped communicating by radio and had disappeared from radar screens after someone turned off its cockpit transponder.

Meanwhile, based on a combination of wrong and misleading information, Major Kevin Nasypany's team at NEADS began to chase a different plane, a phantom jet that no longer existed, supposedly heading south from New York toward the nation's capital: American Airlines Flight 11, which had crashed more than a half hour earlier.

The after-it-crashed search for American Flight 11 represented a striking illustration of the confusion and failed communication between the United States' air traffic control system and the nation's military during the chaotic first hour after al-Qaeda hijackers executed a plan of unanticipated complexity. Whether by design, chance, or a combination of both, the terrorists' simultaneous multiple hijackings vividly and fatally exposed vulnerabilities of America's national defense system on a scale unseen in the sixty years since Japan's attack on Pearl Harbor.

The boondoggle search for Flight 11 kicked into gear when NEADS Master Sergeant Maureen "Mo" Dooley fielded a call from the FAA's Boston military liaison, Colin Scoggins. He'd just taken part in a frenzied conference call with FAA headquarters in Washington and several regional air traffic control centers about the hijackings.

During that FAA conference call, Scoggins heard someone—he wasn't sure who—say that American Airlines Flight 11 remained aloft, flying south. If true, that meant some other plane had struck the North Tower of the World Trade Center. Scoggins consulted with a supervisor, then passed the information to Mo Dooley at NEADS in a phone call at 9:21 a.m., roughly thirty-five minutes after Flight 11 had in fact crashed.

"I just had a report that American 11 is still in the air," Scoggins told Dooley, "and it's on its way towards, heading toward Washington."

Dooley: "Okay, American Eleven is still in the air?"

Scoggins: "Yes."

Dooley: "On its way toward Washington?"

Scoggins: "That was another, it was evidently another aircraft that hit the tower. That's the latest report we have."

Dooley: "Okay."

Scoggins: "I'm going to try to confirm an ID for you, but I would assume he's somewhere over, uh, either New Jersey or somewhere farther south."

The confusion quickly deepened.

Dooley: "Okay. So, American Eleven isn't the hijack at all then, right?"

Scoggins: "No, he *is* a hijack."

Dooley: "He, American Eleven *is* a hijack?"

Scoggins: "Yes."

Dooley: "And he's heading into Washington?"

Scoggins: "This could be a third aircraft."

Dooley pulled away from the call and yelled to Nasypany: "Another hijack! It's headed towards Washington!"

"Shit!" Nasypany answered. "Give me a location."

Two hijacked planes had already crashed into buildings in New York. Hearing this new report of a possible third hijacked plane, Nasypany wanted to throw all available assets toward preventing a catastrophe in the nation's capital.

"Okay," he told his team, "American Airlines is still airborne—Eleven, the first guy. He's heading towards Washington. Okay, I think we need to scramble Langley right now. And I'm, I'm gonna take the fighters from Otis and try to chase this guy down if I can find him."

Colonel Robert Marr at NEADS approved Nasypany's plan to launch more fighters. The two F-16s from Langley and an unarmed training jet scrambled into the air.

As they focused on an airliner that no longer existed, neither Nasypany nor anyone else in the U.S. military knew that a different disaster was developing. A third passenger jet had in fact been hijacked: American Airlines Flight 77 out of Dulles Airport.

AROUND 8:51 A.M., the five Saudi Arabian men aboard Flight 77 executed their plan. They used swift, coordinated takeover methods similar to those used during the previous half hour on Flight 11 and Flight 175.

Twenty minutes later, the phone rang in the Las Vegas home of Ron and Nancy May. Nancy was getting ready for work as an admissions clerk at a community college and she missed the call. The phone rang again a minute later, and this time Nancy May heard the voice of her flight attendant daughter, Renée. They'd last spoken two days earlier, and Renée and Ron had talked the previous day. Renée had sounded happy on both of those calls.

Now Renée sounded serious. She calmly, but erroneously, told her mother that six men had hijacked her flight and forced "us" to the rear of the plane. Renée didn't say how she arrived at the number six, and she didn't explain whether the people crowded together were crew members, passengers, or both. She didn't know the fate of the pilots.

Renée told her mother the flight information and gave Nancy three telephone numbers to call American Airlines.

"I love you, Mom," Renée said. The line went dead.

Nancy yelled upstairs for Ron. Using one of the numbers from Renée, Nancy reached Patty Carson, an American Airlines flight services employee at Reagan National Airport in Washington, who had just returned to her desk from a staff lounge where she watched on television as United Flight 175 exploded into the South Tower. When Nancy relayed Renée's hijacking message, along with Renée's flight number and employee identification number, Patty Carson seemed confused. She told Nancy that she didn't think the plane that struck the World Trade Center was an American Airlines jet.

"No, no," Nancy May interrupted. "We are talking about Flight 77, in the air." She told Carson that Renée had said "We are being hijacked and held hostage."

Ron May took the phone and told Carson that since Renée had just called, it stood to reason that she couldn't have been in a plane that crashed into the World Trade Center.

Carson took the Mays' telephone number and promised to call back as soon as she knew anything. After speaking with Carson, Nancy and Ron May tried to call Renée on her cellphone but the call didn't connect. They turned on the television, hoping for news.

When she hung up with Ron and Nancy May, Carson learned that she had to evacuate the airport, a precaution prompted by reports of hijackings. On her way out of the building, Carson described the call to a flight services manager, Toni Knisley, who called her boss, American Airlines base manager Rosemary Dillard. At first there was some confusion about which plane Renée was on. When they confirmed it was American Flight 77, Rosemary Dillard stumbled backward into a chair. That morning, she'd raced to Dulles Airport with her husband, Eddie, the sharp-dressing, dominoes-playing real estate investor who was going to California to work on a property they owned. She'd kissed him goodbye and told him to come home soon.

Eddie was aboard hijacked Flight 77.

INSIDE THE U.S. DEPARTMENT OF JUSTICE building in Washington, a telephone rang in a fifth-floor office near a mural that depicted a robed figure protecting a cowering man from a lynch mob. Secretary Lori Keyton answered and heard the voice of an operator ask her to accept an emergency collect call from Barbara Olson.

Keyton accepted the charges, and Barbara calmed herself enough to choke out the words: "Can you tell Ted . . ."

Keyton cut her off and rushed into the ornate office of the U.S. solicitor general.

"Barbara is on the line and she's in a panic," Keyton told Ted Olson.

When Barbara reached him from Flight 77, Ted Olson was watching television, viewing a replay of a still unidentified passenger jet hitting the South Tower. When he heard that Barbara was on the phone, Olson's first thought was relief. It meant that Barbara wasn't on either of the planes that had crashed. Then he picked up the call.

"Ted," she said, "my plane's been hijacked."

Barbara told him the hijackers had knives and box cutters. Olson asked if they knew that she was talking on the phone, and she answered that they didn't. She said they'd ordered the passengers to the back of the plane. The call cut off.

Unlike callers from the previous two hijacked planes, neither Barbara Olson nor Renée May mentioned violence against the pilots or anyone else, nor the use of Mace or the threat of a bomb.

Ted Olson tried his direct line to Attorney General John Ashcroft, but Ashcroft was on a flight to Wisconsin. He called the Justice Department's Command Center and reported the hijacking. For some unexplained reason, Olson's call didn't trigger notification of the U.S. military. Olson asked that a security officer come immediately to his office, to offer suggestions if Barbara called again. Before the officer arrived, the phone rang.

Barbara told Ted that "the pilot" had announced that the plane had been hijacked, but it wasn't clear if she knew whether the speaker on the intercom was one of the hijackers or the original cockpit crew. She might have been operating under the old "rules" and believed the terrorists were forcing the legitimate pilots to do their bidding. Barbara said the plane was flying over houses. Another passenger told her they were headed northeast.

"What can I tell the pilot?" Barbara asked Ted. "What can I do? How can I stop this?"

Ted wasn't sure how to answer. He decided that he had to tell Barbara about the other two hijackings and crashes at the World Trade Center. Flight 77 seemed bound for the same fate; the question was where the hijackers intended to crash. Barbara absorbed the news quietly and stoically, though Ted wondered if she'd been shocked into silence.

They expressed their feelings for each other. Each reassured the other that it wasn't over yet, the plane was still aloft, and everything would work out. Even as he said the words, Ted Olson didn't believe them. He suspected that neither did Barbara.

The call abruptly ended.

AT THAT MOMENT, no one at the FAA had any idea what was happening aboard American Flight 77, or where it was.

Shortly after nine, controllers at Indianapolis Center began spreading word that Flight 77 had disappeared from their screens. At 9:09 a.m., controllers at Indianapolis Center reported the loss of contact with the plane to the FAA regional center. Fully fifteen minutes later, a regional FAA official relayed that information to FAA headquarters in Washington.

By 9:20 a.m., after the distress calls from Renée May and Barbara Olson and nearly twenty-five minutes after someone turned off the transponder on Flight 77, Indianapolis controllers finally learned that two other passenger jets had been hijacked. At that point, they doubted their initial assumptions about a crash. They and their FAA supervisors began to consider the evidence that a third passenger plane had been hijacked.

Overall, confusion and uncertainty were almost universal during the first hour after the hijackings, extending far beyond the FAA. At 9:10 a.m., a United dispatch manager wrote in a logbook: "At that point a second aircraft had hit the WTC, but we didn't know it was our United flight." As late as 9:20 a.m., dispatchers from United Airlines and American Airlines were still trying to confirm whose planes had hit the World Trade Center. During one phone call, an American Airlines official said he thought both planes belonged to his airline, while a United official said he believed that the second plane was Flight 175. He reached that conclusion in part because enlarged slow-motion images on CNN showed the plane that flew into the South Tower didn't have American Airlines' shiny metallic skin.

In fairness to FAA and airline officials, these were extraordinarily fast-moving events for which they had never trained. Also, the officials were hamstrung by a mix of incorrect or fragmentary information, as well as by a false sense of security that developed during the years since a U.S. air carrier had been hijacked or bombed. Just four years earlier, a presidential commission on air safety chaired by Vice President Al Gore focused on the dangers of sabotage and explosives aboard commercial airplanes. It also raised the possibility that terrorists might use surface-to-air missiles, and it cited concerns about lax screening of items airline passengers might carry onto planes. The commission's final report never mentioned a risk of suicide hijackings.

Ultimately, though, the FAA bore responsibility as the government agency with a duty to protect airline passengers from piracy and sabotage. Despite that mission, the FAA had significant gaps in domestic intelligence and multiple blind spots. Some of this was attributable to a lack of communication, and perhaps a lack of respect, from federal intelligence-gathering agencies. On September 11, 2001, the FAA's "no-fly list" included a grand total of twelve names. By contrast, the State Department's so-called TIPOFF terrorist watchlist included sixty thousand names. Yet the FAA's head of civil aviation security didn't even know that the State Department list existed. Two names on that State Department list were Nawaf al-Hazmi and Khalid al-Mihdhar, both on board Flight 77. That wasn't the only example of other federal agencies' not sharing information about potential threats with the FAA.

Earlier in the summer, an FBI agent in Phoenix named Kenneth

Williams had written a memo to his superiors in Washington expressing concern about Middle Eastern men with ties to extremists receiving flight training in the United States. Williams's memo presciently warned about the "possibility of a coordinated effort by [O]sama bin Laden" to send would-be terrorists to U.S. flight schools to become pilots to serve al-Qaeda. Among other recommendations, he urged the FBI to monitor civil aviation schools and seek authority to obtain visa information about foreign students attending them. The FBI neither acted on the memo nor shared it with the FAA. The FBI took a similar approach in the case of a French national named Zacarias Moussaoui who'd been receiving flight training in Minneapolis. Moussaoui was arrested less than a month before September 11 for overstaying his visa, and an FBI agent concluded that he was "an Islamic extremist preparing for some future act in furtherance of radical fundamentalist goals." The agent believed that Moussaoui's flight training played a role in those plans. On August 24, eighteen days before the attacks, the CIA described him as a possible "suicide hijacker." But when the FBI told the FAA and other agencies about Moussaoui on September 4, its summary didn't mention the agent's belief that Moussaoui planned to hijack a plane.

In the summer of 2001, the FAA seemed to ignore even its own recent security briefings. A few months before September 11, an FAA briefing to airport security officials considered the desirability of suicide hijackings from a terrorist perspective: "A domestic hijacking would likely result in a greater number of American hostages but would be operationally more difficult. We don't rule it out. . . . If, however, the intent of the hijacker is not to exchange hostages for prisoners, but to commit suicide in a spectacular explosion, a domestic hijacking would probably be preferable."

Now that scenario had come to pass, and the FAA found itself unaware and unprepared.

THE FAA'S INDIANAPOLIS Center controllers continued to search their radar screens to the west and southwest along Flight 77's projected path, having missed the plane's sharp turn back to the east. Although the plane had disappeared from radar at 8:56 a.m., it actually reappeared at 9:05 a.m. But because some controllers had stopped looking when they thought it crashed, and some looked in the wrong direction, they never saw it return to their radar screens. Neither Indianapolis Center controllers nor their bosses at the FAA command center issued an "all-points bulletin" for other air traffic control centers to look for the missing plane.

American Airlines Flight 77 traveled undetected for thirty-six minutes.

The plane's new flight path pointed it on a direct course for Washington, D.C. But yet again, no one told the U.S. military, this time about a threat to the nation's capital.

BY 9:25 A.M., even as American Flight 77 remained missing and a mystery, one top FAA official grasped the severity and growing scope of the crisis.

At the agency's operations center in Herndon, Virginia, FAA national operations manager Ben Sliney knew about the North Tower crash and had seen United Flight 175 hit the South Tower on CNN some twenty minutes earlier. He worried about the disappearance of Flight 77 and feared that more hijackings might be under way. Sliney also had heard about Mohamed Atta's "We have some planes" remark. He felt haunted by the question of how high the hijacking total might eventually reach. He couldn't undo what already happened, but Sliney hoped that he might help prevent the next attack.

Fifty-five years old, with a shock of white hair, an Air Force veteran and a lawyer by training, Sliney concluded that he had both the authority and the responsibility to take drastic action. Accordingly, he declared a "nationwide ground stop," the first order of its kind in U.S. aviation history, which prevented all commercial and private aircraft from taking off anywhere in the United States.

Making Sliney's order even more remarkable, he had only recently returned to the FAA after several years in a private law practice. The morning of September 11 was Sliney's first shift on his first day in his new job as the FAA's National Operations Manager, boss of the agency's command center.

MEANWHILE, BETWEEN ABOUT 9:23 and 9:28 a.m., American Flight 77 dropped from an altitude of 25,000 feet to about 7,000 feet as it continued on its undetected eastward path. By about 9:29 a.m., while controllers fruitlessly searched the Midwest, the Boeing 757 was almost on the East Coast, about thirty-eight miles west of the Pentagon, the physical and symbolic heart of the U.S. military, located a short hop across the Potomac River from Washington, D.C.

Hani Hanjour—or whoever was in the cockpit—disengaged the autopilot destination that he'd previously set to Reagan National Airport and took manual control of the plane.

AS AMERICAN FLIGHT 77 approached the prohibited airspace of the nation's capital, confusion about the plane spread further through the U.S. government.

At 9:31 a.m., an agent from the FBI office in Boston called the FAA seeking information on the planes that had hit the World Trade Center.

An FAA official told him: "We have two reports, preliminary information, ah, believe to be American Airlines Flight 77 and Flight 11, collided with World Trade Center. Also, a preliminary report, ah, United Airlines Flight 175 off radar. Ah, no further information."

AT 9:32 A.M., air traffic controllers at Dulles airport saw a green dot on their radar screens that no one expected, traveling eastbound at the surprisingly fast speed of about 500 miles per hour.

Among those who noticed the unidentified aircraft was Danielle O'Brien, the air traffic controller who for some reason had wished Flight 77's pilots "good luck" when she handed them off an hour earlier. From its speed and how it turned and slashed across the sky, she and other controllers initially thought the object on their radar was a nimble military jet.

O'Brien slid to her left and pointed it out to the controller next to her, her fiancé, Tom Howell, who recognized it as a threat. "Oh my God," Howell said. He yelled to the room: "We've got a target headed right for the White House!"

A Dulles manager called the FAA's control center and controllers at Reagan airport in Washington to warn them. Still no one from the FAA called NEADS or anyone else in the military's air defense system. An FAA supervisor at Dulles, John Hendershot, used a dedicated phone line to alert the Secret Service of the incoming danger. He told the men and women who protect the president and the vice president: "We have an unidentified, very fast-moving aircraft inbound toward your vicinity, eight miles west."

President Bush wasn't in Washington, but Vice President Dick Cheney was in his White House office. Secret Service agents rushed in, lifted Cheney from his chair, and hustled him to a tunnel leading to an underground bunker beneath the White House called the Presidential Emergency Operations Center. The agents also told White House staffers to run from the building.

Simultaneously, Reagan airport officials sought urgent help identifying the mystery jet. They called the closest plane in the sky: a military cargo plane that had just taken off from Andrews Air Force Base in Maryland, fifteen miles from the District of Columbia. The FAA's newly issued ban on takeoffs didn't apply to military planes, and the cargo plane's pilots hadn't heard about it, anyway.

AS THE FAA tried to identify the plane approaching the White House, little more than ten minutes had passed since Major Kevin Nasypany speculated about using a Sidewinder missile "in the face" to stop terrorists from creating another large-scale disaster.

The F-16s from Langley were airborne by 9:30 a.m. with orders from NEADS to fly to Washington. But no one briefed them about exactly why they were scrambled. The pilots defaulted to an old Cold

War plan and flew out to sea, to a training area known as Whiskey 386. The lead pilot, who'd heard about a plane hitting the World Trade Center but knew nothing about hijackings, thought he and his two wingmen were supposed to defend the capital against Russian planes or cruise missiles.

As the Langley F-16s took flight, headed the wrong way, a member of Nasypany's team pressed the issue of how they'd respond if they encountered a hijacked passenger jet being readied for use as a weapon.

"Have you asked . . . the question what you're gonna do if we actually find this guy?" wondered Major James Anderson. "Are we gonna shoot him down if they got passengers on board? Have they talked about that?"

At that moment, the man on whom shootdown authority rested stood before two hundred students, a handful of teachers, and a clutch of reporters in an elementary school in Sarasota, Florida.

PRESIDENT BUSH BEGAN his September 11 at 6 a.m. with a four-mile run at a golf course with his Secret Service protectors. Afterwards he showered, dressed, and sat for a routine, fifteen-minute intelligence briefing from CIA official Mike Morell in the president's suite at Sarasota's swanky Colony Beach Resort. Many of the president's summer 2001 briefings had included mentions of a heightened terrorism risk.

One of those briefings, received by the president on August 6, marked the first time that Bush had been told of a possible plan by al-Qaeda to attack inside the United States. Titled "Bin Laden Determined to Strike in US," the memo read in part: "Clandestine, foreign government, and media reports indicate bin Laden since 1997 has wanted to conduct terrorist attacks in the US. Bin Laden implied in US television interviews in 1997 and 1998 that his followers would follow the example of World Trade Center bomber Ramzi Yousef and 'bring the fighting to America.'"

But there wasn't a word about terrorism in Bush's security briefing on the morning of September 11. Much of it focused instead on the Israeli-Palestinian conflict.

By 8:40 a.m., Bush's motorcade had left the resort for the nine-mile drive to the Emma E. Booker Elementary School. The president intended to use the school as a backdrop to promote his "No Child Left Behind" education policy with a press-friendly event: a reading lesson with a diverse class of second graders.

On his way into the school, Bush shook hands with teachers and students. Meanwhile, senior White House adviser Karl Rove answered a call from his assistant: a plane had hit the World Trade Center. Rove passed the information to the president but said he didn't have details and didn't know what type of plane. Three decades earlier, during the Vietnam era, Bush had served as a fighter pilot in the Texas National Guard. He'd later say that his first thought was pilot error involving a light airplane. Bush also would say he wondered, "How could the guy have gotten so off course [as] to hit the towers?"

Bush ducked away from the receiving line into a classroom. At 8:55 a.m., less than ten minutes after the first crash, he spoke on a secure phone line with National Security Advisor Condoleezza Rice. A half hour had passed since the FAA first learned about the hijacking of Flight 11, but no one had immediately informed the White House, the nation's national security agencies, or the Secret Service. Rice didn't know much more than Bush's other aides.

Bush told the school's principal the situation, then walked to teacher Kay Daniels's classroom. "Good to meet you all!" he said as he entered. "It's really exciting for me to be here." Bush smiled, clapped, and followed along as Daniels led her sixteen students through rapid-fire phonics exercises.

At about 9:05 a.m., two minutes after United Flight 175 struck the South Tower, White House chief of staff Andy Card hesitated a moment at the door to the classroom. He collected his thoughts, then walked to Bush's side. Reporters watching from the back of the classroom perked up, knowing that no one would interrupt the president's event unless something major happened. Card bent at the waist and whispered in Bush's ear:

"A second plane hit the second tower. America is under attack."

Bush's eyes widened, and his expression slackened, a moment preserved for posterity by a photographer for The Associated Press. Card purposely stepped back—he'd say later that he did so so that Bush couldn't ask him a question with cameras and reporters capturing the president's every move and word.

Bush remained silently seated in the classroom for roughly seven more minutes, a decision for which he would be harshly criticized as a deer in the headlights of history. He would explain his reaction as responsible leadership, calculated to project calm and prevent panic. As those minutes passed, his eyes darted left and right, his mouth turned down at the corners. Bush appeared to follow along as Daniels led her students through a story called "The Pet Goat," about a girl who defends her ravenous goat, which by the story's end becomes a hero.

Members of the White House press corps on the other side of the classroom began to receive emergency alerts on wireless pagers clipped to their belts or handbags. Standing among them, Bush's press secretary Ari Fleischer held up a handwritten sign instructing the president: DON'T SAY ANYTHING YET.

At 9:15 a.m., the school's principal told the students to close their readers and place them under their chairs. "These are great readers," Bush said. "Very impressive!"

The president stood and joined Card, Rove, and other top staffers in a secured classroom. There he spoke by phone with Cheney, Condoleezza Rice, and the director of the FBI, Robert Mueller, while his staff arranged for what they expected would be a fast return to Washington. Secret Service agents wanted to hurry Bush out of the school. The president's schedule had been publicized well in advance, and they feared that the event might be a target for a terrorist "decapitation" attack. That is, to kill the head of state.

Bush refused to leave—he first wanted to speak to the nation.

At 9:30 a.m., before television cameras and an audience of students, teachers, and reporters in the school's media center, Bush delivered his first remarks about the attacks of September 11.

"Ladies and gentlemen, this is a difficult moment for America," the president began. "Today we've had a national tragedy. Two airplanes have crashed into the World Trade Center in an apparent terrorist attack on our country. I have spoken to the vice president, to the governor of New York, to the director of the FBI, and have ordered that the full resources of the federal government go to help the victims and their families and to conduct a full-scale investigation to hunt down and to find those folks who committed this act. Terrorism against our nation will not stand."

Bush asked for a moment of silence, then said: "May God bless the victims and their families and America." Then he left.

UNAWARE OF THE crisis, inside the cockpit of a military cargo plane flying over Washington, Lieutenant Colonel Steven O'Brien played tour guide, pointing out the capital's buildings and monuments to his copilot, Major Robert Schumacher. Their unarmed plane was a lumbering gray four-propeller workhorse called a C-130, a model prized since the 1950s for delivering supplies and soldiers anywhere in the world. The plane's call sign was Gofer 06.

After a cargo pickup in the Virgin Islands, O'Brien and his crew had spent the previous night at Andrews Air Force Base in Maryland before returning home to Minnesota. They took flight about 9:30 a.m. without having heard about hijackings or crashes.

As Gofer 06 flew at an altitude of about 4,000 feet, a large silver passenger jet streaked past the cockpit windshield about four miles ahead of them, descending rapidly. At first, O'Brien and Schumacher thought the jet's pilot was in trouble, struggling to keep his aircraft aloft. The hairs on the back of Schumacher's neck stood up. He said a silent prayer: "God, just let them land safely."

Their radio came alive with the voice of a Reagan National Airport air traffic controller: "Do you see an airplane?"

That was an understatement, O'Brien answered. The passenger jet was so close to their C-130 it nearly filled their windshield. If Gofer 06 had taken off even a short time earlier, the two planes might have collided.

"Can you tell me what kind it is?" the controller asked.

O'Brien said it looked like a Boeing 757 or a 767. Its silver fuselage with a red stripe strongly suggested that it belonged to American Airlines.

The controller gave O'Brien an order he'd never heard in more than two decades of flying: follow that plane. He turned Gofer 06 and gave chase.

ALTHOUGH A CIVILIAN air traffic controller had just asked a military C-130 to follow American Flight 77 as it streaked over Washington, still no one at the FAA or any other government agency had informed the U.S. military that a plane that took off more than an hour earlier from Dulles had been hijacked.

Meanwhile, acting on mistaken information they'd been given by the FAA, Nasypany and his NEADS team remained focused on a futile search for the already crashed American Flight 11. Finally, that changed, but only by happenstance, and too late.

At 9:32 a.m., a NEADS technician called the FAA's Washington Center to ask whether controllers there had seen any sign of American Flight 11. Both sides of the conversation ran amok with incomplete or incorrect information.

"There are three aircraft missing out of Boston," the NEADS technician said at one point, repeating the erroneous information about an extra plane that had prompted the search for the phantom American Flight 11. Soon after, she added incorrectly: "They thought that the American 11 was the aircraft that crashed into the World Trade Center with the United 175. However, American 11 is not the aircraft that crashed."

American Flight 11 had in fact crashed into the North Tower roughly forty-five minutes earlier.

The NEADS military technician explained to the civilian Washington Center controller that she called because she "just wanted to give you a heads-up."

"But again," she added, "remember, nothing has been confirmed as far as with the aircraft that hit the World Trade Center. But the other one, we have its information, headed toward Washington." She meant that Flight 11 was still airborne, on an unauthorized flight path to the capital.

After a pause, the Washington Center controller offered news of his own: "Okay. Now let me tell you this. I, I'll, we've been looking. We also lost American 77."

That bombshell remark came at 9:34 a.m., delivered as an afterthought in a conversation initiated by a NEADS technician about American Flight 11. It represented the first time Nasypany's team or anyone else in the U.S. military, other than the cargo crew of Gofer 06, heard about a problem with American Flight 77. That, despite the fact that FAA controllers had been searching for the plane ever since the cockpit stopped communicating forty minutes earlier.

The news of another missing plane rocketed through the NEADS center. Trying to play catch-up, the NEADS technician quizzed the Washington Center controller for information about American 77's origin, destination, aircraft type, last contact, and last known position.

AS THE CONVERSATION between NEADS and the Washington Center ended, from the cockpit of Gofer 06, Lieutenant Colonel Steven O'Brien reported to Reagan National Airport controllers that he had seen the mystery passenger jet make a 330-degree turn, flying at just two thousand feet over downtown Washington.

Meanwhile, Colin Scoggins of Boston Center told Nasypany's NEADS team at 9:36 a.m. that an unknown plane was six miles from the White House, approaching some of the most heavily guarded airspace in the world.

That, and the earlier call with the Washington Center, prompted Nasypany to take an extraordinary step: he seized control of the airspace over Washington from the FAA, to clear a path for the F-16 fighters from Langley to intercept the intruder. Although the question remained unresolved whether the fighter pilots would be authorized to fire a missile at a passenger plane to prevent a potentially greater tragedy, Nasypany wanted them in a position where they might have a chance to protect the White House, the U.S. Capitol, or any other Washington-area target.

"We're gonna turn and burn it—crank it up," Nasypany ordered. "All right, here we go. . . . Take 'em and run 'em to the White House!"

But Nasypany's effort immediately collapsed.

"Where's Langley at?" Nasypany asked. "Where are the fighters?"

The Langley F-16 fighters were nowhere in range. Unknown to Nasypany, they hadn't flown the route he'd wanted, a direct path north to intercept the erroneously reported American Flight 11, which no longer existed. Just when they were needed most, when every second and every movement counted, the two Langley F-16s were about one hundred fifty miles away—the result of an incorrect flight plan generated by misunderstandings, a mistake on coordinates by a military air traffic controller, and an overall lack of information, communication, and coordination.

Still, with the Twin Towers burning and the capital under threat from an unknown aircraft, Nasypany wouldn't quit. He wanted the fighters to fly supersonic, faster than the speed of sound, to Washington.

"I don't care how many windows you break!" Nasypany yelled at 9:36 a.m.

A minute later, a NEADS tracking technician announced that he believed he'd spotted the hijacked plane on radar: "I got him! I got him!" the technician yelled. That meant NEADS might have coordinates to give to the fighters. But just as quickly, NEADS lost track of the plane. It didn't matter—notice of the threat had come too late. Even if the fighter pilots had broken every window in every building en route to Washington, they wouldn't have arrived in time.

American Flight 11 was long gone. And time had nearly run out for American Flight 77.

AT THAT MOMENT, Father Stephen McGraw was running late for a funeral at Arlington National Cemetery, just outside Washington, D.C.

Stressed, trying to make up for lost time, the slightly built Catholic priest had made matters worse by taking a wrong turn off the highway into a traffic jam. McGraw stewed in the motionless left lane of Route 27, alongside an expanse of lawn outside the Pentagon. He pressed both hands to his head and moaned aloud: "Oh God, I'm going to miss the graveside service."

Still nagging at McGraw's conscience was an experience from earlier in the summer. Returning to his Virginia parish on a day off, flustered after missing a turn, he drove by without stopping as ambulance workers tended to a man with a bandaged head who'd been injured in a car crash. McGraw had been ordained only three months earlier, at age thirty-five, after giving up a career as a Justice Department lawyer to follow his calling. Immediately after passing the accident, he felt as though he'd failed to live up to his priestly duties to comfort the injured man. Afterward, in his room in the church rectory, McGraw had dropped to his knees, prayed for the man, and resolved before God to never again bypass someone in need.

Now, as he fretted in his car about missing the cemetery service, McGraw knew nothing of the hijackings or the attacks in New York. Suddenly, the air filled with a whirring, whooshing noise, as though he'd been dropped inside a blender. McGraw looked out the passenger window to see an airplane flash by, so low it clipped a light pole as it approached the Pentagon. From the passenger seat, he grabbed his purple satin stole, his green book of prayers for the sick and dying, and a vial of olive oil blessed by his bishop. McGraw jumped out and ran toward the Pentagon, abandoning his car in the traffic jam to fulfill his promise.

AFTER STEERING AMERICAN Flight 77 in a looping 330-degree turn, Hani Hanjour or another hijacker in the cockpit put the jet on a collision course with a five-sided, five-story, fortress-like symbol of American military power. He pushed the throttle to maximum power and reached a speed of roughly 530 miles per hour. He pointed the jet's nose downward. Moving at about 780 feet per second, the Boeing 757 flew only a few feet above the ground, its wings momentarily level over a grassy field. A hundred feet from the building, its right wing struck a portable generator, triggering a small explosion of diesel fuel, and the right engine wiped out a chain-link fence and posts around the generator. The right wing rose, the left wing dipped. The left engine struck the ground almost simultaneously with the plane's nose touching the Pentagon's limestone-faced west wall just below the second floor.

At 9:37:46 a.m., the Boeing 757 that was American Airlines Flight 77, originally bound for Los Angeles from Dulles International Airport, exploded in an orange fireball and a plume of dense black smoke that rose some three hundred feet into the sky. The immediate toll of the impact with the west wall of the Pentagon was fifty-three passengers and six crew members, along with five murderous al-Qaeda hijackers.

Two passengers, Navy physicist William Caswell and Defense Department economist Bryan Jack, died in the building where they worked. Pilot Chic Burlingame had worked for years at the Pentagon as a Navy reserve officer, but his family would feel certain he didn't die there. The Chic they knew would have fought to the death in the cockpit to save his passengers and crew before giving up the controls.

The terrorists extinguished the life and potential of eleven-year-old "Little" Bernard C. Brown II by crashing the first plane he'd ever boarded. On almost any other weekday, Navy Chief Petty Officer Bernard Brown Sr. would have been inside the Pentagon's newly renovated Wedge One. But "Big Bernard" had taken the day off, so he wasn't there when Flight 77 carried his bright, charismatic son to his death.

The terrorists silenced Barbara Olson's fervent voice, robbing Ted Olson of her companionship and denying her followers her insights.

The Falkenberg-Whittington family would never reach Australia. Newlyweds Zandra and Robert Riis Ploger III wouldn't honeymoon in Hawaii. Yugang Zheng and Shuying Yang wouldn't see their daughter become a doctor. Bud and Dee Flagg wouldn't return to their cattle farm or watch their grandchildren grow. Dr. Yeneneh Betru wouldn't build the first public kidney dialysis center in his homeland. Eddie Dillard wouldn't return home soon to his wife, Rosemary. Mari-Rae Sopper wouldn't save the women's gymnastics team at the University of California, Santa Barbara. Renée May would never surprise her parents with news of her pregnancy.

FROM ABOVE, GOFER 06 pilot Steve O'Brien saw the scene through a haze of smoke and the memory of having witnessed a flight school classmate's fatal crash. He provided the first confirmation to Reagan National Airport controllers about the plane he'd been asked to follow: "Looks like that aircraft crashed into the Pentagon, sir."

Major Kevin Nasypany learned about American Flight 77's fate from CNN, almost ten minutes after it happened. Watching the carnage on television, Nasypany cursed the confusion that had kept the Langley F-16s on the ground longer than he'd wanted and then sent them to a military airspace far from the action.

Nasypany and others could only speculate about what might have been different had the Langley fighters been launched when he first asked for them, at 9:07 a.m., fully a half hour before Flight 77 hit the Pentagon. Some top NEADS officers firmly believed that the fighters, if they had taken flight sooner and been given the correct coordinates, might have been able to intercept American Flight 77. They didn't have authority to use weapons against the passenger plane, but if they couldn't shoot it down, the NEADS officers believed, the fighters might have forced the plane to the ground.

When he learned of Flight 77's loss and the damage at the Pentagon, Nasypany erupted in anger. "Goddammit! I can't even protect my NCA!" he said, using a military acronym for the National Capital Area.

Nasypany turned back to the work at hand, directing the Langley F-16s to create a protective cover over Washington. "Talk to me about my Langley guys," Nasypany ordered. "I want them over the NCA—*now!*" He planned to have the F-16s establish a combat air patrol, or CAP, over Washington, ready to intercept any potentially hostile plane, even though it still remained unknown whether they would have the authority to shoot down a hijacked civilian jet with hostages aboard.

As the Langley fighters belatedly moved toward Washington, unconfirmed reports of other hijacked planes continued to reach NEADS. One report involved another big Boeing 767, Delta Flight 1989, which had taken off from Boston destined for Las Vegas. Another concerned a plane from Canada supposedly headed directly toward Washington.

PRESIDENT BUSH LEARNED about the attack on the Pentagon on his ride to the airport, shortly after leaving the Emma E. Booker Elementary School. By 9:45 a.m., Bush was aboard Air Force One, where he asked the Secret Service about his family's safety. Then he called Vice President Cheney, who several minutes earlier had reached a bench in the underground tunnel to the White House bunker where he could speak on a secure telephone.

"Sounds like we have a minor war going on here," Bush told his vice president, according to notes an aide made of the call. "I heard about the Pentagon. We're at war. . . . Somebody's going to pay."

The attack on the Pentagon scuttled Bush's initial plan to return to Washington, especially with reports of more hijacked planes headed toward the capital. While Bush and Cheney spoke by phone about where the president should go, Air Force One took off at 9:55 a.m. with no set destination. The pilot pointed the nose skyward, determined to get the plane as high as possible as fast as possible.

No one outside al-Qaeda knew how many more planes and targets might be part of the terrorists' plot, and no one other than the plotters yet knew who was responsible, although speculation focused almost immediately on Osama bin Laden and al-Qaeda. If there were more planes, Bush and Cheney had yet to discuss how far the military should go to stop them. Meanwhile, the secretary of defense, Donald Rumsfeld, who normally would receive orders regarding the use of force from the president and then pass them down the chain of command, hadn't yet spoken with Bush or Cheney.

Instead, Rumsfeld had rushed to the parking lot of the Pentagon to help with rescue efforts.

AT 9:42 A.M., five minutes after Flight 77 hit the Pentagon, FAA national operations manager Ben Sliney had seen enough. Determined to prevent any potential hijacking plots not yet hatched, he issued a second unprecedented, nationwide edict.

"That's it!" he cried. "I'm landing everyone!"

Sliney ordered all FAA facilities nationwide to instruct every aircraft flying over the United States to land as quickly as possible at the nearest airport. He marched through the FAA command center and fielded questions from the forty-plus people on his staff, some of whom wondered if their building might be a target and whether the country was at war. Sliney had a direct answer: we're safe, so let's get to work.

"Regardless of destination!" Sliney shouted. "Let's get them on the ground!"

Sliney's emergency order to empty the skies would require compliance from 4,546 planes. It would take hours of effort and precise coordination on the ground and in the air. Ultimately, Sliney's demand would be met by 4,545 of those planes.

All but one.

CHAPTER 9

"MAKE HIM BRAVE"

United Airlines Flight 93

HISTORY REPEATED ITSELF FOR A FOURTH TIME MORE THAN AN hour after the hijacking of American Flight 11. Forty-six minutes after United Flight 93 took off, as it cruised at 35,000 feet over eastern Ohio, the plane abruptly dropped 685 feet. Eleven seconds into the descent, at 9:28 a.m., Captain Jason Dahl or First Officer LeRoy Homer Jr. screamed into the cockpit radio: "Mayday! . . . Mayday! . . . Mayday!"

A second raspy shout from one of the pilots revealed that the source of the emergency wasn't mechanical or electrical, but human: "Hey— get out of here!"

Barely five minutes had passed since United dispatcher Ed Ballinger had warned the pilots of Flight 93 to beware of a cockpit intrusion, followed by Jason Dahl's unanswered ACARS response that asked Ballinger to "cofirm latest mssg plz." The narrow window of opportunity to guard the cockpit against a hijacking had closed.

As the terrorists fought to displace Jason and LeRoy, one or both pilots must have kept his hand pressed on the talk button of the radio microphone: sounds of the struggle in the cockpit were heard by FAA ground controllers and by pilots of planes on the same radio frequency. Thirty-five seconds after the Mayday distress calls, LeRoy or Jason again screamed: "Hey, get out of here—get out of here!"

The messages reached Flight 93's ground controller, John Werth in the FAA's Cleveland Center. Already he was balancing an almost unimaginable burden, far beyond the sixteen flights on his radar screen. He'd heard about the first two hijackings and was calling pilots as they flew through his airspace, asking if they'd seen any trace of a missing plane: American Flight 77.

Some of those pilots had picked up bits and pieces about the World Trade Center and quizzed Werth for more details. Werth had spent three decades guiding planes through the skies over Ohio, but he struggled with what to say, worried that he might panic them. He told the pilots to call their airlines, and purposely never said the words "hijack" or "trip." All the while, Werth kept searching for American Flight 77 on radar and keeping close watch on Delta Flight 1989, which he'd been told was another suspected hijacking. The Delta flight took off from Boston, bound for Las Vegas, around the same time as American Flight 11 and United Flight 175.

Amid the chaos, Werth didn't immediately know the source of the Mayday call he heard, and he could only make out what he thought were "guttural sounds." He responded into his mic: "Somebody call Cleveland?" Then he noticed Flight 93's rapid descent and heard the second panicked call: "Get out of here!" In his headset, Werth heard screams from the cockpit.

"I think we've got one!" Werth called to his supervisor, Mark Barnik, a former police officer.

To be certain which of his flights was the latest hijack, Werth called every plane on the frequency, and he heard back from all but Flight 93. He hailed Jason and LeRoy seven times in less than two minutes with no reply. Aware of the New York crashes, Werth instantly concluded that it was a suicide hijacking. He thought the terrorists' target might be a nuclear plant forty miles from the plane's current position. As other controllers moved nearby flights out of the way, Werth instructed Barnik: "Tell Washington."

Instead, Barnik alerted the United Airlines headquarters in Chicago, where a staffer on the operations desk spoke for everyone: "Oh God, not another one."

ALTHOUGH THE HIJACKERS on Flight 93 had one fewer man and waited longer to act, they followed nearly the same script as the hijackers of Flights 11, 175, and 77. As some stormed the cockpit and attacked Jason Dahl and LeRoy Homer Jr., others moved the passengers to the rear of the plane. The hijackers tied red bandannas around their heads, claimed they had a bomb, and stabbed at least one person other than the pilots.

Three minutes after Jason or LeRoy yelled his final "Get out of here!" controller John Werth heard a new male voice from Flight 93, with a halting command of English and a Middle Eastern accent. The man breathed heavily, apparently from exertion, perhaps from fighting with the pilots or from dragging them from the cockpit. The man said: "Ladies and gentlemen: Here the captain. Please sit down, keep remaining sitting. We have a bomb on board. So, sit."

The voice almost certainly belonged to Ziad Jarrah, the only hijacker aboard Flight 93 with pilot training, who'd been in the first-class seat closest to the cockpit. He had practiced on simulators but had never before flown an actual 757. Like the other hijacker pilots, Jarrah's apparent lack of experience with the cockpit radio resulted in a threat meant to be delivered over the public address system to passengers being conveyed instead to ground controllers. The cockpit's voice recorder soon picked up the voice of a second hijacker, whom Jarrah would address as "Saeed," making it evident that a fellow terrorist at the controls was Saeed al-Ghamdi.

Listening from Cleveland Center, John Werth heard Jarrah say "a bomb." At 9:32 a.m., a Cleveland Center staffer called the FAA's Command Center in Virginia to report that United Flight 93 from Newark had become the morning's fourth hijacked plane and that it might have a bomb on board.

Werth tried to keep the hijacker pilot talking: "Er, uh . . . calling Cleveland Center . . . You're unreadable. Say again, slowly."

Jarrah didn't answer. In the meantime, Werth asked other pilots in his sector to tail Flight 93, to keep eyes on the hijacked plane.

JUST AS WITH the other hijacked flights, it's unknown how the terrorists gained access to the cockpit. One possibility was that they used knives to force a first-class flight attendant, either Deborah Welsh or Wanda Green, to open the cockpit door using a key kept in a storage compartment near the front of the plane. Unlike the other flights, however, one of those flight attendants might have become a cockpit hostage.

Over a thirty-one-minute period starting at 9:32 a.m., the plane's cockpit voice recorder captured the statements of Jarrah and at least one other hijacker, likely Ghamdi. For part of that time, it also recorded comments and pleas by one or two other native English-speaking people who weren't hijackers.

The presence of a hostage or hostages in the cockpit could be discerned at first from a stream of harsh commands Jarrah issued in English after his "Here the captain" announcement. One or two other voices later emerged, along with evidence of resistance, followed by violence. At times, it wasn't clear from the recording who was speaking or what was happening, but the placement of microphones in the cockpit revealed that Jarrah did most of the talking and that the cockpit of Flight 93 became a scene of captors brutalizing a captive or captives.

IMMEDIATELY AFTER JARRAH claimed to have a bomb on board, he turned his attention to a hostage who had evidently bravely refused to comply with the hijackers' orders. Jarrah issued a stream of commands:

"Don't move. Shut up."

"Come on, come."

"Shut up!"

"Don't move!"

"Stop!"

Next came the sounds of a seat being adjusted. Someone in the cockpit apparently continued to resist Jarrah's orders. He resumed his tirade:

"Sit, sit, sit down!"

"Sit down!"

Another hijacker, in the copilot's seat, chimed in: "Stop!"

"No more," someone pleaded, as a hijacker simultaneously ordered, "Sit down!"

In Arabic, a hijacker said, "That's it, that's it, that's it," then switched to English: "Down, down."

Jarrah shouted: *"Shut up!"*

A radio call from air traffic controller John Werth interrupted the rant: "We just, ah, we didn't get it clear. . . . Is that United Ninety-Three calling?"

Seconds later, after several unexplained clicking noises, a hijacker spoke the words of the Basmala, an Arabic verse often recited by observant Muslims before they take an action: "In the name of Allah, the Most Merciful, the Most Compassionate."

After some mumbling in Arabic, someone said: "Finish. No more. *No more!*"

Then came the words, "Stop, stop, stop, *stop!*"

Someone pleaded, "No! No, no, no, *no!*" And again, "No, no, no, *no!*"

A hijacker answered: "Down! Go ahead, lie down. Lie down! Down, down, *down!*"

The back-and-forth continued with shouted demands—"Down, down, *down!*"—and desperate pleas—"No more. . . . no more."

At 9:34 a.m., the recorder captured the voice of a native English-speaking woman, possibly one of the first-class flight attendants. She beseeched her abusers: "Please, please, please . . ."

A hijacker shouted: "Down!"

"Please, please, don't hurt me," she moaned.

"Down! . . . No more."

The woman cried out: "Oh God!"

"Down, down, down!" a hijacker answered.

"Sit down!" said Jarrah.

"Shut up!" said another hijacker.

More commands followed. With them came the sounds of a warning bell that indicated Jarrah was trying to disconnect the autopilot, to change the plane's destination to one of his choosing. A knock on the cockpit door by another hijacker briefly interrupted the abuse. Jarrah answered in Arabic: "One moment, one moment."

Then, "No more."

Then, "Down, down, down!"

Someone begged, "No, no, no, no, no, no . . ."

The answer came: "Sit down, sit down, sit down!"

"Down!"

"Sit down! Sit down! You know, sit down!"

Next came a question from the woman that might have indicated that she wasn't the only hostage in the cockpit. She asked, "Are you talking to me?"

The answer: "Down, down, down, *down!*"

Almost four minutes into the assault, the woman pleaded for her life: "I don't want to die!"

But the hijackers had decided her fate.

"No, no. Down, *down!*" one of the terrorists answered.

"I don't want to die," the woman repeated. "I don't want to die."

"No, no. . . . Down, down, down, down, down, down."

She pleaded again, "No, no, please."

A *snap* reverberated in the cockpit, captured on the voice recorder. The woman cried.

"No!"

Her crying continued. She struggled for her life. Ten seconds passed, then twenty, then thirty. More than a minute went by after the woman began to cry, the longest gap without anyone speaking in the cockpit. Finally, in Arabic, Jarrah broke the silence. He said, "That's it. Go back." Then, in English, "Back."

"That's it!" a hijacker, possibly Jarrah, said in Arabic. Then, in English, he gave an indication that the hostage still refused to surrender: "Sit down!"

Nine minutes had passed since the takeover of United Flight 93. At

9:37 a.m., a hijacker reported in Arabic that they wouldn't have any more problems from their hostage: "Everything is fine. I finished."

The woman's voice wasn't heard again.

In the final section of the four-page instruction letter in the hijackers' possession, a passage read:

> [A]pply to them the prisoners law. Take prisoners and kill them.
> As the Sublime said, "There is not a prophet who takes prisoners
> And goes forth with them on the earth."

BEYOND FLIGHT 93, chaos reigned. The violent silencing of the woman in the cockpit coincided with the moment that American Flight 77 struck the Pentagon. President Bush was en route to the airport in Sarasota, Florida, and Vice President Cheney had almost reached the entrance to the emergency bunker under the White House. Major Kevin Nasypany and his team at NEADS struggled to position fighter jets to find hijacked planes and to protect and defend against more disaster. The North Tower of the World Trade Center had been burning for nearly an hour, the South Tower for more than a half hour. Ron Clifford, the businessman in the yellow tie waiting for his big meeting at the North Tower, had already become a hero, but he didn't yet know that tragedy had struck closer than he could have imagined.

Police and other first responders streamed toward Lower Manhattan, while Twin Tower workers rushed down stairwells and fled uptown or toward bridges and waiting ferries. Grim-faced, determined, burdened by gear and buoyed by a rescuer's code, scores of New York firefighters climbed the stairs of both towers, helping those they could, even as some trapped workers on upper floors jumped or fell to their deaths, or floated a plea for help in a note tossed from a broken window, or made desperate phone calls seeking rescue or solace.

AMID THE APPARENT murder of the woman in the cockpit, terrorist pilot Ziad Jarrah pulled back the control wheel and brought the plane into a climb, reaching 40,700 feet. He dipped the wing and began a left turn, first heading south and then, as the turn continued, southeast. Within minutes Jarrah completed a sharp U-turn, pointing the 757 back toward the East Coast on a heading that would take it north of Washington, D.C.

Jarrah hadn't turned off the transponder, so FAA controller John Werth saw the turn on his radar screen. The turn was so abrupt, Werth thought anyone at the back of the plane who wasn't strapped in must have been tossed like a rag doll. He moved westbound planes out of the way to avoid midair collisions. He called Flight 93 again and again, at one point hoping that the original pilots remained in control and could punch in a transponder code of 7500, to confirm the hijack.

"Ah, United Ninety-Three, if able, ah, squawk 'trip,' please," Werth radioed.

Jarrah ignored him.

AT 9:36 A.M., less than ten minutes after the hijacking began, a supervisor at Cleveland Center followed Werth's original instructions and called the FAA's Command Center in Virginia. He reported that Flight 93 was over Cleveland and asked whether anyone at the FAA had asked the military to scramble fighter jets. If not, he suggested that he'd be more than willing to call a local military base. He was told not to do that, because "that's a decision that has to be made at a different level."

At the same moment, after FAA controllers belatedly and haphazardly told the military about hijacked American Flight 77 as it bore down on Washington, D.C., Major Nasypany ordered the Langley F-16s to fly supersonic—"I don't care how many windows you break"—to be in a position to protect the nation's capital.

Despite Cleveland Center's suggestion about involving the military; despite the crash one minute later of American Flight 77 into the Pentagon; despite the World Trade Center crashes; still no one at the FAA's Command Center in Virginia or at FAA headquarters in

Washington informed Nasypany or anyone else at NORAD, NEADS, or the Defense Department that a *fourth* transcontinental passenger jet had been hijacked and was heading toward the capital.

AT 9:39 A.M., Jarrah delivered another threatening message on the cockpit radio, again thinking incorrectly that he'd be heard by the plane's passengers and crew members: "Ah, here's the captain. I would like to tell you all to remain seated. We have a bomb aboard and we are going back to the airport, and we have our demands. So please remain quiet."

Controller John Werth tried to engage him: "Okay, that's United Ninety-Three calling?" And then, "United Ninety-Three. I understand you have a bomb on board. Go ahead."

Jarrah didn't respond. As Werth tried repeatedly, Jarrah pushed forward on the control wheel. The plane descended sharply, at a rate of 4,000 feet per minute.

"This green knob?" one hijacker asked the other in Arabic.

"Yes, that's the one."

With that move, Flight 93's terrorist pilots followed Atta's script and turned off the transponder. Nevertheless, Werth continued to track the plane using primary radar, although he had to rely on reports from other planes in his sector to estimate its altitude and speed.

As the descent continued, shortly before 9:42 a.m. the cockpit voice recorder captured another person speaking. A native English-speaking man said two words in a low-pitched tone, perhaps a moan: "Oh, man!" Apparently, someone else remained in the cockpit with the hijackers. Before the woman was silenced, she'd asked: "Are you talking to me?" The sudden emergence of the voice of an English-speaking man gave new meaning to her question. The woman's inquiry might have indicated that someone else, most likely either Captain Jason Dahl or First Officer LeRoy Homer Jr., remained alive in the cockpit and also refused to follow Jarrah's commands to "sit down."

FAR MORE THAN the passengers and crew members on the other hijacked planes, men and women aboard Flight 93 almost immediately recognized that their seatback Airfones could be lifelines to call for help and advice. Many also understood that they could use the phones as a source of comfort, for themselves and for the people they cared about most.

Without interference from the hijackers, over a span of a half hour, passengers and crew members attempted to make at least thirty-seven phone calls to United Airlines, to authorities, and to their loved ones and friends. Two calls were made using cellphones, but the rest were made using built-in Airfones from the last twelve rows of the plane. The technology allowed only eight outgoing calls at a time, and poor reception caused twenty Airfone calls to drop immediately or within seconds. The calls that connected formed a spoken tapestry of grace, warning, bravery, resolve, and love.

The content of many of the calls from Flight 93 reflected the fact that the hijackings were no longer nearly simultaneous. The forty-two-minute delay before takeoff, plus the forty-six minutes of flight prior to the hijacking, meant that word of the earlier attacks and the terrorists' suicidal tactics had spread widely on the ground. Almost as soon as telephone calls began to flow from Flight 93, passengers and crew members learned that their crisis wasn't unique. They also learned how the earlier hijackings had ended.

That knowledge became a powerful motivator. It transformed them from victimized hostages into resistance fighters.

AROUND 6 A.M. Pacific Time, or 9 a.m. Eastern, Deena Burnett padded around her home in San Ramon, California, wearing the robe of her husband, Tom, as she always did when he traveled as a top executive for a heart pump manufacturer. She watched television as she made cinnamon waffles for their five-year-old twin daughters, Halley and Madison, and three-year-old Anna Clare.

When the screen showed a second plane hit the World Trade Center, her mind raced to her husband, who she thought was still in

Manhattan on business: "What hotel is he staying in, anyway?" she wondered. "Is he at the Marriott in Times Square this time? How far is Times Square from the World Trade Center?"

Deena's mother saw the same terrifying scenes and called her with the same fears about Tom. Deena reassured her, but then Deena remembered that Tom said he'd take an earlier flight if he could, to be home by noon. Creeping worry set in. Tom's mother called next, then the call-waiting beep sounded. Deena answered the pending call.

"Hello?"

"Deena."

"Tom, are you okay?"

"No, I'm not. I'm on an airplane that's been hijacked."

He gave her a few details: the hijackers claimed to have a bomb and had knifed a passenger. Tom asked her to call the authorities. He hung up and Deena called 9-1-1. Tom called back minutes later. Word that Flight 93 wasn't the only hijacked plane had apparently filtered through the cabin from other calls. Speaking in a quiet voice, Tom went into analytical mode. He asked Deena if she'd heard about any other planes. She said yes, two planes had flown into the World Trade Center. He asked if they were commercial planes, and she replied that specifics hadn't been released. Tom told Deena the hijackers were talking about flying the plane into the ground somewhere.

AT 9:35 A.M., flight attendant Sandy Bradshaw grabbed the Airfone in Row 33, the second-to-last row in coach. She speed-dialed "f-i-x" and reached the United maintenance center in San Francisco, just as a flight attendant aboard United Flight 175 had done forty minutes earlier.

Sandy's call was the first notification to United from inside the plane. Composed and professional, she told a maintenance manager that hijackers were in the cockpit and had pulled closed the curtain in first class, which was emptied of passengers. She said the terrorists claimed to have a bomb. They had a knife, she said, and had killed a flight attendant whom she didn't name.

IN THE FOOTHILLS of the Santa Cruz Mountains of California, Alice Hoglan woke to a ringing phone in her brother and sister-in-law's house, where Alice was staying after giving birth to triplets as their surrogate. Her sister-in-law ran to Alice's room with the phone. Alice heard her son's voice, clear and strong, but the first words he spoke revealed that something had rattled the former collegiate rugby star.

"Mom, this is Mark Bingham," he said, using his first and last names. "I want to let you know I love you. I love you all." Based on what he'd heard or seen, Mark told her that three men had hijacked his flight and that they claimed to have a bomb.

"Who are they, Mark?" asked Alice, a longtime United flight attendant. Mark didn't answer. A few seconds later, he said, "You've got to believe me. It's true."

"I do believe you, Mark. Who are they?"

After another pause, Alice heard voices and murmurs in the background. The line went dead.

Alice called 9-1-1 and was connected to the FBI. Meanwhile, her brother Vaughn turned on the television, where replays of the South Tower crash played in a seemingly endless loop. Word of the Pentagon attack soon followed. Alice and Vaughn understood: the hijackings were suicide missions, and Mark's plane would almost certainly be next. Vaughn urged Alice to call Mark's cellphone, to let him know the situation and to urge him to take action.

Mark didn't answer, so Alice left a message: "Mark, this is your mom. . . . The news is that it's been hijacked by terrorists. They are planning to probably use the plane as a target to hit some site on the ground. If you possibly can, try to overpower these guys, 'cause they'll probably use the plane as a target." In her fright, Alice couldn't find the word "missile." She told Mark that she loved him and said goodbye. But first Alice repeated her message: "I would say, go ahead and do everything you can to overpower them, because they're hell-bent."

Mark didn't retrieve his mother's voicemail, but the call to action came through nevertheless, from other telephone conversations happening all around him.

AT THE SAME time that Mark Bingham spoke with his mother, former national collegiate judo champion Jeremy Glick called his in-laws' white clapboard farmhouse in upstate New York. He knew that's where he'd find his wife, Lyz, and their infant daughter, Emerson. Lyz and her parents watched the World Trade Center burning on live television when the phone rang. His mother-in-law, JoAnne Makely, answered the call.

"Jeremy," she said. "Thank God. We're so worried."

"It's bad news," he replied. He asked to speak with Lyz.

"Listen, there are some bad men on this plane," he told her. Lyz began to cry as he shared details. They repeatedly told each other "I love you."

Jeremy interrupted and told his wife, "I don't think I'm going to make it out of here. I don't want to die." Through her tears Lyz reassured him that he wouldn't die that day. Jeremy was doubtful. "One of the other passengers said they're crashing planes into the World Trade Center," he said. "Is that true? . . . Are they going to blow up the plane or are they going to crash it into something?"

"They're not going to the World Trade Center," Lyz said.

"Why?"

"Because the whole thing's on fire."

As Jeremy and Lyz spoke, JoAnne called 9-1-1 and reached a New York State Police dispatcher. After confirming Jeremy's name and flight number, the dispatcher coached JoAnne into helping Lyz to pump Jeremy for information. Jeremy told Lyz—who relayed the information through JoAnne to the dispatcher—that the hijackers were "Iranian-looking" men who'd put on bandannas when the hijacking began. He said one had a "red box" that he claimed contained a bomb and had threatened to blow up the plane.

Like Mark Bingham and several other callers, Jeremy said he saw only three hijackers. It's not known why they didn't count all four, but it's possible that Ziad Jarrah didn't participate in the initial attack, to avoid potential injury that would have prevented him from flying the plane. In that scenario, Jarrah might have slipped unseen into the pilot's seat after the three other hijackers seized the cockpit and forced remaining passengers and crew to the rear of the plane before closing the first-class curtain.

Jeremy asked a question, relayed by Lyz, that signaled deliberations had already begun among passengers and crew members about fighting back. "Okay," JoAnne told the dispatcher, "his question to you is—he's a big man; he's thirty years old; he's a big athlete. They want to know whether they should attack these three guys, rather than . . . Hello?"

The dispatcher remained on the line, but he didn't answer the question. If the passengers and crew members chose to strike back against the attackers, they'd have to decide for themselves. To do so, they'd have to be relatively certain that this was unlike any hijacking any of them had ever heard about. They'd also need to believe that the possibility of success outweighed the risk of the hijackers' detonating a bomb or crashing the jet before the counterattack could succeed.

As JoAnne spoke with the dispatcher, she kept one eye on CNN: "Oh, no," JoAnne blurted to Lyz. "Turn off the television. . . . They just crashed one into the Pentagon."

Standing beside her, Lyz shared news of the third crashed plane with Jeremy. JoAnne pulled away from the 9-1-1 call and comforted Lyz, who only three months earlier had become a first-time mother. "I know," JoAnne told her. "Be brave. The police are trying to do what they can."

Jeremy remained calm, but Lyz could hear confusion in his voice. In the midst of a spreading crisis, in the sky, in New York, outside Washington, and now in her own home, JoAnne displayed a presence of mind and a maternal instinct that she used to guide Lyz through her fear so she could give Jeremy the boost he needed.

JoAnne told Lyz: "Make him . . . make him brave."

She repeated the phrase, almost like a benediction: "Make him brave."

Soon the 9-1-1 tape picked up the sound of Lyz's voice, speaking to JoAnne: "Some guys are rallying together, and they want all the men to go and attack these . . ." Lyz said Jeremy wanted to know if she thought that was a good idea. She said she didn't know. Lyz asked him if the hijackers had guns. Jeremy said no, then he tried to ease Lyz's fears. In a joking tone, he told her that he and four other men were "going to get butter knives."

JoAnne told the dispatcher about a possible counterattack, then added: "And Jeremy doesn't know whether it's a good idea." The dispatcher changed the subject. He asked JoAnne to spell her name and to have Lyz ask Jeremy what he could see out the window, land or water, and whether the plane was circling or banking to turn.

Jeremy grew serious. He told Lyz that he and the other men had voted, and they'd reached a decision. Aware of the potential consequences, Jeremy asked for Lyz's reassurance. JoAnne's message about how Lyz could help Jeremy took root just when he needed it most.

Lyz told Jeremy: "I think you need to do it. You're strong, you're brave. I love you."

Jeremy told Lyz he loved her, too. "You've got to promise me you're going to be happy," he said. Jeremy asked Lyz not to hang up the phone.

TOM BURNETT REACHED his wife, Deena, again. He told her the passenger who'd been knifed had died. The victim likely was import consultant Mickey Rothenberg, who'd been seated among the hijackers and was the only first-class passenger who didn't try to make a phone call. When the plane took off, hijacker Ahmed al-Haznawi sat directly behind Mickey. On American Flight 11, virtually the same seating arrangement had placed terrorist Satam al-Suqami behind the tech wizard and former Israeli commando Daniel Lewin.

Tom knew about the Pentagon. He asked Deena, a former flight attendant, if she thought the hijackers had been able to smuggle a bomb onto the plane or whether they were bluffing. She didn't answer. Tom said he doubted they had a bomb. Only knives.

AT 9:43 A.M., software salesman Todd Beamer tried to call his wife, but it didn't go through. He dialed 0 on an Airfone in Row 32 and reached an operator. Todd explained the situation and asked her to get a message to his wife that he loved her. The operator grew upset and called over her supervisor, who coincidentally had the same first name as Todd's wife: Lisa.

"I'll finish the call," supervisor Lisa Jefferson told the operator. She and Todd established an immediate rapport. Todd told Lisa that two people were on the floor of the first-class cabin, injured or dead. Lisa overheard a flight attendant tell Todd they were the captain and first officer. Todd asked Lisa if she knew what the hijackers wanted. Lisa answered that she didn't.

Todd described a hijacker with a red belt and what looked like a bomb strapped to his waist. He said two hijackers with knives had entered the cockpit and closed the door behind them. As the call continued, the plane dived sharply.

"Oh my God, we're going down!" Todd yelled. "We're going down. Jesus help us." Lisa heard a man in the background cry, "Oh my God, Jesus! Oh my God!" A woman screamed. Todd yelled again: "Oh, no! No! God, no!"

The plane leveled off and Todd regained his composure: "Wait, we're coming back up." He asked Lisa if she would say the Lord's Prayer with him. From a call center outside Chicago and from a hijacked plane over Pennsylvania, Lisa and Todd prayed together: "Our Father, who art in heaven, hallowed be Thy Name, Thy kingdom come, Thy will be done, on earth as it is in heaven. Give us this day our daily bread. And forgive us our trespasses, as we forgive those who trespass against us. And lead us not into temptation, but deliver us from evil. For Thine is the kingdom, and the power, and the glory, for ever and ever. Amen." They wouldn't be the last to recite those words this day.

"Jesus, help me," Todd said. "I just wanted to talk to someone, and if I don't make it through this, will you do me a favor? Would you tell my wife and family how much I love them?" Lisa promised she would.

She offered to connect Todd with his wife, but he declined. Todd explained why he hadn't already asked for that. "I don't want to upset her unnecessarily," he said. "She's expecting our third child in January, and if I don't have to upset her with any bad news, then I'd rather not."

TOM BURNETT REACHED his wife, Deena, again, this time with a momentous decision: "A group of us are getting ready to do something," he told her.

Tom said he might not be able to call back.

SOME PASSENGERS TRIED calls that didn't connect, while others reached answering machines. Advertising executive Lauren Grandcolas, pregnant with her first child, chose words she thought would comfort her husband, Jack. In a soothing tone, she told him in a voice message: "Okay, well, I just wanted to tell you that I love you. We're having a little problem on the plane. . . . I'm comfortable and I'm okay for now. I'll, I . . . just a little problem. I love you. Please tell my family I love them, too. 'Bye, honey."

Attorney Linda Gronlund left a message for her sister, Elsa Strong. "Apparently, they, uh, [have] flown a couple of planes into the World Trade Center already, and it looks like they're going to take this one down as well." Her voice choked with emotion. Linda fought tears as she told her sister that she'd miss her. She sent her love to their parents and gave Elsa the combination to a safe containing her important papers. With a deep sigh, Linda said, "Mostly I just love you and I wanted to tell you that. I don't know if I'm going to get the chance to tell you that again or not."

Flight attendant CeeCee Lyles tried using an Airfone to resume her never-ending conversation with her police officer husband, Lorne. Roused from sleep after his overnight shift, Lorne saw UNAVAILABLE on the caller ID, so he rolled over and didn't answer. CeeCee left a message: "Hi, baby. I'm . . . Baby, you have to listen to me carefully. I'm on a plane that's been hijacked. I'm on the plane. I'm calling from the plane. I want to tell you I love you. Please tell my children that

I love them very much, and I'm so sorry, babe. I don't know what to say. There's three guys. They have hijacked the plane. I'm trying to be calm. We're turned around, and I have heard that there's planes that have been—been flown into the World Trade Center." Her voice, steady until then, began to crack. "I hope to be able to see your face again, baby. I love you. Goodbye."

Census worker Marion Britton reached an old friend, Fred Fiumano. He tried to comfort her by saying the hijackers would probably land in another country. Struggling to hold her emotions in check, Marion said she didn't buy it. She told Fred she knew about the World Trade Center, and she predicted that her plane would crash, too. Marion's census coworker Waleska Martinez called a friend's Manhattan office, but the call didn't go through.

Marion shared the Airfone with Discovery Channel Stores' district manager Honor Elizabeth Wainio, whose friends called her Lizz and who looked younger than her twenty-seven years, with big hazel eyes and brown hair framing her fair, lovely face. She reached her stepmother, Esther Heymann, at her home outside Baltimore: "Hello, Mom," Elizabeth said. "We're being hijacked. I'm calling to say goodbye." Through her shock, Esther suggested that they find a way to be together in the moment. "Let's just be in the present," Esther said. "We don't know how it's going to turn out. Let's look out at the beautiful blue sky and take a few deep breaths."

Their call continued for four and a half minutes. Esther told her: "Elizabeth, I've got my arms around you, and I'm holding you, and I love you." Elizabeth said she could feel Esther's embrace, and she loved her, too. Like several others, Elizabeth focused not on herself but on the pain she anticipated among the people she feared she'd be leaving behind: "It just makes me so sad knowing how much harder this is going to be on you than it is for me." They remained silent for a while, then Elizabeth said, "I should be talking. I'm sitting here being quiet, I'm not even talking." Esther reassured her: "We don't have to talk, we're together." After another pause, Elizabeth seemed at peace. She told Esther she knew that her late grandmothers were waiting for her.

JOSEPH DELUCA CALLED his father to say he loved him and goodbye.

Flight attendant Sandy Bradshaw told her husband, Phil, a US Airways pilot, that she saw three hijackers put red bandannas on their heads as the attack began. He told her what happened in New York. "Where are you?" he asked.

"We're over a river," Sandy said.

"Which way are you headed?" Phil asked.

"I don't know."

Phil thought a moment and asked, "Well, where's the sun?"

"It's in front of us," Sandy said.

"All right," Phil said. "You're headed east."

She told him that passengers were getting hot water from the galley as they prepared to take action. She asked him for any suggestions for fighting hijackers, but Phil's mind went blank. But he did have an idea if they regained control: Phil told Sandy to call him back when they seized the cockpit from the terrorists. He knew how to fly a 757, and he'd talk someone through it. They expressed their love for each other. Sandy told him to raise their kids right.

Andrew "Sonny" Garcia, the former air traffic controller, connected with his wife for only one second before the call dropped. He spoke her name: "Dorothy."

The plane descended to an altitude low enough for computer engineer Ed Felt to use his cellphone to dial 9-1-1 from inside a locked lavatory. Felt's call connected to an operator in the Westmoreland County, Pennsylvania, emergency dispatch center. "Hijacking in progress," Felt said, his voice shaking. He provided the flight number, the aircraft type, and its original itinerary, from Newark to San Francisco.

CEECEE LYLES COULDN'T stop trying to reach Lorne. With the plane flying low, she caught a cellphone signal. This time Lorne saw CeeCee's number on the caller ID and picked up, still in bed, half-asleep after his overnight shift.

"Babe, I need for you to listen to me," CeeCee said. "My plane has been hijacked."

"Stop playing," Lorne said.

"I'm not playing."

In the background, he heard people yelling. He thought he must be having a nightmare, but CeeCee's voice made it all too real.

She told him the plane had turned around and that she didn't know what would happen. CeeCee told Lorne she hoped she'd see his smiling face again. She asked him to tell her sons that she loved them. "I love you, I love you, Babe," she said. "Take care of the kids."

Lorne heard an edge of panic in her voice. They prayed together.

But CeeCee wasn't giving up. Before the call ended, she told Lorne, "We've got a plan."

CHAPTER 10

"LET'S ROLL"

United Airlines Flight 93

WHILE FLIGHT 93'S PASSENGERS AND CREW MEMBERS STRATEGIZED and made phone calls, drawing insight and inspiration from their connections on the ground, in the cockpit Ziad Jarrah worked to control the big plane as he pushed it lower and lower, flying eastward.

As the descent continued, the mystery deepened as to whether Captain Jason Dahl or First Officer LeRoy Homer Jr. might have survived the initial attack. During an exchange in Arabic at 9:45 a.m., one hijacker, likely Saeed al-Ghamdi, badgered Jarrah about whether to open the cockpit door to the other two hijackers: "How about we let them in? We let the guys in, now. . . . Should we let the guys in?"

"Inform them," Jarrah answered, "and tell him to talk to the pilot. Bring the pilot back."

The hijackers said nothing more about a pilot, and the native English-speaking man who'd moaned "Oh, man!" in the cockpit several minutes earlier wasn't heard from again. Jarrah wouldn't be getting any expert help; he was on his own at the controls.

ALTHOUGH NO ONE at the FAA had informed the U.S. military about the hijacking of United Flight 93, at 9:45 a.m., controllers at the FAA's Cleveland Center warned officials at the Greater Pittsburgh International Airport that their radar showed that an unresponsive plane would soon pass nearby or directly overhead.

Inside the airport's air traffic facility, FAA watch supervisor Mahlon Fuller had just heard about the crash into the Pentagon. A controller called him to a radar scope. Within two sweeps of the radar, Fuller knew two things: the unresponsive plane was moving very fast, and it was headed right for them. Fuller grabbed the intercom at his desk and announced: "Evacuate the facility. There's a 757, thirty miles northwest of the airport, with a suspected bomb on board."

At roughly the same time, not as a direct result of Flight 93, but in response to rumors of more attacks, lawmakers, staffers, and visitors evacuated the White House and, for the first time in history, the U.S. Capitol, where Congress was in session. A Capitol police officer ran through the marble halls shooing people outside and shouting: "There's a plane coming. Get out!" Women ran out of their shoes to escape the threat. Sirens wailed as some members of Congress clustered under trees. Heavily armed security officers rushed high-ranking U.S. Senate and House leaders to a Cold War–era bunker, where a set of federal lawbooks waited in case they needed to legislate from underground during the emergency. Bags of potato chips awaited them, too, for nourishment.

As news spread, a wave of evacuations swept from New York to Washington to cities across the country as workers and visitors ran from federal buildings, state capitols, and many of the nation's iconic landmarks, including the Washington Monument, the Empire State Building, the Sears Tower in Chicago, the Space Needle in Seattle, the John Hancock Tower in Boston, the Transamerica Pyramid in San Francisco, and many more.

AROUND 9:46 A.M., apparently worried that they were losing altitude too quickly, Jarrah jerked the plane's nose upward, then put Flight

93 into another dive. The herky-jerky roller-coaster effect coincided with Todd Beamer's shout, "Oh my God, we're going down!" Muffled screams from the cabin were heard inside the cockpit.

Jarrah sought divine intervention to compensate for his anemic flying skills. Although he'd impressed the owner of a Florida flight school where he'd trained as someone with the potential to become an airline pilot, Jarrah completed only one hundred hours of flight training, and he didn't have either a commercial pilot certificate or an instructor's endorsement to fly multiengine planes. He was in over his head. The voice recorder microphone at the pilot's seat picked up whispers in Arabic of the Shahada, an expression of faith that is one of the Five Pillars of Islam: "In the name of Allah. In the name of Allah, I bear witness that there is no other God but Allah."

The plane's descent continued, but at a slower rate.

Air traffic controller John Werth tried to reach the cockpit by radio. A half dozen times he asked some version of the same question: "United Ninety-Three, do you hear Cleveland Center?" Still NORDO.

Also, at 9:46 a.m., the FAA's operations center in Virginia updated officials at FAA headquarters that United Flight 93 had turned around and was headed toward Washington. Based on its estimated speed and projected flight path, the hijacked plane, being flown by a terrorist who claimed to have a bomb, would reach the nation's capital in twenty-nine minutes. Three minutes later, as Flight 93 drew closer, an official at the FAA's Virginia operations center spoke again with FAA headquarters.

"Ah, do we want to think about, ah, scrambling aircraft?" he asked.

The answer was vague and noncommittal.

"Ah," the FAA headquarters official said. Then he sighed. "Oh God, I don't know."

The FAA operations center official tried to emphasize the urgency. "Uh, that's a decision someone is gonna have to make probably in the next ten minutes."

The headquarters official shrugged. "Uh, you know, everybody just left the room."

INSIDE THE UPSTATE New York headquarters of the Northeast Air Defense Sector, minutes crept past. Major Kevin Nasypany and his team remained wholly unaware of the hijacking of United Flight 93. As Jarrah flew toward Washington, D.C., and as passengers and crew members called the airline, 9-1-1, and their families and friends for advice and comfort, Nasypany focused on trying to redirect the F-16s from Langley to where he wanted them: over the nation's capital.

Officers and technicians at NEADS also kept watch on the Las Vegas–bound plane from Delta Airlines, Flight 1989, that controllers in the FAA's Boston Center feared might be another hijacking. Those worries continued despite key facts to the contrary: the pilots never stopped responding to ground controllers, never turned off their transponder, and never deviated from their flight path. But after three hijacked planes had already crashed into terrorist targets, NEADS couldn't take any chances.

They spotted the Delta plane on radar at 9:41 a.m. NEADS Senior Airman Stacia Rountree called out: "Delta [19]89, that's the hijack. They think it's possible hijack."

Master Sergeant Mo Dooley: "Fuck!"

Rountree: "South of Cleveland. We have a code on him now."

Dooley: "Good. Pick it up! Find it!"

All the NEADS-controlled fighter jets were already engaged, so Nasypany and his team sought help from military bases in the Midwest. NEADS staffers contacted the Selfridge Air National Guard Base in Michigan, which diverted two unarmed fighters on training missions to intercept Delta Flight 1989. Calls to other bases produced offers of more fighters from Toledo, Ohio; Sioux City, Iowa; and Fargo, North Dakota.

A half hour earlier, before American Flight 77 struck the Pentagon, Nasypany had mused aloud about what he'd tell fighter pilots if they engaged a hijacked commercial airplane that terrorists intended to use as a guided missile: "[I]f we have to take anybody out . . . we use AIM-9s in the face." Since then, he'd already replaced one young NEADS technician who'd hesitated when Nasypany asked if he'd be capable of relaying a shootdown order. Now, he wanted clearly defined answers from his superiors, specific orders known as rules of engagement, on whether they were still operating under the old policy of "ID, type, and tail," which called on military fighter pilots to only identify, intercept, and escort a hijacked plane.

"What are we gonna do?" Nasypany asked. "I got to give my guys direction."

In the midst of tightly controlled chaos, Nasypany wanted an immediate, actionable, military answer to a question that had long vexed ethicists. Decades earlier, the British philosopher Philippa Foot formulated the dilemma as the Trolley Problem: Would it be morally acceptable for a trolley driver to divert a train directly onto a track on which stood one rail worker, knowing that that worker would be killed, if the driver felt certain that his action would save five workers on the other track?

From an ethical standpoint, Nasypany's question could be phrased this way: Would it be permissible for U.S. fighter pilots chasing hijacked planes to shoot down a passenger jet filled with innocent people, if they believed that such action would prevent a potential catastrophe with greater loss of life?

Defense had been Nasypany's life work. Now he kept pressing for an official answer, even as he personally believed that he and his team would do whatever needed to be done, no matter how shocking, to prevent another disaster.

WITH EACH PASSING second, Flight 93 flew lower in the sky and drew closer to Washington, D.C. At 9:48 a.m., it reached 19,000 feet. Two minutes later, 16,000.

At 9:53 a.m., the terrorists either overheard or sensed a revolt brewing among their hostages. Speaking in Arabic, one hijacker, apparently Ghamdi, suggested that the hijackers congregate inside the cockpit and use the fire ax to hold off the passengers. That was, of course, the very action that United dispatcher Ed Ballinger had hoped his ACARS warnings would prompt among the legitimate pilots.

"The best thing," the hijacker said. "The guys will go in, [you] lift up the [unintelligible word] and they put the ax into it. So everyone will be scared." After some confusion among his collaborators, he explained that they should hold up the ax to the peephole in the cockpit door: "Let him look through the window. Let him look through the window."

Soon after, Jarrah revealed with certainty that the fourth hijacked plane's target was in the U.S. capital, although he didn't specify the exact location. Instead, he dialed a navigational code into the flight computer for Reagan National Airport, located within five miles of landmarks including the U.S. Capitol Building and the White House. Both buildings were targeted by al-Qaeda leaders and operatives at different times during the lead-up to the Planes Operation. Bin Laden reportedly preferred the White House, but Atta thought the home of the U.S. president would be too difficult to reach. He focused instead on the hilltop Capitol, its dome crowned by the bronze Statue of Freedom.

EVEN AFTER HE knew that Flight 93 had been hijacked, United flight dispatcher Ed Ballinger kept trying to reach the cockpit and to alert his other flights. He sent pilots Jason Dahl and LeRoy Homer Jr. more warnings, with higher degrees of alarm. At 9:32 a.m., four minutes after the terrorists acted, he wrote ACARS messages to several flights including Flight 93: "High Security Alert. Secure Cockpit."

As the terror unfolded, Ballinger could only wonder if a few minutes or a few alternative words of warning might have made a world of difference. Perhaps he could have phrased his first cautionary message with more urgency, or with more specifics about the threat of hijacking. Maybe Jason and LeRoy would have barricaded the cockpit door. Or maybe they would have been inspired to grab the fire ax before the intruders entered the cockpit, to fight them off.

The "what if" questions seemed almost endless. What if United Flight 93 had been delayed longer in Newark? Maybe the plane would have been caught in the FAA ground stop and would never have taken off. Or maybe a longer wait on the ground would have delayed the start of the terrorists' attack, giving Ballinger more time to warn the pilots to secure the cockpit. What if the United maintenance employees in San Francisco, who first learned of United Flight 175's hijacking from a flight attendant, had immediately relayed word to United headquarters instead of waiting for nearly ten minutes? Or what if the flight attendant had used a different telephone code, to dial Ballinger directly at United's Chicago headquarters? What if American Airlines and the FAA had immediately spread word about the hijacking of Flight 11? What if Mohamed Atta's statement about "planes" had been deciphered sooner?

Ballinger knew he wasn't to blame, but he'd be tormented by the questions, especially one: What if he'd sent his first ACARS cockpit warnings about intrusions a few crucial minutes earlier?

FIFTEEN MINUTES INTO Todd Beamer's call with Airfone supervisor Lisa Jefferson, he told her: "A few of us passengers are getting together. I think we're going to jump the guy with the bomb!"

Lisa asked if Todd was sure. Todd said he didn't have much choice, so he'd rely on faith. Lisa got the impression that Todd believed someone among the passengers and crew members could fly the plane. That is, he said, if the bomb supposedly on board didn't detonate, if they overpowered the knife-wielding terrorists, and if they recaptured the cockpit. Lisa told Todd that she'd stand behind him, whatever he and the others decided.

Jeremy Glick mentioned a vote to his wife, Lyz. Do nothing or do something? With life or death at stake, "something" won.

Strangers on a plane minutes earlier, the men and women aboard United Flight 93 called upon their survival instincts and discovered a warrior spirit that some might not have known they possessed. They boarded the flight as airline employees and everyday travelers, bound for home or business meetings or vacations or memorial services, but they emerged from a chrysalis of terror as a fighting band of brothers and sisters. They were at a distinct disadvantage, but they had numbers, they had one another, they had people on the ground who loved them, and they had a collective will that their captors had foolishly underestimated. If they went down, it would be on their terms. Their flight wouldn't follow the catastrophic model they'd heard about in New York and outside Washington, D.C. Whatever the cost, the hijackers wouldn't decide their destination.

Several minutes before ten, Todd Beamer pulled away from the phone. Lisa Jefferson heard him ask, "Are you guys ready?"

Lisa didn't hear the reply. But next she heard three words, a command that set in motion men and women who'd become their own cavalry: "Okay. Let's roll."

ABOUT ONE HUNDRED feet separated the rear of the plane from the murderers in the cockpit. The insurgents' only approach was through an aisle twenty inches wide. They'd have to counterattack single file. Fearsome yells accompanied their charge.

The hijackers heard the racket from inside the cockpit.

"Is there something?" one hijacker asked.

"A fight?" said another.

"Yeah?" said the first.

From the first-class section, a hijacker banged on the cockpit door and sought refuge inside. Sounds of a fight and a man's screams reverberated off the plane's cabin walls. At the controls, Jarrah knew he needed more time to reach their target. He tried to rally his fellow fanatics.

"Let's go guys!" he yelled in Arabic. "Allah is greatest, Allah is greatest. Oh guys! Allah is greatest."

Jarrah turned the control wheel to rock the plane, left and right, left and right, to throw the hostages turned rebels off balance. The struggle outside the cockpit continued with grunts and yells. Jarrah yelled: "Oh Allah! Oh Allah! Oh, the most gracious!"

A hijacker shouted *"Stay back!"* perhaps while threatening the onrushing fighters with a knife, or the cockpit ax, or the box they claimed contained a bomb. Still they came.

"In the cockpit!" a native English-speaking passenger or crew member yelled. "In the cockpit!"

Jarrah stopped beseeching Allah and turned to his accomplices for protection. At 9:59 a.m., he said in Arabic: "They want to get in there. Hold [the door]. Hold from the inside. Hold from the inside. Hold."

Jarrah wagged the plane's wings more sharply. But to no avail.

A passenger or crew member yelled "Stop him."

A hijacker answered "Sit down! Sit down! Sit down!"

The fight continued, with yells and actions that became hard to distinguish on the cockpit voice recorder. In Arabic, someone said: "What? There are some guys. All those guys." A passenger or crew member yelled: "Let's get them!" A hijacker said: "Trust in Allah, and in Him." From some distance away from the pilot's seat, a hijacker insisted, "Sit down." But it was too late for that. Passengers and crew members were done following anyone's commands but their own.

Jarrah continued his erratic flying in an attempt to frustrate the counterattack. Sounds of metal snapping rang out in the cockpit. "Ah!" yelled a hijacker. And again, louder: "Ah!" Glass or plates smashed loudly. Then silence. Warning tones sounded in the cockpit. Then came more crashing sounds. A third time someone yelled: "Ah!"

At ten o'clock, as the plane flew at 5,000 feet, it dawned on the terrorists that they couldn't hold the cockpit indefinitely. But they weren't yet sure what to do.

One hijacker said in Arabic "There is nothing."

Another, apparently Jarrah, asked "Is that it? Shall we finish it off?"

"No. Not yet."

"When they all come, we finish it off!"

The other listened for more sounds from outside the cockpit: "There is nothing."

Maybe they'd have enough time to reach Washington after all. Jarrah pulled back on the control wheel, and the plane climbed. But the fight outside the cockpit resumed. Then came the voice of an English-speaking man: "Ah! I'm injured."

The sound of metal striking metal followed, and then, some distance from the cockpit, "Ah!" From the pilot's seat, Jarrah resumed his spiritual pleas: "Oh Allah! Oh Allah! Oh gracious!"

The rebellion geared up for another push. A leader among the insurgents shouted: "In the cockpit. If we don't, we'll die!"

Jarrah tried a new tactic. He toggled the control wheel forward and back, making the jet rise, then dive, then rise again. He repeatedly pitched the plane's nose up and down. He instructed Ghamdi to work with him using the copilot's control wheel. Switching between Arabic and English, Jarrah commanded: "Up, down. Up, down, in the cockpit." Accompanying the hijackers' exertions was a soundtrack of crashes, thumps, shouts, and breaking glass.

"Up, down. Saeed—up, down!"

But rocking wouldn't stop the passengers and crew. Still trying to gain entry, they apparently repurposed a beverage cart to use as a battering ram. "Roll it!" a male passenger shouted.

Around that time, a pilot in a small plane spotted Flight 93 streaking toward him at 8,000 feet, its landing gear down, flying erratically, banking hard left, then hard right, rocking its wings over farmlands and former coal mines southeast of Pittsburgh.

The Flight 93 cockpit recorder captured a loud crash. Passengers and crew members might have pulled back the beverage cart and smashed it again into the cockpit door.

"Allah is the greatest!" Jarrah yelled. "Allah is the greatest!"

Jarrah stopped wagging the wings and ended his seesawing with the nose. He stabilized the plane. The battle resumed.

At 10:01 a.m., an exchange between two hijackers revealed that they knew their time had nearly run out. They wouldn't be able to hold the cockpit long enough to complete their murderous mission. Months of planning and training would be thwarted by passengers and crew members, outwardly ordinary men and women who revealed extraordinary strength in an effort to save themselves and perhaps others.

Hijacked United Flight 93 wouldn't reach Washington, D.C., some one hundred twenty-five miles away. Jarrah wouldn't slam the fuel-laden 757 into one of his two most likely targets, the U.S. Capitol Building or the harder-to-reach White House. No one would die inside or around either building. Neither landmark would burn on television screens around the world. The ultimate goal of the fourth hijacking would fail.

Instead, Jarrah defaulted to an option that Mohamed Atta created as a fallback: any hijacker who couldn't reach his target should crash the plane into the ground.

"Is that it?" Jarrah asked in Arabic. "I mean, shall we pull it down?"

"Yes," Ghamdi answered. "Put it in, and pull it down."

Jarrah tried to act quickly, before the passengers and crew breached the cockpit. He called out "Saeed!" Then, in Arabic mixed with English, he tried a desperate measure to slow the ongoing insurgency: "Cut off the oxygen! Cut off the oxygen!" He resumed his erratic flying, calling out "Up, down. Up, down."

Still the revolt continued. A confusion of crashes and sounds filled the plane, with metallic snaps mixed with loud grunts and shouts by Arabic and English-speaking men. At 10:02 a.m. came a series of yells: "Shut them off! Shut them off!" followed by numerous metallic clicking sounds.

The passengers wouldn't relent: "Go! Go! *Move! Move!*" one rebel exhorted fellow fighters. An English-speaking man shouted: *"Turn it up!"*

A command sounded in Arabic: "Down, down. . . . Pull it down! Pull it down! *Down!*" Then in English, someone said, "Down. Push, push, push, push. . . . push."

Flight 93 nosedived toward the hills and streams of rural southwestern Pennsylvania. As the earth rushed closer, passengers and crew members apparently made a last-ditch effort to reach Jarrah and Ghamdi in the pilots' seats. Although some aviation experts considered it unlikely that they succeeded, sounds from the cockpit recorder revealed some kind of last-second struggle. One possibility was that a passenger or crew member grabbed the pilot or copilot's control wheel in a desperate effort to pull the plane out of its dive. A hijacker, apparently Ghamdi, shouted in Arabic: "Hey! Hey! Give it to me. Give it to me."

The hijacker repeated "Give it to me" six more times.

Someone turned the control wheel hard to the right. The plane flew sideways, then turned completely over and flew upside down, its belly to the blue sky as it descended. Grunts and loud noises mixed with the voice of a hijacker, who shouted in Arabic: "Allah is the greatest!"

Near the end of the instruction letter the terrorists carried were directions for the last moments of the operation:

> And if possible, when the time of truth and the Zero Hour arrives, then rip open your clothes, and bare your chest to embrace death for the sake of Allah! And you must continue to pronounce His name. And you either conclude with a prayer that, if possible, you begin seconds before the target, or your last words should be: "There is no god but Allah.
>
> "Mohammed is His messenger!"

The struggle continued. A male passenger shouted *"No!"*

Amid screams, loud noises, and whispers of "Allah is the greatest!" the cockpit recording ended. Having turned almost fully upside down at the last moments, flying at an estimated 563 miles per hour, pointing nose-down at a 40-degree angle, the Boeing 757 with more than 5,000 gallons of jet fuel cut through power lines and reached its termination point.

United Flight 93 exploded in flames as the cockpit broke off, plowed forward, and shattered into countless pieces. The rest of the plane burrowed more than fifteen feet deep into the soft earth of a grassy field that had once been a coal strip mine known as the Diamond T. Originally bound for San Francisco, hijacked toward Washington, D.C., the flight ended near tiny Shanksville, Pennsylvania, population 245. The nation's capital was about fifteen minutes' flight time away.

It was 10:03 a.m.

AFTER AMERICAN FLIGHT 77 crashed into the Pentagon, Gofer 06 pilot Steve O'Brien asked his crew if they were too rattled to continue flying. They assured him they were fine, so O'Brien pointed the C-130 toward their home base in Minnesota, on a northwesterly route that took them over Pennsylvania.

Several minutes after ten, a crew member looking out a window in the back of the plane spotted black smoke rising in the distance. He told O'Brien, who reported to Cleveland Center, that the plume, some twenty miles away, reached some five thousand feet into the air.

"You say black smoke in sight?" asked a surprised air traffic controller.

"That's affirmative, black smoke," O'Brien said. "That's not a cloud, it's black smoke, sir."

The crew of the unarmed military cargo plane Gofer 06 had witnessed history for a second time on 9/11; they had seen the aftermath of the crash of Flight 93.

AT ALMOST PRECISELY the same moment as the crash of United Flight 93, Major Kevin Nasypany's team at NEADS learned that a different plane they were worried about, Delta Flight 1989, wasn't a hijacking after all. With an escort from fighters, it landed without incident in Cleveland. They also searched their screens for a jet from Canada that hadn't been hijacked.

At 10:07 a.m., four minutes after NEADS had heard the good news about Delta 1989, the defenders of a large swath of U.S. airspace first learned that United Flight 93 had been hijacked. That notification came not from FAA headquarters, but from controllers in the agency's Cleveland Center. Yet despite the report of black smoke from Gofer 06, the Cleveland controllers didn't tell NEADS that Flight 93 had already crashed.

The delay in the notification was significant and baffling. More than a half hour had passed since the FAA confirmed that terrorists had seized Flight 93. Multiple calls from the plane had reached 9-1-1 operators, United Airlines, and family members and friends of the hostages. But again, no one had informed the military. Unaware that Flight 93 had already burrowed into a Pennsylvania field, NEADS technicians engaged in a wild goose chase, working their phones in an attempt to find out more about a hijacked passenger jet supposedly aiming for Washington.

Amid the confusion, unrelated to Flight 93, radar technicians spotted an unknown plane that appeared to be headed for the White House. Within minutes, Nasypany received an answer to his question about whether American fighter jets would have shootdown authority if they encountered a hijacked plane. He called out: "Negative. Negative clearance to shoot." Nasypany instructed his team to tell fighter pilots to continue standard procedures: only identify and tail any reportedly hijacked planes. As it turned out, the unidentified plane was a fighter from Langley, flying over the White House to protect it.

Shortly after ten, Nasypany had no reason to think that all the planned hijackings had been carried out. He wanted more assets, and he wanted to keep his options open, in the event that orders changed and it became necessary to instruct fighter pilots to force down a commandeered jet heading toward a heavily populated area.

An official at the Air National Guard base in Syracuse, New York, told Nasypany that he had two fighters that would launch within fifteen minutes. Thinking that Flight 93 remained airborne, and planning ahead for other potential hijackings, Nasypany asked one question: "Are they loaded?"

"We've got hot guns," the commander said. "That's all I've got."

Nasypany would have preferred missiles. He answered: "Hot guns, well, that's good enough for me, for the time being."

AT 10:14 A.M., more than ten minutes after the crash of Flight 93, Nasypany pressed his team for more information about the lost plane. NEADS Tech Sergeant Shelley Watson called the FAA's Washington Center for an update.

Watson: "United Ninety-Three, have you got information on that yet?"

Washington Center: "Yeah, he's down."

Watson thought they'd finally caught a break and that a hijacked plane had landed safely. She asked excitedly: "He's down?"

Washington Center: "Yes."

Watson: "When did he land? Because we have confirmation—"

The FAA official became agitated: "He did *not* land."

The message registered. Watson's voice drooped: "Oh, he's *down* down."

AT 10:15 A.M. on September 11, no one outside al-Qaeda knew whether United Flight 93 was the last of the "planes" mentioned by Atta or whether other hijackings might still be in progress or about to occur. No more planes were taking off, and the FAA's Ben Sliney had ordered every plane to land at the closest airport. But neither he nor any other American official knew whether any of the several thousand planes still flying might be in danger. With so much uncertainty, debates continued among military and government officials about potentially shooting down hijacked passenger jets.

President Bush remained aboard Air Force One, soon to be headed toward Barksdale Air Force Base in Louisiana and from there to another base, in Nebraska. When American Flight 77 hit the Pentagon, Secret Service officials had convinced Bush that it would be best to stay clear of more possible attacks on Washington until they were certain the threat had passed. Meanwhile, Vice President Cheney, in the shelter beneath the White House, received continuing, erroneous reports from the Secret Service stating that planes with terrorist pilots were approaching Washington. One report, between 10:10 and 10:15 a.m., claimed that a hijacked plane was eighty miles from the capital. Cheney gave the order for military pilots to engage, then repeated his authorization minutes later when he was told that a hijacked plane was sixty miles away.

At the direction of the Secret Service, outside the normal military chain of command and without the knowledge of NEADS, fighter jets were instructed to launch from Andrews Air Force Base in Maryland. Those fighter pilots received orders to protect the White House and to take out any plane that threatened the Capitol. They launched with permissive rules of engagement that gave pilots discretion to use deadly force if they encountered an airborne version of the few-or-many Trolley Problem.

News reached Cheney in the bunker about an aircraft that had crashed in Pennsylvania. But soon after, around 10:30 a.m., Cheney was told about another supposedly hijacked aircraft five to ten miles away. An aide conveyed Cheney's reaction to a multiagency conference call about airborne threats: "The vice president's guidance was we need to take them out."

That lethal authorization order reached NEADS at 10:32 a.m., delivered in the form of a message from General Larry Arnold, commander of the Continental Region for NORAD. It read: "Vice president has cleared us to intercept tracks of interest and shoot them down if they don't respond." (By "tracks of interest" he meant hijacked passenger jets that refused to heed commands.)

Nasypany finally had clear authority. Nevertheless, no one at NEADS shared that order with the fighter pilots circling high over Washington, D.C., New York City, and elsewhere across the country. In fact, the lead fighter pilot from Langley still hadn't heard anything about hijackings. He still thought he was protecting the capital from Russian cruise missiles. As he flew over the damaged Pentagon, the pilot said to himself, "The bastards snuck one by us."

With no other confirmed hijackings, and with continued uncertainty about the implications of a shootdown order, NEADS commander Colonel Robert Marr, Major Nasypany, and weapons director Major James Fox decided that it would be prudent to wait before telling fighter pilots about the vice president's authorization. They figured they'd have time to give the shootdown order, if it became necessary.

AT 10:39 A.M., Cheney updated Secretary of Defense Donald Rumsfeld, who'd been helping the wounded at the Pentagon.

Cheney: "There's been at least three instances here where we've had reports of aircraft approaching Washington—a couple were confirmed hijack. And, pursuant to the president's instructions, I gave authorization for them to be taken out. Hello?"

Rumsfeld: "I understand. Who did you give that direction to?"

Cheney: "It was passed from here through the [operations] center at the White House, from the [bunker]."

Rumsfeld: "Okay, let me ask the question here. Has that directive been transmitted to the aircraft?"

Cheney: "Yes, it has."

Rumsfeld: "So we've got a couple of aircraft up there that have those instructions at this present time?"

Cheney: "That is correct. And it's my understanding they've already taken a couple of aircraft out."

Rumsfeld: "We can't confirm that. We're told that one aircraft is down but we do not have a pilot report that did it."

In fact, no hijacked flights had been shot down, and none would be, and neither Major Kevin Nasypany nor anyone else at NEADS or NORAD had passed along shootdown authority to the fighter pilots from Otis or Langley. As it turned out, it wasn't necessary. But Cheney's comments revealed the depths of confusion, misinformation, and chaos at the highest levels of the U.S. government fully two hours after the crisis began.

AS FIGHTERS PATROLLED the skies over the United States, it remained unknown whether the ground stop ordered by the FAA's Ben Sliney had interrupted plans for more attacks. One plane that raised questions was United Flight 23, seventh in line for takeoff from John F. Kennedy International Airport, bound for Los Angeles, when flights were halted.

Aviation and law enforcement officials told reporters that when the captain announced over the intercom that they were returning to the gate, four young men sitting in first class who appeared to be Middle Eastern became agitated, stood, and consulted one another. They reportedly refused flight attendants' orders to return to their seats. When the plane reached the gate, the men apparently bolted before they could be questioned.

Flight dispatcher Ed Ballinger, who'd sent one of his cockpit intrusion warnings to United Flight 23, said he was told a similar story about the passengers' strange behavior by airline officials. In that version, the men initially refused to get off the plane. In either case, federal officials never commented publicly about the incident.

ANY POSSIBLE THREAT in the air ended by shortly after noon. Every one of the thousands of planes ordered to land at the nearest airport had done so without incident, an extraordinary feat of coordination by air traffic controllers, pilots, and airport officials. Thirty-eight of those planes landed in the tiny community of Gander, Newfoundland, where they deposited nearly seven thousand passengers and crew members from more than a hundred countries, plus seventeen dogs and cats. An outpouring of kindness, hospitality, and generosity from the people of Gander became a bright spot on a dark day.

As the skies cleared, the FAA and NEADS established an open conference call that would be named the Domestic Events Network. From that moment on, it would operate around the clock, seven days a week, never to be interrupted to this day. As one investigator put it, its creation acknowledged the fact that the FAA never notified the nation's military air defenders at NORAD, NEADS, or anywhere else of "any of the four hijacked flights in time to enable them to respond to the threat before the planes crashed."

The most notice given to Major Kevin Nasypany and his team at NEADS was eight minutes before American Flight 11 hit the North Tower. They had less than four minutes' notice about American Flight 77, and they were told that it was missing, not hijacked. NEADS was notified about United Flight 175 only eleven seconds before it hit the South Tower. The U.S. military's guardians of the sky received no advance warning and had no knowledge of United Flight 93 before it crashed.

In the aftermath, some high-ranking military and political leaders made bold claims, in testimony to Congress and statements to the media, in which they suggested that fighter jets were close to shooting down hijacked jets. Those statements were false.

NO ONE COULD say with certainty which passengers and crew members spearheaded the counterattack on United Flight 93. Phone conversations and personal histories strongly suggested that Tom Burnett, Jeremy Glick, and Todd Beamer were among the leaders. As a bold man who'd wrestled a gun from a mugger, cracked heads on rugby fields, and run with the bulls in Pamplona, Mark Bingham seemed likely to have been among the rebels.

Yet they weren't the only fighters on Flight 93. Joey Nacke had a Superman tattoo on his muscular shoulder for a reason. Toshiya Kuge, Richard Guadagno, William Cashman, and Joe Driscoll didn't make phone calls announcing their intentions, but they weren't men who shied from battle. If they had retaken the cockpit, licensed pilot Donald Greene could have been relied upon to grab the controls, presumably helped by former air traffic controller Sonny Garcia. Flight attendant Sandy Bradshaw could have called her pilot husband, Phil, to guide them to the ground. Sandy had ended her last call to Phil by saying she needed to go because "everyone" was running up to first class.

Sandy wasn't the only woman on board with a brave streak. Cee-Cee Lyles, desperate to return to her family, had mastered an advanced survival course as a police detective. At the end of her final phone conversation with her husband, Lorne, she said: "We're ready to do it now." After Lorne heard a loud boom and raised voices, Cee-Cee had yelled: "Okay, baby! It's happening! It's happening now!" Honor Elizabeth "Lizz" Wainio's last words to her stepmother, Esther Heymann, suggested that she'd be somewhere close to the action: "They're getting ready to break into the cockpit. I have to go. I love you. Goodbye."

Others among the passengers and crew members didn't have their last thoughts or words preserved during phone calls, or didn't mention the counterattack. Some might have been killed or incapacitated earlier. But they are no less worthy of tributes: Christian Adams, Lorraine G. Bay, Alan Anthony Beaven, Deora Frances Bodley, Marion R. Britton, Georgine Rose Corrigan, Patricia Cushing, Jason M. Dahl,

Joseph DeLuca, Edward Porter Felt, Jane C. Folger, Colleen L. Fraser, Kristin Osterholm White Gould, Lauren Catuzzi Grandcolas, Wanda Anita Green, Linda Gronlund, LeRoy Homer Jr., Hilda Marcin, Waleska Martinez, Nicole Carol Miller, Donald Arthur Peterson, Jean Hoadley Peterson, Mark David Rothenberg, Christine Ann Snyder, John Talignani, and Deborah Jacobs Welsh.

The forty heroes of Flight 93 couldn't save themselves. They couldn't return home to their loved ones. But they were all that stood between the hijackers and the destruction of the U.S. Capitol or the White House. All deserved to be honored and remembered as civilians turned combatants, the saviors of countless lives during the first battle of a new war. If there is a heaven, Lizz Wainio's grandmothers were waiting there to greet every last one.

LATE IN THE afternoon, President Bush returned to Washington, D.C., after a brief stop at Offutt Air Force Base in Nebraska, home of the U.S. Strategic Command.

After landing at Andrews Air Force Base, he flew to the White House aboard the Marine One helicopter. The pilot stayed low and zigzagged, in case a terrorist on the ground had a shoulder-launched missile. The short trip gave Bush a clear view of the Pentagon. To no one in particular, the president said: "The mightiest building in the world is on fire. This is the face of war in the twenty-first century."

Later, wearing a dark suit and a gray tie fit for mourning, Bush appeared on television seated at his Oval Office desk. The time was 8:30 p.m., twelve hours after American Flight 11 crew member Betty Ong first reported, "I think we're being hijacked." In a somber tone, his fingers knitted together as though in prayer, Bush expressed sympathy for the dead and their loved ones; praise for first responders; faith in the nation and its military; trust in a higher power; and resolve to punish those responsible.

Without his knowing it, his address echoed the closing line of President Grover Cleveland's speech a century earlier at the dedication of the Statue of Liberty. Bush said: "America was targeted for attack because we're the brightest beacon for freedom and opportunity in the world. And no one will keep that light from shining."

Later that night, the president wrote in his diary: "The Pearl Harbor of the 21st century took place today. . . . We think it's Osama bin Laden."

BY THE TIME darkness crept across the country, some one hundred fifty fighter jets patrolled over the United States. They dealt with numerous false alarms, but nothing else fell from the sky on September 11, 2001. The toll from inside the four hijacked planes stood at 246 men, women, and children, killed by nineteen suicide terrorists. Attention had shifted from the sky to the ground, where the terrible toll had yet to be tallied, and where smoke and ash still obscured extraordinary stories of heroism and sacrifice, survival and loss.

PART II

FALL

To the Ground

CHAPTER 11

"WE NEED YOU"

September 10, 2001

RON CLIFFORD
Ground Level, World Trade Center

ON THE NIGHT OF SEPTEMBER 10, RON CLIFFORD PACED ANXIOUSLY inside the oversized master bedroom he'd converted into an office in his house in Glen Ridge, New Jersey, an affluent suburb of quiet streets lit day and night by flickering gas lamps.

A native of Cork, Ireland, who'd spent two decades in the United States, at forty-seven Ron looked like everyone's favorite priest: comfortably built, with gentle blue eyes, thinning blond hair, and wire-rimmed glasses. He had a storyteller's warm voice, complete with Celtic lilt, and his mouth curled naturally into a smile. But on this night, Ron felt burdened by worry. The company for which he ran East Coast sales was hemorrhaging money. With an unforgiving home mortgage and a wife and daughter to support, Ron believed that his future depended on a high-stakes meeting he'd scheduled for the next day twenty miles away in Manhattan.

Over and over again Ron checked his research and rehearsed the pitch he'd planned for a West Coast competitor in the field of Internet analytics, a business that helped companies understand how customers used their websites. Ron hoped to convince his rival to hire him and buy his bosses' cash-strapped company. Seeking every edge, real or imagined, Ron found synchronicity in the meeting's timing: Tuesday, September 11, would be his daughter Monica's eleventh birthday. Another comfort was Ron's confidence that he'd look every inch the desirable partner—several days earlier he'd called his elegant sister, Ruth Clifford McCourt, for a pep talk and fashion advice.

Ron and Ruth's bond was forged in their youth after a younger brother died in a motorcycle accident and their parents' marriage crumbled. Ruth came to the United States at sixteen with their mother, while Ron and two other brothers stayed with their father in Ireland. When Ron followed, Ruth steered him to an apartment on the Upper East Side of Manhattan and later encouraged him to move to Boston to study architecture. For a time, they lived together in the Boston neighborhood of Brighton, supporting each other as Ron studied and Ruth built her career.

Ruth was forty-five but looked a decade younger, a head-turning beauty with strawberry blond hair. She trained in London as a skin care specialist, then spent a decade running her own day spa and hair salon in the Boston suburb of Newton, Massachusetts. Now Ruth lived in a large waterfront home in New London, Connecticut, with her husband, David, owner of a gas distribution company, and their four-year-old daughter, Juliana, a sprite with her mother's charm in miniature.

Ruth and David had married six years earlier at the Vatican, a privilege that resulted from a celebrity favor trade with a local priest: Ruth introduced the priest to her friend and neighbor the actress Katharine Hepburn, and the priest arranged an audience with the pope. Ruth and Ron's father, Valentine, died days before the wedding, so Ron read the toast he'd planned to give. Ruth gave Juliana the middle name Valentine.

After Ron's wife, Brigid, his sister, Ruth, was Ron's closest confidante. Their families vacationed together, and rarely did more than a few days pass without Ron and Ruth talking by phone. In recent weeks, Ron, Brigid, and Monica had spent long summer days at Ruth's house.

When Ron called Ruth before his big meeting, Ruth boosted his confidence and assured him that he'd close the deal. Meanwhile, Ruth had some news of her own. She, Juliana, and Juliana's godmother, Paige Farley-Hackel, would be flying to California in a few days. Paige had launched a radio program on spirituality, and she had arranged a meeting with Deepak Chopra. Ruth told Ron that after spending several days at Chopra's Center for Well Being in La Jolla, she and Paige planned to visit friends in Los Angeles and take Juliana to Disneyland.

During the call, Ruth urged Ron to wear a bright tie to his big meeting, to make a strong impression. "You always want to stand out," Ruth said. Ron went out and bought a yellow silk tie as sunny as an egg yolk, paired with a new blue suit and a crisp white shirt.

The night before the meeting, a thunderstorm raged over Glen Ridge. An avid sailor, attuned to wind and weather, Ron listened to the rain as he laid out his Ruth-approved wardrobe. He peeked in on Monica then joined Brigid in bed.

ELAINE DUCH
North Tower, World Trade Center

Elaine Duch swam leisurely, her hands caressing the water with each stroke, her legs churning a modest wake. After a busy Monday at work, Elaine had dragged her twin sister, Janet, to adult swim at the Lincoln Community Pool in their hometown of Bayonne, New Jersey. Whether in a pool or at the Jersey Shore, since childhood Elaine had always loved how her body felt in water, buoyant and sensuous, her skin and muscles relaxed and recharged.

With Elaine taking charge, as usual in their twinship, the two sisters had resolved months earlier to get into eye-catching shape. Elaine and Janet, whose surname was pronounced "Duke," spent summer nights and weekends swimming, with long walks and bike rides on the side, while planning for their first yoga class. Both single, closing in on their shared fiftieth birthday, they wouldn't go quietly.

Janet had a boyfriend, and Elaine wouldn't have minded one, but she didn't consider it essential. She'd built a comfortable, fulfilling life highlighted by monthly overnight visits to the casinos of Atlantic City. Occasionally they took longer trips, including a weeklong bus tour in July to Niagara Falls.

Elaine's workdays revolved around hourlong journeys via train between the underdog city of Bayonne and glittering Manhattan, two distinct worlds separated by New York Harbor. Thirty-one years earlier, a week after she graduated from high school, Elaine had followed her older sister Maryann to a job at the Port Authority of New York and New Jersey, in time rising to senior administrative assistant in the real estate department.

Elaine's workplace was the Port Authority's pride and joy, the World Trade Center, where she relished the responsibilities and the people but never warmed to the buildings' immense height. Her feelings hardened in February 1993, when her office was on the 35th floor of the North Tower. Terrorists had packed a rental van with fifteen hundred pounds of explosives and parked in the underground garage. The blast killed six people, injured more than a thousand, and caused an estimated $300 million in damage. The evacuation down soot-filled stairwells ruined Elaine's white pants and her black-and-white winter coat. Soon Elaine discovered that the real cost had been far greater: five of the six people killed were Port Authority employees she knew and liked. The death of secretary Monica Rodriguez Smith, seven months pregnant, on her last day at work before a maternity leave, upset Elaine most of all. Elaine attended yearly memorials and cherished her memory of Monica's gentle nature.

In the years that followed, Elaine's office migrated skyward to the 88th floor of the North Tower. She liked that floor even less, especially on windy days when the elevators creaked and moaned as the building swayed a foot or so in either direction.

In September 2001, Elaine's work life in the North Tower was near an end. That summer, the Port Authority had entered into a ninety-nine-year, $3.2 billion agreement to lease the World Trade Center to two private firms, Silverstein Properties and Westfield America. With management of the real estate changing hands, Elaine had accepted a new job in the Port Authority's audit department, located across the river in Jersey City. She was supposed to have begun working there already, but paperwork for the new lease hadn't come through, so her move was pushed back to later in September.

Their swimming exercise complete, Elaine and Janet toweled dry, ate dinner at a Chinese restaurant, and returned to the apartment they shared. Before bed, Elaine prepared lunch and laid out her clothes for the next day: a favorite gold skirt with a blue paisley print and a new sleeveless cream top. She'd wear her blond hair down, letting it drape over the shoulders of a navy jacket. White canvas sneakers would carry her to work, then she'd change at her desk into strappy leather sandals.

FDNY CAPTAIN JAY JONAS
North Tower, World Trade Center

Eight and eighteen.

The two numbers rattled around Jay Jonas's mind as he drove his aging Subaru south along the Hudson River, heading from his home in the sleepy village of Goshen, New York, to his second home, a busy firehouse in Lower Manhattan. At forty-three, Jay carried a solid 240 pounds on his six-foot-one frame. He had ruddy cheeks, catcher's-mitt hands, steady blue eyes, and a big man's quiet confidence.

Rain splashed musically on his windshield as Jay's numbers kept the beat. Eight: the years he'd been a captain in the Fire Department of the City of New York, the FDNY. Eighteen: Jay's place on the civil service list awaiting promotion to battalion chief.

The raise that came with the higher rank would be welcome, as Jay and his wife, Judy, looked ahead to sending three bright kids to college. But the money was secondary. The jump to battalion chief would pin two gold bugles on Jay's collar and place him among his mentors and heroes: the top brass of "New York's Bravest."

When that might happen was anyone's guess. The FDNY promotional pipeline had long been clogged, with no indication when the spigot might reopen. As 6 p.m. neared on September 10, Jay filed away his numbers and parked behind the squat Canal Street headquarters of Ladder Company 6, nicknamed the Dragon Fighters for the unit's Chinatown territory. Stepping from his car, Jay focused on his immediate future: twenty-four hours on duty, keeping his men sharp and a slice of the city safe.

Jay's career goal had its roots in boyhood. His father, a telephone installer by day and a volunteer firefighter by desire, let Jay tag along to the fire station on weekends. By sixteen, Jay was a junior firefighter. By nineteen, he was studying fire science under veteran FDNY instructors at Orange County (N.Y.) Community College, paying for his classes by working as an ambulance driver and emergency medical technician. Upon joining the FDNY in 1979, Jay was assigned to a New York City borough so synonymous with arson that it had a catchphrase: "The Bronx is burning."

During more than two decades of service, he'd fought too many fires to count. But one stood out. In May 1982, Jay followed a lieutenant into a fourth-floor apartment of a burning tenement building on East 182nd Street. Flames rose through the floorboards and licked at the windows. Choking on acrid smoke, Jay heard a moan. He crawled inside a bedroom, felt around, and found a semiconscious man on the bed. Jay dragged sixty-year-old Al Pecchinenda from the apartment and outside to safety. The next day, the *New York Post* ran a photo of a grateful young woman throwing her arms around Jay's neck and kissing his cheek.

Over time, Jay worked in a rescue unit, rose to lieutenant, earned a bachelor's degree, and made captain in 1993. As a new captain, Jay bounced among fire companies until one day he stood in line at a wake for a lieutenant killed in the line of duty. The unit's captain was on the verge of being promoted, and its troops were reeling. A group of firefighters approached Jay, and Firefighter Tommy Falco delivered their message: "We need you." Soon after, Jay took over as captain of Ladder 6.

Jay was determined to make Ladder 6 the best company in the city. He repurposed materials he studied for promotional tests into drills for his men. He showed them videotapes of building collapses and taught advanced rope techniques. Even on a slow day, they went home with the feeling that Jay loved most: "I was a fireman today."

In the summer of 2001, Jay was one of 11,336 firefighters employed by the FDNY, plus another 2,908 in Emergency Medical Services. Yet to Jay the department felt like a small town: everyone eventually crossed paths. Now and then, Jay ran into Orio Palmer, who'd followed him up the stairs at the 1982 tenement fire. Orio was a probationary firefighter at the time, and afterward he quizzed Jay, soaking up information about how to navigate danger. Orio rose quickly after that and had already been promoted to Jay's dream job of battalion chief.

Rain fell on Jay's shoulders as he entered the Chinatown firehouse to begin his double shift. He'd have his usual Ladder 6 crew, plus firefighters Scott Kopytko and Doug Oelschlager, assigned to the company for the night. Soon the alarm sounded, and they joined multiple companies answering a call for collapsed scaffolding at the base of the nearby Manhattan Bridge. Like neighbors running into one another at a yard sale, Jay caught up with two friends who were lieutenants in other companies, Peter Freund and William "Billy" McGinn, whom Jay had trained when he was a lieutenant in Ladder 11.

Back at the firehouse, Jay ducked into his office to get some rest, happy to be a firefighter, but still in his eighth year as a captain, still eighteenth on the promotion list.

CHRIS YOUNG
North Tower, World Trade Center

Walk into any New York City restaurant or temporary employment agency and yell "Places!" or "Action!"—then stand back as the wait-staff or the clerks spring to life, revealing their true selves as actors awaiting big breaks. Those two sides of Chris Young's life came into stark relief on September 10, 2001.

Tall and lean, thirty-three years old, an introvert in life and an extrovert on stage, Chris had expressive features that casting directors might describe as borderline between "best friend" and "leading man." With his glasses on, Clark Kent; with them off, who knows? To pay his rent between acting gigs, Chris took a temporary administrative job in the Midtown office of Marsh & McLennan, a global insurance and financial services firm. He'd done too many temp jobs to count, from data entry to reception, but Chris liked the people at Marsh & McLennan, where he worked in the training department.

Chris especially enjoyed working for a young manager with a dragonfly tattoo named Dominique Pandolfo. A bubbly brunette, Dominique described herself as "a pizza bagel" to explain her Italian Catholic–Russian Jewish heritage. She quietly allowed Chris to keep a flexible schedule so he could duck out for auditions. After several weeks, Dominique "lent" Chris to Angela Kyte, a hyperorganized manager who in two decades at the company had risen to managing director. Angela needed Chris's help preparing printed materials to accompany a PowerPoint presentation she planned to deliver on September 11 at the company's other New York offices, in the North Tower of the World Trade Center.

As September 10 wound down, Chris sneaked peeks at the clock, eager to move on to his "real" life. Shortly before Chris left work, Angela told him to return to the Midtown office early the next morning. She instructed him to pick up a final box of materials and deliver it to her on the 99th floor of the North Tower.

When the workday ended, Chris hurried to his apartment in the unfashionable Brooklyn neighborhood of Gowanus. Awaiting him was a fellow actor named Ted deChatelet, a friend from their days as theater majors at Wake Forest University, in Chris's native North Carolina. Ted had moved to Oregon, but he'd returned to New York for the September 10 world premiere of the first movie in which either had a role: a low-budget, gender-bending farce called *Macbeth: The Comedy*. Set in modern times, Ted played Shakespeare's hero Macduff, while Chris played a flamboyantly gay witch. Chris still shivered at the thought of being stuck in a New Jersey field on a frigid day to shoot the cauldron scene, chanting "Double, double, toil and trouble!"

As he dressed for his star turn, Chris felt as passionate as ever about acting, though he'd tired of the countless auditions with rare callbacks. A dozen years had passed since his greatest theatrical triumph, playing Don Quixote in a college production of the musical *Man of La Mancha*. Yet the fading glow of that success sustained him. When his confidence flagged, Chris drew upon memories of the ovations and the raves. He used the show's monologues for auditions and often sang its score, especially Don Quixote's rousing cry for courage, "The Impossible Dream."

In April 1999 he'd followed friends to New York for a shot at Broadway. Since then, he'd sung in the chorus with the New York Gilbert & Sullivan Players, but beyond that and his fledgling film career, Chris remained the very definition of a struggling modern actor.

Chris and Ted raced to a theater on the Upper East Side for their movie's opening at the New York Independent Film Festival. Even on their big night, they were reminded of their true status. Security officers held them outside in the rain until Mayor Rudy Giuliani and his entourage swept out of the theater after seeing a different movie.

When the lights came up, the festival audience cheered *Macbeth: The Comedy*. Afterward, the cast and crew celebrated at a nearby bar. But when the party reached full tilt, Chris reluctantly found Ted. "I've got to go home," he said. "I have to be up early in the morning and go down to the World Trade Center."

FDNY PARAMEDIC CARLOS LILLO
Ground Level, World Trade Center

CECILIA LILLO
North Tower, World Trade Center

Relaxing poolside at a Caribbean resort in May 2001, newlyweds Carlos and Cecilia Lillo fell into conversation with other couples, including a married pair of Chicago police officers. Over frosty drinks, Cecilia mentioned that she worked for the Port Authority of New York and New Jersey, with an office in the World Trade Center.

Their new friends asked about the 1993 bombing, so Cecilia briefly told them that she'd evacuated the North Tower down a smoky stairwell. Did she fear another attack? No, she told them, strict new security precautions made her feel safe.

Carlos, a paramedic with the FDNY, chimed in: "If something *was* to happen again in that building, my honey works there. I'll be the first one there, and I'll go looking for her." The poolside discussion moved on, but Carlos's vow bothered Cecilia.

Back in their room, Cecilia told Carlos the story of her harrowing escape eight years earlier, how she'd been enveloped in black smoke and considered hurling herself out a window before she stumbled down to the street. Then she laid down the law.

"Listen," Cecilia said, "whatever you do, promise you won't come inside looking for me. I'm *not* waiting for you on the sixty-fourth floor." When Carlos reluctantly agreed, Cecilia softened. "I promise I'll make it out, because someone who loves me will be waiting for me outside."

Carlos agreed, on one condition: "If you're ever hurt," he said, "find an FDNY ambulance. Tell them who I am. They'll find me." Cecilia promised, and their vacation resumed.

The challenge of finding each other was a running theme for Carlos and Cecilia. At thirty-seven, Carlos had been married once before, unhappily. Cecilia, who was thirty-five, had ended a previous relationship to search for "the perfect man." Half a lifetime earlier, they nearly crossed paths when both attended Long Island City High School in Astoria, Queens. Charming and athletic, Carlos was captain of the gymnastics team. Cecilia, studious and pretty, noticed how kind and protective he was to his then girlfriend, walking her to every class. They shared friends, but never met. A decade later, Cecilia spotted Carlos looking dashing in his blue FDNY uniform outside Elmhurst Hospital. "So *that's* what became of him," Cecilia told her mother, who urged her to say hello. Cecilia demurred: "He doesn't know who I am."

More years passed. In 1997, Cecilia attended a housewarming party for a work friend, Sandra Lillo. While leafing through Sandra's wedding album, Cecilia spotted a photo of Carlos. "That's my brother-in-law," said Sandra, who'd married Carlos's brother Cesar. After months of missed connections, Carlos picked up Cecilia for a New Year's Eve party. By then, he knew that she remembered him from high school.

"What do you think?" Carlos asked when Cecilia met him at the door.

"Oh, you look bigger," she said.

"Are you calling me fat?" he said playfully.

"No! You were a kid, on the gymnastics team. Now you're bigger, a man!"

Carlos smiled and opened the car door for Cecilia, who felt a spark of recognition for the young gentleman she'd admired as a girl. Carlos stayed glued to her side the entire night. After the party, they talked in a café until dawn. Later that day, he drove his ambulance to her house to show her the lifesaving tools of his trade. In time, Cecilia would learn that Carlos loomed large as a mentor to his colleagues, among them a quiet EMT named Moussa "Moose" Diaz, who graduated from the same high school they did.

As romance bloomed, Cecilia worried about Carlos bounding into danger. In June 2001, when she heard a story about a paramedic who ran toward injured firefighters without a helmet, she exploded: "You would have done the same thing! I don't want to get a call that you're in a coma because you weren't wearing a helmet. I'll go to the hospital and finish you off!" Carlos promised he wouldn't be that reckless, then told her: "If you're ever in a situation like that, get into a fetal position and cover your head."

Before they married in 2000, Carlos and Cecilia purchased a comfortable home on Long Island and invited her parents to live with them. As they dreamed of their future, Carlos made plans to take the lieutenant's exam. He intended to retire after twenty-five years at the FDNY, then start a computer business. With a degree in marketing, Cecilia set goals to rise from her job as a Port Authority human resources administrator. As encouragement, for their anniversary Carlos bought her two pairs of professional-looking black flats, to replace the open-toed sandals she favored.

Above all they wanted children, but months of trying hadn't worked. The sight of a father sharing a milkshake with his son gnawed at Carlos's heart. He took a second job, partly to save for a pool their hoped-for children could use in their backyard. Sometimes his job took an especially heavy toll. In early September 2001, he answered a call from a neglectful young mother whose baby had died. That night Carlos vented his frustration with Cecilia: "Kids having babies they don't care for, and here we're people who want to be moms and dads!"

After a series of tests, Cecilia and Carlos made an appointment with a fertility doctor for September 26, 2001. In the meantime, they spoiled their nieces and nephews. After work on Monday, September 10, they visited Cecilia's youngest sister and her three-year-old daughter, Casandra. While the sisters talked, Carlos and Casandra tossed balled-up pieces of paper at each other, squealing with laughter.

BRIAN CLARK AND STAN PRAIMNATH
South Tower, World Trade Center

Day after day, year after year, Brian Clark and Stan Praimnath kept the same hours in the same building. They rode the same elevators to offices three floors apart in the South Tower of the World Trade Center. They lived the same American dream, two immigrant strivers who arrived in New York City as young men, built careers in finance, and went home at night to happy suburban families.

In a small town, they might have been friends. In teeming Lower Manhattan in September 2001, Brian and Stan were a matched set of perfect strangers.

A Toronto native, Brian was fifty-four, with kind blue eyes, graying hair, and a poise that could be mistaken for detachment. An only child who'd longed for a sibling, Brian earned an engineering degree and an MBA, then became the tenth employee of Euro Brokers, a fledgling capital markets brokerage firm. In 1974, a year after hiring

Brian, the firm relocated from Toronto to Manhattan. Brian and his wife, Dianne, who'd been his high school sweetheart, moved with their son and daughter to nearby New Jersey, where another son and daughter completed the family.

At the time of the 1993 bombing, Euro Brokers was on the 31st floor of the North Tower, on the eve of a planned move to the 84th floor of the South Tower. Afterward, Brian became a civilian fire safety warden, trained in evacuation and entrusted with a red flashlight, a whistle, a reflective vest, and a red hat he kept in a credenza above his desk. Under no circumstances would he wear the hat, which he thought made him look foolish.

Brian rose to executive vice president at Euro Brokers, making him one of several top company officials who managed a staff of more than 250 employees, including hotshot brokers and traders who turned knowledge into money. Amid shouting men in shirtsleeves and women in power skirts, their faces lit by computer screens, Brian remained the prototypical unflappable Canadian. A video of the trading floor captured him looking like a bemused math teacher watching raucous teens at recess.

On Monday, September 10, Brian arrived at work before the markets opened. He dived into what promised to be a busy week in a privileged life: days spent wrangling Type A traders, nights with Dianne for dinner, a weekend of church and family events.

Meanwhile, Stan Praimnath took September 10 off, an uncharacteristic indulgence for a man who'd worked his way up from nothing to the South Tower's 81st floor.

As a barefoot boy in a small village in Guyana, in South America, Stan grew up devouring books, pilfering mangoes from nearby groves, and dreaming of life in the United States. While still in his teens, Stan passed a test to become a high school teacher, but he saw little hope for advancement in an impoverished, socialist country. An uncle who'd emigrated years earlier to New York offered to sponsor Stan's family, so in September 1982, Stan, his parents, and his five siblings landed in the borough of Queens.

After six months of consistent effort and constant rejection, Stan found two jobs: as a shipping clerk in a New Jersey textile company and as a data entry clerk in a Manhattan bank, which together meant a sixteen-hour workday, not including commuting time. On cold days, he'd meet one of his brothers in a train station to hand off a winter coat they shared.

On one of his endless subway rides, Stan met a fellow Guyanese immigrant, a shy minister's daughter named Jenny whom Stan courted for years and married in 1989. The same year, Stan became a loan officer for Fuji Bank, coincidentally one of the same institutions served by Brian Clark's company, Euro Brokers.

Twelve years later, in the summer of 2001, Stan was forty-four, with neat black hair, soulful brown eyes, a high forehead, and a cleft chin. Nighttime karate classes kept him in fighting trim. He and Jenny had two daughters and a cozy home on Long Island. He'd risen to assistant vice president of Fuji Bank, where he worked long hours, collected friends, and routinely ate lunch at his desk, where between bites he'd read his Bible and stare out the windows at a priceless view of the Statue of Liberty.

Anticipating a relaxing day ahead, at home on September 10 Stan fixed breakfast for his daughters before school. He shared a cup of tea with Jenny before she left for work in Brooklyn, in the legal department of the New York Stock Exchange. Later, he'd cook dinner, watch the news, and say a prayer of thanks for the bounty of his life.

The next day, September 11, Stan would return to his routine. He'd stride through the South Tower lobby, perhaps alongside a fellow immigrant, a stranger named Brian.

ALAYNE GENTUL
South Tower, World Trade Center

As the summer of 2001 neared its end, Alayne Gentul and her husband, Jack, rode a pair of old Schwinn bicycles up and down the sand-splashed streets of Barnegat Light, a beach town named for a lighthouse on a narrow island along the New Jersey coastline.

Beaches occupied an outsized place in Alayne and Jack's life. Their romance began twenty-four years earlier with a moonlit kiss on another New Jersey beach, surrounded by an armada of amorous horseshoe crabs that had come ashore to spawn.

In their early years together, Alayne and Jack scrimped for weekends at the shore and dreamed of pricey, weeklong summer rentals. Time passed, their careers thrived, and they shared their ocean love with their sons, Alex and Robbie, twelve and eight. Their late-summer bicycle tour was as much for the boys as it was for them. Maybe now they could afford to buy a seaside family retreat a hundred miles from their suburban home.

Alayne and Jack appraised properties as they biked past scraggly trees and crushed-seashell driveways, turning around as each easterly block dead-ended at a winding dune path to the Atlantic. One house stood out, an abandoned ranch-style cottage four lots from the water. A teardown, to be sure, but in its place they could build something all their own. It wasn't for sale, but maybe someday. They had time.

At forty-four, Alayne still looked to Jack like the undergraduate he'd spotted at Rutgers University when he was a young residence hall director. Alayne had walked past him on campus, smiling and self-assured, with shiny brown hair, blue eyes, hoop earrings, and a model's gait. Jack sighed to a friend: "If only I could find somebody like that." When Alayne knocked on his dorm room door seeking a job, he'd found her. After keeping a respectful distance while she worked for him, they married in June 1978, a year to the day after their first beach kiss among the crabs.

While Jack pursued a master's degree in education, Alayne worked at a bank and waited tables at a Friendly's restaurant. She noticed that a male waiter promoted to assistant manager wore a white shirt and tie, but her promotion carried only a new nametag for her polyester uniform. Alayne challenged the policy to her corporate bosses and won, in the process gaining momentum for a career in human resources.

In 1982, Alayne went to work as a personnel assistant for Fiduciary Trust Company, an investment firm with New York offices in the

South Tower of the World Trade Center. Every day she rode elevators built by Otis Elevator Company, where her engineer father, Harry Friedenreich, had spent his career. During the towers' construction, Harry rode to the top of one and withstood the elements, to see how his colleagues tackled the colossal vertical challenge. Harry didn't like heights, so he didn't stay long. Years later, when he visited Alayne in her 90th floor office, Harry felt a rush of pride.

By the time of her parents' visit, Alayne had earned a master's degree at New York University and become Fiduciary Trust's senior vice president of human resources. Jack treasured her ability to deftly balance work and home life. Alayne managed the family finances, cared for the boys, invited Jack's widowed mother to live with them, and every Sunday, after teaching Bible classes, cooked a week's worth of dinners.

Occasionally, Jack caught glimpses of a steely side, as when Alayne shut down a home contractor who tried to overcharge them, or when she told Jack about reprimanding senior executives for acts of sexism. But with her family, Alayne was a font of good humor: helping Alex with homework, baking with Robbie, tending to Jack and their garden. She even tolerated Jack's gentle pranks, laughing when he handed her a watermelon in the supermarket, balanced other fruits on top, then walked away.

After their beach property bike tour, Alayne, Jack, and their sons returned to their busy lives. School started for the boys, Jack began a new job as dean of students at the New Jersey Institute of Technology, and Alayne got wind of a possible promotion that would require them to move to California. A house at Barnegat Light might have to wait.

On the second Sunday in September 2001, the Gentul family joined Jack's new students at a minor league baseball game. Afterward, as the summer sun stretched the day, Alayne and Jack sat on their back porch with a bottle of wine and a tray of cheese. They talked about how much they loved their kids and how happy they made each other. They weren't on a beach, but Alayne closed her eyes and leaned in for a kiss.

LIEUTENANT COMMANDER DAVE TARANTINO, MD
The Pentagon

Dave Tarantino plunged toward earth, spinning and swaying. He looked up to see his parachute dangerously creased by a nylon line draped over the canopy.

The year was 1983. Dave was nearly eighteen, a lanky six-foot-four freshman at Stanford University with blue eyes and a place on the crew team. Born in Guantanamo, Cuba, the eldest of three sons of a U.S. Navy officer and a devoted mother, Dave lived all over the world before graduating at the top of his high school class in Michigan. Introverted but adventurous, Dave arrived at Stanford with plans to become a doctor. Only weeks later, he'd need a team of them—if he survived the fall.

Dave and some dorm friends had talked themselves into a what-the-hell, you're-not-scared-are-you road trip to a slipshod skydiving outfit in Antioch, California. The college kids simply had to pay a fee, nod their way through basic safety training, climb into a little plane, and clip on to a cable that would automatically pull their ripcords when they jumped. When they reached 3,500 feet, one by one they stepped out the open door and enjoyed what they expected would be an exhilarating four-minute fall. A bravery test, with training wheels. The problem was, their parachutes had been packed, or in Dave's case mispacked, by a distracted jump school worker. Dave jumped last.

Immediately, he knew he was in trouble. As he spun, Dave tugged at the chute's steering lines but that didn't help. He reached up and behind his head, hoping to disengage the cord bisecting his parachute, but the canopy twisted into a bow-tie shape. Gravity gained an edge. Dave's mind raced through a risk-reward equation. He could deploy the emergency chute, but first he'd have to cut loose the main canopy, which was still partially slowing his descent. How could he trust that the backup would even open? He abandoned that idea. Gaining speed, Dave fought to calm his nerves. As he corkscrewed toward the ground, Dave focused on executing the tuck-and-roll landing technique he'd learned an hour earlier.

Dave plowed into the field. He lay in a puddle of blood, a crumpled mess on the grass, his body tangled like the limp chute next to him. Stunned and afraid, Dave's friends and the jump school staff hesitated before racing to him in a van, certain he was dead.

Somehow youth, soft ground, luck, and his roll on impact onto his right side conspired to keep Dave breathing. His brain registered a symphony of pain from broken bones—foot, tibia, fibula, ribs, wrist, elbow, and jaw—plus a dislocated hip, a separated shoulder, a lacerated liver, and a punctured lung. On the brink of passing out, he moaned for help.

Flash forward eighteen years to September 10, 2001. On a break for lunch, U.S. Navy Lieutenant Commander Dr. David Tarantino sprinted across a soccer field as a midfielder on the Pentagon's recreational team. After the parachute fall, he had endured multiple surgeries, ate weeks of blenderized meals delivered through a syringe, and spent months entombed in casts. He caught up on classes and refused to leave school, motivated by stubbornness and by Stanford crew coach Jim Farwell's promise to save Dave's place on the team. After months of work on the torture devices that turn rowers' legs into muscular pistons, Dave earned the coach's award as "Most Inspirational Oarsman."

After college, Dave backpacked around the world for a year, immersing himself in foreign cultures and finding himself in books. Joseph Conrad's *Lord Jim* made an indelible impression. Dave was particularly affected by the title character's shame at saving himself and failing to perform his duty as an officer on a sinking ship. The next year, Dave followed family tradition and accepted a Navy scholarship to attend Georgetown Medical School. He valued the financial support, but a bigger motivator was the Navy ethos of "honor, courage, and commitment."

After med school, Dave enjoyed more than two years as a flight surgeon with a squadron of F-18 fighters. He met his wife, Margie, a teacher of English as a second language, and switched his specialty to family medicine. Later, he deployed to Haiti for a humanitarian mission, then worked with the Indian Health Service in Nome, Alaska, where he helped to direct the rescue of more than a dozen survivors of a plane crash during a blizzard.

A posting in Naples, Italy, led to missions in Africa and a focus on global health and disaster preparedness. In the summer of 2001, Dave was stationed at the Pentagon in the Office of the Secretary of Defense, planning for humanitarian missions the American military might undertake in conflict regions. At thirty-five, he remained lean and fit, determined to keep his rebuilt body strong by playing soccer and competing in triathlons.

On September 10, Dave felt at a crossroads, unsure whether to remain in the Navy. He and Margie were considering having kids, and he thought he might settle into private practice and leave globetrotting military medicine behind. As Dave chased a soccer ball in the shadow of the Pentagon, a decision loomed in his near future.

LIEUTENANT COLONEL MARILYN WILLS
The Pentagon

A light rain dappled the windows of her comfortable Maryland home as Marilyn Wills summoned her flock the morning of September 10. Before they rushed off to work and school, she and her husband, Kirk, joined hands with their two young daughters in a prayer circle.

They opened with a Bible verse, then Kirk and the girls asked for wisdom, guidance, and support for good deeds and hard work. Marilyn ended with a flourish: "Lord God, please be with Portia, Percilla, mom, and dad, and see that we're safe when we're apart and that we come back home safely together." Hugs all around, bag lunches in hand, and everyone out the door, a family ritual performed almost every day.

Marilyn sang along to gospel music as she drove sixteen miles to the Pentagon's vast north parking lot. She hustled along a crowded walkway, carrying a garment bag into the military monolith, to make a superwoman-like change from stylish suburban mom into no-nonsense Lieutenant Colonel Marilyn Wills. Emerging from a ladies' room, forty-year-old Marilyn looked crisp in her Class B uniform: dark green pants, light green shirt with silver oak leaves, and Army-issued black cardigan sweater, a necessary defense against an overachieving office air conditioner.

Marilyn had moved several months earlier into a newly renovated section of the Pentagon in her role as a congressional liaison officer for the Army's personnel department. It marked the latest stop in a nearly twenty-year career that shaped her in every way and took her around the world.

Marilyn Wills grew up in Monroe, Louisiana, the second of three daughters of a mother who taught school and a supportive stepfather. Her professional climb began at Southern University in Baton Rouge, where she found a work-study job in the school's ROTC office. No one in her family had served in the military, and she had vowed not to be the first. But then an instructor invited her to a no-commitment summer program at Fort Knox, Kentucky, where she was paid and fed during a tryout for military life. The uniform fitted: Marilyn had a talent for leadership, and the Army prized leaders.

Marilyn graduated with a criminal justice degree and was commissioned a second lieutenant. Sent to Fort McClellan in Alabama, she was the only African American woman in her basic military police officers' class. Less than one hundred pounds soaking wet, she stuffed herself with potatoes at every meal to reach the minimum weight requirement and endure the physical training to become an MP. She felt targeted by white instructors who required only black soldiers to take remedial writing classes. She never complained except to her mother, who encouraged her to persevere.

One morning, Marilyn was pleasantly shocked by the sight of a black female officer, the first one she'd ever seen. The captain became a role model and a mentor. When Marilyn completed training and reported to a base in Louisiana, the older woman asked a fellow captain named Kirk Wills, an old classmate stationed there, to keep an eye on Marilyn. Kirk did one better: he married her. Portia, now twelve, was born when Marilyn and Kirk were stationed in Germany. Percilla arrived six years later.

In the years that followed, Marilyn rose to the rank of lieutenant colonel. She commanded a military police company in Germany, served in Honduras, acted as a force protection officer at President Bill Clinton's second inauguration, and spent four years on staff at West Point, among other postings. In 1999, she traded commanding soldiers for representing the Army on Capitol Hill.

Marilyn thrived in the new role with help from a civilian Pentagon employee, a native New Yorker named Marian Serva. Generous with her expertise, Marian trained Marilyn in the delicate art of managing relations between the Army and Congress on personnel matters. Marian's daughter, Christina, was eighteen, a college freshman, so the two women bonded and swapped stories and advice from adjacent cubicles about the challenges of raising successful girls.

When her workday ended on September 10, Marilyn stepped through a little-used exit door and found three olive-skinned men in civilian clothes standing outside the Pentagon, apparently trying to get inside. "You can't go in this door," she told them. "This is 'exit only.' You have to go around." The men nodded and walked away.

The exchange left her queasy, as her mind raced to the bombing of the USS *Cole* and other attacks on American facilities. That night, after fixing dinner, supervising homework, and getting the girls to sleep, Marilyn told Kirk she felt horrible about having leapt to assumptions about the men's intent. Before she went to sleep, Marilyn vowed she'd never do anything like that again.

CAPTAIN JACK PUNCHES, U.S. NAVY, RETIRED
The Pentagon

Jack Punches had waited long enough. Two weeks had passed since his daughter Jennifer's twenty-fourth birthday. After a business trip, she'd returned to her apartment ten minutes from her parents' sub-urban Virginia home. The night of September 10, Jennifer told her father by phone that she wanted to go for a run and hang out with her boyfriend. Jack had other ideas.

"We're coming over," he told her.

Armed with the carrot cake for which his wife, Janice, was famous in the family, Jack burst through Jennifer's door. He sang "Happy Birthday" in his rich baritone and collected his spoils: a hearty slice washed down with milk, followed by a kiss from Jennifer. Satisfied, Jack rounded up Janice and marched out, bellowing the theme song from *Monday Night Football*. The Denver Broncos would soon play the New York Giants, and Jack planned to watch with his son, Jennifer's twenty-year-old brother, Jeremy.

Jack was fifty years old, solidly built, dark-haired and hazel-eyed. He stood five foot eight in his socks, five foot eleven in his mind. Jack regaled his children with stories of his boyhood in rural Illinois, where he was a standout athlete and homecoming king. He laughed when Jennifer and Jeremy eventually discovered that his stardom resulted largely from the fact that he had been the only boy in his high school class, and that his bighearted mother had ordered him to take all eleven of his classmates to the prom.

Janice grew up in the next town over. She knew she was in love with Jack by their second date, on a summer night when Jack was home from college. They married at twenty-two, after he graduated from the University of Missouri on a Navy ROTC scholarship. A waterskiing accident a year later cost Jack the ring finger on his left hand, but the amputation gave him a lifetime of comedy material. He'd joke that he cut off the finger to avoid wearing a wedding band; later he perfected a sight gag in which he'd pretend that the phantom finger was jammed up his nose. And whenever Jennifer or Jeremy did something great, Jack held up his outstretched hands and shouted, "A perfect nine!"

During more than twenty-six years in the Navy, Jack accumulated a chest full of ribbons and more than seven thousand flight hours, many of them piloting a submarine-hunting aircraft called the P-3 Orion. On his rise to the rank of captain, Jack commanded hundreds of sailors at a U.S. base in Sigonella, Italy, then took command of a logistics and supply squadron in Jeddah, Saudi Arabia, supporting battle groups during the Gulf War in 1990–91. Along the way, he earned a reputation as the best kind of officer: one who puts his sailors first. Off duty, Jack oversaw cookouts, March Madness basketball pools, and crossword puzzle contests. He presided over slip-and-slide antics, using a garden hose to snag sliding sailors like jets landing on an aircraft carrier. Near the end of his service career, he worked at the Pentagon running the Navy's efforts to counter drug smuggling.

At a ceremony when Jack retired in July 2000, the lights dimmed, and a screen filled with a scene from the 1955 movie *Mister Roberts*. As the title character, Henry Fonda plays a junior Navy officer on a World War II cargo ship who shields his men from a tyrannical skipper. After suffering one too many abuses, Mr. Roberts tosses the skipper's prized palm tree overboard. Before leaving for combat, Mr. Roberts is honored by his men with a brass medal shaped like a palm tree as they induct him into the fictitious Order of the Palm, for actions beyond the call of duty against "the enemy."

The lights in the auditorium rose, and a young officer called Jack to the stage: "Thanks for your patience, your understanding, and most of all, your wisdom. This is from the heart." He gave Jack a perfect replica of Mr. Roberts's palm tree medal. Clenched with emotion, Jack told his family, his comrades, his friends: "I want all of you to know and understand that I've loved this uniform, the Navy, the nation. And there has not been a single day that I was not proud to put this uniform on and serve with the best and brightest this nation has to offer."

After a few months in the private sector, Jack knew where he belonged. He returned to the Pentagon in a civilian role, as the top deputy in the Navy antidrug office he formerly ran. He worked under a friend and fellow civilian, a retired Vietnam War aviator whom Jack had once supervised named Jerry Henson.

On the night of September 10, his belly full of carrot cake and milk, Jack watched with Jeremy as the Broncos beat the Giants. Jeremy went to sleep, then roused himself when he realized that he hadn't kissed his father good night. His cheeks fresh with kisses from both of his grown children, Jack climbed the stairs to join Janice in bed.

TERRY AND KATHIE SHAFFER
Shanksville, Pennsylvania

Terry Shaffer was in trouble. A blue-eyed bear of a man, Terry buried evidence of his offense deep in the pocket of his worn jeans to hide it from his wife, Kathie, petite and normally placid, and the love of Terry's life since they met as teenagers at a church camp.

Approaching their twenty-fifth wedding anniversary, blessed with three children and good health, Terry and Kathie lived what they considered an ordinary, happy life, built on family, faith, and their rural community of Shanksville, Pennsylvania. But money was tight, as they had just sent their eldest, Adam, to college, with their daughter Rebecca and son Ben not far behind. Plus, they were struggling through a home renovation to rebuild a big stone fireplace, a costly project that had wrecked the kids' bedrooms and forced them to sleep in the living room.

The proverbial straw that broke Kathie's calm announced itself with an annoying chime. "Who do you think you are," Kathie demanded, "that you are *so* important that you need a cellphone?" Nobody they knew had one. Kathie considered it a frivolous expense, an accessory for people in cities like New York, a five-hour drive east that neither of them had ever made.

Terry scrambled to justify his cellular commitment, explaining that the phone would help him to devote more time to family while juggling his occupation, forklift driver at a Pepsi-Cola bottling plant, and his calling, chief of the Shanksville Volunteer Fire Department. Pepsi paid the bills, but the camaraderie and mission of the fire service satisfied something deeper.

With their marital row still simmering, Terry woke early on September 10 for his 5 a.m. to 1 p.m. shift. Kathie would rise later, to watch their two youngest cross the street to school, before going to work as a registered nurse for a family medical practice. A fire radio with him, as always, his new cellphone charged and ready, Terry pulled the family's Dodge minivan out of his driveway and into the sleeping community he dedicated himself to protect.

The Shaffers were among Shanksville's 245 residents. Some traced deep roots to founder Christian Shank, a German immigrant who arrived in 1791 with his wife, seven sons, and plans for a gristmill. Shank laid out three parallel streets, and so they remained: North, South, and Main. Doomed to isolation when railroad planners turned elsewhere, Shanksville didn't even appear on many maps. In the summer of 2001, the little working-class borough had three churches, a fire hall that was Terry's home away from home, a brick post office, a service station with no gas pumps, the school, and a general store where the owner's wife made $1.99 egg salad sandwiches. The store's owner, Rick King, was Terry's friend and assistant fire chief. As Terry drove out of town, he navigated winding country roads lined by tilting fences and houses needing paint. He passed old barns and covered bridges, abandoned coal mines and busy scrap yards.

After a day of loading trucks with cases of Pepsi, Dr Pepper, and Hawaiian Punch, Terry returned to Shanksville and dropped by the fire station, where he and two dozen other volunteers answered some two hundred calls a year. Major fires were rare. More common were vehicle crashes, water rescues, and medical emergencies. They still talked about the time a horse on an Amish farm fell into a manure pit. Between calls, Terry spent countless hours preparing and strategizing for the worst he could imagine in their sixty-two-square-mile territory, which included settlements smaller than Shanksville and a fourteen-mile stretch of the Pennsylvania Turnpike. That's where he worried they'd someday be tested to their limits or beyond.

Occasionally, tragedy called. Sometimes, Terry knew the victims well, such as the time one of his volunteer EMTs lost control of his car on a bridge and was flung to his death. Or when a local man on a weekend leave from the Navy wrapped his car around a tree, killing himself and two others. Terry took an ax to the tree to spare the families the sight of hair embedded in the bark. On those dreadful days, when no rescue was possible, Terry turned over control of the scene to Wally Miller, the county coroner. Wally was a second-generation funeral home director who lived by a simple creed: "You've got to remember that everybody that dies, that's somebody's favorite guy, whether it's a prisoner or the richest guy in town or somebody else."

Even on quiet days, Terry dived into work at the fire station, doing paperwork, planning a barbecue fundraiser for a new tanker truck, mapping water sources in an area with no hydrants or municipal lines. Eventually dinner beckoned, so Terry went home the night of September 10 to Kathie and the kids.

CHAPTER 12

"HOW LUCKY AM I?"

Ground Level and North Tower, World Trade Center

September 11, 2001

UP BEFORE DAWN, RON CLIFFORD DRESSED IN HIS NEW BLUE SUIT and knotted his bold yellow tie. He gathered his thoughts and scanned his notes for the meeting he hoped would secure his financial future. As Ron prepared to leave his New Jersey home, his cellphone rang: the meeting had been moved, from a hotel in Times Square to the Marriott World Trade Center, tucked between the Twin Towers in Lower Manhattan.

Ron considered the change a good omen, signifying a homecoming of sorts. Before shifting his career to computer analytics, he had spent eight years as an architect for the New York Housing Authority, in the city's Financial District, with a corner window view of the World Trade Center.

Ron liked the bustling neighborhood, though he never cared for the towers, which he thought looked like blocky structural supports of a suspension bridge. Occasionally he ate lunch in Windows on the World, the restaurant and catering complex on the 106th and 107th floors of the North Tower, for a spectacular view of the city unblemished by the towers themselves. Now and then his dislike edged into ridicule. While visiting a friend in the South Tower, Ron ran a strip of masking tape down a window, as a point of reference to show the North Tower swaying in high winds. Ron laughed out loud when he read an interview in which the towers' architect claimed he designed them mindful of the "human scale."

By coincidence, Ron had discussed the World Trade Center while out sailing the previous weekend. When Ron and a friend ran into engine trouble, they called the friend's handy cousin, a recently retired engineer for the Port Authority of New York and New Jersey. Soon they fell into conversation about the 1993 bombing. While he worked on the engine, the engineer confided that the attack was personal: the pregnant woman killed that day was his secretary, Monica Rodriguez Smith.

RON'S COMPLICATED FEELINGS weren't unique.

Three decades after the Twin Towers of the World Trade Center reshaped Manhattan's skyline, they remained enigmas. On one hand, New Yorkers admired the swagger it took to erect the planet's two tallest buildings only a short distance apart, even if they held the title only briefly. And no one could deny the moxie of designing them to look like colossal exclamation points on Gotham's greatness. But size matters only so much. Signature skyscrapers need panache. Souvenir models should look like dream castles, not Kit Kat candy bars. It's hard to love a 110-story monolith, even one with an identical twin.

Doubts, or worse, were the towers' birthright. Criticism accompanied every step of the decadelong process of planning the sixteen-acre World Trade Center complex, located near the southern tip of the thirteen-mile-long island that is Manhattan. The loudest protests arose in the early 1960s from small business owners who faced displacement by construction, and real estate titans who worried that the huge towers would tilt the city's power balance away from Midtown to the Financial District. They felt especially piqued that the developer was a public entity, the Port of New York Authority (soon to be renamed the Port Authority of New York and New Jersey). Private developers didn't mind the agency as an overseer of infrastructure such as bridges and airports, but not as a landlord competing for high-end tenants to fill an unprecedented ten million square feet of vertical office space.

An opposition leader, developer Lawrence Wien, tried scare tactics to ignite a public outcry. He repeatedly invoked the crash of a fog-bound B-25 bomber into the 79th floor of the Empire State Building, an accident in 1945 that cost fourteen lives. Wien's point was sharp and personal: he co-owned the Empire State Building, an Art Deco cathedral to capitalism complete with a spire pointing to heaven, and the longtime holder of the "world's tallest skyscraper" crown. As part of his campaign, in 1968 Wien and his allies bought a large display ad in the *New York Times* with an artist's rendering of a passenger jet bearing down on the north face of the proposed North Tower.

The Port Authority expressed outrage, insisting that a structural analysis had determined that each tower could withstand a direct hit by a Boeing 707, the largest passenger jet of the day, traveling at 600 miles per hour. One of the agency's outside consultants insisted that such a plane crash would trigger "only local damage which could not cause collapse or substantial damage to the building and would not endanger the lives and safety of occupants not in the immediate area of impact." The claim sounded comforting, but in fact no such detailed analysis had been conducted, and no one calculated the risk if a plane's fuel exploded on impact, a predictable result that had, in any case, already occurred in the B-25 crash.

Along with air traffic concerns came withering design complaints. The towers' boxy severity led critics to deride them as oversized filing cabinets. Some said they resembled leftover shipping crates that had once contained more elegant skyscrapers. Even moderate assessments had some sharp edges. In 1966, the powerful *New York Times* architecture critic Ada Louise Huxtable reviewed the blueprints and wrote a lukewarm semiendorsement headlined WHO'S AFRAID OF THE BIG, BAD BUILDINGS? Her conclusion: "On balance, the World Trade Center is not the city-destroyer that it has been popularly represented to be, its pluses outweigh its minuses in the complex evaluation that must be made, and its potential is greater than its threat." She ended the review with a line that read like a cross between a backhanded blessing and a voodoo curse: "The trade center towers could be the start of a new skyscraper age or the biggest tombstones in the world."

Port Authority officials swatted away their opponents and broke ground. In the background, meanwhile, engineers hired to carry out the architects' plans pioneered inventive construction and safety techniques that made the towers models, for good and ill, of countless tall buildings that would follow.

At the outset, the engineers came up with a novel strategy to battle gravity, the bane of all human-made structures from sandcastles on up. Previously, super-tall buildings relied on internal "bones" of heavy steel, upon which hung the structural equivalent of muscle and skin made of stone. In this case, though, the towers' engineers designed each one essentially like a box within a box. The external walls, the outer boxes, were made entirely from thin bands of structural steel. Like the exoskeleton of a crab, those outer walls minimized the need for heavy, bulky internal steel support columns.

The external columns gave the towers a look reminiscent of pin-stripe power suits, but it was more than a stylistic choice. Fewer interior steel columns meant more rentable space on each acre-sized office floor. Some internal columns were still necessary, so the engineers clustered them in the inner boxes, known as the central core, among the elevators, stairwells, and utility shafts. The result was an extraordinary thirty thousand square feet of rentable, customizable space on nearly every office floor, uninterrupted by columns or walls, with incredible views to boot. Also, the narrow windows between the closely spaced exterior columns tended to reduce dizzying vertigo among people afraid of heights, which oddly enough included the architect, Minoru Yamasaki.

More clever innovations were needed to move thousands of people up and down the towers each day without turning the buildings into giant elevator shafts. The Otis Elevator Company solved the problem by pioneering a design that allowed workers and visitors with business on lower floors to ride local elevators up from the lobby, stopping on multiple floors, as they would in any tall building. But people destined for midlevel and upper floors took express elevators to a "sky lobby" on either the 44th or 78th floor, depending on their final destination. From the sky lobby, they boarded local elevators to their desired floor. This arrangement sharply reduced the size of the towers' central core, which meant even more rentable office space. Among other construction advances, the towers' engineers crafted revolutionary solutions to minimize swaying and vibration caused by powerful winds from the Hudson River.

All told, the design and structural innovations lowered the buildings' weight, sped the pace of construction, dropped the cost of materials, and increased the anticipated return on what became a $1 billion investment, more than triple the initial cost estimates. Yet those and other advances came with an unwanted, largely overlooked price: they collectively made the Twin Towers of the World Trade Center more susceptible to fire, especially when compared with older buildings whose exteriors were clad in fire-resistant masonry, whose floors were divided into compartments like the hull of a ship, and whose skeletons contained thicker and more abundant steel.

Exacerbating the potential fire risk was a quirk of timing in revisions to the New York City Building Code. As a public agency, the Port Authority wasn't required to comply with the code, but its top officials promised to meet or exceed the city's standards at the trade center. During initial planning, that meant applying strict rules adopted in 1938. But in the mid-1960s, as the towers took shape, a revised, less stringent code moved toward enactment. Even before it took effect, Port Authority bosses told the engineers to follow the new standards' more lenient, cost-saving rules.

The old code would have mandated six emergency exit stairwells in each tower. The Port Authority interpreted the new rules as requiring only three stairwells per tower. However, even under the new code, each tower should have included at least a fourth stairwell, to accommodate visitors to public spaces on the highest floors. Also, fire safety experts generally urge that stairwells in tall buildings be spaced as far apart as possible. But in each of the Twin Towers, the three stairwells were bunched relatively close to one another in the central core. That left them collectively more vulnerable to fire or other damage affecting the core and made them harder to reach for tenants and visitors working in desirable offices near the windows.

In addition, the old construction code required tall buildings to have a "fire tower," one stairwell encased in masonry, with an entranceway that trapped and vented smoke away from the stairs. The new rules didn't require fire towers, so the World Trade Center didn't have them. Instead, each tower's three central stairwells were encased in lightweight gypsum wallboard, making them far more susceptible to damage.

Also worrisome were the techniques used to stop or at least slow a fire from weakening the spindly steel frames that supported the towers' floors. Because the floor system was so original, neither the new nor the old New York City codes included regulations that addressed the engineers' plans to use sprayed-on fire retardants. Special tests could have determined those answers, but no one conducted them. In the end, Port Authority officials essentially guessed at what type of fire-resistant material to use and how much to apply to prevent the steel floor supports from buckling in a blaze. Initially, they insisted that the fireproofing was adequate, and that each floor was built to be airtight. If a fire did break out, they said, it would be locally contained and cause limited damage. Later, however, they installed a sprinkler system, too.

Construction of the towers took five years, slowed by strikes among elevator builders and tugboat operators, which delayed the delivery of steel. Occupancy began even before completion, although at first the twin giants primarily served the Port Authority and other public agencies. Over time they gained grudging acceptance and by 2001 attracted more than four hundred companies as tenants, from financial giants like Morgan Stanley, with more than eight hundred thousand square feet of office space in the South Tower, to one-person firms that enjoyed the prestige of the address but were crammed into nooks and crannies barely larger than a janitor's closet.

As the towers rose into the clouds, their size demanded that attention be paid. Positioned on a diagonal from each other, the buildings stood 131 feet apart, about the distance of a third baseman's throw to first. Each exterior wall spanned 208 feet. The North Tower rose 1,368 feet, an imperceptible six feet taller than its twin, and its flat roof sprouted a 360-foot television and radio antenna. On clear days, visitors to an indoor observation deck on the 107th floor of the South Tower could see parts of New Jersey, Connecticut, Pennsylvania, and Delaware. Windows on the World, in the North Tower, came with similar views plus fine wine and pricey food that earned mixed reviews. Some critics complained that the menu never lived up to the height.

The Twin Towers officially opened with a ribbon-cutting ceremony in April 1973, only to be dethroned as the world's tallest by the Sears Tower in Chicago a month later. The towers loomed over a plaza that would be named in honor of Austin J. Tobin, the Port Authority's longtime executive director. Eventually, the complex also would include four smaller, conventional office buildings, plus the Marriott hotel that was Ron Clifford's meeting destination the morning of September 11.

Below the plaza was an underground shopping mall called the Concourse that connected the buildings in the complex. Deeper still were parking levels and a train station that served New Jersey commuters and provided connections to New York City subway lines. Surrounding the six underground stories were walls of concrete three feet thick and seventy feet deep, affectionately called "the bathtub." The nickname was a misnomer: the walls didn't contain water, they held back the Hudson River.

In August 1974, sixteen months after they opened, the towers had their true coming-out party. Tightrope artist Philippe Petit captivated the world with a dazzling, forty-three-minute, thoroughly illegal high-wire walk on a cable strung between the roofs. By making the twins his costars, casting them as strong, silent types in his death-defying show, Petit gave them the personality their design lacked. Soon indifference among hardboiled New Yorkers evolved into nodding familiarity and even grudging affection. Two years after Petit's walk, a Hollywood remake of *King Kong* showed the great ape ignoring his old haunt, the Empire State Building. This time he leapt from the North Tower to the South Tower before his demise. Over time, the twins appeared in scores of other movies, instantly setting the scene in Manhattan. They graced countless photos and postcards, often paired with the Statue of Liberty as their leading lady.

After the towers withstood the 1993 bombing, Port Authority officials boasted about their durability, even as the agency upgraded and replaced fireproofing, added an air pressure system to limit smoke rising through the core, installed backup power for emergency lights, and improved stairwell lighting.

By the summer of 2001, occupancy remained high and the buildings' future seemed assured. The Twin Towers of the World Trade Center had endured early trials and a terrorist attack to become icons approaching a comfortable middle age. Still uninspiring, perhaps, but undeniable symbols of American ingenuity and financial might, as synonymous with New York as the Eiffel Tower was to Paris or the pyramids to Egypt.

DRESSED AND READY, shortly before 7 a.m. Ron kissed his wife, Brigid, goodbye, hopeful that when he returned they'd celebrate their daughter Monica's birthday and his new business venture. The changed meeting location gave him extra time, so Ron indulged his love of the water, riding a commuter train to Hoboken, New Jersey, then taking a fifteen-minute ferry trip across the Hudson. The thunderstorm of the night before had passed, so Ron stood on the deck under a cloudless late-summer sky, a leather bag containing his sales pitch hanging from his shoulder. He basked in the cool breeze and watched the rising sun illuminate the towers.

"How lucky am I?" Ron thought. "Who gets to do this?"

The ferry docked, and Ron strolled past the exclusive North Cove Yacht Harbor. He admired the gleaming vessels of the super-rich, who paid berthing fees that topped $2 million a year. Crew members in T-shirts swabbed teak decks as though they feared a captain's lash.

As Ron walked along a promenade by the yachts, a well-dressed man enjoying breakfast alfresco raised a glass to him and called out: "Nice suit."

SHORTLY AFTER RON began his morning journey, a silver PATH commuter train from New Jersey squealed to a stop inside the cavernous rail station five stories beneath the World Trade Center. Out poured work-bound men and women, young and not so young, a diverse and divergent group from every rung on the corporate ladder.

Jostled by the crowd, Port Authority senior administrative assistant Elaine Duch stepped onto the platform in her white canvas sneakers. She held a purse in one hand and a tote bag in the other. Her dark blond hair fell on the shoulders of her smart blue jacket. Elaine's cherished gold skirt with a blue paisley print, the one she'd laid out the previous night after swimming with her twin sister, Janet, swished with every hurried step.

Emerging from the subterranean gloom, Elaine ordered her morning coffee, with a whisper of milk, at a pushcart whose owner knew that she would circle back, frowning, if she opened it at her desk and found it wasn't just right. Sunlight streamed through the cathedral-like windows inside the North Tower lobby as Elaine boarded an express elevator that rocketed skyward and deposited her at the 78th floor sky lobby. There she caught a local elevator up to her destination: the 88th floor.

Elaine reached her desk in the Port Authority's real estate department a few minutes before eight. Ringing telephones and the hustle of colleagues rushing to meet deadlines heralded her arrival. Elaine said quick hellos, dropped her bags, and turned on her computer. Swamped by work, Elaine had no time to change out of her sneakers; her strappy black leather sandals stayed buried inside her tote bag with a new gadget: her first cellphone.

Elaine sipped her coffee, the perfect shade of mahogany, and dived into her day.

A MILE AND a half to the east, FDNY Captain Jay Jonas wolfed down a bowl of Wheaties and gulped black coffee in the kitchen of the Ladder 6 firehouse in Chinatown. He'd been awake nearly all night, busy with runs, and now Jay could only hope for a quiet day ahead.

Around eight thirty, a half hour before the changeover to the day shift, Jay joked around with the two younger firefighters from a different Manhattan ladder company, Scott Kopytko and Doug Oelschlager, who'd worked the overnight shift with Jay and his men. The pair said goodbye, leaving Jay to finish his breakfast.

Alone with his thoughts, Jay prepared for another shift as a captain, eight years into the role, still number eighteen on the promotion list.

WORKING ON SHORT sleep after his movie premiere, aspiring actor and temp worker Chris Young arrived by subway at the World Trade Center shortly after eight. He'd already swung by the Midtown office of Marsh & McLennan to grab the box of materials he had to deliver.

Chris blinked at the precise instant a guard took his photo for a visitor ID badge. Clipped to his shirt, it allowed him access to the North Tower's 99th floor, one of eight floors where the giant insurance and financial services company rented space. Chris had previously worked a different temp job in the South Tower, so as he pushed through a lobby turnstile toward the elevators he anticipated the stunning views awaiting him.

The 99th floor was already a hive of activity at the start of the workday. Chris quickly found his temporary boss, managing director Angela Kyte. He handed her the box, but his job wasn't done. Angela told him that a separate shipment of presentation materials hadn't arrived, so he should track it down.

With a few phone calls, Chris discovered that a planned delivery the previous night had gone awry, but the materials were now on their way. He volunteered to wait, knowing that both Angela and his other supervisor, Dominique Pandolfo, intended to spend the entire day in the North Tower. Their absence from the Midtown office meant that he'd have nothing to do all day if he left now and went back uptown.

Angela surprised Chris by saying that she'd deal with the late arrival herself. He could take the subway back to Midtown.

AT 8:30 A.M., Cecilia Lillo was hungry.

The Port Authority administrator had lately grumbled to her paramedic husband, Carlos, that she'd been gaining weight while he kept fit by jogging after work. In the semiuseful way of husbands everywhere, that morning Carlos had executed a plan. During their shared commute, he bought one bagel for them to split, instead of their usual order of a full bagel each. Carlos had chosen her favorite, plain with butter, but it wasn't enough. Now, hours before lunchtime, Cecilia's stomach growled.

In her office on the 64th floor of the North Tower, Cecilia decided that she'd head up to the 86th floor to deliver a stack of ID cards to colleagues there, shoot down to the 43rd floor to graze through the public cafeteria, then return to her desk fully fueled.

First, though, Cecilia bumped into Nancy Perez, a vivacious Cuban-born Port Authority supervisor. Cecilia admired Nancy, whose nature was to look after people. Among other outside pursuits, Nancy learned sign language to teach karate to deaf children. In a hallway outside a ladies' room, the two friends made lunch plans for a Cuban restaurant and strategized about how Cecilia could balance her desire to become pregnant with her ambitions for promotion.

Before heading upstairs, Cecilia circled back to her desk to check an email.

BY 8:30 A.M., Moussa "Moose" Diaz had already put in a full day.

He awoke at the usual awful time: 2:40 a.m. That was the price he paid to work as an emergency medical technician in New York City while raising his family atop a mountain in upstate Monroe, New

York. This would be Moose's second day back at work after a three-week vacation, part of which he spent visiting Virginia with his wife, Ericka, a waitress, and their sons, eleven-year-old Greg and five-year-old Harrison.

Moose was thirty-six, nearly six feet tall, with a shaved head and soulful brown eyes. He had olive skin, inherited from his Cuban father and his Palestinian/Haitian mother. His mother had chosen his name, the Arabic equivalent of Moses. Calm and thoughtful, happiest with his family, Moose showered and dressed in his dark blue uniform with EMT in white letters over his heart and FDNY across his broad back. He moved silently through the darkened house to avoid waking Ericka and their boys.

Fortified by a protein shake, Moose slid into his 1993 Toyota Corolla and turned on the news radio station 1010 WINS. Still sleepy, he settled into his hour-and-fifteen-minute country-to-city commute to Crescent Street and Thirty-First Avenue in the New York City borough of Queens, home to Battalion 49, Astoria Station, ten miles across the East River from the Twin Towers.

Moose arrived early for his 5 a.m. to 1 p.m. shift at the EMTs' cramped underground workplace, nicknamed without affection the Submarine. The good news was that Moose and his colleagues rarely spent much time inside; they served a working-class district of housing projects and factories with one of the heaviest emergency call volumes in the city. Some days it seemed as though everyone in Astoria dialed 9-1-1.

With more than an hour until sunrise, Moose went through a mandatory routine of making sure his ambulance, 45 Adam, had enough gas, plenty of bandages, and a working defibrillator. His trauma bag was stuffed with stethoscopes, gauze, and airway kits. At 5 a.m. sharp, Moose and his partner of two years, Paul Adams, logged on to the radio network so dispatchers would know they were ready, willing, and available.

Paul was thirty-five, a powerfully built five foot nine with a crew cut and a wild edge: the yin to Moose's tranquil yang. After his father's death two decades earlier, Paul had emigrated to Queens from

Glasgow, Scotland, with his mother and two younger sisters. Although he still spoke with a slight burr, Paul had become a full-throated New Yorker with a ready supply of profanities. Single, a city EMT for ten years, when he wasn't working or playing pool, Paul could be found in the air, piloting small planes.

Moose and Paul piled into their ambulance for the day's first call, a pregnant woman suffering from blood loss and a possible miscarriage. As a precaution, they called for backup from another Astoria crew, 49 Victor, an ambulance staffed by two paramedics: Roberto Abril and Cecilia Lillo's husband, Carlos Lillo.

Moose felt especially glad to see Carlos, whom Moose considered a mentor. Carlos had been two years ahead of Moose at Long Island City High School, where Moose was a wrestler and Carlos captained the gymnastics team. When Moose first arrived on the job, other EMTs and paramedics kept a cool distance. Then one day in the station locker room, a high-pitched voice announced to all within earshot: "Oh my God. I can't believe it—Moose is here!" Moose immediately became part of the squad.

Having delivered the pregnant woman to Elmhurst Hospital by 8:30 a.m., Moose and Paul stood near the hospital's emergency room, waiting to get their paperwork signed. Carlos lingered nearby.

AS HE NEARED the Marriott, Ron Clifford felt the streets around him pulse with controlled chaos. On any given workday morning, a million or more people rushed about Lower Manhattan to command, serve, live in, or visit the main engine of the world's financial system. Stock traders and executives, secretaries and technology whizzes, public servants and messengers, food servers and custodians, retail clerks and tourists jostled for position, all under the watchful eye of police officers, "New York's Finest," who patrolled in cars, on foot, and on horseback, and firefighters, "New York's Bravest," ready for whatever emergency the day might bring.

If money had an aroma, Lower Manhattan would have been as fragrant as a bakery. Instead it smelled of fast-moving people and slow-moving vehicles, asphalt and steel, coffee and steam, cologne and sweat, with salty high notes wafting from New York Harbor. Air brakes hissed in complaint as city buses disgorged passengers. Horns blared as taxis avoided men in polished brogues and sneaker-wearing women who, like Elaine Duch of the Port Authority, carried heels in their work bags. Another among the sneaker set was Jennieann Maffeo.

At forty years old, Jennieann stood five feet one, with luxuriant brown hair held in place by a metal clip. She wore blue pants, a pretty blouse, and a zippered sweater. New Balance running shoes cushioned her steps; her briefcase contained her work shoes alongside her wallet, a book, and a knitting project.

As Ron Clifford approached the Marriott, Jennieann waited nearby for the second leg of her commute in the shadow of the North Tower, at a bus stop on West Street. She intended to grab a New York Waterway shuttle bus to a Hudson River ferry pier. A brief cruise would leave her on a dock in Weehawken, New Jersey, close to her job as a computer systems analyst at the financial firm UBS PaineWebber.

Jennieann's younger sister and best friend, Andrea, often teased her that the long commute would be the death of her. But Jennieann tolerated the ninety-minute trek every morning and every night so she could live with Andrea. Together, they cared for their mother, Frances, a cancer survivor, and their father, Sam, a stroke victim, in a three-family house in a working-class section of Brooklyn.

Single, a gifted soprano in her church choir, Jennieann threw herself into childcentric volunteer work. She raised money to fight juvenile diabetes, supported Make-a-Wish, and spent lunch hours reading to impoverished children. Earlier that morning she had interrogated Andrea, the literacy director for the New York City public schools, about the most economical way to buy art supplies for a needy New Jersey elementary school. Before leaving for work, the sisters made plans to shop that night at a discount store.

As Jennieann waited for the shuttle bus, she stood alongside a quiet colleague, Wai-ching Chung, a thirty-six-year-old UBS PaineWebber vice president. Wai-ching's colleagues knew him as a man so devoted to protecting the firm's databases that he rarely took a day off. A Hong Kong native, painfully shy, Wai-ching would get flustered if anyone at work so much as said hello. He lived with his parents and younger brother in Brooklyn. He spent his little free time with his sister and her family, including his niece Maurita Tam, a recent graduate of Amherst College.

When Maurita was a child, Wai-ching had amused her by blowing sheets of tissue paper into the air so she could watch them float gently to the ground. Now twenty-two, Maurita enjoyed the sight of rainbows that spanned the Manhattan skyline. At that very moment, Maurita was headed skyward, to her job as an executive assistant for the Aon Corporation on the 99th floor of the South Tower, a thousand feet above the shuttle stop where Wai-ching and Jennieann stood waiting.

RON CLIFFORD CLIMBED the stone stairs to the Marriott, a twenty-two-story hotel dwarfed by the adjoining towers. A favorite of business travelers, the 843-room Marriott boasted meeting rooms fittingly named Dow, Stock, Bond, and Trader.

Ron stepped into the beige marble lobby. With some time to kill before his 9 a.m. meeting, he ducked into a restroom for a peek in the mirror. He couldn't explain why, but Ron took appreciative note of the bathroom's antiseptic cleanliness. His hair in place, his bold yellow tie straight, Ron returned to the lobby. Still too early to call his meeting partner's room, he pushed through revolving doors that connected the Marriott to the North Tower.

Despite his ambivalence about the towers, Ron enjoyed the soaring grandeur of their light-filled seven-story lobbies. He gazed through the windows onto the five-acre plaza. It occurred to him that the steel trident columns at the towers' base resembled an upward branching design in stained glass windows created by Frank Lloyd Wright. The master architect called his pattern the Tree of Life.

Invigorated, Ron returned to the Marriott lobby. Diners clinked silverware and spoke in muffled tones over breakfast in the hotel's

Tall Ships Bar and Grill. Guests checking out bustled toward the front desk. Ron reminded himself of everything at stake and inhaled deeply.

BY 8:41 A.M., on the North Tower's 88th floor, Elaine Duch had caught up with her pile of work. She took a break to send a ritual morning email. She wrote her twin sister, Janet, that she'd arrived safely and reminded her of their yoga class that night. Elaine included an exasperated complaint about the first leg of her morning commute: "bus soooooooo crowded. . . . no a/c—i was roasting on bus, then i finally cooled."

Minutes later, a receptionist called Elaine with momentous news: a messenger had arrived with ten brown cardboard boxes whose contents foretold the future of the World Trade Center. Inside the boxes were overdue lease documents that would enable the Port Authority to give two private real estate companies control of the complex for the next ninety-nine years. This was the paperwork that had delayed Elaine's planned move to a new job in the agency's audit department, based in an office across the river in Jersey City. The messenger's arrival signaled not only the trade center's historic turnover to private control, but also the final days of Elaine's quarter century of work inside the North Tower. She'd miss her Port Authority friends, but not the building, whose magnificent views of Manhattan never compensated Elaine for its dizzying height.

Elaine grabbed a set of keys and walked briskly through a glass door that led to a vestibule down the hall from a bank of elevators. The messenger's flatbed cart couldn't fit through the glass door, so Elaine pointed him toward double-wide doors down the hall that served as the main entrance to the Port Authority real estate department.

IN THE MARRIOTT lobby, Ron Clifford made final preparations for his big meeting. In his Chinatown firehouse, Captain Jay Jonas spooned his Wheaties. At Elmhurst Hospital, paramedic Carlos Lillo, EMT

Moose Diaz, and their partners awaited their next call. On the 64th floor of the North Tower, Cecilia Lillo plotted her second breakfast. The World Trade Center hummed with the usual activity of a normal Tuesday morning in September. Roughly 8,900 people were at work or visiting the North Tower.

After being dismissed by his supervisor, Angela Kyte, on the 99th floor, temp worker and actor Chris Young retraced his steps to a local elevator. He rode it down to the 78th floor sky lobby, then switched to one of ten giant express elevators to the ground floor.

Alone in an elevator car built to carry up to fifty-five people, Chris felt tired from his movie premiere the night before, but upbeat about an easy workday ahead. He recognized a rare opportunity to test a childhood theory that claimed a person who jumps inside a high-speed elevator feels as weightless as an astronaut in space.

Chris jumped once. Nothing. A second jump, higher. Still nothing. Chris jumped a third time, higher still. The time was 8:46 a.m.

AT THAT MOMENT, terrorist pilot Mohamed Atta gripped the controls inside the cockpit of American Airlines Flight 11. Thirty-two minutes had elapsed since the takeover began. After flying the Boeing 767 the full length of Manhattan island, Atta pointed the hijacked jet toward his target: the North Tower of the World Trade Center.

Flight attendant Amy Sweeney huddled in a rear jump seat. Using an Airfone, she described the hijacking to Boston-based flight service manager Michael Woodward.

"Something is wrong," Amy told Michael. "We're in a rapid descent. . . . We are all over the place."

Michael asked her to look out the window and describe what she saw.

"Oh my God!" Amy said. "We are way too low!"

CHAPTER 13

"GOD SAVE ME!"

Ground Level and North Tower, World Trade Center

FORTY-SEVEN MINUTES AFTER TAKEOFF, CARRYING EIGHTY-SEVEN hostages, five tons of cargo, ten thousand gallons of fuel, and five terrorists, American Airlines Flight 11 completed its forced conversion from a passenger jet into a 283,600-pound guided missile. Its nose aimed slightly downward, its right wing tipped upward, the silver Boeing 767 with red, white, and blue stripes and "AA" on its tail smashed into the north face of the North Tower at 8:46:40 a.m. Its violent arrival carved an airplane-shaped gash in the steel and glass that stretched at an angle from the 93rd to the 99th floor.

As it entered the building, what remained of Flight 11 sliced through thirty-five exterior steel columns and heavily damaged two more. It severed six core columns and damaged three others. It shattered at least 166 windows. It broke the concrete floor slabs of the 95th and 96th floors eighty feet deep into the building. It launched a fusillade of flying debris that knocked or scraped fire-retarding insulation from forty-three core columns. It stripped the insulation from sixty thousand square feet of steel floor supports over several floors. It severed pipes that fed water into the fire sprinkler system. It stopped elevators in motion and cut off elevator service to at least the sixty upper stories. It sent glass and metal and office contents and body parts raining down a thousand feet to the plaza and the streets below.

It altered the path of American and world history.

All that damage took less than one second.

A wheel from the left wing landing gear crashed through the North Tower's central core, embedded itself in an exterior column on the far side, ripped the steel beam from the building, and flew with that piece another seven hundred feet to the south, landing on Cedar Street. Another wheel also passed entirely through the tower; unencumbered by building parts, it flew twice as far to the south.

The crash immediately killed everyone on board Flight 11 and an unknown number in the plane's path. But that was only the beginning. As the plane blasted through the tower's core, it crushed the walls of all three emergency stairwells in its path, cutting off stair access to everyone on the 92nd floor and above. At the moment of impact, an estimated 1,355 people were inside those nineteen uppermost floors. That included roughly two hundred people dining or working at Windows on the World and attending a technology conference on the 106th floor.

The survivors on those floors had no way down and no way out, although many didn't yet know it. Scores called 9-1-1 as well as family members and friends, while others sent emails. They delivered oral and written messages that spanned the emotional spectrum, from panicked pleas to calm appeals. Some expressed their greatest concern not for themselves, but for the loved ones whom they worried they would soon leave behind.

Yet despite the damage, despite the death and destruction, in the immediate aftermath of the assault, the North Tower still stood. It absorbed the unthinkable blow, bending and swaying but not breaking. Even with its relatively spindly design and sparing use of steel, the tower had what engineers call "reserve capacity" that allowed it to support a far greater load than its own weight plus the weight of people and furnishings. When Flight 11 severed more than forty exterior and core support columns, the building instantly and automatically redistributed the load to undamaged neighboring columns. That kept the North Tower upright and prevented the immediate deaths of survivors in and above the impact zone, as well as more than seventy-five hundred men and women on lower floors who streamed toward undamaged stairwells to try to escape.

With its load redistributed, the tower could have remained standing indefinitely, potentially allowing rescue workers to reach everyone who survived the initial damage. If not for the fires, that is.

The North Tower's external steel columns cut through the Boeing 767's fuel-filled wings like the blades of an egg slicer. Fireballs visible for miles exploded from the entrance wound and from blown-out windows on the east and south sides of the tower. More fireballs raced up and down elevator shafts, blowing out doors and walls as far down as the basement levels. Toxic clouds of hot, thick smoke poured up and down the central core and gushed out of the broken building. No longer was the morning sky an unblemished blue.

Despite the explosions, less than half of the ten thousand gallons of jet fuel from Flight 11 burned in the initial fiery blasts. The rest sprayed through the impact floors and nourished fires that consumed combustibles from the plane and the office furnishings. Those fires fed off fresh air that flowed into the torn-open building. As flames gathered strength and spread, trapped survivors rushed toward sealed and broken windows in desperation. At the same time, the fires began to threaten the remaining support columns that already carried a heavier-than-normal load from their broken neighbors. Meanwhile, fires licked at the exposed steel of floor supports that were shorn of their fire-retarding insulation.

The secondary effects of the crash were in full swing.

Building fires typically don't get hot enough to melt structural steel columns, even relatively thin ones. But long before steel reaches its melting point, it loses strength. The weaker a steel column gets, the less able it is to carry its assigned or reassigned load. Similarly, fires could make unprotected steel floor supports sag, adding stress and pulling down on the exterior and core columns to which they were attached.

Ultimately, if the structural steel in the impact zone became hot enough for long enough, or was forced to carry too much added weight, it would buckle. If that happened, everything above the impact zone would come crashing downward, creating an enormous moving mass that could overwhelm the entire North Tower. Put simply, it could cause total collapse.

Although the impact of the B-25 bomber on the Empire State Building in some ways resembled the crash of Flight 11, fundamental distinctions between the buildings' designs made the two events vastly different. No building like the North Tower had ever experienced such an assault, so no one could say for certain what might happen next. Like everything else about the morning of September 11, this was uncharted territory.

WHEN FLIGHT 11 struck, Elaine Duch had just stepped through the glass door outside her office on the 88th floor to meet a messenger, Vaswald George Hall, and his document-laden flatbed cart. Hall was fifty, the father of four, a police officer in his native Jamaica before arriving in the United States seventeen years earlier. He'd recently scored high on a civil service exam and hoped to start a new career.

Before Elaine could guide him through the larger, double doors of the Port Authority's real estate department, an explosion roared from above. The building swayed and moaned as though threatening to dive into the Hudson. The floor rippled and rocked beneath her sneakers like the deck of a ship in high seas.

Before Elaine could think, before she could act, a fireball of ignited jet fuel traveled down an elevator shaft and burst through the elevator doors. It illuminated the hallway with a brilliant orange flash of dragon's breath. It consumed Elaine. The fire seemed to touch every part of her at once, as though she'd leapt into a cauldron, a scorching immersion that bathed her in unspeakable heat. Certain that she was about to die, Elaine screamed: "God save me!"

The fire considered her open mouth to be an invitation to scorch Elaine's lungs.

All around her, ceiling tiles popped from their frames. Light fixtures crashed to the floor. Gypsum drywall boards burst from their anchors. Elevator doors ruptured off their tracks. The fireball from Flight 11 passed as quickly as it arrived on the 88th floor, with the growl of an engine and the shush of a snuffed-out candle. It left behind a smoky haze and small fires in the far reaches of the corridors.

And it left behind Elaine.

In shock, yet still on her feet, Elaine glanced down and saw charred tatters of her cream-colored top melted to her skin. Blackened remnants of her skirt made gruesome tattoo marks on her exposed legs. Her face and arms shone bright red, as though she'd fallen asleep on a broiling beach. Her jacket seemed to have burned away entirely. The sulfurous smell of her scorched hair mixed with the lingering odor of jet fuel. Smoke rose from her like mist from a morning lake. When the fire embraced her, it cut through her watchband and sent her watch skittering to the floor. Her key case leapt from her hand. Elaine's glasses were askew but intact, having miraculously protected her eyes. Her only other unscathed body parts were her feet, shielded by her sneakers. Vaswald Hall, the messenger who'd been just feet away from her, was gone; Elaine didn't know where.

The fireball burned more than three quarters of her skin, from her scalp to her ankles. Her burns were mostly third-degree, which destroyed Elaine's nerve endings. For the moment, that was a blessing. The absence of nerve endings blocked her ability to feel pain and allowed her to only partly comprehend the severity of her injuries. Elaine's immediate concerns were embarrassment at her disheveled seminakedness, confusion about what just happened, and worry about the strange sizzling sounds all around her.

Elaine walked zombie-like into the real estate office, her arms outstretched, stepping across shattered glass from the door she'd passed through less than a minute earlier.

Startled by the plane's impact, the rising smoke, and the building's pronounced sway, Port Authority workers and several employees of the trade center's new leaseholder, Silverstein Properties, scurried through the office, some invoking the memory of the 1993 bombing.

Everyone halted at the sight of Elaine. People she'd worked with for decades asked one another with alarm, "Who's that?"

Two longtime colleagues, Joanne Ciccolello and Gilbert Weinstein, rushed over. They patted Elaine's head and body with their bare hands to extinguish smoldering embers. Elaine saw the horror in their eyes as they tapped and brushed away sparks. Alarmed by her exposed, damaged skin, Elaine ran to her desk and grabbed an off-white cardigan and tied it around her waist as an improvised wrap skirt. Her modesty partially restored, Elaine grabbed her purse. Years earlier, someone stole her pocketbook when she worked on the 63rd floor. She still winced at the hassle of losing her license and belongings, and she was determined not to let it happen again.

TEN MILES AWAY, at a chemical company in Bayonne, New Jersey, Elaine's twin sister, Janet, looked up from her desk to see her boss standing over her. He said that a small plane had slammed into the World Trade Center. Probably no big deal.

Janet sprinted outside, to a parking lot where she could see across the sparkling bay to Lower Manhattan. She spotted flames and smoke bursting from the World Trade Center tower with the giant antenna on top. Janet knew that Elaine worked on the 88th floor of the North Tower, which looked to be around the affected level. But, seized by fear, Janet couldn't remember which of the buildings had an antenna on top.

Frustrated that she didn't know which Twin Tower held her twin sister, Janet rushed back inside. She called Elaine's desk phone, but the call wouldn't go through. Janet tried Elaine's new cellphone, but it rang unanswered, buried deep in Elaine's abandoned tote bag.

A HOWL FROM an engine followed by a powerful *boom* launched FDNY Captain Jay Jonas to his feet.

For Jay, sudden loud noises were almost as motivating as fire alarms. He abandoned his Wheaties and ran outside as his mind spun through a catalog of horrendous sounds. Jay guessed that a freight truck must've driven off the nearby Manhattan Bridge, where he and his ladder company had been only hours earlier on the overnight shift in response to a scaffolding collapse.

Out on the street, firehouse watchman Ray Hayden already understood that the awful noise hadn't come from something as ordinary as a truck or a bridge. Hayden had seen a passenger plane screech overhead, heading toward the World Trade Center. Other buildings in the cramped Chinatown neighborhood blocked his sightline, but Hayden heard enough and rushed back inside to alert the troops.

The watchman's assumptions proved correct. New York's emergency airwaves burst to life with calls between FDNY dispatchers and a battalion chief who'd been responding to a gas leak on a street corner less than a mile uptown from the Twin Towers. Within seconds, the chief radioed: "We just had a plane crash into the upper floors of the World Trade Center. Transmit a second alarm and start relocating companies into the area."

The radio reports quickly escalated.

"The World Trade Center, Tower Number One, is on fire," said an officer from FDNY Engine 6, a half mile west of the sixteen-acre complex. "The whole outside of the building. There was just a huge explosion."

The first report of victims came from Captain Eugene "Jack" Kelty of Engine 10, located across the street from the towers: "World Trade Center, ten-sixty," he said, using the code for a major emergency with the possibility of multiple casualties. "Send every available ambulance, everything you've got, to the World Trade Center *now.*"

As Jay reached the sidewalk on Canal Street, he heard watchman Ray Hayden yell over the firehouse intercom: "A plane just crashed! A plane just crashed into the World Trade Center!"

Jay turned west to face toward the towers, about a mile downtown. He couldn't see them, but as Jay scanned the sky, he saw an ominous plume of rising black smoke.

"How big a plane?" Jay called to the watchman.

"It was a *big* fucking plane!" he answered.

Jay rushed inside and told Ray to sound the alarm to turn out Ladder 6 and also Engine 9, which shared the firehouse. Neither company had yet been ordered to respond, but Jay felt certain that both would soon be called to duty. In simplified terms, ladder companies climb into buildings to find victims and create ventilation, while engine companies pump water. But those rules applied more neatly to a six-story tenement fire; skyscraper fires had their own rules and made different demands on all firefighters, regardless of company assignments. Jay understood instinctively that a fire fed by jet fuel and a wealth of combustibles, caused by an airplane crashing into an upper floor of one of the Twin Towers, would make unique and vicious demands.

Jay raced into his office to pull on his bunker gear: fire-retardant pants, thick rubber boots, and a heavy black-and-yellow turnout coat. Stuffed in its pockets were gloves, a flashlight, and a smoke hood, along with items more typical for a mountain climber: rolls of nylon webbing and steel carabiners, which had as many potential uses as there were possible disasters. Jay had employed similar gear during the rescue of two firefighters in 1995, and he'd written an FDNY training bulletin on improvised rope rescues. He grabbed his air mask, a twenty-pound device as essential to a firefighter as scuba gear is to a diver. His black helmet, with a bold red 6 and CAPTAIN above the brim, awaited him in the truck.

As Jay dressed, the battalion chief who witnessed the crash into the North Tower added more chilling details, and another alarm, to the initial report that he had issued only twelve seconds after impact. "We have a number of floors on fire," Chief Joseph Pfeifer told the Manhattan FDNY dispatcher. "It looked like the plane was aiming toward the building. Transmit a third alarm throughout; the staging area [is] at Vesey and West Street"—the intersection at the northwest corner of the trade center property.

A third alarm meant a call for a dozen engine companies, seven ladder companies, an elite rescue unit, communications teams, multiple chiefs, and assorted support crews. As Jay anticipated, that included his unit, Ladder 6. Eventually, more than two hundred fire units, with more than a thousand firefighters, would swamp the scene, along with more than one hundred ambulances and the FDNY's Hazmat team. Some would come without being called, desperate to help however they could. Because the crash occurred close to the 9 a.m. shift change, many firehouses had double their usual complement of firefighters. Few if any of "New York's Bravest" wanted to avoid the fight. Trucks that normally carried six men zoomed toward the trade center with twice that number, "riding heavy" in firefighter parlance.

The FDNY responders also would soon include a charismatic, pious, sometimes joyously profane sixty-eight-year-old chaplain named Father Mychal Judge. His lifesaving skills included leading scores of people to Alcoholics Anonymous, where he, too, had found rescue. A day earlier, the silver-haired Father Mychal had rededicated a renovated firehouse in the Bronx, where he reminded firefighters of an essential truth about their work: "You show up, you put one foot in front of another. You get on the rig and you go out and you do the job, which is a mystery and a surprise. You have no idea when you get on that rig, no matter how big the call, no matter how small, you have no idea what God's calling you to. But He needs you. He needs me. He needs all of us. . . . So *keep going.*" When Mayor Rudy Giuliani reached the trade center, he'd call out to Father Mychal: "Pray for us." The priest assured the mayor that he would, as always.

Captain Jay Jonas, Chief Joe Pfeifer, and every other firefighter who sped toward the scene understood that this would be a big one. Maybe the biggest one ever. For his part, Pfeifer wanted some semblance of order from the outset. At a minimum, he wanted to steer rescuers into the tower as quickly as possible. Pfeifer continued his radio call: "As the third-alarm assignment goes into that area, the second-alarm assignment report to the building!"

As Jay Jonas mustered his troops, he knew that they'd be among the firefighters going in.

INSIDE THE MARRIOTT lobby, Ron Clifford heard an explosion, followed by a deep rumble that shamed the previous night's thunderstorm. Vibrations rose up from the soles of his polished shoes, reverberating through his body and everything around him, as though he and the entire hotel were tuning forks. Ron thought a storage tank in the basement might have ruptured. Yet the lobby walls, fixtures, and ceiling remained intact. He looked around and saw people trying to regain their bearings.

Confused, Ron glanced toward the revolving doors that led to the North Tower lobby. On the other side of the glass, he saw a chaotic scene. Ron smelled what he thought was kerosene. He heard screams. His mind flashed to the 1993 attack and his conversation on a sailboat the previous weekend with the retired World Trade Center engineer. Still, he didn't want to jump to conclusions.

As Ron stared at the North Tower lobby, he saw smoke fill the soaring space where moments earlier he'd admired the view. People rushed toward the blown-open revolving doors he'd just come through. Not fully processing the situation, Ron briefly fixated on the broken doors. "My god," he thought, "it must've been some kind of pressure to do that."

Fleeing people entered the hotel lobby and rushed past him, eyes wide, faces gripped by fear. Ron's gaze settled on one person who moved differently, more slowly and clumsily than the others. Through a foggy veil of fuel particles and fumes, a short, stout figure marched awkwardly yet determinedly toward him. His first thought: a homeless person in tatters. Then his mind sharpened: it was a woman, nearly naked, horribly burned.

Fire had consumed the woman's pants, blouse, and underwear. A zipper ran up her blackened chest, fused with her ruined skin. Dark tufts of fried hair sprouted from her head. A metal clip had melted against her scalp. Her eyelids were slits, clamped shut from burns or swelling. She shuffled forward. Her hands reached out, her twisted fingernails scorched white from heat.

The woman moaned in agony. Ron resisted an urge to recoil and banished an impulse to join the fleeing herd. As Ron stared at the woman, her burned lips formed a word that rooted him in place: "Help."

As everyone else ran to save themselves from the explosion that rocked the building and filled the North Tower lobby with smoke, fumes, and debris, Ron stepped toward the oncoming figure. Without medical training, with no idea what happened, Ron didn't know what to do for her. But he knew he had to do something.

"Okay," Ron told the woman, "we'll get some help."

He eased her down onto the marble floor. Her ravaged skin and what remained of her hair still smoldered. Heat and vapor rose from her body. Ron wished he could call his sister, Ruth, who'd spent a decade running a day spa, to draw upon her skin care expertise.

Ron thought water might help. He told the burned woman to wait as he ran into the restroom where minutes earlier he'd straightened his yellow tie. Remembering how clean the lavatory was, Ron snatched an empty garbage bag from its container and partly filled it with cool tap water. He ran back to the woman, still on the floor, alone in the crowd as people coursed toward an exit like deer from a forest fire.

Ron knelt and gently poured water on her burned arms, legs, and body.

"Help!" Ron called to passers-by. He stood and shouted: "Emergency! Help! Can anybody help us?!" No one stopped. No one replied.

He dropped back to the woman and the wet floor. A closer look showed that she had been burned head to ankle, as though she'd stood under a molten waterfall. Only the ragged tops of her New Balance running shoes were partly intact. The rubber soles were melted to the bottoms of her feet.

Despite her shock and pain, somehow the woman remained coherent and able to speak. She told Ron her name: Jennieann Maffeo.

He pulled out a notebook as she told him where she worked. She said she'd been waiting outside the North Tower for a shuttle to the ferry when she burst into flames. She didn't know how or why. As flaming debris and fireballs rained down, Jennieann had been swept into a frightened crowd, unable to see, on legs burned to the muscle. She'd somehow stumbled into the North Tower lobby, then passed through the blown-open doors to the Marriott lobby, and into Ron Clifford's path.

AT ELMHURST HOSPITAL in Queens, FDNY paramedic Carlos Lillo sprinted toward EMT partners Moose Diaz and Paul Adams, his eyes wild and brimming.

"A plane hit the towers!" Carlos yelled. "We gotta go!"

As they raced to their ambulances, Moose heard Carlos shout: "My wife's in there!"

Moose remembered a locker room conversation during which Carlos said he'd drilled Cecilia about how to find an escape route in any emergency, and how he'd made her promise to find an FDNY ambulance to summon him. Carlos had also told Moose about the promise he'd made to Cecilia about not risking his life to look for her if anything like the 1993 bombing happened again. Moose understood that those promises might soon be tested.

As they tore away from the hospital, the EMTs and paramedics radioed dispatch to say they were leaving their assigned district to cross the river to Manhattan. Sirens blaring, lights blazing, the two ambulances raced along Queens Boulevard. Through the passenger seat window, Paul Adams caught a glimpse of smoke pouring from the upper floors of the North Tower. The radio barked a report: a code 1040—airplane crash—at the World Trade Center. Chatter came across the airwaves about a "small airplane" striking the building. Paul thought about the single-engine Cessnas he piloted. He stared at the multistory gash and turned to Moose. "That ain't no small airplane," he said.

Moose fell silent. Fighting nerves, he reminded himself of all he knew about responding to a major catastrophe with multiple casualties. As he ran through the protocols in his head, Moose thought the North Tower looked like a lit cigarette standing on its filter end.

ALONE INSIDE AN express elevator descending toward the North Tower lobby, Chris Young landed from his third unsuccessful anti-gravity test. As he did, he heard a roar. A blast of warm, dusty air and a sickly-sweet scent flooded his senses. Lights popped from ceiling frames but remained lit. The elevator car screeched and shuddered to a stop, knocking Chris off his feet.

Curled into a ball, Chris briefly wondered if his jumping up and down had somehow derailed the elevator as it descended. Then he remembered 1993. He couldn't be sure, but the noise he had heard sounded as though it came from below. "Another bomb under the World Trade Center," he thought.

Chris saw the "L" illuminated on the elevator car's digital control panel, but he didn't know how close it had stopped to the lobby. Heart racing, breathing fetid air, he stood and took stock. Relieved that he wasn't hurt, Chris gathered his wits. He pressed a red emergency button that set off a ringing alarm.

After several minutes with no response, he noticed a second button with the outline of a firefighter's helmet. He pulled the alarm button to silence the ringing and pressed the fire hat. An automated computer voice assured him that his emergency call had been received and would be answered shortly.

The voice repeated itself without pause for the next fifteen minutes.

DRESSED AND READY, Captain Jay Jonas climbed into the front seat of the red-and-white firetruck, a tractor-trailer model called a tiller rig, designed for the tight turns of Chinatown streets. He looked around to see a fire crew he'd match against any in New York City, or anywhere else for that matter.

The men of Ladder Company 6, Jay's men, held job titles that evoked an earlier era of firefighting or the credits from a superhero movie. "Roof Man" Sal D'Agostino brandished a long rope and hand tools; "Can Man" Tommy Falco, who years earlier had asked Jay to lead the company, carried a loaded extinguisher; "Chauffeur" Mike Meldrum drove the big truck; "Tiller Man" Matt Komorowski steered the rear wheels from a raised perch atop the back end; and aptly named "Irons Man" Billy Butler, the strongest among them, wielded special pry bars to get them into and out of trouble.

Sirens screaming, lights flashing, hearts pumping, Mike Meldrum and Matt Komorowski pulled the truck out of the firehouse and pointed it west on Canal Street. The street rose as they approached an entrance to the roadway that led toward the Manhattan Bridge. The higher elevation gave Jay a panoramic view of Lower Manhattan that had never been more breathtaking.

As he stared toward the Twin Towers, Jay felt the strange and unfamiliar sensation of being overwhelmed. He'd spent years studying fire, learning its destructive ways and plotting creative, disciplined responses to every kind of catastrophe he could think of. Now, looking through a windshield the size of a big-screen television, Jay had a view that shamed his imagination: orange flames and gray-black smoke blasted from an enormous, angled hole in the topmost quarter of the North Tower, polluting a sky that had been as crystalline as a mountain lake.

"Buckle up!" he called to his men. "We're going to work."

In the silence that followed, Jay wondered: "How many people need our help right now, right at this very minute?" As he calculated the possible toll and considered the job ahead, Jay couldn't conceive the disaster's true cause: "My god," he thought, "what a horrible accident."

Others in the FDNY understood immediately, with various degrees of precision, that this was no accident. In one of his earliest radio transmissions, Chief Joe Pfeifer used the word "aiming" to describe the route he saw the plane take to the North Tower. Less than three minutes after impact, at 8:49 a.m., Lieutenant William "Billy" McGinn from Squad 18, a Special Operations Command unit based in Greenwich Village, displayed even deeper insight.

Without knowing about Mohamed Atta's "We have some planes" comment a half hour earlier aboard American Flight 11, or that United Flight 175 had also been hijacked and at that moment was pointed toward Manhattan, or that two F-15 fighter jets from Cape Cod were racing toward New York, Billy McGinn told a dispatcher: "Looked like it was intentional. Inform all units coming in from the back it could be a terror attack." An FDNY dispatcher acknowledged the new reality: "Ten-four. All units be advised."

As Jay gaped at the burning tower from the front seat, he silently counted the number of upper floors that appeared to be belching smoke. *We have twenty floors of fire,* Jay thought. As Chauffeur Mike Meldrum wove through traffic, and as rapid-fire voices on the truck's radio confirmed the horror awaiting them, Jay's mind raced. He didn't know the exact point of impact, but he estimated that the lowest fire floor was about a thousand feet in the air, and he knew that each floor of the tower covered about forty thousand square feet, or nearly an acre.

Jay leapt to an inescapable conclusion, one that would be over-whelmingly shared among his colleagues: from a firefighting stand-point, the numbers didn't compute. Jay felt confident that the men of Ladder 6 and the rest of the FDNY would do whatever they could, whatever their bosses asked of them, whatever the public needed of them, but they couldn't defeat this fire. At most, they could limit its spread until, in a best-case scenario, it burned itself out while they res-cued as many people as they could. Jay also understood that the great vertical distance between the ground and the fire meant that many, perhaps most, of the thousands of civilians inside the tower would be forced to rely on themselves and one another if they were to escape.

Jay's instant assessment wasn't idle speculation or defeatism. Two years earlier, one of Jay's firefighting heroes had published a prescient manual that seemed to eerily anticipate the situation awaiting Ladder 6 as it sped down Canal Street.

When Jay was a young firefighter, FDNY Deputy Chief Vincent Dunn was the highest-ranking officer he'd ever met. Awestruck at the time by Dunn's encyclopedic knowledge and his unbridled con-fidence, Jay told his wife, Judy, "I want to be like him." In the two decades that followed, Dunn became a nationally acclaimed expert on high-rise fires, unafraid to tell inconvenient truths. In 1999, Dunn wrote: "The best-kept secret in America's fire service is that firefight-ers cannot extinguish a fire in a twenty- or thirty-thousand-square-foot open floor area in a high-rise building. A fire company advancing a 2½-inch hoseline with a 1¼-inch nozzle discharges only three hun-dred gallons per minute and can extinguish only about twenty-five hundred square feet of fire."

Dunn's calculations meant that a team of firefighters like Jay's unit could, at best, hope to defeat a fire in one corner of one upper floor of a building like the North Tower. That is, if they could reach it and had enough water to spray on it. Multiply that by a hundred, or a thousand, and the impossibility of the situation came into focus.

Dunn took his reasoning one step further: "City managers and department chiefs will not admit this to the public if they want to keep their jobs, but every fireground commander is aware of this vulnerability. What really occurs at a high-rise fire involving an entire floor or more is a controlled burn rather than a suppression operation."

Even more disturbing and more prophetic were Dunn's conclusions, published on the same pages, about how firefighters should expect panicked civilians to react: "People trapped in a burning high-rise who can't be reached by your tallest ladder will leap to their death; they'll try to escape by climbing down ropes or knotted bedsheets, and fall while doing so; they'll scribble notes in desperation, telling of their location, and drop them from smoky windows; they'll leave their last cries for help recorded on the telephones of dispatchers."

As a student of Dunn's work and of firefighting in general, Jay knew all of this. He also suspected that he couldn't count on the building's sprinklers or water supply pipes to be much help, if any, because a plane that cut through the tower likely damaged or destroyed the systems that delivered water. Jay's assumption on that score was correct.

Jay knew that the drive to the scene wasn't the time to instruct his men on the overwhelming, to some extent theoretical, problems they faced. Jay controlled his breathing. He applied a trick that he'd learned from his mentors: the more excited he became, the slower and quieter he'd force himself to speak.

"This . . . is . . . going . . . to . . . be . . . a . . . big . . . operation," Jay told his men. "There's going to be guys that are operating all day, and we'll have a small part of it. And we'll just do the best we can."

Jay sensed the adrenaline coursing through his men. Mike Meldrum's intensity expressed itself as a heavy foot on the gas as they crossed Broadway in the Tribeca neighborhood, speeding toward the Financial District.

"Mike, slow down," Jay told the chauffeur, a beefy twenty-year veteran with a handlebar mustache. "We'll get there. But if you're going this fast, we might *not* get there, you know?" Mike eased off the pedal.

Even as Jay displayed outward calm, anxiety crept up his spine. As the North Tower loomed larger and larger, as smoke and flames licked higher, Jay thought back to a conversation he'd had two weeks earlier. A friend who was a lieutenant knew that Jay had been awaiting his overdue promotion to battalion chief. He asked if Jay was nervous. "Well," Jay answered, "after the busy places that I've worked, there isn't too much they're going to throw at me that I've never seen before."

As Mike Meldrum and Matt Komorowski turned the truck onto Vesey Street, Jay mentally revised his answer. He told himself: "This is really bad. . . . I never saw this before."

ON THE 64TH floor, before Cecilia Lillo began her multifloor delivery and food run, she circled back to her desk to check her email. An explosion announced the arrival of American Flight 11 some thirty floors above.

"Mommy!" she screamed and burst into tears.

Remembering 1993, Cecilia feared that a bomb had split the tower vertically, like an ax through a piece of firewood. The building swayed, lights flickered, a secretary at a nearby desk shrieked her boss's name. Another Port Authority worker struggled for balance as she stood changing into her work shoes. Cecilia saw a large white block, maybe a piece of concrete, or a computer, or something else entirely, plummet past a window.

"Let's go!" one of her coworkers shouted as she rushed with several others to a stairwell. Cecilia lurched toward a cabinet to clip her cellphone to an inside pocket of her purse. When she turned around, the other women were gone. Running through the floor in the new black flats Carlos had bought her for their anniversary, Cecilia searched for her boss but couldn't find her. Neither did she see her Cuban-born friend Nancy Perez or another friend, Arlene Babakitis, a bighearted mother of two sons who'd spent nearly thirty years at the Port Authority.

On the far side of the floor were several other Port Authority workers, including one of Cecilia's favorite managers, Patrick Hoey, a silver-haired father of four who oversaw bridges and tunnels. He and the others around him made no move to leave. After speaking to a police sergeant, Pat Hoey announced to his staff: "Listen up, everybody. They told us not to leave. They are sending police up, and we need to wait here. We are going to be fine."

Cecilia respected Pat Hoey, but she had other plans. Clutching her purse, she scrambled toward the center of the floor to find a stairwell, just as she'd vowed to Carlos during their Caribbean vacation four months earlier. She repeated it to herself like a mantra: someone who loves me will be waiting outside.

ON THE 88TH floor, two of Elaine Duch's colleagues, Dorene Smith and Anita Serpe, led the badly burned woman gingerly toward the empty office of Alan Reiss, director of the Port Authority's World Trade Department, which oversaw the trade center property. He was on street level when the plane hit, meeting with other agency officials in a delicatessen to consider the anticipated effects of the new property lease.

In Reiss's absence, about two dozen men and women congregated with Elaine in his office, in the building's southwest corner, on the far side of the tower from Flight 11's point of impact. Still, smoke from the fires ignited by the fireball seeped in through cracks between the office door and its frame.

As Elaine and the others awaited instructions, several workers scrambled through the debris and the overturned furniture of the 88th floor, looking for anyone who might be hurt and seeking an open stairwell. Leading the searchers was architect Frank De Martini, a Port Authority construction manager. When the plane hit, Frank had been enjoying coffee with his wife, Nicole, a fellow architect who worked in the South Tower. While Frank led the search, Nicole joined Elaine and the others taking refuge in Alan Reiss's office.

Coincidentally, Frank De Martini had been featured a few months earlier in a History Channel documentary, boldly describing the towers' strength. "This building was designed to have a fully loaded 707 crash into it," Frank had told the filmmakers, unwittingly citing the flawed decades-old report that didn't consider damage from a fuel explosion. "I believe that the building probably could sustain multiple impacts of jetliners because this structure is like the mosquito netting on your screen door—this intense grid, and the jet plane is just a pencil puncturing that screen netting. It really does nothing to the screen netting." If only that had been true.

Another man scouring the 88th floor for an escape route was Port Authority architect Gerry Gaeta. When the plane hit, Gerry leapt over his desk and ran out of his office. "That's a bomb," he told his staff. "Let's get out of here." He found the halls filled with smoke and fire. Windows on the building's northeast corner were blown out completely. At the centermost stairway, Stairwell B, Gerry and others found a hole where the stairs used to be. At Stairwell A, the southernmost stairway, he saw gypsum wallboards, crumpled and on fire, blocking the door. Gerry entered the real estate department office and saw furniture tossed around as though a hurricane had blown past.

With no apparent way out and the situation growing desperate, someone on the floor used a two-way radio to call for help from the Port Authority police.

"Uh, we're on the eighty-eighth floor," the caller said. "We're kind of trapped up here and the smoke is, uh, is—" The transmission cut off, but another radio call followed quickly, using the building's designation as the "A" tower and making a clear reference to Elaine.

"We also have a person that needs medical attention immediately."

"What's the location?" the dispatcher asked.

"Eighty-eighth floor, badly burned."

"Eighty-eight?"

"Tower A, south side, eighty-eighth floor."

As Elaine sat in Alan Reiss's office with her colleagues, a lens from her glasses suddenly cracked, a delayed reaction to the heat blast. She took the glasses off, handed her purse to her friend Anita Serpe, and made a close examination of her injuries. As Elaine looked at her legs, she thought the peeling red skin on her calves resembled melting candle wax, dripping toward her sneakers.

CAPTAIN JAY JONAS and the men of Ladder 6 reached the staging area at the corner of Vesey and West Street less than ten minutes after Flight 11 hit. With help from Tiller Man Matt Komorowski, Chauffeur Mike Meldrum parked in a semicircular driveway at the foot of the North Tower. Several other fire companies had already arrived, and more pulled in simultaneously.

Hail-sized pieces of building debris rained on the truck as Jay, Mike, Matt, Sal, Tommy, and Billy jumped out. With their silver emergency air cylinders bouncing against their backs, they ran to a pedestrian bridge that spanned West Street, taking cover under it as flames and smoke erupted from floors a quarter mile in the sky above them. Three times the Ladder 6 crew ran back to the truck to fetch tools, rope, and other equipment, then each time retreated from the falling junk under the pedestrian span. Finally, a break came in the debris storm. Jay called out, "Okay, ready . . . set . . . go!"

As they ran toward the North Tower, Jay glanced to his left. Sprinting alongside him was New York City's civilian fire commissioner, Thomas Von Essen, who'd previously been a longtime firefighter and a top union official. As they approached the tower doors, Jay saw two badly burned people with no hair and melted clothes. Their injuries were so severe that he couldn't tell their gender. Every instinct urged Jay to halt and help them, but then he thought, "We stop and help these two people, or we go upstairs and help a hundred." He spotted two paramedics and waved them toward the burn victims, then rejoined his men and entered the lobby as frightened evacuees poured out.

Inside, Jay saw slabs of decorative stonework and marble tiles smashed on the floor, enormous windows shattered from their frames, and a bank of elevators destroyed by fireballs that had blown down the shafts and incinerated everything in their path. At a melted desk beside the elevators sat the charred remains of a security guard, his badge still visible on his burned jacket, his body fused to his chair. Other firefighters stepped over piles of debris in the lobby that they only later realized were human remains.

Amid the horror and the wreckage, Jay saw a heartening sight: an assembly of firefighting all-stars, lining up for assignments from the chiefs at a security desk being used as the FDNY command post. As he fell into line among them, Jay spotted Captain Terence "Terry" Hatton, leader of elite Rescue Company 1, which specialized in saving trapped or injured firefighters. Six foot four, with a boyish face and a booming laugh, Terry Hatton's can-do vocabulary relied heavily on the word "outstanding." He'd never apply it to himself, but others often did: over two decades, he'd earned nineteen commendations for bravery. His wife, Beth Petrone-Hatton, was at that moment working alongside Mayor Giuliani, whom she'd served as an executive assistant for eighteen years. The mayor had officiated at their wedding three years earlier.

Clustered around Terry Hatton, Jay saw other Rescue 1 members he knew, including Lieutenant Dennis Mojica and firefighters Kenneth Marino, Dave Weiss, and a good friend named Gerry Nevins. When Gerry wasn't saving lives in New York City, he raised pigs, goats, and chickens with his wife and two young sons on a small farm a few miles from where Jay lived. Nearby was another friend, Lieutenant Pete Freund, whom Jay had seen hours earlier at the scaffolding collapse at the Manhattan Bridge.

Jay saw Commissioner Von Essen lean close to catch the ear of Deputy Chief Peter Hayden, the highest-ranking FDNY officer at the scene. Soon, authority would shift to Chief of Department Peter Ganci Jr., who was still in transit. Ganci would establish a new command post outside the tower, across West Street. In the meantime, Jay admired how cool Pete Hayden seemed amid the chaos. Jay heard

Hayden tell Von Essen, "There's no way we're putting this out. This is strictly a rescue mission."

Hayden had been incorrectly told that as many as fifty thousand people worked in the two buildings and as many as seventy thousand visited daily, including shoppers and commuters in the PATH train station below. In fact, the combined number of people inside the two towers at the moment Flight 11 struck was far lower, somewhere between 14,000 and 17,400. Yet even those numbers represented an overwhelming fire rescue challenge for the FDNY.

Standing alongside Hayden was the first boss on the scene, Battalion Chief Joe Pfeifer, who'd called in the initial report. Shortly after he arrived, Joe Pfeifer noticed his only brother, Lieutenant Kevin Pfeifer, in the lobby. Joe quickly briefed Kevin, who nodded and led five members of Engine 33 toward a stairwell. Chief Pfeifer returned to the task at hand, juggling a two-way radio and a telephone as he took reports from firefighters and Port Authority officials around him, as well as calls from workers and safety wardens on upper floors, and people trapped on elevators pleading for rescue.

Numerous calls for help poured in from workers at Cantor Fitzgerald, an investment bank with offices spread across floors 101 to 105. More than 650 Cantor Fitzgerald employees had arrived to work early that morning, as usual, to get a jump on the markets. Calls also came from employees of commodities broker Carr Futures, a smaller firm tucked among the giants on the North Tower's upper floors. In addition, more than 350 employees and consultants had already arrived in the offices of Marsh & McLennan on the 93rd to 100th floors, where actor and temp worker Chris Young had just dropped off materials for a presentation. The huge insurance and financial company was the sole occupant of all but one of the floors directly in the path of Flight 11. The exception was the 93rd floor, which Marsh & McLennan shared with Fred Alger Management, an investment firm with thirty-five employees at work. An unknown number of people died instantly or soon after Flight 11 hit, as fire engulfed those impact floors.

Four emergency calls to Port Authority Police came from Christine Olender, assistant general manager of Windows on the World.

Thirty-nine, born on the Fourth of July, despite the crisis Christine retained the polite effervescence of the pompom girl and homecoming queen she'd once been.

"We're getting no direction up here," Christine told Officer Steve Maggett. "We're having a smoke condition. We have most people on the 106th floor—the 107th floor is way too smoky. We need direction as to where we need to direct our guests and our employees, as soon as possible."

"We're doing our best," Maggett answered. "We've got the fire department, everybody. We're trying to get up to you, dear. All right, call back in about two or three minutes, and I'll find out what direction you should try to get down."

Christine called back three more times, each time speaking to Officer Ray Murray. With each call she grew more frightened. On her fourth and final call, Christine said: "The situation on 106 is rapidly getting worse. . . . We . . . we have . . . the fresh air is going down fast! I'm not exaggerating. . . . What are we going to do for air?" In desperation, Christine asked permission to break a window.

Ray Murray sat in a windowless, plaza-level office at Five World Trade Center, yards from the unfolding disaster, but his only view was a video monitor that showed bodies in the driveway outside the North Tower. He suspected that Christine and hundreds of others on the North Tower's upper floors were beyond saving, but during each call he offered hope. "You can do whatever you have to, to get to the air," he told her.

Ray Murray was affable and well liked, thirty-two years old, six years on the Port Authority Police Department. In addition to Christine's, he answered more than fifty emergency calls during the first half hour after Flight 11 hit the North Tower. Ray juggled calls from his bosses and worried family members, including his own wife; he fielded inquiries from NBC, ESPN, a CBS affiliate in Seattle, New York's Channel 11, a German radio reporter, and a local rabbi; and he answered security companies whose alarms rang wildly inside the towers. Above all, Ray offered reassurance to men and women calling from upper floors, advising every one to find a stairwell if possible. "If you can get to the stairs, go down the stairs," he told a woman on the North Tower's 83rd floor. "If not, you get on the ground. We're going to get people up there."

Ray Murray didn't know if that was true, but it was all he could offer.

AS THEY STRATEGIZED to rescue as many survivors as they could, FDNY chiefs Hayden and Pfeifer concluded that none of the North Tower's ninety-nine elevators appeared to be working, although a battalion chief later used one to reach the sixteenth floor. Without elevators, firefighters bearing upwards of eighty pounds of equipment would have to walk up fifteen hundred or more steps to reach people in desperate need of help. That is, if time, stairway access, and physical strength permitted.

Adding to the difficulties, communication and coordination immediately proved to be critical problems for most of the firefighters sent upstairs. That shouldn't have surprised anyone. FDNY handheld radios had worked poorly, sometimes not at all, during the response to the 1993 World Trade Center bombing. In 1999, the department purchased new ultra-high-frequency Motorola radios designed to vastly improve communications inside steel and concrete buildings. But shortly after the new radios went into service, a New York firefighter lost in a house fire couldn't be heard calling for help. Disputes arose about whether the radios were faulty or if they weren't being used properly. Either way, in early 2001, the FDNY reissued its old analog radios, the same ones that were ineffective in 1993. Those radios had the same shortcomings now.

After the 1993 bombing, the Port Authority had installed what was known as a "repeater" system at the World Trade Center. The system was designed to boost FDNY radio signals in emergencies by amplifying and rebroadcasting them. Chief Pfeifer tried to activate the repeater almost as soon as he arrived in the North Tower lobby. He tested it with help from Battalion Chief Orio Palmer, an authority on radio communications, who nineteen years earlier had worked

alongside Jay Jonas as a probationary firefighter during a Bronx tenement rescue. After several unsuccessful attempts to use the repeater,
Pfeifer decided that they couldn't count on it. The outdated radios,
though proven inadequate, would have to suffice.

Not only couldn't firefighters reliably talk to one another inside the
building, they didn't have access to information from outside sources.
The FDNY had a notoriously fraught relationship with the New York
Police Department, built on more than a century of competition and
cultural differences. The Finest and the Bravest used different radio
frequencies and different equipment. As a result, FDNY chiefs inside
the North Tower couldn't communicate with police helicopters hovering above the building. Independent of one another, fire chiefs and
police officials determined almost immediately that helicopter rooftop rescues would be impossible because of fire and smoke. Still, police chopper pilots could have shared information in real time about
the fire's progress and the building's structural integrity. Fire chiefs in
the North Tower lobby also didn't have access to the extensive local
and network television coverage of the crisis scene, rendering them
almost blind to the destruction directly above them.

Making matters even worse from a communications standpoint,
Flight 11's impact had wiped out the North Tower's public address
system, preventing firefighters in the lobby from offering survivors
hope of help on the way. Yet even if that message had reached the
uppermost nineteen floors, it wouldn't have made a difference. With
all three stairwells destroyed, and no possibility of rooftop rescues,
people on the upper floors were beyond help.

In their desperation, office workers clustered at the windows at the
highest floors to escape the fire, heat, and smoke. Within six minutes
of the crash, the first person fell or jumped. At least 110 more lives
would end that way from the upper stories of the North Tower. Each
thousand-plus-foot fall, each flailing or graceful ten-second descent,
would end with a thud that sounded like a gunshot. With each loud
bang, emergency responders in the lobby involuntarily flinched as one.

The falling or jumping victims fulfilled the awful predictions of Captain Jay Jonas's mentor, high-rise firefighting expert Vincent Dunn. They also echoed a tragedy that had occurred ninety years earlier, one mile uptown. In 1911, 146 garment workers, most of them women and teenage girls, died in a fire at the Triangle Shirtwaist Factory. Trapped above the flames in that ten-story building, roughly fifty women and girls leapt or fell to their deaths. Nearly a century later, the calamity remained seared in the collective memory of FDNY members and many New Yorkers.

Unaware that the upper stairwells were impassable, or that the public address system was broken, Chief Joe Pfeifer tried to offer reassurance. He pleaded with people trapped in the North Tower to await rescue. "Please don't jump," he said into a public address microphone at the lobby command post. "We're coming up for you."

No one on the highest floors could hear him.

ON THE MARBLE floor of the Marriott lobby, the air around them a greasy haze, burn victim Jennieann Maffeo told Ron Clifford she worried about a coworker who'd been standing beside her at the shuttle stop. She felt certain he'd been killed. Jennieann didn't tell Ron the man's name, but she almost certainly meant Wai-ching Chung, the shy database executive whose niece worked in the South Tower.

Jennieann listed her medications and told Ron she had asthma and a severe allergy to latex. Ron recorded it all in his notebook.

Jennieann also gave Ron her boss's name and phone number. She instructed Ron to tell her boss to call only her sister, Andrea. She begged him, "Please don't call my mother and father. They're old and sick and they'll be very upset." Ron promised.

"I'm going to die," Jennieann said.

"No, you're not," Ron answered. "You're going to be fine."

Even as he said it, Ron wondered with rising anger why no help came. Several more times he called for aid, but a seemingly endless stream of people rushed past, headed toward an exit at the south end of the hotel lobby.

BY 8:57 A.M., little more than ten minutes after the crash in the North Tower, FDNY commanders had asked a Port Authority police officer and building workers to immediately evacuate the unaffected South Tower, based on their belief that the entire trade center property was unsafe. At 9 a.m., the commanding officer of the Port Authority Police ordered an evacuation of all civilians in the entire World Trade Center complex. But communication problems thwarted the evacuation order. Sent out only over a radio channel used by Port Authority Police commanders, it didn't reach Port Authority Police officers or other emergency responders.

Almost like the children's game of telephone, in which a message gets garbled or lost entirely when passed from one person to the next, evacuation messages didn't reach at least some of the people who needed them most. The orders weren't passed to 9-1-1 operators, who fielded calls not only from people trapped in the damaged North Tower, but also from workers in the unaffected South Tower seeking guidance whether to stay or leave. As a result, at least some 9-1-1 operators and FDNY dispatchers told callers in both buildings not to evacuate, and to instead wait until emergency workers reached them.

For instance, shortly after Flight 11 hit the North Tower, a man on the 92nd floor of the South Tower called Port Authority Police: "We need to know if we need to get out of here, because we know there's been an explosion. I don't know what building."

An officer told him: "Do you have any smoke . . . smoke conditions up in your location at [Tower] Two?"

The caller answered: "No, we just smell it, though." After several interruptions and crosstalk, the caller persisted: "Should we stay or should we not?"

The officer told him: "I would wait till further notice. . . ."

"Okay, all right," said the caller. "Don't evacuate."

Callers received different advice from other Port Authority Police dispatchers, some of whom consistently told civilians to leave.

ALTHOUGH A MAJORITY of emergency responders entering the North Tower along with Captain Jay Jonas and the men of Ladder 6 were firefighters, they weren't alone. Also responding were members of the NYPD, which sent elite officers from its Emergency Service Unit. In addition, at least three plainclothes police officers without radios or protective gear climbed high into the North Tower. They checked floors for trapped civilians and refused orders by their chiefs to leave.

Port Authority Police officers responded in droves as well. Among them was Officer David Lim, who'd been working in his office in the South Tower when Flight 11 hit next door. When he heard the explosion, Dave worried that the explosion marked a repeat of the 1993 bombing. He turned to his bomb-sniffing dog, a yellow Labrador retriever named Sirius that Dave considered the smartest dog he'd ever known.

"Maybe they got one by us, Sirius," Dave told his partner, who moonlighted as the pet of Dave, his wife, and their two children.

Dave locked Sirius in a cage in the South Tower basement and ran toward the North Tower. On the way, he spotted a body next to a bandstand set up for a scheduled concert on the Austin J. Tobin Plaza. As he radioed a report to dispatch, another person landed fifty feet from him on the plaza's pink granite. On impact, the body disintegrated into a puddle of flesh, bone, and blood. Dave kept moving, into the North Tower.

Most of the people who fell, jumped, or were swept from the building died instantly. But not all. Amid smoking debris on the plaza, Ernest Armstead, an FDNY emergency medical specialist, found a well-dressed woman in her fifties, with brown hair and tasteful earrings, who'd suffered catastrophic injuries that left only her head and right torso intact. Somehow, she remained conscious. He hung a black triage tag around her neck, to signify to other responders that she was beyond help.

"I am not dead," the woman insisted. "Call my daughter. I am not dead."

Shocked, Armstead stammered, "Ma'am, don't worry about it. We will be right back to you."

He knew he was lying. Her wounds were incompatible with life. Armstead wondered if an air draft had somehow cushioned her fall. He wished he knew if she'd been a passenger on the plane, or a worker swept through an office window by the crash, or someone who'd leapt to escape flames, but he didn't ask. He had to keep moving.

As Armstead stepped away, the woman yelled: "I am not dead! I am not dead!"

"They're coming," he told her as he moved on. "They're coming."

AS JAY JONAS and the men of Ladder 6 awaited their assignment, a rakish FDNY captain named Patrick "Paddy" Brown urged Jay to immediately follow him into a stairwell on the hunt for survivors. "Jay, don't even bother reporting in," Brown said. "They're just gonna send you upstairs. Come with me."

Jay knew that his fellow captain, five years his senior, was among the most decorated officers in FDNY history. Paddy Brown had movie-star looks and a bigger-than-life reputation: savior of a baby from a burning building, supervisor of a daring rope rescue, Golden Gloves boxer, marathon runner. He had a black belt in karate and a devoted following of yoga students. Before fighting fires, he served as a Marine sergeant in Vietnam. Decades later, Paddy Brown remained eager to take the next hill.

"This is a little too big not to follow procedure," Jay told him. "Let me get my name on paper that I'm here." Paddy shrugged and led his Ladder 3 crew to Stairwell B, the only one of the three North Tower stairwells that opened onto the lobby level. The other two opened onto the mezzanine level, just above.

At that moment, a New York Police helicopter circled the Twin Towers, some 1,700 feet above where Jay stood. Officer Timothy Hayes, a pilot for the police aviation unit, had already despaired that thick smoke made rooftop rescues impossible. Now he spotted a large aircraft speeding toward his copter.

"Jesus Christ!" he told his partner. "There's a second plane crashing."

They pulled up and United Flight 175 flew beneath them. Seconds later, at 9:03 a.m., Jay Jonas heard a thunderous explosion.

CHAPTER 14

"WE'LL BE BROTHERS FOR LIFE"

South Tower, World Trade Center

WHEN THE DAY BEGAN, BRIAN CLARK SAT SILENTLY AT HIS DESK AT Euro Brokers, his back to the windows on the west wall of the 84th floor of the trade center's South Tower.

A Canadian immigrant who'd risen to become the firm's executive vice president, Brian pecked at a keyboard, writing emails and updating spreadsheets. Outside his office, more than two hundred fifty men and women prowled the trading floor, angling for an edge, bantering among coworkers who'd become friends. Brian loved that about Euro Brokers. He considered it an enchanted workplace, with a culture of camaraderie that defied the cutthroat stereotype of Wall Street firms.

At 8:46 a.m., Brian's reverie was broken by a double explosion, a *boom-boom!* to his ears, that caused the lights in his office to flicker. He spun toward the noise. Swirling flames pulsed against his windows, then fell from sight. Brian registered something strange about the fire: it had mass, not like the dancing flames in a fireplace, but a molten weight subject to gravity's pull. Brian leapt to his feet, thinking that a welder must've hit a gas line on a higher floor of the South Tower.

Brian transformed instantly from financial executive to civilian fire safety warden. He reached into the mounted credenza above his desk and grabbed his flashlight and whistle, leaving behind the ugly red hat and the reflective vest.

AT 8:46 A.M., Stan Praimnath rode a local elevator up from the South Tower's 78th floor sky lobby, briefcase in one hand, a brown paper bag with a toasted raisin bagel in the other. He wore new rubber-soled work shoes that he considered stylish and his wife, Jenny, deemed hideous.

The elevator doors opened onto the 81st floor offices of Fuji Bank, nearly empty as a result of recent job cuts. Stan walked toward his desk in the loan department, near the southwest corner of the building. Along the way, he passed a temp worker, Delis Soriano, and engaged her in a ritual greeting as she toiled at a photocopier.

"How are you, Delis?"

"Fine, Stanley, and you?"

"I'm special," he said, knowing that it would make her smile. A pastor had taught Stan that line, as an opening to share the gospel. Delis had heard Stan use it often. She knew that if she wanted to finish her copying, she shouldn't ask why he'd said it.

"I know you're special, Stan," she said kindly as she resumed her work.

Neither Stan, who'd been encased in an elevator, nor Delis, standing at the noisy copier, knew that a plane had just flown into the north wall of the North Tower next door.

AT THE MOMENT of Flight 11's impact, Brian Clark, Stan Praimnath, and Delis Soriano were among approximately eighty-six hundred people inside the South Tower. Most couldn't identify the cause immediately, but roughly half heard the sound of an aircraft hitting the North Tower, only 131 feet away. About one in ten people inside the South Tower felt the building move as a result of a shock wave from the tower's damaged twin.

Red flashlight in hand, whistle around his neck, Brian rushed from his office. He yelled the emergency instructions he'd learned during training sessions and fire drills held every six months during the eight years since the 1993 bombing: "Go to the center corridor and await further instruction!"

Roughly two hundred of Brian's colleagues followed the first half of those orders. Then they kept going, exiting the 84th floor down stairwells and elevators, at least some without knowing what had happened. The Euro Brokers evacuees were part of a much larger South Tower exodus. Within five minutes of the North Tower strike, acting without formal instructions to flee, about half of the people in the South Tower exited their floors and headed toward the ground via elevators or stairwells.

Brian felt no temptation to join them. As he herded colleagues toward the intersecting hallways at the floor's center corridor, where the building's elevators and stairwells stretched through its inner core, he overheard television reports from the trading floor that said the emergency was next door at the North Tower. He adopted a stereotypical New York attitude: close doesn't count.

While other Euro Brokers workers crammed into elevator cars and stairwells, Brian spotted several dozen traders clustered against the building's north-side windows, looking up at a ring of fire around the 93rd floor of the North Tower. As Brian approached, he heard someone gasp that people were jumping. He didn't want that image burned in his memory, so he hung back, about fifteen feet from the windows. A colleague named Susan Pollio, who volunteered in homeless shelters as she rose from secretary to bond broker, wasn't so fortunate. The sight of a person plummeting from the North Tower sent Susan running to Brian in tears.

"Oh, Brian, it's terrible," Susan wept. "People are dying."

Whispering comfort, Brian embraced Susan as she heaved with sobs. He led her to a ladies' room near the center of the floor and guided her inside. Brian returned to his office and called his wife, Dianne, and his father in a Toronto nursing home, to say he was safe. The damage was next door, he told them.

As Brian set down the phone, strobe lights flashed and a *whoop, whoop* siren sounded. About ten minutes after Flight 11 crashed, shortly before 9 a.m., the voice of a South Tower building safety official boomed over the intercom: "Your attention, please, Building Two [the South Tower] is secure. There is no need to evacuate Building Two. If you are in the midst of evacuation, you can use the reentry doors and the elevators to return to your office. Repeat, Building Two is secure."

Brian recognized the voice: it belonged to a deputy safety director who regularly conducted fire drills on the floor. Confirmation from a trusted source that the emergency was elsewhere smothered any lingering thought Brian had about joining South Tower evacuees cramming into stairwells. Flashlight in his pocket, whistle around his neck, he wandered onto a smaller trading floor on the west side of the 84th floor, where he bumped into Robert "Bobby" Coll, an affable thirty-five-year-old senior vice president, a devoted husband and father, and an avid sailor, skier, and surfer.

"Bobby, what do you know?" Brian asked.

Bobby said that he'd joined the initial evacuation but had run back up to the 84th floor with several others upon hearing the announcement that the building was secure. Together, they decided that they'd figure out what, if anything, to do next.

STAN PRAIMNATH'S IMMEDIATE focus was on his raisin bagel. He'd barely set down the bag on his desk when his phone rang. His mother asked if he was all right. He told her yes and said goodbye. Similar calls followed from his three brothers. Stan gave each the same brisk response. His family assumed that Stan must have known about the North Tower, so none explained the reasons for their concern. The calls left him increasingly hungry and pleasantly confused. "There's a lot of love here," he thought. "It's not even nine o'clock."

Stan's desk, an L-shaped wedge of industrial steel, faced the tower's south wall windows, with a priceless view of the Statue of Liberty on its island two miles away. As Stan got to work, a flash of light caught his eye. He turned right, toward the Hudson River, and saw fireballs raining from the sky. He phoned two colleagues who'd recently relocated to the North Tower, but neither picked up. Stan realized that he and temp worker Delis Soriano appeared to be alone on the 81st floor.

Stan called to her as he rushed toward the elevators: "Let's get out of here."

Stan and Delis took a local elevator down to the 78th floor sky lobby. There they joined eighteen other senior Fuji Bank executives and colleagues from three other floors, most of them Japanese nationals, who had also self-evacuated in the first minutes after the Flight 11 crash. As they awaited an express elevator to the ground floor, Stan silently recited his most reliable daily prayer: "Lord, cover me and all my loved ones under your precious blood, and take me to work and bring us back home in peace and safety."

The group rode down without speaking. Stan stepped off first in the South Tower lobby, where a security guard in a blue uniform asked where they were going.

"We're going home. I saw fireballs falling from the sky," Stan said, still unaware that a plane had hit the North Tower minutes earlier. "Something's wrong."

"No," the guard said, as sirens wailed nearby. "Your building is safe, it's secure. Go back to your office."

The Fuji Bank executives and other employees dutifully reentered the elevator, all but Stan and Delis. The doors began to close, but a hand reached out and held them open. It belonged to John "Jack" Andreacchio, a general affairs staffer who worked on the 80th floor. Stan knew Jack as a jovial, Brooklyn-born prankster, always ready with a quip.

"C'mon, Stan the man, what are you scared of?" Jack asked. "Let's go back up."

Stan hesitated. He thought about his evacuation through smoke after the 1993 bombing. Delis leaned in close: "Stan, I'm scared. I want to go home."

"Take the rest of the day off," Stan told her.

The permission rankled some of Stan's workcentric colleagues. He heard a voice in the elevator call out: "Stan, we've got an operation to run here!"

"No," he said, surprising himself with his rebelliousness. "I'm running the operation, and she is going home." But Stan's defiance only stretched so far, and it didn't extend to himself. He stepped into the elevator and the doors closed, reuniting him with the other Fuji Bank employees on the minute-long ride up to the 78th floor.

Calmed by the security guard's assurances, the coworkers enjoyed a livelier return trip. No one mentioned the North Tower. Stan gently teased a youthful executive, Hideya Kawauchi, a fashionable native of Japan fond of Brooks Brothers clothes, about tucking a sweater into his khaki pants. That drew a laugh from a human resources executive, Alisha Levin, who loved New York but often returned home to Philadelphia to see her parents, her sister, and her two nephews. Stan nodded warmly to a bank vice president, Manny Gomez Jr. Stan turned to Brian Thompson, a vice president for human resources, and half-joked: "You'd better start thinking relocation." After dispersing in the 78th floor sky lobby for local elevators, only Stan and Joseph Zuccala, a consultant who'd started work at the bank several days earlier, exited on the 81st floor.

At roughly the same time, approximately 9:02 a.m., a new announcement sounded over the public address system in the South Tower. This one partially revised the remain-in-place order broadcast several minutes earlier, replacing it with an optional evacuation: "May I have your attention, please. Repeating this message: the situation occurred in Building One [the North Tower]. If the conditions warrant on your floor, you may wish to start an orderly evacuation."

Brian Clark on the 84th floor and Stan Praimnath on the 81st were among more than six hundred people who either remained on the thirty highest floors of the South Tower or had left and then returned to them during the first sixteen minutes after American Flight 11 struck the North Tower. Some might have heard the Port Authority public address announcer's mild suggestion at 9:02 a.m. about an "orderly evacuation." Stan didn't; neither did Brian.

Unlike Stan, Brian, and the others on the upper floors, by 9:03 a.m.,

an estimated thirty-two hundred men and women had already escaped from the South Tower, out of concern about the crisis in the
North Tower. Another forty-eight hundred were somewhere between
the lobby and the 76th floor, in elevators, stairwells, and offices. Some
were in the process of evacuating, headed down toward the street,
while others stayed fixed in place. Still others were in the process of
returning to their desks.

THE INITIAL ORDERS in the South Tower to "remain in place" and
"return to your office" demonstrated that Port Authority officials,
despite good intentions and extensive precautions after the 1993
bombing, possessed the same blinkered outlook as wide swaths of
the U.S. aviation and defense communities. They couldn't imagine
that terrorists would carry out synchronized attacks. This restricted
their thinking, their actions, and their defenses to what had happened
already, ignoring what might yet occur.

Yet not everyone shared that limited vision and faulty imagination.

The South Tower's largest tenant was the financial megalith Morgan Stanley Dean Witter & Co., which had offices from the 43rd to
74th floors, occupied by twenty-seven hundred employees. To those
workers' great fortune, one of their leaders was a big man in a pin-
stripe suit named Rick Rescorla, Morgan Stanley's vice president of
corporate security.

At sixty-two, Rick Rescorla had spent a lifetime honing his innate
gift for knowing the best response to danger. His exploits as a platoon
leader in Vietnam were the stuff of legend, featured in the book *We
Were Soldiers Once . . . and Young*. He appeared on the book's cover, a
bayonet fixed on his M-16, an iconic image of a warrior moving forward. Rescorla had spent the previous eight years feeling certain that
the 1993 bombing was only the first strike.

After hearing the announcement to remain inside the South Tower,
Rescorla invited a Port Authority official to "piss off." He called his
best friend, a counterterrorism expert, and told him: "I'm getting my
people the fuck out of here." He grabbed a bullhorn and ushered hun-

dreds of Morgan Stanley workers into stairwells, two abreast, just as he'd trained them. Then he went in search of stragglers.

Rescorla was remarkable, but he wasn't the only person who insisted upon immediate evacuation for others, regardless of the personal danger.

ALAYNE GENTUL HAD arrived that morning on the 90th floor of the South Tower dressed as though she anticipated an emergency. She wore a bright red blazer that her husband, Jack, had never especially liked. In the weeks since she and Jack had hunted for New Jersey beach houses by bicycle, Alayne had kept up her usual busy pace. She had visited her parents in Florida, weighed a move to California for a promotion, readied her two sons for the start of school, and helped Jack settle into his new job, all without missing a beat at work.

As senior vice president of human resources at Fiduciary Trust Company, Alayne juggled countless responsibilities involving more than six hundred workers, spread across five South Tower floors: the 90th and the 94th through 97th. When American Flight 11 hit next door at the North Tower, Alayne behaved as though each employee were her personal responsibility.

Interrupting a debate among her junior colleagues about whether to stay or leave, Alayne immediately ordered a group of ten people to board the elevators that her father's company had installed three decades earlier. To receptionist Mona Dunn, Alayne resembled a teacher marshaling students through a fire drill. One of Alayne's deputies, training director Ed Emery, a marathon runner with a wife and son, ushered another group to a stairwell, including Anne Foodim, who had just returned to work after treatment for cancer. Anne considered Fiduciary Trust to be like a family, and she credited Alayne for creating that atmosphere. When Anne's energy flagged on the way down, Ed Emery channeled an old track coach to keep her moving: "If you can finish chemo, you can get down those steps!"

On the crowded 78th floor sky lobby, Ed herded Anne and others into an express elevator to the ground floor, then returned upstairs to help Alayne. Rather than follow the others down, she climbed seven floors above her office, determined to evacuate six technology consultants from California who'd begun work that morning on the 97th floor. Their job had instantly turned from precautionary to prescient: they were supposed to create backup computer systems in the event of disaster.

WHEN STAN PRAIMNATH returned to the 81st floor shortly after 9 a.m., he reached his desk and grabbed a ringing phone. The caller was a colleague from Fuji Bank's Chicago office who'd visited the South Tower weeks earlier. "Stan!" she said. "Are you okay?" When he assured her that he was fine, the woman yelled: "Get out!"

The well-meaning colleague was eight hundred miles away. Stan doubted that she knew more than a security guard eight hundred feet below him. Still standing at his desk, looking absently out the window overlooking New York Harbor, Stan pressed the woman for details. She said something about "no time," but Stan had already stopped listening.

In the distance, beyond the Statue of Liberty, he spotted an object rushing toward him. Stan's mind registered that it was gray, large, and loud enough for him to hear its roaring engines through the sealed windows. A second later, he understood that it was a passenger plane, with a U on its tail. Strangely, it seemed to be speeding directly toward him, its wings banking sharply, its nose level with his. It grew larger, then larger. Then it filled the windows.

Stan dropped the phone and dived under his desk. Curling into a ball, he shouted: "Lord, I can't do this! You take over!"

FOUR MINUTES EARLIER, aboard hijacked United Flight 175, former Navy fighter pilot Brian "Moose" Sweeney left the voicemail message for his wife, Julie. His last recorded words: "I just totally love you, and I'll see you when you get there. 'Bye, babe. Hope I'll call you."

As Stan Praimnath lunged under his desk on the 81st floor, as Brian Clark and Bobby Coll plotted their next moves on the 84th floor, as Alayne Gentul reached the 97th floor, al-Qaeda terrorist Marwan al-Shehhi aimed the Boeing 767 full throttle at the south face of the South Tower.

Inside the plane, sales executive Peter Hanson huddled with his doctoral candidate wife, Sue Kim, and their two-year-old daughter, Christine. Nearby were Ruth Clifford McCourt and her four-year-old daughter, Juliana; Ronald Gamboa and Daniel Brandhorst with their three-year-old son, David; the Reverend Francis Grogan; pacifist professor Bob LeBlanc; retired nurse Touri Bolourchi; hockey great Garnet "Ace" Bailey; flight attendants Michael Tarrou and Amy King; and all the other innocents on board.

At 9:03 a.m., Peter Hanson spoke on the phone to his father, Lee. His mother, Eunice, watched on television as the North Tower burned and a plane suddenly approached the South Tower. Lee heard a woman inside the plane scream as Peter spoke his last words: "Oh my God, oh my God!"

AT 9:03:11 A.M., less than seventeen minutes after American Flight 11 devastated the North Tower, United Flight 175 bored deep into the South Tower.

The plane struck the tower's south face, twenty-three feet from the midpoint, toward the southeast corner. The off-center jolt caused the upper floors to rotate like a boxer's torso twisted from an unexpected blow. The entire building vibrated from rooftop to ground. The plane struck on a 38-degree angle, its right wing sharply higher than its left. The nose, pointed slightly downward, hit the slab of the 81st floor, near where Stan Praimnath trembled under his desk. The immediate impact lasted about six-tenths of a second.

Just as parts of American Flight 11 tore through the North Tower, the right engine of Flight 175 passed entirely through the South Tower and blew through the building's northeast corner. It damaged the roof of a neighboring building before landing fifteen hundred feet north of the tower, near the corner of Murray and Church Streets. The right landing gear followed a similar trajectory.

Damage from the fuselage and the 156-foot wingspan stretched across nine floors, from the 77th to 85th floors. The two additional impact floors, compared to Flight 11's damage, resulted from the more banked approach. The impact shattered 433 windows on the south, west, and east facades. It cut the pipes for fire sprinklers. It destroyed nearly all elevator service, severing cables and trapping occupants, although one freight car from the lobby to the fortieth floor remained operable.

Unlike in the North Tower, where Flight 11 destroyed all three stairwells, in the South Tower one exit route from the uppermost floors remained at least partially intact: Stairwell A. Although located in the central core, Stairwell A was positioned to the west of where Flight 175 entered and was shielded by heavy elevator equipment. That left the stairwell potentially usable for anyone who could reach it.

Structurally, the South Tower fared worse than its twin. The plane severed thirty-three exterior steel columns and ten core columns, causing the South Tower to immediately lean slightly to the southeast above the impact zone. The crash stripped fire-suppressing insulation from dozens of core columns and steel trusses that supported the building's concrete floor slabs. Like the North Tower, the building absorbed the impact and remained upright, as promised by the dreamers and designers who built it. But the severed steel columns and the loss of fire-suppressing insulation portended a disaster caused by fire.

Flight 175 carried more than nine thousand gallons of jet fuel, plus fourteen tons of flammable luggage, mail, seats, food, and electrical equipment. The impact floors contained tons of office furniture, carpets, and other combustibles. Fires from the fuel wouldn't melt the intact steel beams, but they would burn hot enough to undermine the strength of the overburdened exterior and core columns, the structural elements that kept the South Tower upright.

Now both Twin Towers faced existential threats.

THE HUMAN TOLL, already brutal from the North Tower crash, radically swelled.

In addition to everyone aboard Flight 175, the impact slaughtered an unknown number of people on the nine impact-affected floors, leaving others fatally or severely injured. The dead and injured included many of an estimated two hundred men and women who congregated in the 78th floor sky lobby during the initial haphazard go-don't-go evacuation. The plane's left wing shredded that floor, killing and maiming men and women as they waited for their places in packed lobby-bound express elevators, as well as others who hoped to catch local elevators to return to their offices.

For a moment all was silent on the 78th floor, the darkness relieved only by the half-light of flaming embers. Moans and cries soon filled the void. A handful of bloodied survivors struggled to their feet amid severed limbs and lifeless coworkers crushed under steel and stone. The stench of scorched flesh and jet fuel mixed with swirls of smoke that sandpapered survivors' throats and singed their eyes. A few shuffled toward an area that seemed better lit, while others remained prone, waiting for help, too stunned or hurt to move.

Onto the 78th floor bounded a young man in a white T-shirt, with a red bandanna tied around his face. Unhurt, he carried a fire extinguisher and exuded confidence and a clarity of purpose. He pointed those who could walk toward the stairwells. He told anyone who could help others to do so. About a dozen survivors followed his directions, in groups of two and three. The man, whose identity wasn't known to the people he helped, hoisted an injured woman onto his back. He steered two others ahead of him into a stairwell. One was a woman named Ling Young, an auditor for the New York State Department of Taxation, who'd been thrown across the sky lobby and badly burned. When the man first arrived, his voice brought Ling to her feet. Until then, she'd sat motionless on the debris-filled floor, stunned, riven by fear, her face bloodied. Inside the stairwell, they reached clear air on the 61st floor. The man gently eased the woman off his back. He told Ling and the others to keep going. Then the man in the red bandanna turned and went back up the stairs.

WHEN THE PLANE hit, Brian Clark and his coworker Bobby Coll felt as though a concussion bomb had struck the Euro Brokers office on the 84th floor. Facing each other, they braced themselves in football stances, legs spread, arms out. Noise buffeted their ears and the building swayed. The false ceiling collapsed around them, bringing down lights, intercom speakers, and air conditioning ducts that dangled from their sockets. The raised trading floor, built of two-foot-square concrete slabs sitting on six-inch pedestals, popped out of place like jumbled cobblestones. Doors burst from frames. The air filled with gritty gray gypsum dust from jagged, broken wallboards.

Terror gripped Brian as he felt the South Tower lean to the west, farther than he imagined possible without falling, as though the building would topple like a chopped oak into the Hudson River. It righted itself with a decisive jerk, which his engineer's mind interpreted as the building's steel bones realigning themselves.

Brian understood now that this wasn't a welder's error but a terrorist act, related to whatever happened to the North Tower. Still he didn't know that a plane had hit his building, or that its right wing had sliced through the far side of the floor, to the east of where he stood. Yet within the first ten seconds, his fear passed. He told himself: "Brian, you're going to be all right."

He returned to his role as a safety warden, shining his flashlight around the dust-darkened room and calling for anyone within earshot to follow him through the wreckage. As Brian rounded up colleagues, unaware of the extent of the disaster, a wave of annoyance washed over him. "Oh crap," he thought. "Look at this mess. We've got to come back in here tomorrow and clean it all up."

Brian led Bobby Coll and five other men toward the 84th floor's center hallway. He could have gone straight to Stairwell B or turned right to Stairwell C, the closest escape routes, each just a few yards away. Yet for reasons he couldn't explain, Brian didn't choose either, despite having no inkling that both those stairwells were destroyed. Perhaps influenced by an internal compass guiding him toward the elevators he normally rode, Brian felt an almost physical presence pushing him to the left, toward the most remote exit. He allowed it to lead him and his group to Stairwell A.

Elsewhere on the 84th floor, unknown to Brian, other Euro Brokers employees remained alive and trapped. A dozen were somewhere on the west side of the floor. Among them was a broker named Randy Scott, a fun-loving, motorcycle-riding, happily married father of three daughters. With no other way to seek help, he scribbled a plea:

84th floor
west office
12 People trapped

Before tossing it out a window, to flutter among countless bits of paper blown from both towers, Randy Scott pressed a bloodied finger against the note, leaving his unique DNA proof of its authenticity.

ON THE SOUTH TOWER'S 81st floor, at the heart of the impact, Stan Praimnath crept out from under his desk. Nothing could adequately explain his survival. Factors might have included the tilt of Flight 175's wings and the location of his sturdy steel desk near the south-west corner of the building, while the plane entered to the southeast. Stan gave at least as much credit to the ever-present Bible on his desk-top and his appeal to the Lord.

Wind blew through what remained of the 81st floor, sucking out papers, fueling fires along the east side, and adding smoke to the dust swirling around Stan's head. He felt lost in a fog. The ceiling drooped, hanging only feet above his desk. Tangled electric wires swung like vines, water spewed from broken pipes, mangled debris and broken walls surrounded him at shoulder height. He realized he'd been deaf-ened, at least temporarily. He saw no sign of anyone else, never spot-ting the only other person he thought had returned to the 81st floor, consultant Joseph Zuccala. Stan smelled jet fuel but initially thought it was sulfur.

Stan's eyes focused on a burning chunk of metal lodged in a doorway twenty feet away: it looked like part of an airplane wing. Small fires burned nearby.

"I'm dead here," Stan thought.

He tried to climb through the debris toward a stairwell but slipped, slicing open his left calf. As blood dripped toward his shoe, Stan screamed: "Lord, send somebody, anybody! . . . I have two small children! . . . I don't want to die! . . . Why am I alone? . . . Send someone, Lord!"

No one answered.

Stan began to crawl. He scrambled from Fuji Bank's wrecked loan department through a ruined office lounge, over mangled desks and broken office dividers. He inched through a jumbled computer room. He lay flat and used a swimming motion to move forward. Somewhere ahead was a stairwell he hoped would bring him closer to his wife, Jenny, and their daughters. Sharp edges of debris shredded his shirt and scraped his skin. His heart and head pounded. His lungs ached. He crept onward.

Bruised, bloody, and filthy, Stan crawled and swam and yelled for help.

BRIAN CLARK AND six other men from Euro Brokers hurried downward inside Stairwell A. They rushed past a door to the 83rd floor, not knowing that people were trapped there, one of them a software company manager who frantically tried to reach 9-1-1.

Melissa Doi was thirty-two, with a heart-shaped face and a five-foot-two frame that foiled her childhood dream of a ballet career. A graduate of Northwestern University, she lived with her mother in a condo she'd bought for the two of them. At 9:17 a.m., a 9-1-1 call recorder captured Melissa muttering in distress, "Holy Mary, mother of God." A New York Police dispatcher picked up the call as Melissa fought panic.

Melissa: "Oh, my God, I'm on the eighty-third floor."

After some confusion and crosstalk, Dispatcher Vanessa Barnes tried to soothe Melissa, her voice alternating between pragmatic and sympathetic.

Dispatcher Barnes: "Hi there, ma'am, how are you doing?"

Melissa: "Is it . . . is it . . . are they going to be able to get somebody up here?"

Dispatcher Barnes: "Well, of course, ma'am. We're coming up to you."

Melissa: "Well, there's no one here yet, and the floor's completely engulfed. We're on the floor and we can't breathe. . . . And it's very, very, very hot."

Dispatcher Barnes: "It's very . . . are the lights still on?"

Melissa grew agitated. She answered: "The lights are on, but it's *very* hot!"

Dispatcher Barnes: "Ma'am, now ma'am . . ."

Melissa: "*Very* hot! We're all on the other side of Liberty [Street], and it's very, very hot!"

Dispatcher Barnes: "Well, the lights, can you turn the lights off?"

Melissa: "No, no. The lights are off."

Dispatcher Barnes: "Okay, good. Now everybody stay calm. You're doing a good job . . ."

Melissa: "Please!"

Dispatcher Barnes: "Ma'am listen, everybody's coming, everybody knows. Everybody knows what happened, okay? . . . They have to take time to come up there, you know that. You gotta be very careful."

Melissa: "Very hot!"

Dispatcher Barnes: "I understand. You gotta be very, very careful. How they approach you, okay? . . . Now, you stay calm. How many people where you're at right now?"

Melissa: "There's, like, five people here with me."

Melissa explained that her companions were conscious, though some were worse off than others. All struggled to breathe. Coughs could be heard in the background of the call. Melissa said she couldn't see fire but believed that it must be somewhere near, as the heat and smoke intensified.

Dispatcher Barnes told Melissa the authorities were doing everything possible to help. Melissa's anguish worsened. She grew upset, then re-signed. Melissa's words became a heartbreaking litany of sorrow.

Melissa: "I'm going to die, aren't I?"

Dispatcher Barnes: "No, no, no, no, no, no, no!"

Melissa: "I'm going to die. I know I am."

Dispatcher Barnes: "Ma'am. Say your . . . ma'am, say your prayers."

Melissa: "I'm going to die."

Dispatcher Barnes: "You gotta think positive, because you gotta help each other get off the floor."

Melissa: "I'm going to die."

Dispatcher Barnes: "Now look, stay calm. Stay calm, stay calm, stay calm."

Melissa: "Please, God."

Dispatcher Barnes: "You're doing a good job, ma'am. You're doing a good job."

Melissa: "No. It's so hot. I'm burning up."

Desperate to offer advice, Vanessa Barnes urged Melissa to get off the hot floor, either forgetting or not realizing that rising heat and smoke would make the floor the safest option. She continued to en-courage Melissa while simultaneously trying to hurry responders to the 83rd floor. As Vanessa Barnes spoke to another dispatcher, Melissa interrupted with a burst of optimism: "Wait! Wait! We hear voices!"

Vanessa Barnes urged her to stay calm.

Melissa screamed: *"Hello! Help! . . . Heeeeeelp! . . . Heeeeeelp!"*

She got no response. Melissa returned to the 9-1-1 call.

Melissa: "Oh, my God. . . . Find out if there's anybody here on the eighty-third floor."

Dispatcher Barnes: "Ma'am, don't you worry. You stay on the phone with me and we'll . . ."

Melissa: "Can you find out if there's anybody on the eighty-third floor, because we thought we heard somebody!"

Dispatcher Barnes didn't know if emergency responders had reached that high. She told Melissa she'd already notified a lieutenant that five people were trapped on the 83rd floor of the South Tower amid smoke and fire. She assured Melissa, "They won't overlook you."

Melissa's spirits flagged: "Can you, can you stay on the line with me, please? I feel like I'm dying."

Dispatcher Barnes: "Yes, ma'am, I'm going to stay with you."

Melissa perked up again, apparently believing that she heard someone coming. She called out the name Karen, likely meaning her colleague Karen Schmidt, a software engineer from Long Island. But again, hope ebbed.

Near the end of their telephone call, Melissa spelled out her mother's name and asked if they could arrange a three-way connection. Vanessa Barnes said that wasn't possible but promised to call Melissa's mother. Melissa gasped for air, sensing that time was running out. She had one more request for the 9-1-1 dispatcher, a message for her mother: "Tell her," Melissa said, halting between words, "I love her, with all my heart and soul, and that she was the best mother a person could ever have."

Melissa Doi stopped talking, but Vanessa Barnes heard breathing. It seemed as though Melissa and her companions might have passed out from smoke inhalation. Just as she promised, the dispatcher stayed on the line. She called Melissa's name more than sixty times. As minutes passed, Vanessa Barnes stopped being an anonymous police dispatcher and became motherly. "Ma'am" and "Melissa" gave way to "dear." Soon Vanessa Barnes began to call her "baby."

Dispatcher Barnes: "Please don't give up, Melissa. . . . Oh, my God. Melissa? Melissa? Melissa? Please don't give up, Melissa. . . . Hold on, baby, hold on. You're going to be fine, baby, can you hear me? You're going to be fine, you're going to be fine."

Despite a long stretch with no response, despite an unprecedented torrent of incoming 9-1-1 calls, Vanessa Barnes refused to disconnect. She was Melissa Doi's last connection to a world that wasn't on fire. The dispatcher turned to a colleague, although it sounded as though she was still speaking to Melissa and even to herself.

"Not dead, not dead," Vanessa Barnes said. "I don't know if she's unconscious or just out of breath. . . . That's why I keep talking to her."

AFTER PASSING THE door to the 83rd floor, Brian Clark and six others from Euro Brokers navigated down two more flights of Stairwell A. They moved in smoky semidarkness, guided by Brian's flashlight and reflective tape applied to the stairs after the 1993 bombing. On the stairwell's 81st floor landing, they encountered a heavyset woman and a frail man, neither of whom Brian knew, laboring as they headed upstairs. Breathing heavily, her arms stretched wide, the woman blocked their way. She insisted that Brian's group turn around and head upstairs with them.

"Stop, stop. You've got to go up! You can't go down!" the woman said. "We've just come off a floor in flames."

Brian illuminated her face with his flashlight, then aimed the beam toward the faces of his colleagues as they debated what to do. Then he turned the light back toward the woman, who spoke for her silent companion. Maybe they could make it to the roof, someone said, or find shelter on a floor above the flames until firefighters reached them. Or maybe they should continue down. No one but the woman felt certain what to do.

As the debate continued, a sound from nearby distracted Brian. He stepped closer to a door that led onto the 81st floor, its frame twisted out of shape by the impact.

Brian heard a banging noise and a muffled call for help:

"I'm buried. Is there anyone there? I can't breathe!"

Brian pointed the light through the ragged opening next to the door but saw no one. The 81st floor looked smoky but not consumed in flames, and the anxious calls for help continued. Brian grabbed the shoulder of the man closest to him, a soft-spoken broker named Ron DiFrancesco, a fellow Canadian who lived with his wife and four children in the same New Jersey suburb as Brian and his family.

"C'mon, Ron," Brian said, "we've got to get this guy."

The two men pushed away the broken drywall, enlarging the entrance enough to squeeze through the opening next to the locked stairwell door. Briefly looking back, Brian noticed that the argument had been settled—Bobby Coll and a Euro Brokers senior vice president, Kevin York, each took one of the woman's elbows and began to climb. "It's okay, lady," one said. "We're in this together. We'll help you."

Leading the group upstairs was Dave Vera, a senior telecommunications specialist at Euro Brokers, carrying a walkie-talkie that he used in his job. The others followed them up as Brian Clark and Ron DiFrancesco forced their way onto the 81st floor.

BRIAN AND RON found themselves amid a quarry of broken office furnishings. Chalky clouds of demolition dust obscured the daylight from blown-out windows. Brian waved his flashlight, calling "Who's there? Where are you?"

Before they heard a reply, Ron DiFrancesco felt overwhelmed by smoke. He pressed the fabric of a gym bag to his mouth as a filter, but it didn't help. Fearing he'd pass out, Ron scrambled back to the stairwell. Knowing that the others in their group had acquiesced to the woman's pleas and gone upstairs, Ron climbed after them.

Brian remained on the 81st floor, yelling and waving the flashlight. With his hearing affected by the crash, Stan Praimnath could barely hear the calls of his would-be rescuer. Fortunately, above his head he saw a small ray of light behaving like a lighthouse beam, sweeping back and forth.

"I can see your light!" Stan yelled, and started to call out "left" and "right" to help his rescuer line up the beam with his location, moving toward it until only a nine-foot wall separated the two men. The office's false ceiling had fallen, so above the wall a gap revealed the steel bones of the South Tower, the support trusses that held up whatever remained of the 82nd floor.

Stan gathered his strength, reached back, and called upon his karate training. He punched a hole through the sheetrock wall, thrust his arm through and waved madly. "Can you see my hand?" he screamed.

Brian followed the waving hand up the length of the arm, through the hole to a soot-coated face. From the other side of the wall, the trapped man's brown eyes locked onto his would-be liberator's blue eyes.

"Hallelujah!" Stan yelled. "I've been saved!"

Yet the wall still separated them. Before they tried to breach the divide, Stan asked Brian what he considered to be the most important question of all: "One thing I gotta know. Do you believe in Jesus Christ?"

Unprepared for a theological inquiry, Brian stammered something about church on Sundays. He wondered if the man he was trying to save had lost his mind. Brian changed the subject to the task before them.

"C'mon, we've gotta get you out of there," Brian said. He dragged a desk to the wall, turned it on one end, and climbed on. He looked over the top of the wall to see a man standing in a wretched pit of debris. "If you want to live, climb over this wall," Brian demanded.

Stan reached up and jumped, but he missed the top of the wall. Brian tried but couldn't catch him. During his jump, Stan grabbed a piece of wood that hung from the collapsed ceiling, driving a black sheetrock screw into his right palm. Still in the debris pit, Stan yelped in pain.

"Bite it out and try again!" Brian said.

"I can't do it!"

Brian told Stan to pound on the wood, to rip the screw from his flesh. It worked, painfully. A golf-ball-sized swelling ballooned at the puncture wound. Convinced that he couldn't use the injured hand to grip the wall, Stan cried: "I'm not going to make it over." He shouted to the heavens, invoking Jenny and their daughters in his pleas: "Lord, if you wanted me to die, why did you bring me all the way here to leave me? Why, Lord, why?"

Fires grew closer. Smoke grew thicker. Worried about himself and his own family, Brian had heard enough: "You've got to think about your children. Think about your family. Climb over or you're going to die!"

Stan tried again, half-jumping and half-scrambling up the wall, his new rubber-soled shoes helping him gain traction. When Stan reached his apex, Brian hooked him and slung him over the wall. They fell backward in a heap, off the upturned desk and into a debris pile, with Brian on his back and Stan atop him. Overcome with emotion, Stan planted a kiss on Brian's cheek.

"Whoa, whoa, whoa!" Brian said, pulling back. He stood, brushed himself off, and straightened the tie he incongruously still wore.

"I'm Brian," he said, holding out his hand.

"I'm Stanley. We'll be brothers for life."

The stranger's comment touched Brian, an only child who'd always wanted a sibling. Brian noticed that during his exertions he'd cut his right palm, too. To his surprise, and Stan's as well, Brian took Stan's wounded hand and pressed it against his.

"In fact," Brian said, "we'll be *blood* brothers for life."

Their bond sealed, Brian draped his arm around Stan's shoulder.

"Let's go home."

AFTER AMERICAN FLIGHT 11 hit the North Tower, Alayne Gentul went higher in the South Tower, to the 97th floor, to evacuate six technology consultants hired to prepare Fiduciary Trust for a theoretical disaster that was suddenly all too real. The company's training director, Ed Emery, joined her there after leading other colleagues into elevators on the 78th floor. Now that United Flight 175 had struck the tower beneath them, they were trapped.

Alayne called her husband, Jack, in his office as dean of students at the New Jersey Institute of Technology. Jack's secretary interrupted a telephone conversation between Jack and a friend who'd called to alert him that a plane had crashed into the North Tower.

"Oh," Jack said when he switched to Alayne's call, "thank God you're okay."

"Well, not really."

Alayne explained her situation, and Jack asked why they didn't try to get out.

"There's smoke coming in," she said, "and it's really hot out there."

"That doesn't make sense," Jack said. "It was the other building."

"I know, but there was an explosion beneath us."

"Jeez," Jack said, "I don't know what that could be."

As Jack began to put the pieces together, he understood that Alayne's reference to "out there" meant outside the office where she and the others took refuge, as fire and smoke made the exits unreachable. Jack grabbed another phone and called the college's police chief, a former New York City detective who promised to find out what he could.

Alayne told Jack the sprinklers wouldn't work, so he called an engineer in the campus facilities department, who rushed to Jack's office and got on the phone.

"Hit the sprinkler head with your shoe!" the facilities engineer told Alayne.

Ed Emery climbed atop a table and tried, not knowing that the crash had severed water pipes. He lost his footing, and Jack heard Alayne call out: "Oh, Ed! Are you all right?"

The engineer asked if there was a water cooler nearby and told them to wet clothes and use them to seal ventilation ducts. Ed tried with his blazer, to no avail.

Smoke grew thicker, Alayne's breaths grew labored.

"Don't breathe so hard," Jack said. "Try to relax."

"We don't know whether to stay or go," Alayne said. "I don't want to go down into a fire. . . . I'm scared."

Jack knew Alayne to be smart and practical, self-assured and steely when needed. After twenty-four years together, scared was something new.

"Honey, it'll be all right. It'll be all right," Jack said. "You'll get down. You'll get down."

"Tell the boys I love them," Alayne said.

"Of course I will, but this doesn't have to be it, okay?"

"Tell the boys I love them," she repeated. "And I love you."

"I love you," Jack replied.

They said "I love you" again and again, until Alayne said goodbye. Jack refused to accept it was the end: "Call me when you get down."

Jack called their pastor, who started a prayer chain. Jack fell to his knees beside his desk and began to pray.

CHAPTER 15

"THEY'RE TRYING TO KILL US, BOYS"

Ground Level and North Tower, World Trade Center

WHEN UNITED FLIGHT 175 EXPLODED INTO THE SOUTH TOWER, A startled FDNY Captain Jay Jonas looked out through the empty frames of shattered lobby windows in the North Tower. He saw flaming debris and large pieces of metal crashing to the ground, but he didn't know the cause. He and other firefighters waiting in line for assignments stared at one another in confusion. Maybe a fuel tank exploded above them, Jay thought.

A frantic civilian rushed into the North Tower lobby and set them straight: "A plane just hit the second tower!"

For a moment, no one spoke. Chief Joe Pfeifer and other early callers to FDNY dispatch had been right from the start: the first plane had *aimed* at the North Tower, and now another one had *aimed* at the South Tower. Any lingering doubt ended. These were terrorist attacks.

Firefighter-farmer Gerry Nevins of Rescue 1 broke the silence: "We may not live through today." Jay and several others turned to one another and shook hands. "I hope I see you later," one said. "Great knowing you," said another.

The men of Rescue 1 were already gone, heading to the stairs behind their captain, Terry Hatton. Before entering the stairwell, Terry embraced one of his closest friends, Firefighter Tim Brown, and said "I love you, brother. It might be the last time I see you."

The chiefs huddled. Assistant Chief Donald Burns, a thirty-nine-year FDNY veteran who'd commanded operations at the 1993 bombing, left the North Tower to establish a command post and take control of the new calamity at the South Tower. In the FDNY command structure, an assistant chief outranked a deputy chief, who outranked a battalion chief. Among those sent to the South Tower with Chief Burns was Jay's old friend, Battalion Chief Orio Palmer.

Jay stepped up to the command post. "Chief," he said, addressing Deputy Chief Peter Hayden, "do you know the South Tower has just been hit with another plane?" The question was Jay's way of letting Hayden know that Ladder 6 was ready and willing to head into the unknown disaster next door, if needed.

Hayden closed his eyes and shook his head: "Yeah, I know. Just take your guys upstairs in this building and do the best you can."

Jay saluted. "Okay, chief." As he walked away, Jay thought Hayden had the air of a general who knew that some of the men he sent into battle wouldn't return.

Jay approached his five-man crew, who clustered nearby. With both towers burning and an overall lack of useful information, rumors took flight. Roof Man Sal D'Agostino heard someone say they were under attack from missiles being fired from atop the Woolworth Building, two blocks away. He told Can Man Tommy Falco: "This is not going to be good."

"All right, here's the deal," Jay told them. "It's a raw deal, but it's what we've got. We got to go upstairs for search and rescue in this building. And we gotta walk up eighty floors. We can't use the elevators because they've been exposed to fire already."

Jay had one more message for Ladder 6, about the unknown enemy who'd thrown two giant daggers into the heart of New York City: "They're trying to kill us, boys. Let's go."

"Okay, Cap," said Chauffeur Mike Meldrum, speaking for them all. "We're with you."

As they walked single file to Stairwell B of the North Tower to begin their upward trek, Roof Man Sal D'Agostino caught up with Jay: "Hey, Cap, I wonder where the Air Force is."

It occurred to Jay that he'd never considered the need for air cover at any of the thousands of fires he'd fought. But Sal's question made perfect sense. In fact, after both towers were struck, the F-15s from Otis Air National Guard Base, piloted by Lieutenant Colonel Timothy Duffy and Major Daniel Nash, were still on their way to New York.

IN THE MARRIOTT lobby, Ron Clifford kept calling for help. Finally, a man stopped—he told Ron that he worked in emergency medical services on Long Island. Ron tore off his blue suit jacket and asked the man to cover Jennieann, for modesty, while he returned to the restroom to fetch more cool water. A woman who Ron believed worked as a nurse at the Marriott joined them, with an oxygen tank and gauze.

When Ron returned to Jennieann's side, he rinsed her wounds again.

"Sacred heart of Jesus," Jennieann said, "please don't let me die."

"Are you Catholic?" Ron asked.

"Yeah, I'm Catholic."

Ron's religious life had ebbed as a young man, but remnants of faith remained deep inside him.

"Well," he told Jennieann, "maybe we can say the Our Father while we're waiting for help."

Ron's brogue harmonized with Jennieann's smoke-strained Brooklynese as the Irish American immigrant and the first-generation Italian American recited words that summarized the Christian gospel: "Our Father, Who art in heaven, hallowed be Thy Name. Thy kingdom come, Thy will be done, on earth as it is in heaven."

Before they reached the prayer's last words, "deliver us from evil,"

United Flight 175 slammed into the South Tower. Again, the building quaked. This time debris fell from the hotel lobby's ceiling. On his knees, Ron felt it even more powerfully than he did the first blast. It pounded through his chest, rumbled through his bones, bounced him against the marble floor. The sailor in Ron felt as though he'd been dashed against the rocks.

He had no idea how much rougher his day would become.

Confused, afraid, Ron still had no idea what happened to either tower, but he knew better than to remain in place. The hotel lobby had grown packed as guests emptied from their rooms and an endless stream of people from the North Tower used it as a refuge or a pass-through, seeking safety from debris and people falling or jumping from the flaming upper stories.

As he rose to his feet, Ron heard the hotel nurse say they'd find help outside, through an exit at the far end of the lobby that would lead them near Liberty and West Streets, at the southwestern corner of the World Trade Center complex.

"Let's get her out," Ron told his fellow helpers.

He asked Jennieann: "Can you get on your feet?"

She agreed to try, so Ron helped her up. The man from Long Island held Ron's suit jacket like a cloak, a few inches from Jennieann's burns, shielding her. Ron handed the man his leather shoulder bag so he could use both hands to steady Jennieann. The nurse held the oxygen mask over Jennieann's mouth as the four of them jostled slowly through the teeming crowd.

Ron sensed that Jennieann felt embarrassed by her nakedness, despite the suit coat. He spotted a large man who looked like a Marriott waiter or busboy and appealed to him for help. The man disappeared for a moment then threw Ron a fresh white tablecloth. Ron wrapped it around Jennieann as they kept moving.

The mass of people trying to squeeze through the outer doors made for slow going. As they approached the exit, Ron roared a demand to let them pass. When people saw Jennieann's ravaged face, her body draped in white, they turned in horror and cleared a path. From somewhere in the crowd, Ron heard a person shout, "It's a plane!" Someone else yelled back, "It's two planes." Ron had his first inkling of what happened.

RACING TO THE scene in their ambulance, EMTs Moose Diaz and Paul Adams finally reached the mouth of the Queens-Midtown Tunnel, which crosses under the East River to connect Queens with Manhattan. Moose tried to enter the tunnel going in the wrong direction, to save time, but a taxicab exiting the tunnel refused to yield, despite the sirens and swirling lights. Paul leaned out the window and launched obscenities like mortars at the cabbie, to no effect. He saw the cab's passenger write down the ambulance's vehicle number, presumably to complain about Moose's driving and Paul's language. That aggravated Paul more.

"Push this fucking cab out of here!" he told Moose. Trying to stay composed, the more placid Moose maneuvered around the cab while Paul flashed the cabbie and his passenger a digital salute.

Oncoming cars cleared an ambulance-sized pathway. At the far side of the tunnel, five miles from the World Trade Center, the ambulance barreled south down Second Avenue at the relatively breakneck speed of 50 miles per hour. The whole city seemed to be heading the opposite direction, away from danger. Only fire trucks, police cars, ambulances, journalists, and intrepid volunteers were headed toward Lower Manhattan.

As they drew near, Moose and Paul watched United Airlines Flight 175 slam into the South Tower. Moose called his wife, Ericka. "Listen, I'm going," he told her. "I can't believe the towers got hit." Ericka made him promise to be careful. They exchanged I-love-yous and hung up.

Moose hit the brakes at a makeshift staging area at the northeast corner of the trade center site, at Vesey and Church Streets. They jumped out near the rear of the city's oldest church, the boxy, beloved St. Paul's Chapel, where George Washington worshipped on his inauguration day.

Carlos Lillo and Roberto Abril took another route through Manhattan, steering their ambulance down Broadway and ending up on the northwest corner of the trade center site. Carlos told Roberto he couldn't wait for him to park at a staging area; he snatched his trauma bag, jumped out, and went looking for injured people to help.

Upon their arrival, they found a mass casualty scene no manual or exercise could have prepared them for. Paramedics and EMTs came upon every imaginable body part, some with clothing or jewelry still attached, some burned or mangled beyond recognition. Among all the pieces, amid all the gruesome human wreckage, one image locked into the minds of several emergency responders who saw it: a girl's foot, inside a pink sneaker. One EMT immediately thought of his own daughter, whose foot was the same size. He looked away, up to the sky to clear his mind, only to see people jumping from the North Tower.

Even as he worked among his colleagues on the wounded, Carlos kept thinking about Cecilia. He asked Kevin Barrett, a fellow Battalion 49 EMT: "Where is my wife—have you seen my wife?" Barrett told him not to worry. "She is going to be okay. She is around. . . . Concentrate on what you're doing."

RUSHING TO LEAVE the 64th floor of the North Tower, Cecilia Lillo ducked inside Stairwell A. She found it crowded but orderly, with no sign of panic among the evacuees. At first Cecilia recognized no one, but then she heard her friend Nancy Perez calling up to her from the stairs below: "Come down! Come with us!"

Cecilia saw their mutual friend Arlene Babakitis with Nancy and wanted to be with them, but she wouldn't cut the line. As Nancy continued to yell, strangers stepped aside and urged Cecilia forward to join her friends. When Cecilia reached them, Nancy shook as she grabbed Cecilia's hands.

"It's a plane!" Nancy said. "Somebody has a portable radio, and they're announcing it's a small plane." Nancy wanted to keep a tight grip on Cecilia's hand as they went down, but only two of them could walk side by side in the 44-inch-wide stairwell. Cecilia told Nancy to hold on to Arlene, who at forty-seven was the oldest of the three friends.

AS HE WAITED inside a motionless elevator, oblivious to what happened, Chris Young's lone companion was the automated voice that incessantly promised that help was on the way. He didn't have a cellphone, only a pager clipped to his belt that alerted him to messages left on his answering machine. Usually they were calls for auditions, but now he grew anxious about the possibility that his predicament wasn't isolated. He wondered if the messages pinging his pager meant that his mother in North Carolina had heard that something bad had happened in New York.

Trapped in a box, with no one answering his emergency signal, Chris turned to his personal security blanket: memories of his college acting triumph as the imaginary knight Don Quixote in *Man of La Mancha*. Forcing himself to stay calm, Chris recited a monologue from the show that he often used at auditions:

> *I shall impersonate a man. . . . Being retired, he has much time for books. . . . All he reads oppresses him; fills him with indignation at man's murderous ways toward man. He ponders the problem of how to make better a world where evil brings profit and virtue none at all. . . .*

The automated voice repeated its only line.

INSIDE ALAN REISS'S crowded 88th floor office, badly burned Elaine Duch listened as about two dozen of her colleagues debated whether to wait for building security officers to lead them to safety or to try on their own to evacuate. Smoke wormed its way between the door and the frame, so several people cleared papers from Reiss's desk and stuffed them into the cracks in an effort to avoid suffocation.

Meanwhile, construction manager Frank De Martini and a few

others, including construction inspector Pablo Ortiz, a former Navy SEAL, continued to scout through the wreckage for an escape route. Soon Frank returned to Reiss's office looking like a chimney sweep, covered in gray soot, his eyes red from smoke. Architect Gerry Gaeta opened the door and Frank delivered hopeful news: Stairwell C, toward the building's west side, appeared to be clear, if they could climb over piles of debris to get there.

Still not grasping the cause or the severity of the damage, several Port Authority workers declared that they wanted to wait on the 88th floor for help. Some recalled the exhausting, smoky, hours-long walk down darkened stairwells during the 1993 evacuation. But finally they were persuaded by their bosses, and by the rising smoke in Reiss's office, that no one would reach them in time.

Frank De Martini, Gerry Gaeta, Pablo Ortiz, and several other men led Elaine, Frank's wife Nicole, and the rest through a scrapyard of upturned furniture, broken wallboards, and office wreckage that was waist high and burning in places. The receptionist who minutes earlier had called Elaine about the waiting messenger loudly quoted scripture as they sought deliverance.

They moved through the real estate office reception area, adjacent to the elevator banks, then turned right into a corridor. Gerry Gaeta was stunned by the sight of the elevators: gaping black holes where the doors had been. Each new sight, each new piece of information, expanded Gerry's understanding of the magnitude of what happened.

When the group reached the steel door to Stairwell C, they discovered that Frank had been right: not only were the stairs passable, they were lit by battery-powered emergency lights and luminescent tape, installed after the 1993 bombing. Gerry went down one flight to be sure it was safe, then returned with news that the smoke seemed less intense below them.

"Let's go for it!" he yelled.

Most headed down, but not all. Frank De Martini, Pablo Ortiz, and several other men from the 88th floor went up to break through walls and pry open doors in an effort to rescue people trapped on higher floors. With crowbars and flashlights, resolve and grit, they were credited with saving at least seventy people, reaching as high as the 90th floor.

Inside the lightly populated stairwell, everyone cleared a path for Elaine. Her friend Dorene Smith stepped in front of Elaine and held an empty arm of the sweater tied for modesty's sake around Elaine's waist. Dorene used it like a rope to pull Elaine along and to catch her if she lost her footing. After a brief separation, Gerry Gaeta fell in behind Elaine and Dorene, gripping the knot tied in the sweater at Elaine's back, to steady her.

With Frank De Martini's wife, Nicole, and most other evacuees from the 88th floor following close behind, the threesome of Elaine, Dorene Smith, and Gerry Gaeta descended step by step down a dozen flights of stairs. Unaware that Flight 11 had cut through all three emergency stairwells, preventing escape from above, Elaine was surprised not to encounter people evacuating from the upper floors.

When they reached the North Tower's 76th floor, the group discovered that the stairway led into what was called a crossover corridor, a passageway roughly one hundred feet long that bypassed mechanical equipment, with fire doors on either end. At the corridor's far end, Gerry Gaeta struggled to open the door. He tugged and kicked at it, to no avail. Gerry wondered if terrorists had locked the stairwell doors to trap survivors, as part of what he mistakenly believed was another bombing plot. More likely, the door was bent inside the frame by the plane's impact on the building.

Elaine and the other 88th floor evacuees reluctantly reversed course, retracing their steps inside Stairwell C and climbing up two flights to the sky lobby on the 78th floor. They found the lobby nearly empty except for several Port Authority security workers and a few men trying to free someone trapped in an elevator. A security guard pointed them toward a potential new escape route: Stairwell B seemed passable from the 77th floor down.

As they moved toward their new exit path, Elaine and the others saw that the ornate sky lobby had been heavily damaged. Green marble panels that had lined the walls near the elevators lay broken on the floor. The group walked down the stairs of a frozen escalator to the

77th floor, where they stopped in their tracks. Facing them was a corridor full of water puddles and live electrical wires shooting sparks. They had two bad options.

"We are going to die waiting here," Gerry Gaeta told the others, "or we are going to die getting electrocuted." He chose to risk the latter.

Dipping one foot gingerly into the water, Gerry discovered that either the wires weren't live or the current wasn't powerful enough to electrocute them.

They reached the door to Stairwell B and resumed their escape march. Down ten steps, turn at a landing, then nine more. Down ten steps, turn at a landing, then nine more. Elaine kept pace, flight after flight, floor after floor, inside the larger, 56-inch-wide stairwell.

After about twenty floors, they caught up with crowds of other evacuees who slowed their progress. One reason for the reduced pace was the clogging that resulted from several thousand North Tower workers funneling into stairwells to escape. Another was the fact that on the lowest floors, firefighters had begun climbing upward inside the same narrow stairwells. The evacuees and their would-be rescuers brushed shoulders as they walked single file in opposite directions inside Stairwell B.

Gerry Gaeta worried that the slowdown would cause Elaine to stop moving entirely. Once she halted, he feared, she might not start again. He knew that he would try to carry her, but he didn't expect that would go well. "Please move to the right!" Gerry yelled at the civilians in front of them. "Burn victim!"

Everyone stepped aside. As Elaine passed, sandwiched between Dorene and Gerry, some gasped; others burst into tears. Elaine gritted her teeth and said nothing. She focused her energy on the stairs—ten down, turn, then nine more, each floor taking a minute or so. She kept her eyes down, focused on her feet. It dawned on her that her sneakers might qualify as lifesavers; if she'd changed into sandals when she arrived at work, her feet would have been hideously burned by the fireball, making it all but impossible to walk down these seemingly endless flights of steps.

Floor after floor they continued, pressed to the right wall of the stairwell as they passed a conga line of red-faced firefighters who trudged upward, burdened by the weight of their equipment and the anticipation of what awaited them. One young firefighter poured water on Elaine from an extinguisher, stimulating what remained of her nerve endings and unintentionally causing pain. She pulled away, fearful that the water would worsen her injuries.

At one point, Gerry noticed two men trying to help a heavyset woman in a dark dress who seemed to be struggling. Something appeared to be wrong with her legs, and he didn't know how she'd endure the rest of the descent. But Gerry kept moving, his attention on Elaine, cheering her along like a trainer.

"Only another ten floors, Elaine," he told her. "Only another ten minutes."

AT 9:11 A.M., twenty-five minutes after American Flight 11 hit the North Tower, eight minutes after United Flight 175 hit the South Tower, some Port Authority Police dispatchers inexplicably continued to tell worried callers not to evacuate.

Cecilia Lillo had already left the North Tower's 64th floor, but others remained there. Port Authority executive Pat Hoey, an engineer who oversaw bridges and tunnels, remembered the chaos of the hours-long evacuation in 1993. Now, he sought expert advice for himself and twenty or so other agency workers alongside him, so he called the Port Authority Police.

Pat Hoey: "What do you suggest?"

A sergeant answered: "Stand tight. Is there a fire right there where you are?"

Pat Hoey: "No, there is a little bit of smoke on the floor."

Sergeant: "It looks like there is also an explosion in [Tower] Two."

Pat Hoey: "Okay."

Sergeant: "So be careful. Stay near the stairwells and wait for the police to come up."

As the building burned, as people fled if they could, or jumped if they couldn't, as firefighters climbed, as civilians helped one another to escape, Pat Hoey and his remaining office mates followed those instructions. Despite clear stairwells and time to leave, they remained fixed in place on the 64th floor, waiting for someone to come get them.

ONCE INSIDE STAIRWELL B, Captain Jay Jonas and the men of Ladder 6 encountered a steady stream of office workers, many frightened, a few burned, but all remarkably calm, coming down the stairs. Evacuees cheered the firefighters as they passed.

"You deserve a raise," one man told Jay's crew.

"God bless you," said another office worker.

"Go get 'em," said a third.

Evacuees smashed the glass of vending machines in stairwell landings and handed Jay and his men bottles of Poland Spring water. One man removed Tommy Falco's helmet and drenched Tommy's head and neck to keep him from overheating. "Good luck," the man said before moving on. The lights were on and the air inside the stairwell was relatively clear, but the smell of jet fuel lingered.

Down in the lobby, Chief Peter Hayden expected firefighters to go no higher than the seventieth floor, but Jay estimated they might need to ascend as many as ninety floors, or more than seventeen hundred stairs, to reach victims awaiting rescue. "We'll take ten floors at a time," he told his men. "Ten floors. Take a quick break, get a drink of water, and press on."

By their first stop, on the tenth floor, the men of Ladder 6 were drenched with sweat, red-faced and breathing heavily. After a brief rest, they trudged up to the twentieth floor, their legs rising more slowly on those 190 steps, with stops along the way to help two firefighters from other units suffering chest pains.

As Ladder 6 and other companies struggled to reach the fires and survivors on upper floors, numerous mobility-impaired civilians received help from friends and colleagues. Some of the impromptu rescuers offered a guiding hand, some a calming word, and some a strong back. Ten coworkers of accountant John Abruzzo, a quadriplegic who relied on a wheelchair, banded together to save him. They carried him from the 69th floor using an evacuation chair that worked like a sled on stairs. When they crossed paths with firefighters, the men declined help. They'd get their friend John out on their own.

Jay and his men encountered several such scenes of civilians helping one another. At one point, they stepped aside for a terribly burned woman, her clothes melted to her skin. Jay watched as she slowly but steadily descended the stairs with a man following close behind making sure she didn't fall. The woman said nothing. She continued downward, and Ladder 6 persisted upward. From the description of the woman's burns and her escort, the timing of the encounter, and the location inside Stairwell B, it seems likely that Jay and his men had crossed paths with Elaine Duch and her Port Authority colleague Gerry Gaeta, with Elaine's friend Dorene Smith somewhere nearby, clearing the way.

ANDREA MAFFEO STARED across the East River, watching horrified from her fifth-floor office on Fifty-Ninth Street in Brooklyn as flames and smoke poured from both Twin Towers. She couldn't shake a strong premonition about her big sister.

"Oh my god," Andrea told her secretary, "Jennieann is there. She's there, and she's hurt badly."

Her secretary said that made no sense. Jennieann Maffeo didn't work at the World Trade Center, she only stopped there on her commute, to hop on a shuttle bus to a ferry. Even if Jennieann had been close by when the planes hit, the fires were near the top of each tower. Andrea couldn't argue with that logic, but she had a bad feeling in her bones.

Both single, the Maffeo sisters lived together, laughed together, went to movies together, cared for their elderly parents together, and vacationed together. A month earlier they had been to Cancun, Mexico, where they'd sat on a bed in their hotel room and talked about the afterlife. Jennieann had been deeply affected by a book she'd just finished, *Crossing Over*, written by the self-professed medium John Edward.

"You have to read this book," Jennieann told Andrea. "It made me feel so comfortable. It talks all about how, when you go, everybody you love who's gone before is going to be there."

Sitting in the hotel room, Andrea had dismissed her. "What the hell are you talking about? Where are you going? You're forty years old."

"I'm not afraid to die anymore," Jennieann answered.

Now, watching the towers burn, Andrea felt chills of fear. She grabbed her cellphone to call Jennieann, but when she pressed the talk button she got a surprise: her mother was already on the line.

"Where's your sister?" demanded Frances Maffeo, a retired seamstress who survived cancer but remained in poor health.

"Ma, she's fine."

"Where. Is. Your. Sister?" Frances Maffeo repeated, louder with each word.

Andrea decided not to share her own fears.

"I'm sure she's at work already. I'm sure she's fine. Don't worry about it."

Frances wasn't buying it. She had intuition, too, and she'd already called Jennieann's employer, UBS PaineWebber, in New Jersey. "She's not there."

Andrea pivoted, trying to keep her mother from panicking. "She's not answering the work phone because she's probably on the bus."

Andrea said goodbye and repeatedly tried Jennieann's cellphone. She called UBS PaineWebber, but no one there had seen Jennieann. Anxiety rising, Andrea looked out the window at the towers, watched television, kept calling her sister, and prayed.

ABOUT TWENTY MINUTES after the North Tower express elevator shook then stopped on its descent to the lobby, Chris Young heard a human voice on the intercom.

"Is anybody in there?" a man asked.

"Yes!" Chris said.

"Is anyone in there with you?"

"No, I'm by myself."

The disembodied voice likely belonged to Dave Bobbitt, a Port Authority elevator supervisor who contacted about seventy-five elevator cars from his post in the North Tower lobby and estimated that he spoke to people in ten stalled cars. Flight 11 severed an unknown number of elevator cables in the North Tower. Some cars followed their programming, returned to their lowest floor, and opened their doors. But many didn't. Some elevator cars stopped on their lowest floors but didn't release passengers, while others plunged dozens of stories, killing or injuring those inside. Elevator shafts also became deadly chimneys, channeling smoke and flames, and blowing out doors and walls from the impact zone down to the basement floors.

Some people trapped in elevators would die, some would be saved, and some would save themselves. Almost halfway up the North Tower, six men in a smoke-filled elevator pried open the doors but found themselves facing a sheetrock wall. Jan Demczur, a Port Authority window washer, reached into his bucket and pulled out a squeegee. With help from the other men, he used its metal edge and a second brass squeegee handle to carve through three layers of sheetrock. They punched a hole through white wall tiles, squeezed through, and landed on the floor of a men's restroom on the fiftieth floor. From there, the six men rushed down a stairwell and escaped.

The voice on the intercom told Chris Young not to try to force his way out. Instead, he should wait for someone to rescue him. The man didn't tell Chris what caused the car to stall, perhaps assuming that everyone already knew about the planes. He didn't ask the elevator number, so Chris suspected that an electronic indicator showed which car or cars had sustained mechanical failures. Chris didn't know that his was one of ninety-nine elevators in the North Tower, with the same number in the South Tower, nearly all of them stuck or worse. Chris also didn't know that emergency responders were focused on helping people trapped on high floors, so they never systematically checked the towers' elevators. Dave Bobbitt made a list of elevators with trapped survivors, but no one followed up on it.

Comforted by his contact with a live person, still unaware of what happened, Chris resumed his dramatic monologues and ran through song lyrics in his head. After about ten minutes, he heard a man's voice shouting.

"Anybody there?"

"Yes!" Chris called, thinking he'd been found by firefighters. "I'm in an elevator!"

"We're in an elevator, too," the man said.

Chris and the man yelled back and forth, straining to hear each other through the walls separating them. They made a pact to tell anyone who came for them about the other. The man told Chris that he was one of seven people stuck in the car. The four men and three women had just entered the elevator to ascend to the 78th floor sky lobby when it rocked and stopped. No one had answered their emergency calls, and like Chris, they didn't know what had happened.

Chris repeatedly pressed the fire helmet button. When someone responded after several minutes, Chris told him about the other elevator. "We've got people working on it," the man said. "We'll get you out, and we'll get them out, too." The man on the intercom again told Chris not to try to free himself.

Chris thought the voice belonged to a different person than his earlier contact. It worried him that the new person didn't seem to remember him. Still, the news seemed hopeful. Chris yelled to the man in the other elevator that he'd reached someone on the outside. He waited, then shouted again. No one answered.

AS SHE DESCENDED the stairwell with her friends and coworkers, Cecilia Lillo repeatedly tried calling her paramedic husband, Carlos, on her cellphone, but she got no signal. Somewhere in the low fifties, she passed a beefy, square-jawed man going up: Fred Morrone, superintendent of the Port Authority Police Department. Cecilia liked Fred, who had always been kind to her. As Fred climbed toward the fires of the impact zone, the former New Jersey state trooper encouraged the civilian evacuees headed down.

"It's above the eighty-something floor," Morrone said. "You guys will be okay."

Morrone's confidence calmed Cecilia. The slow downward progress had been making her uneasy about whether she, Nancy Perez, and Arlene Babakitis would get out. They stopped again when they heard someone from a higher floor yell that everyone should stand aside to let injured people pass. A woman who halted next to Cecilia screamed and cried when a burned woman shuffled past. Cecilia turned the distraught woman toward the stairwell wall. "It's okay," Cecilia told her. "We're going to be okay. You don't have to look."

As they waited, Cecilia overheard a man nearby yelling and cursing on his cellphone, ordering his wife or girlfriend not to drive into New York from New Jersey. Cecilia wanted to ask to borrow his phone, to call Carlos, but the man sneaked in behind the walking wounded to jump the line and speed his exit.

A dozen floors lower, at around 9:30 a.m., Cecilia thought she heard a man's voice in the stairwell call "Sessy!"—pronounced to rhyme with "Bessy"—the nickname Carlos used for her.

"I think my husband's here!" Cecilia told her friends.

"Huh? I didn't hear anything," Nancy said. Whoever it was, it wasn't Carlos. They kept going.

At the 23rd floor, firefighters directed them out of Stairwell A without explaining why. Cecilia and her friends waited in a 23rd floor hallway for about ten minutes, then the firefighters directed them to Stairwell C and told them to squeeze in among the evacuees already inside.

At the tenth floor, Cecilia and Nancy started a floor-by-floor countdown—"Ten! . . . Nine! . . . Eight!"—to bolster their spirits and encourage Arlene, who'd grown weary.

DURING THE FIRST half hour of the disaster, 9-1-1 operators passed word to the fire chiefs in the North Tower lobby that locked doors had foiled efforts by some people trapped on upper floors to reach the roof. Roof access was routinely limited to hinder suicides, vandals, and anyone who might attempt to repeat Philippe Petit's 1974 wire walk between the towers. Although rooftop rescues had already been ruled out, a "lock release order" was transmitted to the building's Security Command Center, so people trapped on burning, smoky floors might have at least a chance at fresh air. But the impact of Flight 11 had damaged the computerized system, and the rooftop doors remained locked. Survivors on the nineteen floors above the North Tower impact zone had nowhere to go.

At 9:32 a.m., an assistant FDNY chief radioed an order requiring all firefighting units in the North Tower to return to the lobby. He issued the order either because of reports of a third inbound plane to New York, which proved false, or because of concerns about the building's structural integrity. With radio communication inside the building ranging from weak to nonexistent, few or no firefighters apparently heard him. Captain Jay Jonas and the men of Ladder 6 never did, so they continued climbing. By all accounts, no FDNY units returned to the lobby as a result of that order.

It soon became clear that a third plane had in fact been hijacked a half hour earlier. But the hijackers weren't headed for New York.

CHAPTER 16

"THEY DONE BLOWED UP THE PENTAGON"

The Pentagon

CONCEIVED DURING THE RUN-UP TO U.S. ENTRY INTO WORLD WAR II, the Pentagon was designed in a rush to house what was then called the War Department. Its groundbreaking took place on September 11, 1941, sixty years to the day before 9/11.

Built atop thirty-four acres across the Potomac River from Washington, D.C., the Pentagon was a stone, brick, and steel monument to military strength, unadorned by decoration except for a layer of Indiana limestone as its outer skin. Only 71 feet tall, each of its five sides stretched 921 feet, or longer than three football fields. Among the world's largest office buildings, it boasted 6.6 million square feet of floor space and more than seventeen miles of polished corridors.

To passersby, the long, low outer walls gave the impression of a fortress-like monolith, serious and solid to the core. But when viewed from above, the Pentagon revealed a hidden grace, like a geometric flower. The five-story building consisted of five concentric pentagonal rings that surrounded a verdant five-and-a-half-acre courtyard of grass and trees. According to Pentagon lore, Russian war planners used the coordinates of the courtyard's hotdog stand as a potential target for nuclear missiles during the Cold War. Less dramatically, the courtyard, known as the Center Court, served as an outdoor respite for the twenty thousand people who worked in the Pentagon, half of them members of the military and the rest civilians and Defense Department contractors. As a bonus, the courtyard was among the largest areas in the world where U.S. military members weren't required to salute their superiors.

Each of the building's five rings had an alphabetic label: the A Ring was closest to the Center Court, then the B, C, and D rings, and finally the E Ring at the outer edge. An open-air breezeway between the B and C rings, confusingly called AE Drive, served as a utility road that allowed vehicles to move equipment or supplies around the building. The Pentagon's five rings were connected by ten long corridors, predictably designated One through Ten, arranged at equal intervals, like spokes of a giant wheel. The corridors contributed to the Pentagon's hyperefficient design: despite the building's immense size, an able-bodied person could walk between any two points in under ten minutes. Office labels followed a simple alphanumeric code that made them easy to locate: floor, ring, corridor, then room. So, 2D491 was an office on the second floor of the D Ring, near corridor Four, room 91. Another way to think of the building was as five connected wedges, each one a slice of a pentagonal pie.

NAVY DOCTOR DAVE TARANTINO spent the first hours of September 11 puttering around his makeshift Pentagon office, a temporary assortment of desks squeezed into a hallway on the fourth floor of Wedge One, in the innermost A Ring, near the Fifth Corridor. Tarantino had already picked out a desk in a newly renovated area of Wedge One, on the building's outermost E Ring. But the planned August move for him and a half dozen workmates had been delayed.

At thirty-five, still the slender athlete who'd survived a crushing fall beneath a twisted parachute as a college freshman, Dave had spent the previous year working in the Office of the Secretary of Defense. The Pentagon job was a highlight of a career that sent him around the world into crisis situations caused by natural and human-made disasters in Africa, Turkey, and elsewhere. Now he worked for a program whose name struck some as oxymoronic in a building synonymous with war: Peacekeeping and Humanitarian Affairs. In fact, modern military doctrine included a "you-break-it-you-bought-it" attitude toward armed conflict; after bombs and bullets came relief and rebuilding, overseen by men and women focused on healing. People like Dave.

During the first hours of the workday, Dave checked emails, made calls, and worked on reports about migrant refugee issues in Haiti and Cuba. Shortly before nine, Dave heard a colleague say something strange was happening in New York. They turned on a television and watched the burning North Tower. When Dave saw the strike on the South Tower, two thoughts sprang to mind: "I'm surprised they haven't hit a military target" and "This has to be Osama bin Laden." Then Dave jumped ahead two moves to what he expected would come next: a U.S. military response in bin Laden's adopted homeland of Afghanistan, followed by a humanitarian crisis and years of cleanup.

As Dave watched the burning towers on CNN and considered the work ahead, the Pentagon shuddered beneath his feet.

SHORTLY AFTER NINE o'clock on the first floor of Wedge One in the Pentagon's D Ring, a distracted Jerry Henson walked past men and women clustered around large television screens bolted to a wall of the new Navy Command Center, whose staffers managed daily naval operations around the clock, around the globe.

Jerry was a few months shy of sixty-five, but he looked a decade younger: square-jawed and graying at the temples, trim enough to fit into the commander's uniform he'd hung up more than two decades earlier. As a Navy flier during the Vietnam War, Jerry had flown seventy-two combat missions. Back then, a normal day might involve photographing enemy positions in a reconnaissance plane, avoiding antiaircraft fire, and landing on the deck of an aircraft carrier in high seas with his fuel gauge on *E*. Retirement from active service gave

Jerry more time with his family and less dangerous work: now he piloted a Pentagon desk as the civilian chief of a seven-person staff that oversaw Navy antidrug efforts and relief operations.

As he reached his office, Pentagon room 1C466, a cubbyhole adjacent to the Command Center, Jerry carried bad news from a meeting he'd just attended in the office of the Secretary of Defense. Budget cuts meant he needed to cancel a planned task force gathering the next day at a Hampton Inn in Norfolk, Virginia. Jerry entered the cramped six-by-ten-foot room he shared with a jovial retired Navy captain: Jack Punches, the nine-fingered former submarine hunter who'd received the Order of the Palm medal upon his retirement. Jack had been up late the previous night, watching *Monday Night Football* with his son, Jeremy, after celebrating his daughter Jennifer's birthday.

Jack and Jerry had worked together for more than six years. Jerry came from a generation of men more comfortable rebuilding engines than expressing emotions, but he considered Jack to be "one of those guys you loved immediately." As he entered the office, Jerry saw Jack staring stone-faced at a wall-mounted TV.

"What's happening?" Jerry said.

Jack gave him a rundown of the terror in New York, but the images spoke louder.

"That was no accident," Jack said.

An aide arrived with a pile of paperwork, then dropped an offhand remark on his way out: "I guess we'll be next on their list."

Jerry and Jack watched multiple replays of the South Tower crash, two retired Navy aviators bearing witness to the start of a new war. In 1941, Jerry turned five years old within days of Pearl Harbor. Now he understood how the adults had felt back then.

Jerry answered a phone call from his wife, Kathy, who managed a dance studio in Virginia. "Are you aware?" she asked.

"Yes, we're watching it on television."

Kathy imagined that everyone in the Pentagon would be on high alert, with red lights flashing, ready to spring into battle. Or, even better, evacuating the pentagonal target with its bullseye courtyard.

"What's the plan there?" she asked.

Jerry laughed her off. "It's a normal working day. We're working."

Simultaneously, Jack had nearly the same conversation with his wife, Janice.

Jerry turned away from the television and called in two Navy petty officers on his staff, Christine Williams and Charles Lewis. They stood by his desk as Jerry dialed the Hampton Inn to cancel their reservations. The call connected as Jerry heard a booming *crump-thump* noise.

Then the line went dead and the lights went out.

AT 8:45 A.M., in a second-floor cubicle on the outer E Ring of Wedge One, Lieutenant Colonel Marilyn Wills made last-minute tweaks to a presentation she needed to deliver about an upcoming conference.

Marilyn's day had begun hours earlier in a rush, with no time for her family's usual morning prayer circle or Marilyn's typical closing plea: "Lord, God, bring us all back home safely together." Instead, Marilyn swayed and sang along in her car with gospel singer Fred Hammond's rousing rendition of "Jesus Be a Fence Around Me." She ignored fellow commuters staring through her windows, hit repeat, and gave a full-throated, full-bodied encore until she reached the Pentagon parking lot.

Nearly two hours into her workday, Marilyn rehearsed her presentation for her friend and cubiclemate Marian Serva, who pronounced Marilyn ready. She gathered her files and marched toward a conference room some twenty feet away. Along the way, Marilyn glanced across the front lines of her new professional home: the sprawling workspace of the Army's Office of the Deputy Chief of Staff for Personnel, under the command of Lieutenant General Timothy Maude.

The general and his immediate staff occupied a suite of offices at the outer edge of the Pentagon's E Ring, with windows facing west toward Arlington National Cemetery. Marilyn and 275 colleagues, two thirds of them military and the rest civilian, had moved two months earlier into a newly renovated, acre-sized plot of carpeted real estate that stretched uninterrupted from the E Ring through the D and C rings. Some ranking officers had private workspaces, but most worked in a "cubicle farm" of 138 partitioned workstations under acoustic ceiling tiles. Still smelling of fresh paint, and with all the charm of a robocall center, the area was officially called the Bay. Its inhabitants called it Dilbertville, a not-so-fond reference to the comic strip that satirized the fluorescent grind of office life.

Dilbertville was humming. On her way to the meeting, Marilyn passed a cluster of cubicle dwellers that included Specialist Michael Petrovich and civilians Tracy Webb and Dalisay Olaes, all checking email, working the phones, and organizing reports. Tracy and Dalisay hoped to slip away for a coffee break with their friend Odessa Morris, but Dr. Betty Maxfield, an Army demographer, delayed their departure when she dropped by to talk about personnel matters and a scarf she'd bought from Tracy, who had a sideline as an Avon representative.

Along with Marilyn, eleven officers and civilians headed toward the conference room for a bimonthly executive officers' meeting. Colonel Philip McNair, executive officer under General Maude, would run the show. Also on their way were Major Regina Grant, a human resources officer who earlier in her career worked as a fire safety educator; Lieutenant Colonel Robert Grunewald, an information technology expert who'd helped to design the new personnel offices; Martha Carden, General Maude's assistant executive officer and the office's resident mother hen; and Lois Stevens, a civilian action control officer who'd be sitting in for a major away on leave.

Another participant was a minor Army legend: the affable sixty-nine-year-old Max Beilke, who'd served in the Korean and Vietnam Wars, and who now looked after fellow veterans as deputy chief of retirement services. A lifetime earlier, on March 29, 1973, Master Sergeant Max Beilke walked to a transport plane in Saigon and took his place in history as the last American combat soldier to leave Vietnam. As Max approached the plane, a North Vietnamese lieutenant colonel stepped forward and handed him a parting gift: a straw placemat decorated with a serene image of a pagoda.

Upon entering the conference room, Marilyn did a sharp about-face and returned to her desk.

"Quick meeting," Marian Serva said.

"I wish," Marilyn replied. "It's freezing in there." She snatched her black cardigan from her chair and hustled back to the conference room.

One by one, starting at 9 a.m., the Army personnel staffers delivered their reports. Bob Grunewald ribbed Max Beilke about how great life sounded for retirees. Max left after his presentation, to attend a simultaneous meeting across the hall, in General Maude's office on the outer E Ring.

Outside the conference room, word spread through Dilbertville about the attacks in New York. A handful of officers and civilians crowded around a television above Marian Serva's desk. Among them was John Yates, fifty years old, a strapping former staff sergeant who served as the department's security officer. He'd spent the previous eight years planning for the Army personnel team's move to its new quarters. John watched for a while, then returned to his desk to call his wife, Ellen, whom he'd married the previous year.

"Honey, do me a favor," Ellen said. "For the rest of the day, work from underneath your desk."

John laughed: "Sure, babe, no problem." He told her he loved her and returned to the horrifying images on Marian Serva's TV.

No one disturbed the executive officers' meeting with news of the terrorist hijackings, so the presentations continued with the participants in high spirits. As her turn to speak neared, Marilyn sensed that the scheduled half-hour 9 a.m. meeting had run long. She glanced at her watch: 9:36 a.m. Colonel Phil McNair gave her the floor.

Marilyn took a deep breath and began: "Well . . ."

The room rocked. McNair leapt to his feet and shouted: "What the hell was that?" Everything went black.

THE ROAR OF jet engines startled facilities manager George Aman as he worked in Arlington National Cemetery, the vast military burial ground beside the Pentagon. A Vietnam War veteran, Aman had spent a quarter century tending to the cemetery's hallowed hills, pampering the thick green grass dotted with more than four hundred thousand marble tombstones in eternal formation. He was accustomed to planes passing overhead, but never this loud, never this low.

The noise rose to a growling crescendo. Aman feared that the plane would crash into the maintenance building where he stood terrified. But it flew past him, so close that Aman swore he could see the faces of American Airlines Flight 77 passengers in the windows. Aman couldn't say whether he spotted author and commentator Barbara Olson, sixth-grade student Bernard C. Brown II, pregnant flight attendant Renée May, or retired Rear Admiral Wilson "Bud" Flagg. All he could say with certainty was that the faces bore empty looks of despair. Overcome, confused, Aman worried that he'd wet his pants.

As fast as it came into view, the plane disappeared.

"Good God almighty," he said, an exclamation that doubled as a prayer.

Aman heard another worker's radio broadcasting urgent updates about the terror attacks in New York City, with frenzied reports about hijacked planes and burning towers some two hundred fifty miles to the northeast. Now, at 9:37 a.m., seconds after the plane streaked past the cemetery, a thunderous explosion shook George Aman's world. Yet he didn't immediately connect the horror he heard on the radio to the low-flying plane or the blast he had heard nearby.

Then a coworker screamed: "Goddamn, they done blowed up the Pentagon!"

WITH TERRORIST PILOT Hani Hanjour pushing the throttles, Flight 77 pierced the west wall of the Pentagon's E Ring at the first-floor level, between corridors Four and Five, only a few feet off the ground. Because the plane struck at an angle, part of the upturned right wing slashed through the concrete slab of the second floor. At the moment of impact, the jet was traveling at roughly 530 miles per hour, or 708 feet per second, loaded with 5,300 gallons of fuel.

The bone-rattling impact, felt several blocks away by George Aman and untold others, sent a fireball two hundred feet above the building's roof. Whirling plumes of dense gray-black smoke were visible for miles. The explosion consumed much of the plane's tail in the milliseconds before it entered the building: although the Boeing 757's tail was forty-five feet high, the initial damage to the Pentagon reached only twenty-five feet up the outer wall.

When it pierced the Pentagon's limestone skin, the jet acted like an incendiary bullet entering a body between two ribs, tearing through the vital organs within. The plane's six hundred thousand bolts and rivets and sixty miles of wiring became superheated, superdeadly shrapnel. Momentum and multiple explosions of fuel enlarged the swath of destruction to more than an acre of office space on each of the first and second floors, with less intense damage on the three upper floors. The plane and its contents fragmented into countless lethal pieces, slashing and burning through people and furniture and equipment and load-bearing columns in the newly rebuilt section called Wedge One, where some thirty-eight hundred men and women, military and civilian, were at work at 9:37 a.m. on September 11.

Slightly minimizing the destruction was the fact that Wedge One was the first section of the building to undergo a major renovation, with work five days from completion. Newly installed structural steel tubing reinforced the outer wall of the E Ring, where the plane entered. Blast-resistant windows with inch-and-a-half-thick glass had been mounted on the exterior wall, reducing flying shards. The new windows were a response to the 1995 domestic terrorism bombing of the federal building in Oklahoma City, where fractured glass caused some of the 168 deaths and numerous injuries.

In addition, Wedge One had the building's only fire sprinklers, and its renovated exterior walls were backed by Kevlar cloth, similar to the material used in bulletproof vests, to prevent shattered pieces of masonry from becoming deadly projectiles. The Pentagon's four unrenovated wedges would have been more vulnerable to damage on September 11. On the other hand, Wedge Two was nearly empty, in anticipation of the next phase of construction, so even if the building damage had been greater, casualties there likely would have been fewer.

When it entered Wedge One of the Pentagon, what remained of Flight 77 took an angled path, penetrating the E, D, and C rings. Airplane parts punched holes through thick walls, reaching as deep as AE Drive, about 270 feet from where the plane entered. The front portion of Flight 77 disintegrated as it moved through the building, creating a hole through which rear sections of the plane traveled farther into the Pentagon. As a result, remains of the five hijackers, who were in the front section, would be found relatively close to the impact point, while remains of passengers and crew members, who'd been herded toward the rear, would be recovered farther inside.

It took less than one second for Flight 77 to stop moving after it hit the building. But, just as in the World Trade Center towers, the damage had only begun: fires raged, debris rained down, water lines burst, electrical wires snapped, and acrid black smoke spread and rose to the floors above.

Several dozen Pentagon workers died instantly. Thousands more headed for escape routes that ranged from straightforward to treacherous. Others, some horribly burned, faced death as they sought safety or awaited rescue as air ran out in darkness, intense heat, and fire-scorched mazes of chest-high wreckage. Some military members and civilians, thrust by circumstance or impulse, became impromptu first responders, risking their own lives to save others.

WHEN THE PENTAGON shook, Navy doctor Dave Tarantino thought it was a construction accident from the renovations. Maybe a welder had hit a gas line, or a crane had toppled over. Then he and his coworkers, deep inside the building and three stories above where Flight 77 entered, thought about New York.

Maybe it was a bomb.

They had no time to speculate further, though: in less than a minute, alarms blared and loudspeakers ordered an evacuation. Black smoke snaked into their makeshift workspace. They rushed to seal secure files and shut down computers. Military and civilian workers poured from their offices into the hallway around the humanitarian assistance group's desks, heading toward a main corridor. Dave and several others guided them through rising smoke to a stairwell that led down and out to the central courtyard. Dave could have gone outside with them, to set up an open-air triage station and wait for any wounded that might arrive. But a unique combination of forces acted upon him in ways no one could have predicted, not even Dave himself.

As a doctor with trauma care experience, Dave believed that he should immediately help seriously injured people still trapped, just as he'd done years earlier when a plane crashed into an Alaska hillside. A third-generation Navy officer, he'd absorbed the principle of prioritizing ship and shipmates over oneself. The Pentagon was his ship, and aboard a burning ship every sailor is a firefighter, bound by duty and honor to fight flame. As a young man, Dave had vicariously suffered the shame felt by Conrad's fictional Lord Jim for saving himself before others. And as the survivor of a near-fatal parachute fall, Dave knew how it felt to need help. While others rushed toward safety, Dave remained inside the Pentagon.

The explosions had knocked out the lights, plunging the hallways into darkness, and thickening smoke obscured emergency exit signs mounted high near the ceilings. Dave half-crawled along the corridor walls to find a bathroom, where he wet paper towels to make crude breathing masks. He surmised that the explosion, whatever had caused it, had originated outside the building. Dave needed to help his Wedge One coworkers overcome their natural inclination to rush toward the outer edge of the Pentagon, the E Ring, and from there to the parking lots and roadways beyond. Dave had quickly realized that the smarter though less intuitive path would be to head deeper inside the Pentagon, to the central courtyard, through which survivors could then enter unaffected wedges of the building to reach safety.

Exiting the bathroom into the darkened fourth-floor corridor, Dave called out to make sure no one was lost in the smoke. Staying low, he raced down a stairwell to the floor below. Choking and retching, unable to see his hand in front of his face, Dave led confused workers, some of them bruised and dazed, toward a stairway leading to the courtyard. He did the same on the second floor, at times crawling or crab-walking as he helped to guide and prod several dozen Pentagon colleagues through the smoky darkness.

As Dave helped the fleeing survivors, he handed out wet paper towels to help people with breathing problems, leaving none for himself. Dave felt his throat and lungs burn from the bitter smoke of jet fuel. He needed fresh air, and he knew he'd find it in AE Drive. He wove his way there, moving against the crowd of people heading for safety. When he reached the breezeway, Dave drank in the clean air and regained his bearings.

When Dave's smoke-strained eyes cleared he saw military men and women from all branches, some heading toward the courtyard but others like him, looking for ways to help. He moved toward them and saw several holes in the cinderblock and brick outer wall of the C Ring. The largest hole was nearly the size of a Volkswagen Beetle, blackened at its rough edges. Smoke poured out and climbed the wall toward the clear blue sky. As his eyes followed the smoke upward, Dave saw several men and women scrambling out a second-story window farther down AE Drive, then falling or jumping into the arms of several stout men who stood waiting below. One of the rescuers, a muscular Navy SEAL named Commander Craig Powell, soon moved from catching jumpers toward the holes in the C Ring where Dave stood.

Dave looked around the debris and realized he was looking at an oversized tire, still attached to twisted metal mounts: clearly part of a commercial jet's landing gear. Reality dawned on him that the damage wasn't from a construction accident or an earthquake or a bomb. He'd wondered minutes earlier if terrorists would choose a military target—now he understood they'd picked the biggest one of all, the one where he worked. As Dave stared at the wreckage in AE Drive, he realized the debris wasn't just mechanical parts and office junk. He recognized part of a foot, part of a leg, part of a skull.

Then he heard voices pleading for help from inside the building.

SITTING AT HIS desk, a split second after he heard a crunching sound and the lights died, retired Navy flier Jerry Henson felt a crushing blow to the head. As the plane devastated the first floor of the E Ring, it tore through the Navy Command Center adjacent to the small office Jerry shared with his friend Jack Punches.

Flight 77 didn't strike their warren directly. Its wake created a tsunami of broken walls and ceilings, crushed masonry, splintered furniture, torn electrical wires, and disemboweled plumbing and heating pipes. Jerry felt as though all of it landed on him.

The top of his desk had detached from its base and slid across the arms of his chair, at once immobilizing and protecting him, like the tray of a child's high chair. Hundreds of pounds of rubble pinned Jerry off kilter in his chair. Part of the load pressed his head hard against his left shoulder, squeezing his face like a vise grip. The weight twisted his neck so painfully he thought it might be broken. Blood from a head wound dripped down his shoulder and right arm. More blood drained from cuts on his chin and cheek. Yet he remained conscious, and he believed that his torso and lower body were uninjured. To his surprise, Jerry could move his legs.

"Help!" he called, coughing, unable to see in the sudden darkness. The smell of burning jet fuel, familiar from his flying days, convinced Jerry that his building had been hit the same way as the World Trade Center. "All this stuff is on top of me because an airplane has come through the Pentagon," he thought. Black smoke at first seeped, then poured into what remained of his office. A fire burned behind his chair, and others burned nearby, adding to the smoke and casting a flickering glow that served as the only light.

The plane's entrance had lifted petty officers Christine Williams and Charles Lewis off the ground and thrown them to the floor. Both were trapped under mounds of debris. They answered Jerry's calls but couldn't free themselves from the wreckage. The two petty officers joined Jerry's cries for help as they tried to dig themselves out. Even as Williams and Lewis slowly worked their way free, they had nowhere to go. Jerry and Jack's windowless office was misshapen and filled with rubble. Much of the debris had come from the direction of the only doorway. If there was a way out, it was obscured by broken walls, upturned furniture, darkness, and smoke.

Battling pain, inhaling smoke, Jerry fought to keep from passing out. Williams and Lewis were accounted for, but where was Jack Punches? Jerry had last seen Jack sitting by the television set after speaking with his wife, Janice. In his mind's eye, Jerry could see Jack flanked on one side by heavy bookcases and on the other by a tall filing cabinet. Either or both might have crushed him.

"Jack?" he called. "Jack? You all right?" He got no reply but kept calling. Maybe, Jerry hoped, Jack was alive but had been knocked out cold. Maybe he'd awaken to the sound of his name. "Jack!"

With no one answering his or the petty officers' calls, Jerry tried to save himself. He snaked his left hand free. Little by little, he pushed away pieces of debris within reach. He'd survived combat in Vietnam and too many close calls to count. His family wanted him home. He didn't want to die at his Pentagon desk.

After a few minutes of clearing, Jerry repositioned his head just enough to ease the pain in his neck, although he remained trapped. He stretched and flexed his legs, fearing they'd fall asleep and make it hard to walk if he somehow got free. As more time passed, that became a fading hope. Jerry remained imprisoned by his desk and the collapsed walls and cabinets of his office. Weakened, his throat and lungs burning, Jerry knew that it wouldn't matter if his legs worked if rescuers didn't arrive soon.

ALTHOUGH FLIGHT 77 struck the west face of the Pentagon at the E Ring's ground-floor level, the impact sent a volcanic force into the Army personnel offices directly above. Explosions of jet fuel and dense smoke burst upward as the plane's wreckage plowed through three rings of the first floor toward AE Drive.

In the Army personnel conference room, acoustic tiles lifted from their frames as a fireball belched from the ceiling behind Martha Carden's seat. Walls cracked and peeled away. Lieutenant Colonel

Marilyn Wills flew from her chair to the other side of the room. She banged her head and blacked out momentarily, then awoke in the pitch darkness with no idea how she'd gotten there. Her left cheek was burned, her hair singed, her nose and mouth coated with the smell and taste of jet fuel. She heard Colonel Phil McNair: "Get under the table! Get below the smoke!"

Lieutenant Colonel Bob Grunewald, like Marilyn a former military police officer, disobeyed that order. Worried about his friend, he cried out: "Where is Martha?"

"Help!" Martha Carden called from the other side of the room. "Somebody help!"

"I'll come get you!" Bob Grunewald yelled as he dived across the table. He crawled ten feet to reach Martha, still in her chair, disoriented and covered with pieces of broken ceiling tile, but unhurt. When Bob reached her, he pulled her to the floor and told her to hang on to his belt. With Martha clinging to him, Bob crawled toward what he hoped would be an exit.

Orienting herself after her blackout, Marilyn Wills scrambled on hands and knees several feet to a door that led into a corridor that circumnavigated the E Ring. She grabbed the handle and yanked, but the door was jammed inside its casing.

Marilyn couldn't see anyone else in the room, but she yelled to alert them: "The door's locked—we can't get out of here!" Marilyn thought of her family's prayer to be reunited safely at the end of each day, which she hadn't had time to say that morning. "Oh Jesus, it's so hot," she thought. "I'm going to die."

When no one answered her calls, Marilyn wondered if she was alone. Yet that didn't make sense. Moments earlier, after Max Beilke's departure, she'd been one of eleven people in the meeting room. Perhaps because she was stunned by the initial blow, Marilyn didn't hear Phil McNair, Bob Grunewald, and others yelling orders and encouragement as smoke filled the room.

"Stay low!" "Keep together!" "Move fast!" "We've got to get out of here!"

In the suffocating darkness, determined to make it home to her family, Marilyn knew that her only hope would be the door she'd entered through, at the opposite end of the room, which opened onto Dilbertville. The conference room wasn't large, but Marilyn felt as though she crawled for a long time. Heat baked her uniform and skin, and Marilyn could no longer imagine the frigid air that had prompted her to grab her sweater. She pulled one arm out of the black cardigan, half-dragging it as she crawled. As she neared the door, Marilyn felt a hand lock onto her ankle.

"Who is it?" she asked.

"It's Lois," said the civilian control officer, Lois Stevens, a petite woman in her sixties who'd spent a quarter century working at the Pentagon. She sounded calm.

"Okay, good," Marilyn said. "Come on, you hold on to me. Don't let go. We're going to get out of here."

DAVE TARANTINO STOOD in the breezeway that was AE Drive, where the ground was wet from broken pipes and littered with human remains, ruined furniture, and airplane wreckage. He joined an impromptu squad of about twenty people, a mix of civilians and military, officers and enlisted personnel, from the Army, Marine Corps, Air Force and Navy, engaged in a spontaneous joint rescue effort. Among them were Lieutenant General Paul Carlton Jr., the surgeon general of the Air Force, and the oversized Navy SEAL Craig Powell, who had reported for work in the Pentagon only days earlier.

Another first responder was Navy Captain Dave Thomas. At forty-three, smart and articulate, balding and bespectacled, he was the son of a career Navy officer and one of four brothers to attend the Naval Academy. He'd spent a dozen years at sea as a surface warfare officer, most recently as commander of the guided missile destroyer USS *Ross*, named for an American hero of Pearl Harbor who was the first Medal of Honor recipient of World War II. Wounded and blinded by a bomb aboard the USS *Nevada*, Donald K. Ross repeatedly braved smoke and fire to keep the battleship's dynamos spinning, singlehandedly preventing the big ship from sinking and blocking the harbor.

At the Pentagon, Dave Thomas had spent almost eighteen months leading a team preparing a massive report on Navy strategy and policy required by Congress every four years, part of the painstaking work that set him on a path to becoming an admiral one day. Every morning at 9:30 a.m., four hours into his workday, he had a standing appointment for coffee with his best friend, Captain Robert Dolan, a rising Navy star who worked in a strategy office within the Navy Command Center. Their bond spanned their adult lives, starting with three years as roommates at the Naval Academy. Bob Dolan had served as Dave Thomas's best man and later became godfather to one of Dave's two children. Dave loved Bob like a brother and felt the same love in return.

When Dave Thomas heard about the New York attacks, he called Bob, who agreed that they'd be too busy for a coffee break. Minutes later, when Flight 77 hit the Pentagon, Dave Thomas wanted to get outside to safety as quickly as possible. Instead he was driven by instinct to move toward the flames, a throwback to his shipboard days. He hustled from his fourth-floor office, located two corridors away from the point of impact, down the stairs and out onto AE Drive. He looked up and saw black smoke rising from farther down the breezeway, so he ran toward the flames, just as Dave Tarantino had done. As Dave Thomas moved against the crowd toward the smoke, he realized that the trail led toward his best friend Bob Dolan's office.

Dave Thomas skidded to a stop when he reached the ruins of Flight 77 on AE Drive. His gaze settled on what looked like five black dots on the ground. Then he recognized that the dots were the polished nails of a detached foot. He wondered why strands of seaweed waved in a creek flowing through the Pentagon breezeway. Then he realized he was looking at human hair and part of a scalp in water pouring from a broken main. Nearby he saw what looked like a field-dressed deer. No. Minutes earlier the butchered flesh had been a living, breathing person.

Dave Thomas spotted an elderly custodian he knew, a small man, always neatly dressed, dragging two heavy fire extinguishers. The man trudged toward the smoking holes in the wall of the C Ring.

Beyond those holes was whatever remained of the Navy Command Center. Dave Thomas's friend Bob Dolan wasn't among the rescuers, which meant that he was somewhere inside, trapped, hurt, or worse.

Dave Thomas took a fire extinguisher and crawled into one of the holes. He found himself inside what had been an electrical closet, now a mess of tangled wiring. He reached a metal door where he could hear people pounding and screaming for help from the other side. He grabbed a metal pole and tried to break through, his efforts prompting the people on the other side to yell louder in hope of deliverance. But the door held, and the people inside grew quiet.

As he searched for another way around, Dave Thomas could only hope that they found a way out. The extinguisher ran out of water and he returned to AE Drive, his lungs aching. Someone gave him a penlight. Guided by the tiny white beam of light, he crawled back into the building through a different hole. It felt like entering a kiln.

"Anybody in here?" he yelled. "Bob, you here?"

Molten metal from welded seams and liquefied plastic from the sheaths of electrical wire dripped from what remained of the ceiling. The drops burned his skin, so Dave Thomas took off his khaki uniform shirt and wrapped it around his head as an improvised pith helmet. His glossy dress shoes began to melt, baking his feet. He went outside, splashed water on his head and feet, and crawled back inside through another hole in the C Ring wall. That route was even hotter, with fires burning and smoke boiling, but Dave Thomas saw a path that led to where Bob Dolan's desk had been. He went as far as he could, but the desk was no longer there. He screamed for Bob, for anybody. An Army officer covered in blood and dust ran past him, out to AE Drive, but Dave Thomas found no sign of his best friend.

STILL TRAPPED IN his chair, Jerry Henson felt air and time running out. Perhaps fifteen minutes had passed since the explosion. The smoke was so thick he felt as though he could grab chunks of it. He grew listless, and the skin around his nostrils and mouth became blackened by soot. His breaths grew shallow and his calls grew faint, then collapsed into coughs. His throat burned. Jerry felt as though he was being strangled by unseen hands. He felt rising heat at his back from a fire behind his chair.

Petty Officers Christine Williams and Charles Lewis weren't talking anymore, and he worried they'd been asphyxiated. Jack Punches still hadn't answered his calls.

In the darkness and the silence, his mind fading, his lungs aching, the burden across his lap and on his head unyielding, Jerry felt certain that he had less than five minutes to live. He didn't think about God or family or whoever did this to him and his friends and colleagues. He thought only of his situation at that very moment: "Well, this is it. If somebody doesn't get you out, you're not getting out."

Jerry tried again: "Help!" he croaked. "Help!"

WITH CIVILIAN LOIS STEVENS clinging to her belt, Lieutenant Colonel Marilyn Wills crawled through the conference room door into what remained of Dilbertville. The normally bustling workspace was unnaturally quiet. Smoke obscured her view, and noxious fumes stung her mouth, nose, and eyes. Marilyn squinted through the darkness to her right, toward a door that led out to the E Ring corridor. She hoped it would be an easy way to escape. But through the crack at the bottom of the door she saw red and felt the heat of flames consuming the Pentagon's outer edge.

"Let's go left!" she told Lois. "To the windows. I know there are windows."

Unbeknownst to Marilyn, two officers who'd been with them in the conference room, Lieutenant Colonel Dennis Johnson and Major Steven Long, both unfamiliar with the new office layout, had gone through the door. Both perished in the inferno of the E Ring corridor.

Marilyn set herself and Lois on a path in the opposite direction, toward the second-floor windows that overlooked AE Drive. On a normal day, it would be a two-minute walk, including stops at friends' cubicles to chat. Now it was an uncertain, serpentine crawl in pitch darkness through more than two hundred feet of poisonous smoke, rising heat, and overturned furniture, cabinets, and office partitions. Marilyn reached for her wrist, to illuminate her watch, but she'd lost it somewhere along the way. The Pentagon alarm siren rang in her ears. She heard the system's recorded voice on an endless loop: "Attention! A fire emergency has been declared in the building. Please evacuate!"

As Marilyn led Lois on the journey toward the windows, her head bumped into a woman's rear end.

"Talk to me," Marilyn said. "Who are you?"

It was Major Regina Grant, a human resources officer who'd been with them in the conference room. She identified herself, and Marilyn said they'd be okay as long as they stayed together. She had a plan.

Moments earlier, when Regina escaped from the conference room into Dilbertville, she'd spotted John Yates, the security officer who had jokingly promised his wife that he'd spend the day underneath his desk. Regina saw John on his feet, towering over her, but he was in shock. Burns lashed his face, winding around his ears and down his neck to his back, buttocks, and legs. His arms and hands were blanched as white as surgical gloves.

Yates had been watching Marian Serva's television when the plane hit and the lights went out. An orange fireball coursed through the office just below the ceiling, consuming some people and missing others altogether. It knocked Yates off his feet. A mist of jet fuel coated his glasses and the remains of his tattered clothing. Regina thought he looked as if he'd been blown out of a cannon. She'd grabbed the back of his leg just before Marilyn bumped into her rear.

Despite John's injuries, his security training kicked in. "Get up and come with me," he told Marilyn, Lois, and Regina as he stumbled slowly away.

"No!" Marilyn yelled. "You get down. Get down! The smoke is too thick!"

John Yates either didn't hear or didn't listen. Regina Grant started to follow him. But Marilyn snatched her by the skirt. "No! Stay down."

Marilyn led Lois and Regina on the crawl through debris-strewn Dilbertville toward the C Ring windows over AE Drive. Heat rose from fires in the Navy Command Center below them, burning Regina's knees through her nylon stockings. Somewhere on the floor beneath them were Jerry Henson, still trapped in his chair, Petty Officers Christine Williams and Charles Lewis, Jack Punches, and Dave Thomas's best friend, Bob Dolan, among dozens of others.

Regina stopped crawling to rest, so Marilyn halted their flight. As they caught their breath, emergency sprinklers doused them, providing relief from the heat. Yet the smoke grew thicker and the air hotter, and Marilyn knew they didn't have much time before they baked or choked to death. She urged them to get moving.

Lois Stevens tugged on Marilyn's leg. Marilyn turned to see Lois patting her throat. "I can't breathe, Colonel Wills," Lois said. "Just go ahead."

"No—here, suck on my sweater," Marilyn said, handing Lois the empty sleeve of the black cardigan, moist with water from the sprinklers. "We got to keep going!"

But Lois was spent. Even with Marilyn's sweater, her lungs felt scorched from smoke inhalation. She feared that she might prevent Marilyn and Regina, both younger and stronger, from reaching the windows.

"I am too tired," Lois said. "I just can't go any farther."

"Please," Marilyn begged. "Please come on. We've got to get out of here."

Lois refused. She pressed down against the hot carpet and prepared to die.

AS NAVY CAPTAIN Dave Thomas searched for his best friend, Bob Dolan, Dave Tarantino and other rescuers, strangers to one another, took turns using fire extinguishers to beat back flames. Someone passed around flashlights. They worked their way from AE Drive into the first floor through the blown-open holes in the C Ring walls.

Several developed a strategy they hoped would lead them to survivors. One took the point, staying low to the ground for as long as he could bear, then he'd switch places with the man behind, leapfrogging deeper into the Pentagon's wrecked Wedge One. Drops of melting metal and plastic burned tiny holes in their skin. Secondary explosions made them fear that the building would collapse around them.

The group cleared the jagged debris as much as possible, trying to avoid live electrical wires that sparked in the smoky gloom. Twice, Dave Tarantino recoiled from shocks, then kept moving. His uniform shirt, a no-iron synthetic style banned aboard ship for its fire-friendly properties, began to melt from the heat, so he tore it off and threw it onto AE Drive. Someone handed him a wet T-shirt to breathe through, and he went back in. One word rattled around his brain: "Apocalypse."

Working separately, still hoping for any sign of Bob Dolan, Dave Thomas took a slightly different route deep into the burning Navy Command Center. He scanned the darkness, through smoke and flames. Perhaps twenty yards away, he saw what appeared to be a disembodied face. It looked like a misshapen Halloween fright mask, its eyes fixed open. Then it blinked.

CHAPTER 17

"I THINK THOSE BUILDINGS ARE GOING DOWN"

South Tower, World Trade Center

ON THE SOUTH TOWER'S 81ST FLOOR, NEW BLOOD BROTHERS BRIAN Clark and Stan Praimnath returned to Stairwell A battered and covered in soot. Brian remembered the heavyset woman's insistence that fire would block their way down, but Stan knew he needed medical help, so they decided to try. The two men pushed away the broken drywall that in places covered the stairs like playground slides and stepped through water from broken pipes that cascaded underfoot. Smoke fouled the air, but the stairwell was passable.

They passed the stairwell door to the 80th floor, unaware that some of Stan's Fuji Bank colleagues huddled in the northwest corner of the floor. Jack Andreacchio, who urged "Stan the man" to return to work a half hour earlier, told a 9-1-1 dispatcher that he, Manny Gomez, and three others were trapped by heat and smoke, unable to reach a stairwell.

"I think we should break the window," Jack Andreacchio said during a call logged at 9:23 a.m. The operator told him that would only bring fire closer.

"Fidnee," the operator said, using dispatch jargon for FDNY, "is on the way. They have it. People are on their way to you, okay?"

"You've gotta help me," Andreacchio said.

INSIDE THE STAIRWELL, at the 78th floor sky lobby level, Brian and Stan skirted past flames that spurted through cracks in the wall. They didn't cross paths with the man in the red bandanna or the people he led to safety. By the 74th floor, the air in the stairwell began to clear. Lights were on and the stair treads were dry. Brian thought they'd be safe. Stan grew steadier on his feet.

The first person they encountered in the stairwell was a colleague of Brian's from Euro Brokers: José Marrero, who'd risen from the kitchen staff to become the firm's facilities manager. José, a married father of three who served with Brian as a fire safety warden, had led other Euro Brokers workers down partway through the building from the 84th floor. Now, breathing heavily, he climbed back up.

"José, what are you doing?" Brian asked.

José nodded to his walkie-talkie. He'd spoken with another member of their safety team, Dave Vera, who was one of the half dozen people Brian originally led down from the 84th floor. At the urging of the woman in the stairwell, Dave had reversed course and gone back upstairs with Bobby Coll, Kevin York, and the others.

"Dave's a big boy," Brian told José. "He can get out. We've just come through hell to get here. Come with us."

"No, no, no. I'm okay," José said as they parted. "I'll be okay."

Brian and Stan continued downward, trading tidbits of what they knew about what happened. Stan told Brian he'd seen the plane that hit their tower, and the enormity came into clearer focus.

At the 44th floor sky lobby, they exited Stairwell A, hoping to phone their families, unaware that the South Tower's structure was growing weaker by the minute. They found a security guard, a man in his sixties, standing watch at his desk.

"Do you have telephones?" the guard asked.

Brian and Stan noticed a semiconscious man on the floor behind the security desk, moaning from massive head injuries. Stan had occasionally seen the injured man in the building but didn't know his name. Stan heard him say: "Please tell my wife and our baby that I love them."

"When you get to a phone," the guard said, "you must get a medic and a stretcher up to the forty-fourth floor." Stan and Brian promised and returned to the stairwell.

They exited again on the 31st floor and walked through the deserted offices. They found a working phone in a conference room of the investment firm Oppenheimer Funds, whose nearly six hundred employees on four floors had escaped unhurt. At shortly after 9:30 a.m., Brian called home. He'd last spoken to Dianne less than forty minutes earlier, after Flight 11 struck the North Tower. After they hung up, Dianne saw United Flight 175 hit the South Tower on television. Neighbors and church friends had rushed to her side, to keep her and their children from panicking.

"Hi, honey," Brian said, breezy as ever. "I'm on the thirty-first floor. I got a good story to tell you. Got to go now, want to get out of the building. But I thought I'd check in." He hung up, and Dianne shared the news that Brian was alive, but still inside the burning tower.

Stan tried to reach Jenny, but she'd fled from work when she heard about the South Tower, certain that Stan was dead. Stan left a message with her coworkers.

Brian then called 9-1-1 for the injured man on the 44th floor. During the next three minutes and seventeen seconds, the depth of chaos in the 9-1-1 system became evident. Brian was transferred twice, put on hold, and talked to three different people. His frustration rising, Brian told the last woman he spoke with, "I'm only going to say this once— don't put me on hold." He again relayed details about the man, then hung up.

NONE OF THE 9-1-1 operators Brian spoke with asked him how high he'd been inside the building, conditions on the upper floors, or details about which stairwell he'd used. As a result, they never learned potentially lifesaving information that might have helped firefighters or trapped people in the South Tower who survived the crash and then called 9-1-1 for guidance.

Amid the chaos, overwhelmed 9-1-1 operators never learned that Stairwell A remained relatively intact above the South Tower impact zone, at least to the 91st floor and possibly higher. No evidence exists that any person who called 9-1-1 from above the impact zone was urged to seek out Stairwell A as a potential path to survival.

One 9-1-1 operator, besieged by urgent calls for help from people trapped in the towers, vented her frustration to a colleague: "How can you have a big building and no way to get out of it? That's ridiculous."

Emergency operators also weren't told that rooftop rescues wouldn't be attempted because of smoke, or that exit doors to the roof remained locked, even as some 9-1-1 callers mentioned plans to try to reach the roof. Some of those heading upstairs might have recalled the daring rescues of twenty-eight people from the North Tower roof by a New York Police Department helicopter after the 1993 bombing.

One person who tried to reach the South Tower rooftop knew the route well. Roko Camaj was an immigrant from Montenegro, a sixty-year-old husband, the father of an adult daughter, and a man with an eagle's comfort with heights. He'd spent twenty-eight years working as a window washer at the World Trade Center—squinting against the sun's reflection off the towers had carved deep crows' feet around his eyes. Roko operated a machine that automatically cleaned the lower ones, but he climbed into a harness to soap the highest windows by hand. Roko was the subject of a children's book in which he declared: "It's just me and the sky. I don't bother anybody and nobody bothers me."

Now, unable to reach the roof, Roko went down several floors and called his wife. He said he was stuck on the 105th floor with about two hundred others. He radioed a fellow Port Authority worker: "Don't let no people up here . . . there's big smoke!"

Meanwhile, following traditional rules that didn't apply to this unprecedented situation, at least some 9-1-1 operators told survivors on high floors of the South Tower to remain in place, regardless of conditions. How many people received those instructions isn't clear.

When Brian Clark made his 9-1-1 call, nearly seven thousand people had already escaped the South Tower, while more than a thousand remained between the lobby and the 76th floor, heading downward if they could. More than six hundred people were on the 77th floor and above. Those numbers didn't include the emergency responders who flooded into the South Tower after United Flight 175 struck.

Time was becoming an enemy, but not everyone understood that. A 9-1-1 caller on the 73rd floor, below the impact zone, told an operator that oxygen was running out, only to be instructed not to leave the floor. A police dispatcher gave a similar order to a man on the 88th floor, in the offices of Keefe, Bruyette & Woods: "You have to wait until somebody comes there."

Meanwhile, the Stairwell A escape route remained nearly empty.

AND AT ALMOST the same time as Brian Clark made his 9-1-1 call, the FDNY suffered its first fatality of the day.

Firefighter Danny Suhr, a thirty-seven-year-old captain of the FDNY football team, had just arrived on the scene. Debris littered the ground and people fell from the sky. As he neared the South Tower, Danny told his captain, "Let's make this quick."

Before they reached the building, a person who fell or jumped from an upper floor clipped Danny's head on the way down. EMTs tried to help, and two fellow firefighters from Engine 216 jumped into the ambulance with their friend, who had a two-year-old daughter and a love affair with his wife that dated from grade school.

"Danny! Danny! Danny!" they yelled.

EMT Richard Erdey knew that Danny Suhr was gone, a victim of catastrophic head injuries, but he kept trying.

"Look," Erdey told the firefighters, "we're doing the CPR for that small glimmer of hope, but I'll tell you what they're going to do. They're going to call it at the hospital. Please stop staring at him.

You're going to burn this image in your head. I want you to remember a better image."

AS EVACUEES RAN from the building, as upper-floor survivors called for help, and as firefighters rushed inside, the South Tower experienced ominous structural changes.

Shortly after 9:30 a.m., bursts of smoke spouted from the building's north side, on the 79th and 80th floors, possibly from shifting floor slabs or the sudden ignition of pools of unspent jet fuel. Simultaneously, threats to the building's integrity worsened. Intact core columns strained under the added burden previously borne by the severed columns. External columns on the tower's east side supported added loads that had shifted from the severed columns. The remaining external columns also struggled to hold up sagging floors.

The strain caused the exterior columns to bow inward, like straws trying to support a brick. All the while, unchecked fires undermined the strength of steel columns throughout the impact zone, contributing to a threat that almost no one imagined. The South Tower neared its breaking point.

TIRING AS THEY continued inside Stairwell A, Brian Clark urged Stan Praimnath to slow down. Stan had begun bounding down the steps two at a time. "I'd hate to break an ankle and have to walk thirty more floors," Brian said, "then come in on crutches tomorrow."

As they slowed their descent, they encountered no one else, neither evacuees nor firefighters, on the final floors of Stairwell A.

Brian and Stan emerged on the north side of the South Tower, facing the sprawling Austin J. Tobin Plaza. Both men had long cherished its sparkling fountain, with a twenty-five-foot bronze sculpture called *The Sphere* at its center. They liked how the plaza served as an urban oasis for workers, tourists, and vendors. Now they saw it buried in ash and debris. Brian thought it looked like an abandoned archaeological site.

Stan and Brian stood shoulder to shoulder, silently absorbing the ruins for twenty or thirty seconds, then walked down a nonworking escalator to the Concourse level. A female officer at the bottom of the escalator told them it would be safer if they didn't exit onto Liberty Street. She directed them through the Concourse mall under the plaza, past the Victoria's Secret shop and out by the Sam Goody music store. The route snaked underneath Four World Trade Center, to an exit at the southeast corner of the complex.

Brian and Stan walked through the Concourse, feeling safe and in the clear. They passed firefighters pulling on boots and equipment, and police officers who seemed busy but under control. Although the two men were the only civilians in sight, covered in dust that should have made it apparent that they'd come from affected floors, no one asked them about conditions inside the tower or which stairwell they had used to escape.

When they reached the exit, a firefighter told them, "Whoa, whoa. If you're going to leave through these doors, you've got to run for it."

"Why?" Brian asked.

"There's debris falling from above."

"Should I look up?"

"No! Just go for it."

Brian couldn't follow that order. He slowly opened the door and poked his head out. "It looks clear," he told Stan. "Ready?"

"Ready!"

AS FIREFIGHTERS SET up shop inside the South Tower lobby, Assistant Chief Donald Burns led relatively few troops compared to the throngs in the North Tower. But Chief Burns had at least one advantage: he'd brought along Battalion Chief Orio Palmer.

Forty-five, married with three children, Orio commanded an area that stretched across Midtown Manhattan, including the Empire State Building, Penn Station, the Garment District, and scores of high-rise office and apartment buildings. A marathon runner and three-time winner of the department's fitness award, Orio had a brushy mustache, a thatch of comb-resistant brown hair, and boundless energy that hadn't subsided since his first days on the job. Almost twenty years earlier, as a probationary firefighter, Orio had followed Jay Jonas up the stairs of the burning Bronx tenement. Later, he had interrogated Jay to learn all he could. Now, while Jay guided the men of Ladder 6 up the North Tower, Orio led FDNY's charge up the South Tower.

An elevator mechanic before becoming a firefighter, Orio found the one functioning freight elevator and took it to the fortieth floor. He also discovered that, despite earlier problems, the FDNY's radio repeater channel somehow worked inside the South Tower, amplifying his signal and enabling him to remain in contact with his bosses and his troops as he moved from the elevator to the South Tower's Stairwell B.

As Orio climbed, a firefighter trailing him, Scott Larsen of Ladder 15, radioed that "the director of Morgan Stanley" had told him: "Seventy-eight seems to have taken the brunt of this stuff—there's a lot of bodies." With that information, almost certainly from Morgan Stanley security chief Rick Rescorla, Orio knew exactly where he was needed most. As he rose higher, Orio displayed no concern that the building wouldn't stand. When another firefighter radioed that he'd stopped to catch his breath, Orio told him: "Take your time."

At 9:32 a.m., Orio radioed down from the 55th floor to Battalion Chief Ed Geraghty that they should set up a command post on the 76th floor, just below the impact zone. Among the firefighters rushing up the stairs behind Orio were Scott Kopytko and Doug Oelschlager, who'd worked the overnight shift with Jay Jonas and his Ladder 6 crew.

By 9:45 a.m., Orio had sprinted up thirty-four flights from where he exited the freight elevator, reaching the 74th floor. He'd climbed nineteen floors in thirteen minutes despite being loaded down with gear. Breathing heavily, he cautioned the men behind him that the walls of Stairwell B on the 73rd and 74th floors were ruptured.

On the 75th floor, at around 9:49 a.m., Orio met Fire Marshall Ronald Bucca, a twenty-three-year FDNY veteran who also was a registered nurse and a reservist in the U.S. Army Special Forces. Fifteen years earlier, Ron earned the nickname the Flying Fireman when he fell five stories from a fire escape, surviving thanks to telephone and cable wires he struck on the way down. After Flight 175 hit, Ron had walked up from the lobby with Fire Marshal Jim Devery. On the 51st floor, Jim Devery came across a burned woman who needed help: Ling Young, who'd first been rescued from the 78th floor sky lobby by the man in the red bandanna. Jim Devery helped Ling down to the lobby, while Ron Bucca kept going up.

Blocked from climbing higher inside Stairwell B, Orio found a new path: Stairwell A. He and Ron Bucca switched to the intact stairwell only minutes after Stan Praimnath and Brian Clark had used the same route on their way down.

At 9:52 a.m., gasping for breath, sounding stressed but in control, Orio reached for his radio. Roughly forty-five minutes had passed since he rushed into the burning South Tower, rode an elevator up forty floors, then scaled more than seven hundred steps. Now, he faced the twin forces driving his professional life: a fire to fight, and people to help.

"We've got two isolated pockets of fire!" Orio radioed Lieutenant Joe Leavey of Ladder 15, who'd reached the 70th floor. "We should be able to knock it down with two lines." Encountering the blood-bath in the sky lobby, Orio described what the man in the red ban-danna, Ling Young, and others experienced more than a half hour earlier: "Seventy-eighth floor, numerous ten-forty-five, Code Ones." In FDNY terms, Orio was describing a floor filled with fatalities. But others, still alive, needed rescue.

"Numerous civilians," Orio told Joe Leavey. "We're gonna need two engines up here."

Orio offered guidance to the men coming up behind him, who were busy fighting a fire in Stairwell B: "I'm gonna need two of your firefighters, Adam [A] Stairway, to knock down two fires. We have a house line stretched, we could use some water on it. Knock it down, 'kay?"

Meanwhile, another problem emerged, as the freight elevator Orio found earlier had stalled, trapping ten injured people sent down by members of Ladder 15. "We're chopping through the wall to get out," a firefighter radioed. He advised his commanders that they'd have to find another way up and down.

As Orio surveyed the 78th floor, he eyed an escalator in the sky lobby: "We have access stairs going up to the 79th floor." He announced that he wanted firefighters up there and higher. Some six hundred people remained on the upper floors, some calling 9-1-1 and their loved ones, some jumping or falling to their deaths. Orio intended to help as many as possible.

As they awaited more firefighters, Orio and Ron Bucca apparently tried to free a group of people in an elevator who'd tried to evacuate after the North Tower crash, only to become imprisoned for nearly an hour when Flight 175 hit the South Tower. At 9:57 a.m., a Port Authority security guard named Robert Gabriel Martinez radioed a call for help: "We need EMS over here! On the double! Two World Trade Center!" He told the dispatcher that he was in the 78th floor sky lobby, then said, "The firefighters have eighteen passengers stuck, and they're going to try to get them out! They're trying!"

One minute later, Orio made a final radio transmission, an unfinished, unanswered call from a commander to his troops. It cut out after the first words: "Battalion Seven to Ladder 15."

BRIAN AND STAN ran across Liberty Street, avoiding or hurdling debris, and jogged a block and a half south down Greenwich Street. Shopkeepers in doorways cheered as the two men passed. Brian stopped outside a deli to catch his breath. The deli owner gave them each a bottle of water, then rushed back inside and reemerged with an unclaimed breakfast platter of sliced fruit and pastries. "Nobody's coming for this today," he said.

Brian carried the platter as they continued south. They turned east and bumped into two priests. Stan felt an outpouring of built-up emotion. He wobbled and broke down.

"This man saved my life!" Stan wept to the priests. "He called to me in the darkness!"

Brian welled up. "I think you saved my life, too," he told Stan. "You got me out of that argument about whether I should go up or go down."

In the middle of the street, one priest placed his hands upon them and led a prayer. Brian and Stan embraced. The priest mentioned that Trinity Church remained open, a block away. Brian and Stan headed for the sanctuary, walking along the upward-sloping street on the church's south side, stopping beside a wrought-iron fence a few feet from the grave of Alexander Hamilton. They looked up at the burning South Tower, roughly a thousand feet away. At that angle, they couldn't see its twin. In two minutes, it would be ten o'clock.

"I think those buildings are going down," Stan said.

An engineering student in college, Brian dismissed him: "No way. Those are steel structures." He described the great strength of steel and assured Stan that the fires were from furniture, paper, carpeting, and other combustibles. Not knowing the true extent of damage in the South Tower, Brian felt certain that the fires would burn out and the buildings would remain standing.

AS BRIAN AND Stan debated, and as Orio Palmer, Ron Bucca, and other firefighters went above and beyond, inside the South Tower an insurance company vice president named Kevin Cosgrove dialed 9-1-1. He was forty-six years old, a caring husband, and the kind of father who indulged his three children with dessert before dinner.

Kevin worked for Aon Corporation, a giant insurance brokerage, on the 99th floor, far above the impact zone. After the plane hit, he walked down to the 79th floor, but he couldn't get down through the stairwell he'd chosen. He climbed back up, beyond his office floor, apparently headed toward the roof. His 9-1-1 call connected from the 105th floor, where window washer Roko Camaj also took refuge.

Kevin coughed on smoke as he sought help and reassurance from a male firefighter and a female police operator on the call.

Kevin: "What floor are you guys up to?"

FDNY: "We're getting there. We're getting there."

Kevin: "Doesn't feel like it, man. I got young kids."

FDNY: "I understand that, sir. We're on the way . . ."

Kevin: "Come on, man."

The female operator came on the line: "We have everything, sir."

Kevin: "I know you do, but it doesn't seem like it. You got lots of people up here."

Operator: "I understand."

Kevin: "I know you got a lot in the building, but we are on the top. Smoke rises, too. We're on the floor. We're in the window. I can barely breathe now. I can't see!"

Operator: "Okay, just try to hang in there. I'm going to stay with you."

Kevin: "You can say that. You're in an air-conditioned building. . . . What the hell happened?"

Operator: "Sir, I'm still here . . . still trying. . . . The Fire Department is trying to get to you."

Kevin: "Doesn't feel like it."

Operator: "Okay, try to calm down, so you can conserve your oxygen, okay? Try to . . ."

Kevin: "Tell God to blow the wind from the west! It's really bad. It's black. It's arid [sic]. . . . We're young men! We're not ready to die!"

The operator's voice dropped to a whisper. She seemed on the verge of tears. "I understand," she said.

Kevin: "How the hell are you going to get my ass down? I need oxygen."

Operator: "They're coming. . . . They have a lot of apparatuses on the scene."

Kevin's voice grew raspy, his breaths shallow. His words became at once labored and frantic: "It doesn't feel like it, lady. You get them in from all over. You get 'em in from Jersey. I don't give a shit. Ohio."

Seeming unsure what else to say, the operator asked his name again.

"Name's Cosgrove. I must have told you about a dozen times already. C-o-s-g-r-o-v-e. My wife thinks I'm all right. I called and said I was leaving the building, and then—bang!"

Kevin said he was with Doug Cherry, a colleague at Aon and a married father of three, and another person he didn't name. He described his location again: "We're overlooking the Financial Center. Three of us. Two broken windows!"

Suddenly the building shifted. Before the call disconnected, Kevin screamed: "Oh, God! . . . Oh!"

THE FORCES OF catastrophe and heat had gnawed at the South Tower impact zone for fifty-six minutes. The fires weakened and added stress to its remaining core columns, its exposed steel floor supports, and its load-bearing exterior walls.

The clock read 9:59 a.m. Passengers and crew members of United Flight 93 over Pennsylvania fought to prevent a fourth building strike. The Pentagon and the North Tower burned. Countless millions watched on live television.

Thick gray smoke gushed with greater intensity from the South Tower. The weakened east wall, where fires had been the most intense, lost the strength to withstand the inexorable pull of gravity. As the east wall failed, its multi-million-pound burden shifted through the building's core to the adjacent north and south walls. But those walls were compromised, too. The damage from Flight 175 and the resulting fires made the load too much to bear. Sapped of their fortitude, the steel wall columns bowed farther and farther inward. Floors sagged deeper and deeper.

The end began.

Above the impact zone, twenty-five stories tilted as one, to the east and south, then went into free fall. The plunging upper floors overwhelmed the undamaged lower floors with a mass impossible to resist. Down it all went, almost straight down. The roar of steel, concrete, furnishings, and so many lives crumbling to the ground was a seismic event. It registered as a minor earthquake on sensors throughout the Northeast. It lasted about ten seconds.

A cloud of grayish smoke and dust rose like the ghost of a vanished civilization. The collapse took the lives of everyone still inside.

It left behind a hole in the skyline, a mound of rubble, and a question: Would its twin follow suit?

ROUGHLY EIGHT THOUSAND people escaped the South Tower. But 619 people remained on the 77th floor and higher, and eleven stood in the lobby, when it fell. That didn't include emergency responders and others whose exact final locations were unknown. Among the dead were men and women who almost certainly would have lived if they'd evacuated immediately or soon after but who remained inside because they were told not to leave or because they stayed to help others.

One was Rick Rescorla of Morgan Stanley. Of the firm's twenty-seven hundred employees in the South Tower, all escaped safely except the former Vietnam platoon leader and five others, several of whom were members of Rescorla's security team.

Another fatality was the civilian rescuer who appeared in the 78th floor sky lobby, his features masked by a red bandanna. He'd later be identified by several people whose lives he saved as Welles Crowther, a driven, charismatic twenty-four-year-old equities trader with Sandler, O'Neill & Partners. A volunteer firefighter and college athlete whose bandanna was his trademark accessory, Welles had recently told his father and friends that he wanted to give up Wall Street to become a New York City firefighter. His actions spoke even louder than his words. Welles Crowther died in the South Tower lobby, surrounded by members of the FDNY, among them Assistant Chief Donald Burns.

The South Tower collapse claimed the lives of the responders who reached higher in the building than anyone else: Battalion Chief Orio Palmer and Fire Marshal Ron Bucca. Others who sacrificed themselves included Battalion Chief Ed Geraghty; Lieutenant Joe Leavey; firefighters Scott Kopytko, Scott Larsen, and Doug Oelschlager; and sky lobby security guard Robert Gabriel Martinez. Outside the tower, among those killed was Police Officer Moira Smith, who made the

first NYPD report of the Flight 11 crash. She guided dozens of evacuees, many of them hurt and bloody, away from the South Tower, each
time returning for more. Victims trapped on upper floors included
9-1-1 callers Melissa Doi and Kevin Cosgrove, and window washer
Roko Camaj.

The World Trade Center's security chief, John O'Neill, was last
seen ten minutes before the collapse, walking from the North Tower
command post toward the South Tower. O'Neill had begun the job
only weeks earlier, after retiring from a storied career as the FBI's
foremost authority on Osama bin Laden. He'd publicly warned of
the hidden dangers of militant terrorist groups four years earlier.
At the time, he said: "A lot of these groups now have the capability
and the support infrastructure in the United States to attack us here
if they choose to." O'Neill had been proven right.

If Alayne Gentul had evacuated instead of going higher in the building to tell others to leave, she wouldn't have found herself trapped on
the 97th floor. When the building collapsed, as Jack Gentul prayed in
his New Jersey office, he felt as though his wife's spirit passed through
him. On his knees, Jack suddenly smelled Alayne's perfume.

FOR SEVERAL SECONDS as the South Tower fell, Brian Clark and Stan
Praimnath stood side by side on the sidewalk outside Trinity Church,
mouths agape. When each floor pancaked downward, window glass
burst into the bright blue sky. The shards looked to Brian like spar-
kly confetti in the morning sunlight. In their shock and disbelief, neither Brian nor Stan yet registered the human toll. Seconds later, the
surreal beauty yielded to horror when dust as thick as malted milk
stormed toward them.

Brian and Stan sprinted down Broadway. Trinity Church's brownstone walls shielded them, absorbing the blow. They glanced back
to see swirling dust rise over the church, obscuring the steeple that
once marked the highest point in New York City. The ash-filled cloud
threatened to crash down onto them.

Before the dust could overtake them, Brian and Stan ducked into a century-old stone building at 42 Broadway. As Brian pushed through the doors, he realized that he'd run the entire way carrying the breakfast platter. He stripped off the cellophane and invited several dozen strangers in the lobby to share the fruit and sweet rolls.

Brian and Stan remained there for forty-five minutes, filthy and disheveled, resting and talking, unaware of the full extent of what had befallen the South Tower. They didn't know it had collapsed entirely. In their partial awareness, they knew about the hijacked planes, but not that there were four. As they waited for sunlight to return, Stan handed Brian a business card, with his home address and a phone number for a clothing company that Stan and his wife, Jenny, ran on the side.

When the dust cleared, Stan and Brian left via a back entrance, behind the New York Stock Exchange, and made their way onto Broad Street. Winter seemed to have arrived three months early. Gray-white ash coated buildings, cars, mailboxes, parking meters, and anyone caught outside in the unnatural blizzard.

Somehow Brian and Stan became separated as Stan rushed toward the Brooklyn Bridge, to reach Jenny and their daughters. In the crush of people trying to leave Lower Manhattan, Stan caught a ride from a stranger in a pickup truck who was driving on sidewalks to avoid pedestrians in the streets. When the driver saw Stan's battered and bloody condition, he offered a cigarette. Stan's wry humor returned: "No, man, I had enough smoke for one day."

When Brian realized that he'd lost Stan, he scrambled through the crowd, yelling, trying to find him, just as he'd done on the 81st floor. But Stan was gone.

Brian walked on alone, in disbelief of all that had happened. The building where he'd worked for twenty-seven years was gone, and with it who knew how many friends and colleagues, other workers and visitors, emergency responders and plane passengers and crew. Half in shock, Brian wondered: Had Stan been a figment of his imagination? A guardian angel who prevented him from climbing higher in a building doomed to fall?

Brian paused. He reached into his shirt pocket and pulled out the business card. It read FROCKS & TOPS INC., STANLEY PRAIMNATH, PRESIDENT & CEO. Brian relaxed. He tucked away the card, comforted by the certainty that amid all the loss and madness, his new blood brother was real.

CHAPTER 18

"TO RUN, WHERE THE BRAVE DARE NOT GO"

Ground Level and North Tower, World Trade Center

AROUND 9:30 A.M., INSIDE STAIRWELL B OF THE NORTH TOWER, Elaine Duch, her colleagues-turned-caretakers Gerry Gaeta and Dorene Smith, and their fellow evacuees couldn't tell how badly the building had been damaged. Most still didn't know the cause.

When they emerged in the main lobby, they saw blown-out windows, broken walls, and water from severed pipes streaming down from the mezzanine level. Gerry thought it resembled Niagara Falls. As they sloshed through six inches of water, Gerry still gripped the knot in Elaine's sweater while Dorene held the empty sleeve.

"Burn victim!" Gerry yelled. "Where do we take her?"

Someone pointed to the entrance to the Concourse-level shopping mall, toward the building known as Five World Trade Center. The subterranean route would take them under the outdoor plaza, where bodies and burning debris continued to fall, and where high above, the South Tower had begun its death spiral. Soon, two medics spotted Elaine. Each took an arm and sped her through the Concourse. Gerry's legs ached from the seventeen-hundred-stair descent, and he knew he couldn't keep up. He told the others to run ahead. Dorene jogged alongside, staying as close as she could to Elaine as they reached an exit onto Church Street, where ambulances awaited them.

"I'm hurt," Elaine announced to anyone in earshot. "Help me."

Dorene helped her hobble toward the street. Elaine turned to her and asked: "How's my hair?"

"It's fine, Elaine," Dorene lied. "It's fine."

AS THEY MOVED toward the fleeing masses, EMT Moose Diaz turned to his partner Paul Adams: "This is the big kahuna, man."

They headed past St. Paul's Chapel toward the Millennium Hotel, where scores of men and women from the towers gathered outside. The moment anyone left the North or South Tower, their status changed from evacuees to terrorism survivors.

The injured streamed toward Moose and Paul, bleeding, badly burned, some hyperventilating, some with chest pain, some with broken bones. Protocol called for the EMTs to triage the wounded, to prioritize treatment by issuing green tags to those with relatively minor injuries, yellow for more serious needs, and red to those who needed immediate care to survive. Black was for hopeless cases.

Paul told Moose they had no time for triage tags, blowing past a lieutenant who insisted otherwise. Convinced that the implausible might happen and the burning towers might fall, Paul yelled to Moose: "These people should be getting the fuck out of here!"

Two women approached them, one whose terrible burns served as their own triage tag. Her blouse was melted to her peeling skin. A sweater partly covered her ruined flesh. For the moment, she'd been spared from pain by the fireball's destruction of her nerve endings. The woman had a single polite request for Moose and Paul, issued in a hoarse, smoke-strained voice.

"Could you help me?"

It was Elaine Duch, with Dorene Smith still at her side.

Moose and Paul dropped everything and eased Elaine onto their stretcher, with Moose talking softly to her all the while. "What happened?" Moose asked.

"I felt a hot flash," Elaine answered. She asked him about her burned and matted hair. Moose told her not to worry.

Elaine looked for Dorene. "Please don't leave me," Elaine begged. "Come with me to the hospital." Dorene promised she would.

A priest from St. Paul's Chapel seemed to appear from nowhere. He asked Elaine, "Are you Catholic?"

"Yes."

"Do you mind if I give you your last rites?"

"Please do," Elaine said.

The priest knelt next to the stretcher and delivered the sacrament as quickly as he could, so Elaine would be properly reconciled with God. At the prayer's end, Moose and Paul lifted Elaine into the back of 45 Adam, as Dorene scampered aboard. Elaine made Dorene promise that she would call her twin sister, Janet. Elaine repeated Janet's phone number multiple times, like a mantra, to be sure.

Paul jumped behind the wheel and drove two miles uptown to St. Vincent's Catholic Medical Center on Twelfth Street. In the back, Moose cut away whatever he could of Elaine's clothing and gently bandaged her wounds.

"Am I going to die?" Elaine asked. A hush fell over the ambulance. No one answered.

RON CLIFFORD AND his newfound helpers, the hotel nurse and the man from Long Island, led burn victim Jennieann Maffeo from the Marriott lobby. As they reached the street, a photographer spotted their little group. The resulting image captured Jennieann swaddled in the white tablecloth and Ron holding her up. His yellow tie remained bright. He stood out from the crowd, just as his sister, Ruth, had advised.

They emerged to find a burning Federal Express truck, broken glass, scattered debris, ash-covered people and vehicles, with a hellish soundtrack of sirens and screams. Ron glanced up and thought the towers looked as if they were melting. He heard loud thumping noises, not like gunshots but strange, awful thuds. Another glance revealed the source: the impact of bodies falling or jumping from great heights.

A firefighter with the white helmet of an officer stepped toward them out of the smoke and ash. The towers groaned as fires warped their insides. "Run!" the firefighter told them. "Run, run, run!" Yet he ignored his own advice; instead of leaving, the chief or captain led a group of fellow firefighters toward the dying buildings. Ron marveled at their bravery.

"Can you run?" Ron asked Jennieann.

"I'll try."

The group tottered in unison across West Street to a line of waiting ambulances. Medics rushed to help Jennieann, as Ron passed along information she gave him about her latex allergy, her asthma, and her boss's telephone number. As they loaded her aboard, Ron told Jennieann: "You have to make it now, after all we've been through."

Then Ron Clifford stood at the curb, watching the ambulance pull away.

STILL CLIMBING HIGHER inside Stairwell B of the North Tower, Captain Jay Jonas remained focused on the steps ahead and his men around him. Determined to keep Ladder 6 together, Jay took regular head counts. "This is the mother of all high-rise buildings," he told his crew. "I can't afford to be looking for you guys."

During their third ten-floor march, Jay realized that two of his men had fallen behind. They'd been delayed when they stepped aside to let evacuees helping one another pass two abreast on the stairs. "Wait here," Jay told the men still with him. He found the missing men and reassembled his team on the 27th floor.

By then, the number of descending civilians in Stairwell B had slowed to a trickle. Roughly an hour had passed since Flight 11 hit the North Tower, and nearly all the able-bodied men and women below the impact zone had fled.

By one estimate, 1,250 people evacuated the North Tower in the first few minutes, even before United Flight 175 hit next door. Another 6,700 men and women left before ten o'clock. Yet along with more than nine hundred civilians still heading for the exits, more than two hundred firefighters remained inside the North Tower, either climbing or resting for the next upward push.

"All right, everybody take a knee, get some water," Jay said. They'd try to make up for lost time and shoot for the fortieth floor on their next effort. They exited Stairwell B through a door that led into a hallway on the 27th floor.

As they rested, they were joined by a friend of Jay's, firefighter Andrew Fredericks, a nationally known fire service instructor from Squad 18 whose obsession with hose techniques earned him the nickname Andy Nozzles. Andy was still mourning three fellow firefighters who died three months earlier in a Father's Day fire he'd helped to fight in a Queens hardware store. He'd written a tribute to the men that laid bare his emotions that night. "Emptiness is the only way to describe the way I felt," Andy wrote. "I kissed my kids and hugged them and watched the news and cried."

While they caught their breath, Jay, Andy, and Ladder 6 also encountered the leader of Engine 21, Captain William "Billy" Burke Jr., tall and lean, the son of a retired deputy chief. In his free time, Billy Burke was a photographer, a Civil War aficionado, and a summertime lifeguard on Long Island. Jay and Billy went back years, to when Jay was a lieutenant and Billy served under him.

At 9:59 a.m., as Jay greeted his old friend, an earthquake seemed to rock Lower Manhattan. The floor heaved, and the North Tower swayed. A whoosh of air crested into a roar, then subsided. The lights switched off. The North Tower righted itself and the men and women inside regained their balance, if not their equilibrium.

ALONE ON THE floor of his elevator car, oblivious to everything beyond the walls of his box, Chris Young occasionally heard loud cries amid sirens and fire alarms. Nothing sounded like human agony, only some kind of vague emergency unfolding nearby. He knew nothing of planes or fires or people falling and jumping to their deaths. Several times he tried calling out to the people in the other elevator but got no response. As his anxiety rose, Chris tried to pry open the doors, but they wouldn't give. To pass the time and calm his nerves, he alternated between sitting and pacing.

More than an hour had passed. At 9:59 a.m., Chris's enclosed world shook. The rumbling intensified, accompanied by a roar. The lights flickered and went out, then a few seconds later came back on. Gusts of smoke and dust blew into his elevator car. Still suspecting that the crisis had been caused by a bomb under the trade center, now Chris thought a second bomb had detonated. Terrified that he would soon die, he again rolled into a ball. Seconds felt like minutes.

The rumbling stopped. Struggling to breathe, Chris took off his blue dress shirt and wrapped it around his head as a filter against the smoke and dust. He rose and pressed the fire hat button, but this time the electronic voice didn't answer. He toggled the alarm button on and off in three short spurts, three long spurts, then three short ones again: the S-O-S distress signal in Morse code. No answer. He shouted: "Is anybody there?"

No one was. Emergency workers had evacuated the North Tower lobby. The seven people who'd been trapped in the nearby elevator had freed themselves. Not knowing any of this, Chris grew afraid. He thought about his mother. He comforted himself again with monologues, but they weren't enough. He thought again about the moment when he'd felt most self-assured: on a college stage, performing *Man of La Mancha*.

Chris's voice bounced off the elevator walls as he sang its most famous lyrics, about an ordinary man who faces overwhelming odds yet wills himself to be a hero:

To dream the impossible dream,
To fight the unbeatable foe,
To bear with unbearable sorrow,
To run where the brave dare not go.

To right the unrightable wrong,
To love pure and chaste from afar,
To try when your arms are too weary,
To reach the unreachable star.

AROUND 9:30 A.M., after guiding Jennieann Maffeo into the ambulance, Ron Clifford went looking for refuge and a safe place to phone home. Ron ran across West Street, about a hundred yards to Three World Financial Center, the American Express Tower. He curled up on the lobby floor, reeking of fuel and smoke. Ron looked down and saw charred pieces of Jennieann's skin clinging to the elegant clothes that his sister, Ruth, had helped pick out for his meeting.

Ron shifted his view to the horrors outside. He watched a couple jump together from the North Tower, holding hands. In the woman's other hand, she gripped her purse, for comfort or eventual identification, or by habit, perhaps. It confused Ron. "That's weird," he thought. "Why is she carrying her purse if she knows she's going to die?" The image replayed itself on a loop in his mind.

Ron reached Brigid, who was racked by worry as she watched CNN. "It's pretty bad here," he told her, his voice on the verge of cracking. "I'm on my way."

"As long as you're okay," she said.

"I'm alive. I'm okay. I love you."

Ron thought about returning to the towers to see if he could help anyone else. Then he thought of his family, and a mantra took shape in his mind, one he repeated multiple times: "Monica's birthday. Got to get home to Monica's birthday."

Looking out at the Twin Towers, even with their damage Ron couldn't imagine that the giants might fall. But he worried about the North Tower's teetering communications antenna, a 362-foot mast that was the most prominent feature that distinguished the twins from each other. "That big antenna is going to tip over," Ron thought, "and it's going to take out a couple of streets."

Ron rushed to the ferry terminal, jumped over the gate, and climbed aboard just as the boat left the pier, overloaded with passengers desperate to get home or simply away. A young man showed off a piece of one of the planes, a grisly souvenir that he'd grabbed as he ran away. Other passengers shamed him for it, and he slunk away.

As the ferry reached Hoboken at 9:59 a.m., passengers burst into screams and cries as they watched the South Tower tilt.

"Holy shit!" Ron said aloud.

AFTER DELAYS AND detours, as 10 a.m. neared, Cecilia Lillo and her friends Nancy Perez and Arlene Babakitis exited Stairwell C at the mezzanine level, one floor above the lobby. They gasped in horror as they looked out the windows onto the debris-strewn trade center plaza. Cecilia desperately wanted to exit there immediately, to find her husband, Carlos, or, as they'd planned, to locate an FDNY ambulance and ask another paramedic to radio him.

But emergency workers wouldn't let them out onto the plaza, for their own safety. Cecilia looked longingly out the tall windows and saw an airplane part painted the patriotic colors of American Airlines. She stopped and shuddered as she began to grasp what had happened.

"It wasn't a small plane," Cecilia thought. "That's why the building shook the way it did."

"Keep moving!" someone yelled.

Like schoolgirls on a fire drill, Cecilia held Nancy's hand and Nancy reached back to hold Arlene's. Emergency responders and Port Authority officials ushered everyone single file toward motionless escalators, to walk down to the lobby level. The three women and others nearby splashed through puddles as sprinklers drenched Cecilia's hair and her sand-colored suit. At the bottom of the escalator, Cecilia prepared to sprint outside, toward West Street, as she'd done after the 1993 bombing. But a man in a white shirt blocked her way and told her to turn left, toward the blown-open revolving doors to the Concourse mall.

"Walk fast, don't run," the man said, pointing her toward the exit at Five World Trade Center that led onto Church Street.

With Nancy trailing several yards behind, and Arlene farther back, Cecilia quick-stepped down an east-west mall corridor, to a spot near a Banana Republic store. Had she turned right, another corridor would have taken her directly to the South Tower. Instead, she veered to the center of the polished stone floor, toward a security guard she recognized who on normal days checked IDs at the elevator turnstiles. He faced the oncoming pedestrian evacuees and waved them onward. As she passed, Cecilia overheard him say "Five minutes."

Cecilia didn't know what he meant, but the guard's comment disturbed her. She turned around, back toward where she expected to see Nancy and Arlene, to scream that they should run. As Cecilia turned, the clock struck 9:59 a.m. The ground shook. A rumble became a *whump*, then crescendoed into a roar, bass notes of destruction pierced by the treble of human shrieks.

In the distance, Cecilia saw the escalator inside the North Tower lobby disappear in a cloud of smoke and dust. She couldn't see Arlene, but she locked eyes with Nancy, perhaps ten yards away, her arms outstretched, her face contorted in fear. Amid the cacophony, Cecilia didn't hear Nancy scream, but she read her unmistakable body language: "Help me!"

Before Cecilia could respond, a gust of smoke enveloped Nancy, the security guard, and everyone and everything else between Cecilia and the entrance to the North Tower lobby. Cecilia spun around, away from the danger, to run east toward the exit onto Church Street. Before she could take a step, the world went black. Cecilia went down. The force of the collapsing South Tower funneled through the North Tower lobby and channeled underground through the Concourse mall corridors, throwing her to her knees.

"Please don't let me die!" Cecilia prayed.

She feared for her friends, who'd been only steps behind her, but she focused her thoughts and energy on her promise to Carlos that she'd escape on her own. "I have to get up!" she told herself. Cecilia began to crawl when another burst of pressure from the falling tower next door propelled her forward, lifting her onto her tiptoes. She feared the store windows would shatter and flay her with flying glass.

As she fought to keep her balance—"Please, God, please don't let me fall!"—yet another pressure burst hurled her headlong toward a marble pillar outside a Gap store. Somehow, she remembered Carlos's advice: she skidded onto her side and curled into a fetal position, her arms wrapped around her head.

Afraid that the ceiling might collapse, Cecilia lay motionless, knees pulled toward her chest, arms protecting her head, one hand still gripping her purse. She slowly opened her eyes but saw nothing. Gripped by panic in the pitch darkness, unsure if she was buried alive or dead, Cecilia screamed.

"Help! Is anybody out there? . . . Carlos! . . . *Carlos!*"

She heard the voice of a man she didn't know: "Don't worry, I'm going to come for you."

Cecilia unwrapped one arm from her head and held out her hand. She guided the man with her voice as he crawled toward her until he touched her fingertips. Together, they rose unsteadily to their feet. Afraid to take even one step in the dark, worried that they'd fall into a hole in the floor and plummet to the parking levels below, Cecilia remembered having seen several firefighters farther down the mall corridor before everything went black.

"Anybody out there?" Cecilia cried. "Anybody have a flashlight?"

From around a corner, where the Gap store led toward the PATH trains, several lights came on and shone in their direction. The firefighters clustered around Cecilia, the stranger who still gripped her hand, and several others, and formed them into a tight circle. With flashlights pointed to the floor, they led the group to a stationary escalator for the climb up to the street-level exit outside Five World Trade Center.

"My husband is going to be out there," Cecilia told the man beside her. "He's a paramedic."

CARLOS LILLO RADIOED his partner Roberto Abril that he was between two buildings near where they separated, which Roberto interpreted to mean a passageway between the North Tower and Six World Trade Center, near the northwest corner of the complex. Roberto tried to get there but couldn't. A short time later, he ran into a boss, EMS Captain Joseph Rivera.

"Listen," Roberto said, "I'm missing my partner. We got separated for some reason and I need to go back. I need to find him." Rivera told Roberto to care for the injured people all around them while he searched for Carlos.

Before the South Tower fell, numerous emergency responders had seen Carlos at one point or another, helping people even as he continued to ask if anyone had seen his wife. Paramedic Manuel Delgado saw Carlos on the east side of the Trade Center, on Church Street, with tears streaming down his face as he worked. The sight struck Delgado as strange, because Carlos always seemed to be smiling. At first, Delgado thought his friend was overwhelmed by the enormity and needed to get away from the scene. Then Carlos gestured toward the North Tower and told him, "My wife's in there."

"Listen, man," Delgado said, "this is God's will. You've got to help me with the people. Snap out of it. We've got a lot of patients. You've got to help me here." Carlos kept working, and eventually Delgado lost sight of him.

AFTER DROPPING OFF Elaine Duch at St. Vincent's Hospital, Moose and Paul drove their ambulance back to the staging area at Vesey and Church and resumed work. A woman who'd escaped from one of the towers appeared to be having a heart attack. A lieutenant called to them: "You got a cardiac patient—take her!" The time was 9:59 a.m.

As they strapped the woman onto a stretcher, Moose and Paul heard a volcanic rumble, then a boom, then a Dante's choir of screams.

EMT Kevin Barrett yelled, "Moose, *run!*"

The South Tower began its collapse.

Moose tripped over the stretcher as the woman fought to unbuckle herself. Moose mentally downgraded her condition from heart attack to anxiety attack. He helped her off the stretcher and screamed: "Ma'am—just go! Go straight, go straight!"

But there was nowhere to go, no way to outrun the tsunami of smoke and debris. It caught Moose as he ran and slammed him to the ground. It engulfed him, choked him, coated him. He tried to keep moving, but he banged headfirst into a metal scaffolding pole. He lost his helmet and went down. People rushed past, grabbing him, moaning, falling. He helped when he could, even as he kept moving. His mind whirled to a decade earlier, when he had run out of oxygen during a scuba dive at a Pennsylvania quarry. Death beckoned.

Everything grew quiet. The South Tower had come to rest in a smoking pile, and all Moose could hear were agonized cries. But smoke and ash still enveloped him, and Moose felt his lungs running out of air. A hand grabbed his leg and he began to panic, costing him more precious oxygen. Moose reached down to help but feared being dragged down, so he pulled away from the grasping hand. His lungs burning, the world still dark with smoke, Moose fought the urge to lie down and accept his end.

Feeling with his hands, still moving, he looked for a place to curl up and die. But then he thought about his phone call with his wife, Ericka, and how she said she loved him and to be careful. Moose began to cry at the thought of leaving her and their sons. He silently scolded himself: "I told her I was going to be all right, and I'm doubting myself now. I shouldn't have told her." The thought of breaking his word kept Moose moving.

Moose started to run, but he still could barely breathe, so he slowed down. He feared he couldn't keep his promise to Ericka. Then Moose saw a light.

The white beam, strong and bright enough to cut through the smoke and haze, bobbed oddly up and down. Behind it was a man with a white beard and long hair. Moose wondered if he was already dead. He choked out a question through his tears: "Are you Jesus Christ?"

"No," said the bearded man. "I'm not Jesus. I'm a cameraman."

Moose looked more closely: the bobbing light sat atop a shoulder television camera.

"I'm going to die," Moose said.

"No, you're not," the cameraman said.

"Damn, you saved my life."

They hugged, and the cameraman began to cry, too. "I've had enough of this job," he said as he grabbed Moose. "We'll do this together."

Moose and the cameraman felt their way down the street and came upon a city bus. They pushed open the door and went inside. Trying to turn on the ignition, Moose pressed a button and compressed air from the brakes created a brief blast that cleared the smoke. Seeing people all around, Moose and the cameraman pulled more than a dozen people onto the bus: men in suits, women in business skirts, a group of construction workers, anyone they could reach. They huddled for a few minutes, hugging one another.

When the South Tower began to collapse and the woman who thought she was having a heart attack ran from Moose and Paul's stretcher, Paul ran, too. He tripped over a fire hose, and by the time he rose, he couldn't find Moose. Paul sprinted to a tall gate at the entrance to the Millennium Hotel's parking garage. A small overhang shielded him from the rain of debris, but he felt smothered by smoke and ash. Alongside him was a man with a small gold shield who Paul thought was an FBI agent.

Paul covered his face with his helmet, then used it to protect the man at his side. His lungs filled with soot. It felt as though he'd plunged his face into a barrel of dust and breathed deep. It occurred to him that he was inhaling a pulverized building, concrete, glass, carpet, chemicals, and who knew what else. Plus whatever remained of the people who'd been inside, the emergency responders and executives, the secretaries and messengers, the custodians and security guards, along with the passengers, crew, and whoever had flown a Boeing 767 into the South Tower on that beautiful summer morning.

Minutes passed. Paul's eyes burned. Choking, desperate for air, Paul began to hyperventilate. On the verge of collapsing, he forced himself to slow his breathing until the plume dissipated. Seeing an opening, Paul grabbed the man with the gold shield and ran to an ambulance parked about twenty feet away on Church Street. Using a master key he kept on a chain attached to his work belt, Paul helped the man inside and gathered about a dozen other people needing help. He drove to St. Vincent's Hospital and dropped them off.

WHEN ITS TWIN fell, the North Tower lobby filled with blinding dust and debris, plunging everyone into pitch darkness and forcing FDNY commanders to flee. Flashlights guiding their escape, they climbed frozen escalators from the lobby to the mezzanine, where Chief Hayden stumbled on a body. A beam of light revealed it was Father Mychal Judge, the senior FDNY chaplain, last seen in his clerical collar and his white helmet, praying for the horror to end. Judge had suffered an apparent heart attack during the collapse. Efforts to revive him failed, and five men carried his slumped body to St. Peter's Church, a scene captured by Reuters photographer Shannon Stapleton.

As they abandoned the North Tower command posts for safer locations, fire bosses didn't immediately know what happened. A fireboat on the Hudson River reported the South Tower's total collapse, but none of the FDNY commanders heard the announcement. Nevertheless, as they searched in the artificial darkness for an exit, Chief Joe Pfeifer used his handheld radio to order all firefighters to immediately evacuate the North Tower: "Command to all units in Tower One, evacuate the building!" But again, communication failed.

Jay Jonas and his men didn't hear the order, and neither did scores of other firefighters still responding to the last order they had received: go upstairs and help everyone you can. The evacuation message apparently did get through to at least some firefighters, but the rest had to figure out for themselves what had happened and decide what to do next.

Inside Stairwell B, the lights flashed back on within thirty seconds after Jay heard an explosion and felt the North Tower shudder. Still on the 27th floor, Jay told Captain Billy Burke: "You check the south windows, I'll check the north windows. We'll meet back here."

Jay rushed to the north side of the North Tower. He pressed close against the narrow windows but could see nothing but smoky white dust. He returned to find his friend Billy Burke with a strange look on his face, his head tilted, as though he doubted what he'd seen. Jay expected him to say that part of the North Tower had broken off high above them.

"Is that what I thought it was?" Jay asked. But he'd underestimated the damage.

"The other building . . . just . . . collapsed," Billy replied.

That made no sense. Jay had fought countless fires and studied countless more, and he'd never heard of a skyscraper collapsing from fire. But the look on Billy's face convinced him it must be true. Jay instantly understood that hundreds, perhaps thousands, of people had just died in the South Tower, including some of his brother firefighters. He also grasped that the building where he and his men stood, which had been burning more than fifteen minutes longer than its twin, was in grave, imminent jeopardy.

Jay hesitated. He put great stock in following orders and maintaining discipline, especially in the midst of a chaotic fire response. Jay hadn't heard a command on the radio ordering them to evacuate, and he didn't want to break ranks. But Jay also knew that he couldn't rely on his radio, so maybe he'd missed a Mayday order to evacuate.

Jay made his decision. He walked over to his men, all still catching their breath: "It's time for us to go home." He didn't explain his reasoning, or even that the South Tower had fallen. At that moment, just after 10 a.m., Jay and his men were among roughly eight hundred people still inside the North Tower stairwells.

Nobody moved. The men of Ladder 6 stared back at Jay, teetering between confusion and defiance. As far as they were concerned, the South Tower still stood, civilians in the North Tower still needed rescuing, and they still had a job to do. Leaving wasn't an option. Moments like these were why they became firefighters. Jay didn't realize that his men hadn't heard his conversation with Billy Burke, so they didn't know the danger they faced. But he did know that he'd given them an order.

"What?" Jay demanded. "I said, 'Let's go!'"

The men of Ladder 6 rose, grim-faced and reluctant, to retrace their steps downward in Stairwell B, along with Andy "Nozzles" Fredericks and Billy Burke.

At some point in their evacuation, they lost touch with Andy. Jay thought he must have returned upstairs. Separately, Billy Burke radioed the men of his company to head for the exits—"Start your way down," he told them. "We'll meet at the rig." But Billy didn't join them or Ladder 6. He remained inside the North Tower to help two other men on the 27th floor.

Ed Beyea and Abe Zelmanowitz were close friends who worked as computer programmers for Empire Blue Cross Blue Shield. At forty-two, weighing nearly 300 pounds, Ed had relied on a wheelchair since a diving accident as a young man left him paralyzed. Abe was fifty-five, an Orthodox Jew who lived with his older brother and his family. Both bachelors, Ed and Abe exchanged DVDs, shared meals and musical tastes, and played computer games of chess and golf. Ed couldn't get down the stairs, and his pal Abe refused to leave him. Abe had already told Ed's aide to leave because she had children at home. Billy Burke made the pair his responsibility, and so Billy, Ed, and Abe became a trio.

CHAPTER 19

"REMEMBER THIS NAME"

The Pentagon

ON THE PENTAGON'S BURNING FIRST FLOOR, WORKING HIS WAY through the heat and smoke of the debris-filled Navy Command Center, Captain Dave Thomas shined his penlight at what looked like a rubber Halloween mask. He squinted through the gloom and saw that the squished face belonged to a man trapped in a chair, his head jammed against part of a wall and whatever else had fallen on him. It wasn't his best friend, Bob Dolan, it was a stranger: retired Navy aviator Jerry Henson.

"Hey, there's a guy in here!" Dave Thomas yelled. No one replied, so he screamed at the top of his lungs: "I've got a guy! Help!"

Dave Thomas moved closer. The tiny beam of light and his shouts sparked something in Jerry's assistant, Petty Officer Charles Lewis, who'd been trapped nearby. He'd thought there was no way out, but now he saw possibility for deliverance. Lewis scrambled and squirmed through burning pieces of the wrecked office, past Dave Thomas.

In the semidarkness, Lewis bumped into Dave Tarantino and SEAL Craig Powell, who directed him out to AE Drive. On his way toward safety, Lewis told them that others remained trapped inside.

Dave Tarantino and Craig Powell moved in and quickly freed Petty Officer Christine Williams from under a pile of rubble. She told them that her boss, Jerry Henson, remained stuck. As Powell led her outside, Dave Tarantino slogged deeper into the wreckage. He shined his flashlight toward movement and saw Dave Thomas struggling to raise the load of broken furniture and shelves to free Jerry from his desk entrapment. It wouldn't budge.

ON THE SECOND floor, civilian Pentagon worker Lois Stevens had given up. Exhausted, fighting for breath, she couldn't crawl another foot. Every time she opened her mouth she thought she'd spit fire. She curled into a ball, prayed, and thought of her husband. Smoke filled the room, drawing closer to the carpet. Lois urged Lieutenant Colonel Marilyn Wills to leave her behind, to let her die in the rubble of Dilbertville.

"Just get on my back!" Marilyn demanded. "Please. Get on my back. I'll crawl, and I'll carry you out of here!"

Lois remained in her fetal curl, but Marilyn refused to accept it. Even as she feared that they'd soon run out of air, Marilyn worried more about having to tell Lois's husband and grown children that she'd left behind their beloved wife and mother. Burning bits of ceiling tiles melted from their frames and fell like tiny meteors around them.

Marilyn beseeched Lois: "Come on. Please. We've got to get out of here!"

Lois relented. She rose to her knees and hung on as Marilyn resumed crawling like a blind mole, burrowing through debris, feeling her way through the dark.

While Marilyn was encouraging Lois not to surrender, Major Regina Grant disappeared. She'd gone off on her own, generally following the route she had seen badly burned security officer John Yates take when he vanished into the smoke. Still on her hands and knees, Regina kept her head down, focusing on her fingernails, as she scrambled toward an exit. Several times she found herself underneath a sprinkler. Each time, the smoke cleared enough for her to see a man's black shoes, moving slowly through piles of furniture and debris ahead of her. She assumed the shoes belonged to John Yates, so she kept crawling in that direction.

When the shoes disappeared, Regina Grant yelled, "Where are you?!"

Someone called back through the darkness: "Right here. Are you okay?"

It wasn't Yates, but Sergeant Major Tony Rose, an Army career counselor who'd been in the service since he graduated from high school twenty-nine years earlier.

"Follow the light!" Rose told her.

Regina tried, but soon she stopped. From somewhere behind her, she heard Marilyn Wills encouraging Lois Stevens not to give up. Ahead, Regina saw a dim glow, but the smoke and heat became too much. Spent, she stopped and said her prayers. Regina thought about her husband and about the military life insurance policy she'd recently increased from $10,000 to $150,000, with portions dedicated to her brothers and sisters, nieces, and nephews. Knowing that she'd be leaving something to her whole family gave Regina comfort. "I have done everything that I can do," she told herself.

As she prepared to die, Regina saw personnel administrator Tracy Webb trying to rise. The fireball that had torched John Yates as it blew through the room had badly burned Tracy's head. Initially, Tracy followed Lieutenant Colonel Bob Grunewald, who'd crawled across the room with Martha Carden clenching his belt, but Tracy lost sight of them. Now, like Regina, Tracy felt she'd gone as far as she could. She stood up, intending to fully inhale the smoke and accept her fate.

The sight of Tracy rising in the gloom gave Regina a burst of adrenaline. It felt like witnessing a child running toward traffic. Distracted from her own submission, Regina reached up and grabbed Tracy's skirt. She yanked Tracy to the ground and resumed yelling for help.

"Where are you?" Regina called.

"Come on! We're right here!" said Tony Rose.

Tony told Regina that she was close to a door. He urged her and

Tracy to keep moving. She followed his voice, bringing Tracy with her. On the other side of the door was Rose, who'd crawled back to lead them through the Fourth Corridor. He guided them through the building to the interior A Ring and down into the courtyard.

Separately, Bob Grunewald, who'd helped to design Dilbertville's technology layout, led Martha Carden on a serpentine crawling path.

"Don't you leave me here," Martha pleaded, even as she knew he wouldn't.

"I've got you, Martha," Bob said. "We're going to make it."

Until the sprinklers came on, Martha feared that her clothes would catch fire from the heat. Even after, she wondered how long it would take to die and how much it would hurt. Only when they reached the Fourth Corridor did she know she'd survive. After rescuing Martha, Bob Grunewald tried to return to search for other survivors, but the smoke drove him back, so he followed Martha to the A Ring and down into the courtyard.

Burned and disoriented, John Yates also somehow made it to the Fourth Corridor, where he collapsed. After a short time, he stirred and rose. He focused on the office space that he'd spent years helping to design, now ruined but still imprinted on his mind. Although still in shock, Yates realized he knew where he was. He started walking and ran into a lieutenant colonel, Victor Correa, who helped him to the courtyard to begin treatment for burns that covered nearly 40 percent of Yates's body.

Meanwhile, Marilyn Wills grew frantic when she realized that she'd lost track of Regina Grant. She felt for the major in the dark and yelled her name. She heard nothing, so Marilyn crawled forward, with Lois Stevens still hanging on to her for dear life.

ON THE FIRST floor, Dave Tarantino took a halting, zigzag route through the dark, on a narrow path lit by the glow of flames, half-crawling over and around debris. With fires closing in, he didn't want to get closer to where Dave Thomas tried to help free Jerry Henson. Dave Tarantino could hear the damaged building creak with complaint about its broken pillars. The roar of flames grew louder. The smoke felt so dense it seemed to make a sound all its own, and he thought he'd soon be overcome.

Jerry Henson's eyes were dull and glassy, but Dave Tarantino locked onto them.

"Come on, man. Get out of there!" Dave yelled.

Jerry said nothing.

Not realizing that Jerry was pinned, the Navy doctor issued a stream of orders familiar to anyone who'd been through basic training: "Get your ass out of there! Move your ass! Let's go! C'mon, you've got to come to us!"

Pale and unresponsive, struggling for breath, Jerry seemed about to slip into shock. Nearly out of breath, he mouthed two words: "Help me."

Dave Tarantino was out of options. A voice in his head screamed that he might never get out. But once they'd made eye contact, he knew he wouldn't leave Jerry behind. Dave Tarantino went to his knees, then crawled on his belly into the small space between Jerry and Dave Thomas, who continued his exertions. Dave Tarantino placed a wet T-shirt over Jerry's mouth to help him breathe.

"I'm a doctor, and I'm here to get you out," Tarantino said. "You're going to be okay."

Jerry roused from his torpor. He thought Dave Tarantino's promise was the sweetest thing he'd ever heard.

Then Dave Tarantino added a catch: "But you gotta fight. We can't do this alone."

"I'm pinned," Jerry murmured. "I can't move."

Dave Tarantino stood and tried to help Dave Thomas lift the load off Jerry. They removed what they could from the desktop. But even with both rescuers straining against it, the weight wouldn't give. Jerry was nearly a goner. If they didn't get out soon, Dave Thomas and Dave Tarantino knew that they would be, too. Each breath grew more strenuous.

The same was true for SEAL Craig Powell and General Paul Carlton, who'd fought their way back into the burning remnants of Pentagon room 1C466 to help. Carlton and several others formed a line, standing in ankle-deep water, passing out pieces of wrecked furniture and computers to clear a path. When smoke overcame one man at the front of the line, Carlton stepped up and took his place.

Powell stood at the far end of the ruined workspace, tossing burning pieces of debris from their path. He felt the sting of burning metal and looked up to see melting slag from wires dripping from above. The little office shared by Jerry Henson and Jack Punches had been a secure facility, encased by a wire cage to prevent electronic eavesdropping. Powell held up the hot, twisted remains of the cage. He warned Dave Thomas and Dave Tarantino to hurry: "Hey, ceiling's going to come down. Get out!"

Meanwhile, General Carlton tried using a fire extinguisher to knock down the flames all around them, but the water caused flare-ups, which he understood meant that they were fuel fires. The general saw smoke rising from Jerry Henson's clothing, so he doused him instead.

From behind him in the darkness, Dave Tarantino heard Powell shouting: "You gotta get out—now! It's going to go!"

With no time, little air, and no other way to free Jerry Henson, Dave Tarantino took one last shot. It was a throwback move, inspired by grueling days in his college weight room after his skydiving accident, when he rebuilt the atrophied muscles in his broken foot and leg. Stanford University's "Most Inspirational Oarsman" scooted under the desk, into a coffin-like space next to Jerry. His lungs aching, his heart pounding, Dave Tarantino flipped onto his back. He pressed the half-melted soles of his dress shoes against the underside of the desktop stretched across Jerry's chair.

With every last ounce of strength, Dave Tarantino pushed.

ON THE FIERY second floor, Marilyn Wills continued her crawl through the darkness toward the AE Drive windows with Lois Stevens, not knowing that Major Regina Grant had already found another way out. Even as she reassured Lois, Marilyn doubted herself.

"God," Marilyn thought, "I hope I'm going the right way."

She sensed that they were getting close, but she expected to see bright sunlight shining through the windows. Instead she saw more smoke. Marilyn began to pray: "God, please help me. I cannot see anything in here." Marilyn and Lois kept crawling, and Marilyn saw a dim glow, illuminating the outline of a young soldier banging on a window frame. She focused and saw that he was flanked by several other survivors from the conference room and nearby cubicles who had crawled in the same general direction, with the same escape plan. Marilyn had delivered Lois to the windows, as promised. But they still weren't safe.

Fighting to break open the window was Specialist Michael Petrovich. He ignored the pain of second-degree flash burns that had melted the skin on his nose and forehead and swelled his ears to twice their normal size. Despite his injuries, he'd led his friend and colleague, civilian Dalisay Olaes, from their cubicles through the wreckage of Dilbertville. Also at the windows were Lieutenant Colonel Marion Ward, who'd been in the conference room, and demographer Betty Maxfield, who'd been talking with Tracy Webb about her Avon scarf when their world exploded. For a short time after the blast, Betty had clung to Bob Grunewald's heel as he crawled ahead with Martha Carden. She'd lost her grip and become disoriented but bumped into Colonel Phil McNair, who led her to the windows.

To no one's surprise, Wedge One's new blast-resistant windows worked as designed: they hadn't shattered and become flying shrapnel. But they also wouldn't open. Michael Petrovich found one window bowed outward by the explosion, its metal frame bent a few inches out of whack. He grabbed a laser printer from a nearby desk and beat it against the window, to no avail. He threw it against the bent frame, only to have it bounce back into Marilyn's lap. Phil McNair joined Michael on the windowsill and stomped and kicked at the frame. The window held fast.

Marilyn feared they'd all die from smoke inhalation or the encroaching flames. Half in alarm and half in rage, she set Lois aside and climbed on a chair to reach the window. She slammed the frame with her hands and feet. With Marilyn and the two men beating it, the window frame pushed open about a foot and a half, the safety glass inside bowed but still intact.

As smoke poured out, the three stuck out their heads to gulp fresh air from AE Drive. Michael Petrovich tossed out the printer to attract help from other service members farther down AE Drive, who came running.

Marilyn turned to Lois: "Come on—you're going first."

Phil McNair and Michael Petrovich held Lois's wrists as they lowered her as far as they could from the window, a nearly twenty-foot drop. They released her safely into the arms of fellow Pentagon workers below. While they waited, Marilyn shared her moist sweater with rasping Betty Maxfield. Next out was Dalisay Olaes, who clung to the window sill, afraid, until Marilyn commanded: "Jump!" Dalisay broke her leg in the fall, but she understood that her choice was "jump or get toasted" and remained grateful to Marilyn. Next came Betty Maxfield. The commotion drew the attention of more sailors and soldiers in AE Drive who were helping with rescue efforts in the Navy Command Center. Several rushed over yelling "Come on down! We'll catch you!"

His facial burns festering, his lungs scarred by smoke, Michael Petrovich's throat began to close. Struggling to breathe, he went next. Marion Ward jumped out on his own, breaking his leg in the fall.

The last two, both of whom had saved others, were Colonel Phil McNair and Lieutenant Colonel Marilyn Wills. He told her to go, but Marilyn refused. She worried that Major Regina Grant was lost in the smoke, and she feared that Marian Serva and others were trapped inside. She told her boss that she wanted to return to Dilbertville.

"We have to go," Phil McNair said.

"Sir, we can't go," Marilyn answered. "Why don't we just stay here and let's yell and scream and people will come to the window. And we can get them out of here."

Ignoring the pain of their smoke-ravaged throats, both let loose:

"Come this way!"

"Come over here if you can hear us!"

"We can get you out of here!"

No one answered. No one came.

"You've got to go," the colonel told Marilyn.

"Sir," Marilyn said, her mind fixed on Regina Grant. "I'm trying to buy time because I'm thinking she's got to be right there. . . . Sir, let's be quiet. If you and I just be quiet right here, maybe we'll hear somebody crying or wailing and we can go back and get them."

Again, they heard nothing. McNair told Marilyn to wait while he looked for others. But the smoke was too thick, and the colonel quickly returned to the window. He shook his head, "No." As Marilyn broke down in tears, McNair gave her a direct order.

"You'll get out of the window—now!"

By then, the helpers in AE Drive had placed a painter's ladder atop a garbage bin outside the window, and two men had climbed atop the rickety pyramid. With Phil McNair right behind, Marilyn climbed out the window and scrambled onto the backs of the men who'd turned themselves into extension ladders.

"Just jump," someone called from below. "I'll catch you."

Marilyn followed the sound of the voice. Standing in AE Drive was the biggest man she'd ever seen: Commander Craig Powell, his muscular arms outstretched. He encouraged her again to jump. The slender Army lieutenant colonel looked down at the big Navy SEAL and let go.

FLAT ON HIS back in the stifling heat, feet against the desktop, Dave Tarantino pushed. His leg press raised the load of debris that trapped Jerry Henson by an inch, then two, then an inch more. Dave Thomas helped by squatting low like a weightlifter and pushing upward with all his might.

As Dave Tarantino gritted his teeth and held the weight aloft with his legs, he thought, "Well, now you've shifted this debris. What's going to happen when you let it down? Is it just going to keep coming and coming?" He'd worry about that soon enough. First, he needed to get Jerry free. He reached up, grabbed Jerry, and pulled him out of the chair, dragging the older man across his body. Jerry clawed at Dave Tarantino's arm, then his neck, reaching toward Dave Thomas, who pulled him the rest of the way out.

"Are you it?" Dave Tarantino asked, still holding the load with his legs. "Is there anyone else in here?"

"My buddy," Jerry rasped. "Jack Punches."

As Dave Thomas dragged Jerry out, Jerry's foot caught on a thick printer cable.

"Get your ass out!" Dave Tarantino yelled, his legs shaking as he still supported the desktop. "This thing's going to go. I can't hold it."

Dave Thomas pulled off Jerry's shoe and hauled him out toward AE Drive. They moved past SEAL Craig Powell, who'd entered the ruined Navy Command Center after catching several people from the second floor, including Marilyn Wills. He braced open the melting wire mesh in the ceiling to preserve their escape route.

Dave Tarantino eased down the load. He said a silent prayer of thanks when it held steady atop Jerry's empty chair. People were still calling for him and Powell to hurry out, but Dave Tarantino yelled back, "Be quiet! Shut up out there!"

General Carlton repeated the doctor's orders for silence, even as he grew anxious at the sight of flames spreading across the desktop above Dave Tarantino's head.

When the voices silenced, Dave Tarantino listened carefully. He heard nothing, so he shouted again for Jerry's friend Jack Punches: "Is there anyone else in here? Anyone else?"

More silence, except for the crackle of fire, the sparks and pops of live electrical wires, and the ominous creaks of a building losing its integrity. Sapped by the intense heat, Dave Tarantino rolled over and crawled out from under Jerry's desk. He moved past Powell, a tower of strength still holding up the passageway. General Carlton heard someone yell, "Come on, Big John!" The general chuckled at the reference to the old Jimmy Dean song, about a mountain of a man who held up the timbers of a collapsing mine to save his friends.

Dave Tarantino hustled outside with Carlton, Powell, and several others. They huffed like three-pack smokers, trying to catch their breath on AE Drive. Even before he reached safety, Dave Tarantino planned to go back inside to look for Jack Punches and other possible survivors. But just after he and the other searchers exited through the jagged hole in the C Ring wall, a boom sounded. Fire and smoke spouted through the hole like a geyser shooting sideways onto AE Drive. What remained of the first-floor supports collapsed, bringing down the upper floors and preventing further rescue efforts.

Jerry Henson was the last person rescued from inside the Pentagon.

AFTER CATCHING MARILYN WILLS, Craig Powell handed off the lieutenant colonel to another military first responder, who carried her to the Center Court. Her ordeal caught up with her. Even in the bright sunlight and clean air, Marilyn's smoke-filled lungs wouldn't work. She couldn't speak and could barely breathe. Scrapes and burns blistered the skin on her cheek, shoulder blades, knees, and shins. Her biceps muscles refused to relax. Her left shoulder locked into its socket.

Through the haze of pain, she heard the voices of people attending to her:

"Oh my God, she is not going to make it."

"Get some water."

"We don't have any water."

"Get some oxygen."

"We don't have any oxygen."

"We have to do something! She's dying."

Even as she neared losing consciousness, Marilyn refused to believe that. It might take a while, but she knew that somehow, she'd get home to Kirk, Portia, and Percilla.

Through her haze, Marilyn understood that she was being moved to a triage station in the north parking lot where she'd left her car only hours earlier, its CD player ready to resume blasting its gospel plea for divine protection. Then came a bumpy ride to Arlington Hospital in the back of an SUV. As they pulled away, Marilyn opened her bloodshot eyes long enough to see smoke pouring from where her office had been. At the hospital, she wrote Kirk's phone number on a doctor's sleeve. Then she passed out.

WHILE MARILYN WAS being taken to the hospital, civilian survivors Lois Stevens and Martha Carden fell into each other's arms in the Pentagon courtyard. Trying to wring humor out of tragedy, Martha grumbled that her lifesaving crawl with Lieutenant Colonel Bob Grunewald had ruined an especially good pair of pumps.

"Lois," Martha joked, "I want reimbursement for my damn shoes."

Lois answered: "Well, I'd *like* some damn shoes."

Martha looked down and saw Lois's stockinged feet.

"Okay," she said, "never mind."

Soon Bob Grunewald joined Martha and several others from their office. They huddled close on a bench in the courtyard, grateful to be together.

After helping to rescue colleagues including Major Regina Grant, Tracy Webb, and Betty Maxfield, in separate escape routes from the Army personnel office, Sergeant Major Tony Rose and Colonel Phil McNair joined other military first responders in AE Drive, pulling several survivors through the punched-out holes on the first floor. From there, McNair went to the courtyard, where he saw a group of medics huddled over a soot-covered figure on the ground. He overheard someone say a name: "Yates."

John Yates, the Army personnel office security chief, lay on the courtyard grass as medics cut off his pants. His face was charred, his hair burned off. John looked down at his ghostly white hands and saw skin hanging off them like shrouds. Phil McNair walked over with words of encouragement. John heard a female doctor say, "He needs to get out of here—and needs to get out of here *now*."

BACK IN AE DRIVE, Dave Tarantino found his crumpled uniform shirt on the ground and tugged it over his blackened T-shirt. Coughing from smoke inhalation, his eyes burning, his back and leg muscles aching, he walked to the Pentagon's Center Court. He found Dave Thomas standing by a stretcher that bore the bloody, battered, but very much alive Jerry Henson, waiting for an ambulance out.

It was shortly after ten. About a half hour had passed since Flight 77 struck the Pentagon. In that brief time, the South Tower had crumbled, the evacuation of the North Tower continued, the horror of people falling or jumping to the World Trade Center plaza had worsened, and the heroes of Flight 93 had fought their final battle.

Dave Tarantino watched as a nervous medic tried to start an intravenous line in Jerry Henson's arm. Dave thought about taking over, then decided to give the young man a chance. "You got this," Dave gently told the medic. "Just focus and get this done." On the next try, the IV line hit its mark.

As the medic cared for Jerry, Dave Thomas introduced himself to Dave Tarantino. During their rescue efforts, there hadn't been time for the niceties of names. As they shook hands, Dave Thomas looked at Dave Tarantino with something approaching awe. Dave Thomas knew that he entered the burned-out Navy Command Center to find his best friend, Bob Dolan. He wondered what drove Dave Tarantino into that brick oven to crawl through jagged rubble, flip onto his back, and leg-press a load of burning debris, knowing that it might crash down on top of him. Dave Thomas decided that he'd never seen a more courageous act. But it worried him—he feared that Dave Tarantino might be selfless enough to return to the burning building, and the next time he might not get out.

Almost without thinking, Dave Thomas broke from their handshake and reached toward his new friend's chest. Before Dave Tarantino knew what was happening, Dave Thomas snatched the name tag—TARANTINO, STAFF PHYSICIAN—off his tattered shirt. Dave Thomas stashed the tag deep in his pocket for safekeeping. If Dave Tarantino didn't survive the day, Dave Thomas wanted to be certain that he remembered the man's name. He'd tell Dave Tarantino's family and anyone else who'd listen about the young Navy doctor's heroism.

As medics carried away Jerry's stretcher, Dave Thomas called out to the wounded old flier: "Remember this name—Tarantino. That's who saved you!"

DURING THE FIRST half hour after Flight 77 smashed through the Pentagon, scores of life-and-death events and countless acts of heroism, sacrifice, and kindness played out simultaneously on the fire-scorched floors of Wedge One and nearby. Some ended in triumph, some in heartache, some in both.

In an Army office on the first floor, the blast threw Captain Darrell Oliver against a wall, opening a gash above his left eye and briefly knocking him unconscious. When he awoke, Oliver felt certain he wouldn't make it out alive. He felt enraged that he hadn't yet taught his two young daughters all they needed to know about life. Walls crumpled, furniture lay strewn about, and pieces of ceiling rained down on him. Partitions tilted at 45-degree angles, separating him and others in his office from a possible path to safety. A secretary who'd been blown from one office into the next grew frantic. Oliver and another officer dug her out from under debris.

"We're not going to get out of here!" the secretary yelled. "We're going to die in here!" Oliver tried to reassure her, but she wouldn't listen, so he stopped arguing and ordered her to climb onto his back. He scrambled over one collapsed wall, then another. Rather than go outside with the secretary, Oliver handed her off to another officer. He returned to help a janitor who'd curled into the fetal position and sobbed in fear. Oliver knew the man had a severe hearing impairment; every day when the janitor came to empty the trash, Oliver rose from his chair, shook his hand, and chatted with him. Days earlier, the man told Oliver that he'd lost a hearing aid. Now he was confused and frightened, unable to follow shouted instructions from other officers.

As the office filled with smoke, Oliver put the man on his back and again climbed over the two fallen partitions, inches below live electrical wires that hung from the torn-open ceiling. Once outside, Oliver joined a team of volunteers who carried a severely burned woman from the building. The secretary, the janitor, and the burned woman all survived, as did Oliver.

ON THE FIRST floor, Navy Lieutenant Kevin Shaeffer rolled on the floor and ran his hands over his face to extinguish flames from the fireball that had blown through the Navy Command Center and killed most of his officemates. Surprised to be alive, he yelled encouragement to himself: "Keep moving, Kevin! Keep moving!" He clawed through rubble and under dangling electrical wires as burned skin sloughed off his hands and arms.

Shaeffer rose when he saw a dim light obscured by smoke, then followed it to one of the punched-out holes leading to AE Drive. He told himself that he looked like the naked child burned by napalm in an iconic Vietnam War photograph: "You're as helpless as that little girl, Kevin." He made it outside and into the care of Army Sergeant First Class Steve Workman, who shepherded him to a hospital. As doctors prepared to cut off his wedding and Annapolis class rings, Kevin pried them off, then passed out.

AFTER THE PLANE flew past, Father Stephen McGraw abandoned his car in the traffic jam next to the Pentagon. The newly ordained priest ran toward the carnage with his prayer book, his purple stole, and his blessed olive oil. He vaulted over a guardrail and sprinted to the grass that flanked the building's ruined west face. At first, Father McGraw remained away from the action, scared by explosions, worried about trespassing on military property, unsure if he'd know what to do if he encountered anyone who needed help.

Within minutes, Father McGraw saw medics and military volunteers carrying injured people to the soft green grass surrounding the Pentagon. To his surprise, everyone treated him like a frontline chaplain, as though it would have been odd if a priest hadn't been waiting for them. Responders pointed and shouted: "Father, there's someone over there who needs you!"

Father McGraw rushed to Juan Cruz-Santiago, a civilian accounting worker for the Army who was burned over more than 60 percent of his body. Survival seemed doubtful. Juan couldn't see and was in no condition to confess his sins, but he told Father McGraw he was Catholic. The priest anointed Juan with oil and granted him a battlefield absolution, whispering, "May the Lord who frees you from sin save you and raise you up."

Father McGraw rushed to a woman and fell to his knees—to pray, but also because he buckled at the sight of her injuries. Caught in a fireball, the woman's clothes had melted to her skin. She told him her name was Antoinette. Placed facedown on the ground, to avoid aggravating the angry burns on her back, she said her one remaining shoe was causing her great pain. Gently, Father McGraw pulled it off. She made one more request before being placed on a helicopter: "Tell my mother and father I love them."

ON THE SECOND floor, initial news of the New York attacks sent research officer Major John Thurman to the website of the *Washington Post*. Thurman was thirty-four, a veteran of the Persian Gulf War who'd been a military police platoon leader. As he watched a replay of United Flight 175 hitting the South Tower, he felt a whoosh and heard a crunch.

The shock wave tossed his chair backward against his cubicle partition. He dived under his desk as the ceiling collapsed and the fluorescent lights went dark and then crashed to the floor. Flames shot overhead, then raced down the wall of the windowless office. Lockers and filing cabinets crashed down. As smoke filled the room, Thurman suspected that a treasonous construction worker had detonated a bomb in sync with the World Trade Center crashes.

"Who else is here?" he yelled as he crawled under the smoke.

Thurman heard muffled yells nearby. He pushed a fallen file cabinet out of the way and clambered toward the voices. Ten feet away, he found Lieutenant Colonel Karen Wagner, a forty-year-old medical personnel officer who had played basketball at the University of Nevada, Las Vegas. A third-generation soldier, Karen Wagner had an effervescent energy and an unquenchable appetite for hard work. Nearby lay fifty-seven-year-old Chief Warrant Officer William Ruth, who flew helicopters to evacuate the dead and wounded in Vietnam, and who returned to battle in the Persian Gulf War. Both were hurt, Bill Ruth worse than Karen Wagner.

John Thurman pulled each from under debris into what remained of his cubicle. As they huddled on the carpet, the room heated up and the smoke thickened. Overhead sprinklers trickled a weak stream of water onto them. Bill Ruth mumbled incoherently. In the dark, Thurman couldn't see the extent of their injuries, but he knew they needed to move.

"Okay," Thurman said, "we've got to stay down. We have to get out of here *now!*"

John Thurman crawled toward a door with Karen Wagner hanging on to his belt. When he reached the door, tilted off its frame with a broken hinge, Thurman thought back to schoolboy fire drills. He tested the hallway by sticking his hand through an opening at the bottom, then snatched it back from the searing heat. They retreated to Thurman's desk. Bill Ruth stopped speaking. Karen Wagner began to hyperventilate.

John Thurman concluded he was going to die. Lying face-down on the carpet, he felt an overwhelming need for a nap. Then, gripped by fury, he thought of his parents—his father had emailed him that morning to say that John's pregnant younger sister had gone into labor. He realized that his parents might lose their eldest child on the same day they became first-time grandparents. With that, he pushed up from the carpet.

"We've got to get to the back door!" Thurman yelled.

He pulled Karen Wagner along with him, his head inches off the carpet, feeling his way through the dark. He rose to his knees, but the heat drove him back down. He pushed overturned furniture out of the way.

"Karen, come on," he called. "Karen, follow me."

She didn't answer. He hoped she was somewhere close behind him, following his voice. Thurman crawled to the opposite side of the smoky room. He looked up and saw the faint red glow of an Exit sign. He pushed through the door to a corridor that led to a stairwell where the lights still worked. Choking and gasping, Thurman removed a shoe and used it to prop open the door for Karen Wagner and, he hoped, Bill Ruth.

Thurman hobbled down the stairs and ran into his boss, Colonel Karl Knoblauch, along with Lieutenant Colonel William McKinnon and a half dozen other rescuers scouring the building for stragglers. The impromptu team had already rescued several people, including a man they found badly burned and bleeding in the Fourth Corridor. Only after they began carrying the soot-caked man did McKinnon notice his name tag; he hadn't recognized his classmate and friend, Lieutenant Colonel Brian Birdwell.

Thurman told them about the others still trapped in the office, so they helped him back upstairs to the propped-open door. Black smoke poured out, thick and hot.

"We have to go back in!" Thurman pleaded. "I know where they're at—we can get them."

"We can't go in," Knoblauch said.

The colonel and his rescue crew brought Thurman to the courtyard. Shivering, struggling to breathe, he repeated to anyone who'd listen, "Karen and Bill are in the room. Karen and Bill are in the room. We've got to go in and get them." As Thurman spiraled into shock, medics took him to the north parking lot, then to a hospital.

FROM ACROSS THE street at Arlington National Cemetery, facilities manager George Aman surveyed the scene as the Pentagon burned. When he heard rumors of another inbound plane, he and several coworkers piled into a pickup truck. The old soldier in George took over. He sped up Patton Drive to higher ground, a hillside in Section Eight of the cemetery, to stand guard for an unseen enemy.

Another hijacked plane never came, but multiple reports of incoming aircraft sent Pentagon survivors scurrying for cover under nearby overpasses and slowed efforts by firefighters. The exodus sent thousands of people from the Pentagon and other nearby buildings rushing toward George Aman and his crew. After leaving the Pentagon and the nearby Navy Annex, men and women, some in uniform, some not, cut through the cemetery on their way toward safety, or home, or wherever they needed to be.

George watched as the evacuees marched stone-faced among the tombstones of the nation's war dead, where in the days and months ahead some of the victims of the Flight 77 hijacking would be laid to rest.

THE FIRST FIREFIGHTERS from the Arlington County Fire Department arrived at the Pentagon less than five minutes after the crash of Flight 77. With reinforcements, and without pause, they fought the blaze for the next thirty-six hours. Arlington ambulance teams treated the wounded on the Pentagon lawn. Several crews of Arlington firefighters rushed into the flames and led Pentagon workers to safety. Inside, rescue workers found a scene they described as "huge heaps of rubble and burning debris littered with the bodies and body parts of . . . victims [that] covered an area the size of a modern shopping mall."

From the moment they began blasting the blaze with water and foam, firefighters heard the building creak and moan. Around 10:15 a.m., the second through fifth floors of the E Ring of Wedge One collapsed. The crumpled area extended about ninety-five feet in width and fifty feet in depth, a small fraction of the enormous building physically, but a huge blow symbolically.

Determined to demonstrate that the assault on the American military's headquarters was by no means fatal, Defense Secretary Donald

Rumsfeld invited reporters to a dinnertime press briefing in an unaffected area of the still burning building.

"The Pentagon is functioning," Rumsfeld told them. "It will be in business tomorrow." Standing alongside him were General Hugh Shelton, chairman of the Joint Chiefs of Staff, and the top Democrat and ranking Republican on the Senate Armed Services Committee. Shelton called the attacks "barbaric terrorism carried out by fanatics" and vowed that they would be answered with overwhelming force: "[M]ake no mistake about it, your armed forces are ready."

AFTER THE AMBULANCE drove off with Jerry Henson, Dave Tarantino hoped to help more survivors. But when the affected portion of Wedge One collapsed, it became clear that no one else would be brought out alive. He returned to the central courtyard to treat injured survivors awaiting transport.

More than once, Dave Tarantino thrilled at the sight of F-16s flying overhead, each fighter jet reassuring everyone there that there'd be no more inbound hijacked aircraft that day.

When he felt certain that there'd be no more need for his medical skills, Dave Tarantino called his wife and parents to say he was okay. He walked around to the building's west face, weaving through fellow service men and women, firefighters, emergency officials, and the occasional priest, to get his first look at the crumpled walls where Flight 77 terminated.

Dave's mind flashed to the moment he first saw Jerry Henson, trapped and helpless. A thought formed, one he'd refine and rephrase but never forget: "Someone tried to kill us, to kill me. Someone tried to kill all of us, out of blind ideological hatred, in the most brutal way. They tried to kill us by hurtling Americans at us."

A day earlier, busy with Pentagon team soccer practice and routine emails and reports, Dave Tarantino had felt torn over whether to leave the Navy and settle into a family medical practice. Now he knew: he'd continue to serve.

As smoke swirled into the sky, he turned away from the burning

Pentagon. Lieutenant Commander David Tarantino, MD, hurt, sore, pungent as an ashcan, limped several blocks to a Metro rail station. He paid the fare and boarded a train toward home. As he reflected on all that he'd seen and done, Dave noticed a woman staring at him from a few seats away. She studied his scrapes and bruises, the burns on his hands. Her gaze worked its way down his torn, stained uniform to his ruined shoes.

The woman looked up, into Dave's bloodshot eyes, and burst into tears.

CHAPTER 20

"THIS IS *YOUR* PLANE CRASH"

Shanksville, Pennsylvania

JUST AFTER 9 A.M., INSIDE HER HILLTOP HOUSE IN RURAL STOYSTOWN, Pennsylvania, homemaker Linda Shepley watched her television in shock. The screen showed smoke billowing from a gash in the North Tower as *Today* show anchor Katie Couric interviewed an NBC producer who witnessed the crash of American Flight 11.

"You say that emergency vehicles are there?" Couric asked Elliott Walker by phone.

"Oh, my goodness!" Walker cried at 9:03 a.m. "Ah! Another one just hit!"

Linda watched the terror in her living room beside her husband, Jim, a Pennsylvania Department of Transportation manager, who'd taken the day off to trade in their old car. The Shepleys saw a grim-faced President Bush speak to the nation from Booker Elementary School in Sarasota, Florida. Then Couric interviewed a terrorism expert but interrupted him for a phone call with NBC military correspondent Jim Miklaszewski, who declared at 9:39 a.m., "Katie, I don't want to alarm anybody right now, but apparently, it felt just a few moments ago like there was an explosion of some kind here at the Pentagon."

From the home where they'd lived for nearly three decades, the Shepleys could have driven to Washington in time for lunch or to New York City for an afternoon movie. Yet as the political and financial capitals reeled, those big cities felt almost as far away as the caves of Afghanistan. Jim went to the garage, to clean out the car he still planned to trade in that day. Linda hurried to finish the laundry before she accompanied Jim to the dealership.

Forty-seven years old, with kind eyes and three grown sons, Linda loved the smell of clothes freshly dried by the crisp Allegheny mountain air. As ten o'clock approached, she filled a basket with wet laundry and carried it to the clothesline in her backyard, two grassy acres with unbroken views over rolling hills that stretched southeast toward the neighboring borough of Shanksville. As Linda lifted a wet T-shirt toward the line, she heard a loud *thump-thump* sound behind her, like a truck rumbling over a bridge. Startled, she glanced over her left shoulder and saw a large commercial passenger plane, its wings wobbling, rocking left and right, flying much too low in the bright blue sky.

As the plane passed overhead at high speed, Linda saw the jet was intact, with neither smoke nor flame coming from either engine. Linda made no connection between the plane's strange behavior and the news she'd watched minutes earlier about hijacked airliners crashing into the World Trade Center and the Pentagon. Instead, she suspected that a mechanical problem had forced the plane low and wobbly, on a flight path over her house that she'd never before witnessed. Maybe, Linda thought, the pilot was signaling distress and searching for someplace to make an emergency landing. Linda worried that their local airstrip, Somerset County Airport, was far too small to handle such a big plane. And if that *was* the pilot's destination, she thought, he or she was heading the wrong way.

Linda didn't know the plane was United Flight 93, and she couldn't imagine that minutes earlier the passengers and crew had taken a vote to fight back. Or that CeeCee Lyles, Jeremy Glick, Todd Beamer, Sandy Bradshaw, and others on board had shared that decision during emotional phone calls, or that the revolt was reaching its peak, or that the four hijackers had resolved to crash the plane short of their target to prevent the hostages from retaking control.

Linda tracked the jet as sunlight glinted off its metal skin. Its erratic flight pattern continued. The right wing dipped farther and farther. The left wing rose higher, until the plane was almost perpendicular with the earth, like a catamaran in high winds. Linda saw it start to turn and roll, flipping nearly upside down. Then the plane plunged, nosediving beyond a stand of hemlocks two miles from where Linda stood. As quickly as the jet disappeared, an orange fireball blossomed, accompanied by a thick mushroom cloud of dark smoke.

"Jim!" Linda screamed. "Call 9-1-1!"

Her husband burst outside, fearing that their neighbor's Rottweiler mix had broken loose from its chain and attacked her.

"A big plane just crashed!" Linda yelled.

"A small plane," Jim said skeptically, as he regained his bearings.

"No, no, no, no. It was a big one. It was a big one! I saw the engines on the wings."

Jim rushed inside and grabbed a phone.

Heartsick, still clutching the wet T-shirt, Linda stared toward the rising smoke. Soon she'd wonder whether, in the last seconds before the crash, any of the men and women on board saw her hanging laundry on this glorious late-summer day.

THE FIRST 9-1-1 call about United Flight 93 came from inside the plane, when computer engineer Ed Felt used his cellphone from a rear restroom to report "Hijacking in progress." Felt's call reached dispatchers in Westmoreland County, adjacent to Somerset County, where the crash occurred. Jim Shepley got a busy signal on his first several tries, as other witnesses around Shanksville, Stoystown, and the nearby villages of Lambertsville and Buckstown jammed emergency lines.

The first 9-1-1 call from the ground that connected came from hairdresser Paula Pluta, who'd heard nothing about the first three hijackings and crashes. Having taken the day off from work, Paula was lost in a simpler time, watching a rerun of the old television show *Little House on the Prairie*, when the low-flying plane rattled her house and everything in it. She ran onto her porch, which overlooked the grounds of the old Diamond T coal mine, where men and machines had spent three decades extracting the soft black wealth buried there. Paula spotted the plane racing downward at a sharp angle. It disappeared from her sight behind fifty-foot-tall trees a fraction of a second before it exploded, roughly a half mile from her home.

"Oh my God!" she told Somerset County 9-1-1 dispatcher Jeremy Coughenour. "There was an airplane crash here by Shanksville-Stonycreek School." Although Paula had lived in the county her whole life, the school with five hundred children was in fact four miles in a different direction. The error could be explained as a product of distress, or as the distance between what her eyes saw and her heart feared. Paula Pluta's son and daughter were in school that morning.

Moments later, another 9-1-1 caller provided a more accurate location. Bulldozer driver Daniel Meyers was part of an excavation crew that had spent several years returning soil and grass to the grounds of the old strip mine, restoring it to rolling fields of green. Breathing hard, Meyers told the dispatcher, "There was an airplane just went down over by Diamond T!"

Coughenour knew the mine's location, off Lambertsville Road and Skyline Drive, so he moved to answer another call. Before hanging up, Meyers added another detail: "Well, I don't know if you have to tell anybody, but it went down nose first, upside down."

By the time Jim Shepley's call got through—with Linda in the background whimpering "Oh my God"—fire horns and whistles had sounded. Volunteer firefighters and ambulance crews rushed to their vehicles to race toward the scene from Shanksville, Stoystown, and neighboring communities. Citizen responders poured in from near and far, drawn by the smoke and in some cases by glimpses of the falling plane. Soon they'd be followed by scores of local, state, and federal officials, among them state police troopers, the FBI, and the county coroner, funeral home owner Wally Miller.

AT THE FAMILY medical practice where Kathie Shaffer worked as a registered nurse, word spread that Somerset County Hospital had sounded an emergency code for disaster. Doctors would remain at the hospital, regular appointments would be canceled, and Kathie and the rest of the staff would keep the office open to help anyone who wandered in upset or confused.

During the first minutes after the 10:03 a.m. crash, details were scant. The medical office secretary tuned her transistor radio to a staticky news station and told Kathie something had happened involving a plane and the World Trade Center in New York City. Kathie drew a blank, having never been to Manhattan. Her mind formed an image of a generic skyscraper on fire. If she pictured the Twin Towers, the image would have been obsolete—the South Tower had collapsed minutes earlier.

As Kathie pieced together information, her sister, Donna Glessner, called the medical office to say she thought she felt an earthquake. Kathie returned to work, but Donna, who worked at her family's hardware store and lived a mile outside Shanksville, soon called a second time: it wasn't an earthquake, it was a plane crash, apparently a few miles away, in little Lambertsville, one of the communities served by the Shanksville Volunteer Fire Department.

"You need to call Terry," Donna insisted.

Kathie found the number of her husband's costly new cellphone, the contentious device they'd bickered about for weeks. As she dialed the number, Kathie accepted that Terry had been right: the phone wasn't an indulgence, it would help him as the fire chief.

"Another plane has crashed, and they think it's near Lambertsville," Kathie told Terry, raising her voice over the din of machinery at the Pepsi plant where he loaded trucks.

"It's no time to joke like that—it's not funny," said Terry, who'd heard reports about the events in New York from a television in a break room.

"I'm serious," Kathie said. "This is *your* plane crash."

Almost simultaneously, Kathie heard Terry's name called on the plant loudspeaker: he and all other volunteer emergency responders were released from work. Terry rushed outside, called the county 9-1-1 center for confirmation, then jumped into his minivan. His mind whirled as he sped toward home. "Okay, who's in town? Is Rick there?" Terry wondered about his assistant chief, Rick King, owner of Ida's Store, located a hundred yards from the fire station. One question led to another: "How many trucks went? What are they encountering? How are they handling it? Where can we get water if we need it?"

The fire radio on his front seat was nearly useless. So many calls filled the airwaves that Terry heard mostly the garble of responders and dispatchers talking over one another. He tried to envision the scene, to assess the possibility of survivors and determine the size and intensity of the fire, but the images wouldn't come. Terry had always expected that his rural department's big test would occur on the Pennsylvania Turnpike, from a truck rollover or a multicar pileup, not from a passenger plane crashing into an exhausted coal mine.

As Terry broke the speed limit along the winding country roads, an unfamiliar feeling took hold. After two dozen years as a volunteer firefighter, including fourteen as Shanksville's chief, Terry had honed a professional calm befitting his training, experience, and leadership position. Now, though, he felt frightened, scared by the thought of what he and his fellow responders would encounter at the Diamond T.

RICK KING HAD spent the morning shuttling the two hundred feet back and forth between his house and his country store, watching television reports about the New York and Pentagon crashes between paying bills and talking by phone with family members. Compact and affable, a volunteer firefighter for twenty-one years and assistant chief for two, Rick called his sister, Jody King Walsh, who lived in Lambertsville. Her children were watching *Barney & Friends* on television, and she knew nothing about the hijackings. As they spoke, Jody said: "Rick, I hear a plane." He dismissed her at first, but Jody was adamant.

Still holding the phone, Rick went out to his front porch, facing to the northwest of Shanksville. He heard it, too, a screaming jet engine. The porch boards rumbled beneath his feet. Rick couldn't see the plane, but he heard the crash and saw the fireball.

"Oh my God, Rick! It crashed!" Jody said.

"I know. I've got to go."

Rick threw the phone inside and sprinted back to his store. No one there had heard the explosion. A customer thought he was joking when Rick described what had just happened. He bolted back outside and ran a hundred yards uphill to the fire station to answer the dispatch call from Somerset County's 9-1-1 center.

"What stations are dispatched?" Rick asked.

The dispatcher said only Shanksville, Stoystown, and a nearby village called Friedens, then asked: "Do you wish any additional fire units?"

"Affirmative, Somerset," said Rick, who connected the dots to New York and the Pentagon as quickly as anyone. "This is a large jetliner," he said, breathing hard, "probably related to what's going on." He asked for three more departments. Soon he'd ask for more.

Rick geared up and leapt into the driver's seat of "Big Mo," a 1992 fire engine that carried a thousand gallons of water, accompanied by three other Shanksville volunteers: Keith Custer, Merle Flick, and Robert Kelly. As they pulled out of the station, Rick called his wife, Tricia. Fearing that more planes would fall from the sky, he told her: "Go get our kids out of school. . . . I don't know what's happening." Anticipating mass casualties, Rick called the dispatcher again and requested every available emergency unit in Somerset and Cambria Counties.

The four Shanksville firefighters barely spoke as the siren wailed and they wondered what they'd find. Keith said his mouth was so dry he couldn't swallow; Rick realized that he couldn't swallow either. As they approached the crash site, Keith spoke for them all: "Guys, prepare yourself. Because this is something we've never seen before."

WATCHING THE TWIN TOWERS burn on television in the smoking lounge of his Somerset funeral home, Wally Miller turned to his father, Wilbur, who'd preceded Wally as the elected county coroner.

"Hey," Wally asked, "how'd you like to be the coroner in New York City today?"

Wilbur and his wife, Wilma, had founded the Miller Funeral Home in 1953, four years before Wally's birth. Lanky, six foot four, with a passing resemblance to Abraham Lincoln, Wally joined the family business in 1980, after college and a year of mortuary science school. He took over the funeral home in 1995 and won election as coroner two years later, after his father retired. The county job busied him with three or four death certificates a week, mostly for natural causes, plus occasional accidents, drug overdoses, and suicides. Wally determined cause and manner of death, notified next of kin, and shepherded remains to the morgue or a funeral home, his own or a competitor's. His professional life was the sensitive, orderly business and bureaucracy of death and grief, except in the rare cases when it was anything but orderly.

Three weeks earlier, Wally had met members of the Shanksville Fire Department on a case he felt certain would be among the most memorable of his career. A car had struck a deer, sending it airborne and into the path of an oncoming pickup truck. The deer had sailed through the truck's windshield, passing entirely through the passenger compartment and out the rear window to its death. The pickup driver suffered only minor injuries, but as the deer flew through the cab, one of its hooves nearly decapitated a forty-eight-year-old man in the passenger seat. As Wally sized up the scene, he asked the Shanksville fire crew: "Could you ever think of something more bizarre happening than this?"

Wally left his father watching the televised destruction and went into the funeral home's office to catch up on paperwork. Around 10:30 a.m., he answered a call from a secretary in the coroner's office in neighboring Cambria County. She told him their office was available to help however he needed with the plane crash that just took place in Somerset County, an event that Wally knew nothing about. Wally did, however, know that most of his elected colleagues were attending an annual conference of the Pennsylvania State Coroner's

Association. He thought the secretary's call was his fellow coroners' idea of gallows humor. The secretary insisted it was no joke, so Wally called the county 9-1-1 office.

"It's possibly a hijacked aircraft, and it's possibly a 747, although they don't have a lot of verifications yet," the dispatcher told him.

"You gotta be kidding me," Wally replied.

"No, no, Wally. It's just . . ."

"You gotta be kidding me," Wally repeated.

"No, Wally."

The dispatcher handed the phone to Rick Lohr, director of the county's Emergency Management Agency. "Wally," Lohr said, "if I were you, I'd find a place to set up a temporary morgue somewhere."

DRESSED IN FIREFIGHTING gear, steeling himself, Terry Shaffer parked his minivan as close as he could to the crash site, near the end of a dirt road formerly used by trucks to haul away coal. He lumbered ahead on foot, searching for Rick King or anyone else from Shanksville who could get him up to speed. Along the way, Terry saw about fifty people, among them state troopers, other emergency responders, and civilians who'd flocked to the catastrophe. The strangers seemed to float by him, almost in slow motion, their faces pale and expressionless. Terry thought they looked like ghosts.

About forty minutes after the crash of Flight 93, Terry reached his destination, a place al-Qaeda certainly never knew existed: a rolling field of grass the color of dried moss, roughly the size of New York City's Central Park, with a dense grove of hemlock trees at its far edge. On a hilltop in the distance, the field's most prominent features were two rusting red-and-white machines, one as big as a bungalow, the other the size of a minibus. Both had cranelike booms that extended one hundred feet into the air, with steel cables dangling like jungle vines. Called draglines, the machines had spent decades pulling toothed metal buckets across the ground to strip away rock and soil and expose countless tons of coal. They were left behind when the mine played out, like rotted schooners on a dry seabed. The drag-lines were vaguely the size of the commercial airliner that Terry had expected to see crumpled before him.

Instead, all Terry initially saw was a smoking crater. Sixty yards be-yond it, he saw several acres of burned or smoldering hemlocks. From the ground, Terry couldn't see that the plane's wings had left black-ened impressions on either side of the crater, evidence that would soon be filmed by a police helicopter hovering over the site.

Firefighters sprayed water into the crater, onto a smoking airplane tire, and onto the trees, as small fires sparked and flared on evergreen branches. Everywhere Terry looked on the ground, spread like dan-delions as far as he could see, were ragged bits of honeycomb insula-tion, twisted metal, torn wire, crumpled paper, and shredded fabric. As his nose absorbed the sickly-sweet stench of burned flesh and jet fuel, Terry understood that if he looked closely, he'd find more grue-some evidence hidden in the grass.

During his drive, Terry had imagined a mass casualty scene strewn with bodies, limbs, and other large remains. Maybe, just maybe, some survivors. Now, he understood they'd find nothing of the sort. Terry would soon learn that United Flight 93 had struck the ground at a 40-degree angle, nearly upside down, its nose and right wing hitting first. Its fuel-filled wings detonated on impact, sending up the fireball and the plume of smoke. The cockpit and front section blasted for-ward into the hemlocks. The remaining fuselage burrowed into the soft earth and burst to pieces. The explosion annihilated everything, including the forty men and women who fought to save themselves and the four terrorists who resolved to kill them all. Except for a few large pieces of metal, engine parts, and the burning tire near the cra-ter, all that remained of the Boeing 757 were millions of charred frag-ments.

Terry walked past the crater, which stretched some thirty feet across and more than fifteen feet deep. There he found Rick, whose four-man Shanksville crew was the first firefighting team to reach the site. Rick had parked Big Mo near the edge of the crater. He and the others had leapt from the engine and took part in the brief, fruitless search for survivors. Initially the ranking fire official on the scene, Rick had felt besieged by people asking "What do you want us to do?" Now Rick looked wan. Terry thought he might be in shock. Terry put his hand on Rick's shoulder: "Are you okay?" Rick wasn't, at least not yet, but a wave of relief crashed over him as he realized that he could hand off control to the chief.

"This is really big," Rick said.

"This is huge," Terry agreed. "This is international."

They would repeat versions of those phrases to each other for the rest of the day.

No matter how big it was, they had only small fires to fight and no one to rescue. There wasn't much for them to do, not until they could take part in the painstaking work of collecting evidence, organic and otherwise. The United Flight 93 crash site was a crime scene, to be supervised by the FBI, and a monumental challenge for a small-town coroner.

WALLY MILLER SHOWED up in khaki pants and gum boots. He knew the Diamond T site, having fished nearby as a boy. Back when the mine still operated, coal workers excavated old fieldstones carved with the names of local settlers from a farm cemetery. The bodies once interred there had long since turned to dust, but the coal company worked with Wally, his father, and a local historian to find a new location for the headstones, casket handles, buttons, and other items that remained.

Wally felt as surprised by the crash site as Rick, Terry, and other emergency responders, but unlike them he understood immediately that he had months, possibly years, of work ahead. He needed to preserve the integrity of the site, to oversee the recovery and identification of remains, and equally important, to provide information and comfort to the next of kin. Wally understood that these tasks would test his core belief—"Everybody that dies, that's somebody's favorite guy, whether it's a prisoner or the richest guy in town or somebody else"—on a massive scale.

As Wally took his first steps around the field, he recalled the crash of USAir Flight 427, which seven years earlier spiraled into the ground near the Greater Pittsburgh International Airport, killing 132 people. He thought about the reports he'd heard describing a gruesome scene of charred remains, followed by years of acrimony that victims' families directed at the county coroner who had handled it.

As Wally surveyed the scene and walked alongside Rick King among the hemlocks, he heard the *pssss, pssss* sounds of melting plastic dripping from the trees and sizzling when it hit the ground. Someone tossed out an inflated estimate of how many people might have been aboard Flight 93, a number nearly ten times too large. The wild guess rattled Wally, especially considering how little he saw in the way of human remains. "If there's four hundred people on this plane," Wally thought, "we've got a serious problem."

From the very first, he resolved to identify every last passenger and crew member who died. "This is a big country," Wally told himself. "It happened here? Well, okay, we're going to deal with it."

AS LAW ENFORCEMENT officials tried to impose order and remove civilians and rubber-neckers, Terry walked through a blue haze of fumes past the crater, where searchers would find the plane's two "black boxes" that recorded cockpit conversations and flight data. Wandering into the woods, Terry saw a stone bungalow, home of a lumberyard worker named Barry Hoover, blown off its foundation, its windows shattered, the door to a garage turned inside out.

Terry walked toward a pond where he would have drawn water if there had been a fire. He saw shards of metal embedded in trees, clothing in high branches, shoes that may or may not have been empty. He saw papers and U.S. mail and remnants of luggage. Terry saw wires and debris he couldn't identify, most of it no bigger than his fist. He saw seat belt buckles and torn pieces of foam cushions and laminated instructions for oxygen masks and emergency exits. A quarter mile from the impact point, Terry came upon a sapling chopped down by the impact of a footlong piece of the plane's nose axle. In the woods south of the crater he saw the biggest, most recognizable evidence that a plane had fallen there: a crumpled section of fuselage about six feet by seven feet, with two intact window openings.

Searchers would find twisted silverware for first-class service, the blunt end of a spoon implanted in a tree, knives that might have belonged to the hijackers, and a weathered but intact copy of the hand-written terror instructions titled "The Last Night." At the base of a hemlock tree they'd find a SunTrust bank card belonging to terrorist pilot Ziad Jarrah that would help investigators trace the flow of money that financed the attackers and the attacks.

Some of the earliest responders on the scene chased down more than a dozen twenty-dollar bills pinwheeling through the charred grass and handed the cash to the FBI. Soon searchers would find thousands of personal effects, much of it as ordinary as socks and ties, but also passenger Todd Beamer's ID badge and assorted jewelry including passenger Andrew "Sonny" Garcia's wedding ring. The gold band would be returned to his wife still inscribed with their wedding date and "All my love." Passenger Richard Guadagno's credentials, with the badge he'd earned from the U.S. Fish and Wildlife Service, would go to his parents. A state trooper would find flight attendant Sandy Bradshaw's flight log, with a photo of her family inside. Copilot LeRoy Homer's wife, Melodie, would receive the dog tags from his service in the Air Force, which he'd hung on a keychain. She'd be reunited with his wedding band, inscribed with a Bible verse about the blessings of faith, hope, and love: "And the greatest of these is love."

Walking on, Terry saw several drivers' licenses but didn't pause to read the names. One, found by other searchers, belonged to flight attendant and former police officer CeeCee Lyles. Terry thought about how the licenses and other IDs had been in someone's wallet earlier that morning, and now they were singed or half-melted on the ground, their owners gone.

Nearby, on a pile of rocks, were the remains of a coiled snake, its head up and ready to strike. The snake had been flash-burned and preserved as ash, like the ancient volcano victims of Pompeii. The macabre find attracted a crowd of onlookers who knew that if they touched it, it would disappear.

Back out on the field, not far from the crater and Big Mo, Terry looked down and saw an open Bible, its cover charred but its pages pristine and white, untouched by fire. Others who saw it would claim it was pressed open to a particular passage, but Terry saw the pages fluttering in the breeze that always seemed to blow at the Diamond T. The Bible belonged to retired electric company executive Don Peterson, who'd been heading to Yosemite with his wife, Jean, for a family reunion. Tucked safely inside was the handwritten list of names of men struggling with alcohol and drugs whom Don counseled and prayed for.

Terry saw another book, too: a flight manual written in Arabic.

AT FIRST, TERRY didn't want to see the human remains that he knew were all around him. But soon he surrendered to the responsibility and trained his eyes to distinguish between the countless man-made items and the untold pieces of burned flesh, bone, teeth, and cartilage in the grass and soil.

Once Terry became a witness to death, he couldn't stop seeing it. Remains were everywhere, small bits of men and women who at that moment should have been landing in San Francisco. Some larger parts, too, including part of a torso; a charred buttock; a piece of spinal cord with five vertebrae; and a foot with three toes that tree-climbing arborists, sent by investigators, would find in the hemlocks. All the remains would require DNA matching, dental records, or other means of positive identification. Somewhere near where Terry stood was a napkin-sized piece of skin, charred at the edges, whose source was never in doubt: the intact Superman logo tattoo on the flesh of Joey Nacke's shoulder.

Years would pass before Terry spoke of the most disturbing thing he saw: not far from Barry Hoover's ruined house, on the grass near a red bandanna, Terry spotted a detached male face. Something about the misshapen features and the complexion convinced him that it belonged to one of the four men who rained death on the little borough of Shanksville and beyond.

As night fell, the FBI took full control of the scene, with Wally Miller empowered to do his job as he saw fit. Most responders were sent home, but the FBI decided to include Terry and the Shanksville Volunteer Fire Department in the recovery efforts. They'd serve as evidence workers and the primary on-site emergency response team.

Terry returned to his fire station around 10 p.m. Kathie was there, along with two other local women determined to organize a flood of donated food, clothing, and other goods that would provide hot meals and small comforts to investigators and emergency responders. Exhausted, haunted, and changed by what he'd seen, vaguely aware that their little community would never be the same, Terry swept Kathie into a bear hug.

CHAPTER 21

"MAYDAY, MAYDAY, MAYDAY!"

Ground Level and North Tower, World Trade Center

THE EXPRESS ELEVATOR IMPRISONING CHRIS YOUNG HAD TWO SETS of doors, on opposite ends of the car. Each set had interior doors attached to the car and exterior doors connected to the building. Motors on the car's roof were supposed to keep all interior and exterior doors shut until the elevator reached a landing. However, a loss of power like the outage caused by the South Tower collapse might allow people trapped inside to force open the doors to escape. Yet even that might not be enough.

Since 1996, the Port Authority had been installing devices called "door restrictors" that prevented passengers from forcing open stalled elevator doors even if power were lost. Designed to prevent falls down shafts, the restrictors acted like deadbolts. They lifted automatically within eighteen inches of a landing but otherwise could only be disengaged by a technician or a firefighter atop an elevator car's roof. Of the 198 elevators in the Twin Towers, about half had been equipped with restrictors.

As minutes ticked by, the smoke and dust settled inside Chris's elevator, but his fear rose. No one responded to his yells or his S-O-S alarms with the emergency button. Chris fought panic. Desperate to free himself, he pressed against a set of doors he'd tried earlier with no luck. This time, to his surprise, Chris felt them give. The loss of power apparently removed pressure from the motor atop his elevator. The car either didn't have a door restrictor installed yet, or it had disengaged automatically.

Still breathing through his dress shirt, Chris wriggled his fingers into the crack. He pulled open the doors a few inches—to reveal a blank wall.

Chris rushed to the opposite doors to repeat the process. This time, light poured in. Chris pried the doors open farther and discovered that the car had stopped a little more than a foot from the floor of the North Tower lobby. He'd been only a hundred feet from where firefighters, Port Authority officials, and other emergency responders had operated the now empty command post for more than an hour after Flight 11 hit the North Tower.

Chris squinted against the sunlight, his eyes sensitive from more than ninety minutes stuck in a dark elevator. He stepped into a moonscape of broken stonework and shattered glass, coated with a pinkish-gray dust as thick and fluffy as a layer of fresh snow. He turned to his right, where he thought the seven people he'd spoken with had been, and saw the doors to that elevator open and no one inside. The lobby was empty, a modern ghost town after an unspeakable horror.

Confused, Chris pushed through a turnstile similar to the one he'd entered before so much changed in so little time, back when this was a normal day. Back when he was headed to the 99th floor, a floor Chris didn't yet know had been gutted by a hijacked plane, to see Marsh & McLennan managing director Angela Kyte and other colleagues who were killed instantly or trapped by flames and broken stairwells, to deliver a box of materials for a PowerPoint presentation that would never be given.

Half in a daze, treading gingerly through the powdered debris, Chris felt as though his feet never touched the ground. He stepped through a broken window frame and headed toward a fire truck he spotted on West Street. But first he turned and looked toward the sky.

The twisted steel and rising smoke made no sense. "Oh my God," Chris said aloud.

He yelled to two firefighters helping an injured person: "Where should I go?"

One firefighter waved for him to follow them, but for reasons Chris couldn't explain, he continued walking, onto Vesey Street, west toward the Hudson River.

He'd walked less than a hundred feet when he heard someone scream, "Run!"

It was 10:28 a.m.

AS LADDER 6 began to evacuate the North Tower after the South Tower collapsed, Roof Man Sal D'Agostino asked Captain Jay Jonas if they could drop their heavy tools. If they weren't going to rescue anyone, why carry irons and a roof rope? Even as he asked the question, Sal could have guessed the answer. Jay regularly preached to his men: "What's a fireman without his tools? He's a walking lump of carbon. Worthless."

Jay barked, "We bring everything with us."

Down they went, moving double-time for the first few floors, with Can Man Tommy Falco and Irons Man Billy Butler out front.

Around the twentieth floor, they came upon a heavyset woman in a purple dress, standing in a doorway, softly weeping. Her name was Josephine Harris, and she was a bookkeeper for the Port Authority, one week shy of her sixtieth birthday. She had limped down about one thousand steps from the 73rd floor, hobbled by fallen arches, a bad leg from having been hit by a car several months earlier, and assorted maladies. An office manager and others had helped Josephine to get this far, but she could go no farther and had sent her helpers ahead to safety. Now she was alone. Josephine might well have been the same woman Gerry Gaeta noticed as he helped Elaine Duch down Stairwell B.

"Hey, Cap," said Tommy Falco, "what do you want us to do with her?"

Every fiber in Jay's body wanted to speed up and get as far away as possible. Every human instinct told him to run, back to his firehouse then home to his family. But if a firefighter followed raw instincts, he'd never charge into a burning building. Ladder 6 had already moved past two burn victims they saw on the way in. They still hadn't fought a fire or rescued anyone. If Ladder 6 survived this day, Jay wanted his men to look themselves in the mirror and say the phrase he lived by: "I was a fireman today."

Jay turned to barrel-chested Irons Man Billy Butler, as powerful a firefighter as Jay had ever known. "Give your tools to the other guys," Jay told him. "Take her arm, put it around your shoulders. We'll stay together as a unit."

Billy did as he was told, then turned to his new charge.

"What's your name?" he asked.

"My name is Josephine."

"Josephine," Billy Butler promised, "we're going to get you out of here today."

WHEN CECILIA LILLO reached Church Street, on the east side of the trade center, she let go of the stranger's hand and watched him walk away, like a zombie through an apocalyptic landscape. As he disappeared, Cecilia wondered if she should have asked his name. Near the Millennium Hotel, she brightened at the sight of an ambulance from FDNY EMS Station 49: Carlos's unit in Astoria. Eager to put their emergency rendezvous plan into effect, Cecilia waved her arms and called for help. But the ambulance was empty, and she realized that she was alone on Church Street.

Wondering where everyone had gone, Cecilia turned around to see the North Tower still burning. She knew nothing of a second plane, or that she'd survived the fall of the South Tower from inside the mall below the trade center. Cecilia agonized about her friends Nancy and Arlene, and about the security guard and everyone else who had disappeared in the smoke. But now her priority was to find Carlos.

Cecilia reached into her purse for her cellphone but discovered it must have fallen out during her ordeal. She grew frightened, upset that she couldn't reach Carlos. Cecilia worried that he'd be terrified that something awful had befallen her. She zigzagged northeast, caught up with a group of stragglers, and followed them to Broadway, where she saw a police officer.

"I need to get hold of my husband—he's a paramedic!" Cecilia told him.

"Sorry, hon," the officer said, "we're on different frequencies." His offhand comment summarized the massive communication and co-ordination failures that plagued the entire response. He offered one piece of advice: "You gotta find a firetruck."

None was in sight, but she spotted several buses a few blocks up Broadway, near City Hall. Cecilia climbed onto one, unsure where it was headed. A woman who didn't speak English gave Cecilia her seat and handed her a bottle of water. When the woman pantomimed washing her mouth and eyes, Cecilia realized that grime coated her face.

The driver announced that they were going to a hospital. Cecilia began to panic. All she wanted was to find Carlos. She joined a stream of passengers who got off before the bus pulled away. A woman approached her on the street.

"Are you okay?"

"No," Cecilia said.

The woman explained that she'd seen both planes hit the towers and watched the South Tower fall. Only then did Cecilia understand why she hadn't seen the North Tower's twin when she emerged from underground.

"I have to get back down there," Cecilia said. But the woman convinced her that no one would be allowed near the trade center. Still hoping to find an FDNY ambulance, Cecilia and the woman walked toward the Brooklyn Bridge. Halfway across, they turned and saw the North Tower fall.

Once in Brooklyn, they boarded a bus to Woodhull Hospital, where Cecilia gave a social worker telephone numbers for Carlos and her sister. Alone in a hospital room, Cecilia prayed. When the social worker returned, he said he could only reach her sister, who was on the way.

Images of her life with Carlos flashed in her mind. Every day since their first date, she had basked in the certainty that he would do anything to protect her. She expected that to be true for every day of the rest of her life, for herself and for the children they hoped to have together.

On their commute only hours earlier, Carlos was his usual kind self, expressing his love with a small act of thoughtfulness that on any other day she might have forgotten: he'd chosen the type of bagel she preferred for them to share. Cecilia spread a napkin across his lap as he drove, and asked: "Why are you always trying to please me?" He just smiled.

Amid her fear and confusion, Cecilia felt certain about one thing: even as Carlos helped to save lives amid the catastrophe of two collapsing skyscrapers, unless he was grievously hurt or worse, he'd find a way to reach her or her family, to learn if she was safe. When her sister arrived, Cecilia asked: "Have you heard from Carlos?"

"No."

Cecilia felt her heart clench.

LADDER 6 AND Josephine resumed their evacuation, but at a much slower pace, frequently moving aside to let other fire companies pass. Step-step-step-step became step . . . pause . . . step . . . pause . . . step . . . pause. Jay stayed close behind Josephine and Billy, who still hadn't learned the fate of the South Tower.

A voice in Jay's head screamed: "We gotta get outta this building! Move fast!" But he kept that to himself, instead relying on his old habit of forced calm.

"Billy," Jay said in his quiet voice, the one he used when he felt most anxious, "can you move a little faster, please?"

"Okay, Cap," Billy said. Beads of sweat rolled down his face.

"Josephine," Billy told her, "your kids and your grandkids want you home today. We gotta keep moving."

Josephine tried, but even as she clung to Billy's thick shoulder, even as she wanted to get out, Josephine's pain dictated their pace. She placed both feet on each step, rested, then repeated the process. Her bad leg threatened to give out, and she grew shakier with each downward flight.

Around the fifteenth floor, Ladder 6 and Josephine ran into Battalion Chief Rich Picciotto, a longtime study partner of Jay's for promotional exams. Picciotto brandished a bullhorn, a rare tool among firefighters, because he remembered the communications and crowd control problems during the 1993 bombing.

"All FDNY, get the fuck out!" Picciotto boomed on his bullhorn. To anyone who could hear him over the radio, he yelled: "We're evacuating, we're getting out—drop your tools, drop your masks, drop everything. Get out, get out!" Dozens of firefighters heeded his lifesaving command and raced toward the exits.

However, some firefighters continued to linger, including a large group seen resting on the nineteenth floor. By some accounts they numbered as many as one hundred men. On his way out, a fire lieutenant told them, "Didn't you hear the Mayday? Get out." Without moving, one answered casually: "Yeah, yeah, we'll be right with you, Lou." Some straggling firefighters apparently didn't know the South Tower had fallen. Others had become separated from their units and insisted on waiting to regroup. Still others refused to leave while fellow firefighters remained inside the building and rescue work continued. One firefighter who heard an evacuation order made his intentions clear, radioing back: "We're not fucking coming out!"

Jay didn't know what was happening high above him in the North Tower, or even above or below him in Stairwell B, but with each step he imagined that he was inside a ticking bomb, each second bringing them closer to the moment when the North Tower followed its twin to the ground.

JAY'S NERVOUS SUSPICION had merit. The collapse of the South Tower had sent a pulse of air pressure that appeared to fuel and intensify the fire high in the North Tower. Within four seconds of the South Tower collapse, flames erupted from south-facing windows on the North Tower's 98th floor, and flared and brightened on three floors just below. Within two minutes of the South Tower collapse, fire spewed from the North Tower's 104th floor, three floors higher than where it had previously been seen. The fires threatened the building's integrity, weakening the steel of the North Tower's core, floors, and exterior walls.

Observers with unique vantage points watched it happening. At 10:07 a.m., a police helicopter pilot radioed, "About fifteen floors down from the top, it looks like it's glowing red. . . . It's inevitable." A second NYPD pilot added, "I don't think this has too much longer to go. I would evacuate all people within the area of that second building."

But with no clear communication links between police and fire-fighters, FDNY commanders didn't hear those messages. Neither did would-be rescuers, some climbing higher, some catching their breath, and some slowly descending the North Tower. Yet even without warnings of impending collapse, a sense of impending doom spread.

At 10:12 a.m., from the North Tower's 64th floor, engineer Pat Hoey sought fresh advice from a Port Authority Police dispatcher. During a call an hour earlier, Hoey had been told to wait in his office for rescuers, so that's what he and a group of colleagues did, using tape and coats to seal out smoke. Now, he checked again.

Pat Hoey: "I'm in the Trade Center, Tower One. I'm with Port Authority and we are on the 64th floor. The smoke is getting kind of bad, so we are going to . . . we are contemplating going down the stairway. Does that make sense?"

At that moment, eighty-seven minutes had elapsed since Flight 11 hit the North Tower. Thirteen minutes had passed since the South Tower fell. Roughly twelve hundred steps separated Pat Hoey from the lobby and possible salvation. Now, finally, Pat Hoey and his co-workers got different, better advice.

"Yes," the dispatcher told him. "Try to get out."

LADDER 6 AND Josephine Harris continued down Stairwell B. When they reached a landing, Jay ran into another firefighter friend, Faus-tino Apostol, a lanky, easygoing aide to Battalion Chief William Mc-Govern. Apostol made no move to leave.

"Faust, let's go," Jay insisted. "It's time to go."

"That's okay," Apostol replied. "I'm waiting for the chief."

Jay, his men, and Josephine, kept moving. When they reached the twelfth floor, Jay found several firefighters from Ladder 5 gathered around a man suffering from chest pains. Their leader was Lieutenant Michael Warchola, a twenty-four-year FDNY veteran. Years earlier, when Jay was in Rescue 3, he and Mike Warchola lived in the same town and carpooled to work. They'd remained friends, and Jay knew that Mike had recently submitted his retirement papers. His last scheduled FDNY shift was two days away.

"Mike, let's go," Jay told him.

"That's all right, Jay. You got your civilian, we got ours. We'll be right behind you."

"Don't take too long."

As their descent continued, Jay's radio worked intermittently, crackling and squawking with calls for help, updates on fire company locations, and occasional moments of heroic grace under pressure. At one point, Jay heard Paddy Brown, the hard-charging, charismatic FDNY captain who'd urged Jay to skip the lobby check-in line and immediately start climbing.

"Dispatch, Captain Brown, Ladder 3. . . . I'm at the World Trade Center. . . . I'm on the 35th floor, okay? . . . Just relay to the command post, we're trying to get up, you know. There's numerous civilians in all stairwells. Numerous burn injuries are coming down. I'm trying to send them down first. Apparently, it's above the 75th floor. I don't know if they got there yet, okay?"

Paddy Brown intended to find out—Jay listened on the radio as he reinforced his reputation as a human engine with no reverse gear: "Three Truck, and we're still headed up, all right?" Brown said.

"Okay," the dispatcher said.

Ever the gentleman, even in the midst of disaster, Paddy Brown signed off as he kept climbing: "Thank you."

By the time Ladder 6 reached the tenth floor, Jay relaxed enough to start calculating. He knew that Roof Man Sal D'Agostino carried a 150-foot rope, coiled in a bag on his shoulder. If they got trapped, they could use the rope to rappel down the outside of the building. They'd have to rig a harness for Josephine, but if their lives depended on it, they'd make it work. A few floors lower, Jay took a sip of optimism: "We may actually make it out of here."

Before Jay could share that thought with his crew, hope went sideways: Josephine's legs gave out. Wracked by pain, she fell to the floor in tears on the fourth-floor landing.

"Don't touch me!" Josephine cried. "Leave me alone! Go! Leave me!"

As his exhausted men turned to him, Jay gritted his teeth and swallowed his frustration. He had no intention of leaving her there, but with Josephine immobile, they needed a new plan. Jay pried open the steel door from the stairwell to the fourth floor, to search for a chair they could use as a litter to carry Josephine the rest of the way down.

Jay discovered that it was a mechanical floor, not an office floor. He found only a flimsy stenographer's chair and an overstuffed couch. Neither would work. Jay went deeper onto the floor. When he reached the far side of the North Tower from Stairwell B, a chill passed through him. The imaginary ticking bomb in his head grew louder. They'd carry Josephine out by hand, drag her if necessary. Upon deciding to return to the stairwell, Jay remembered the first rule he'd learned as a young firefighter: "Never run at a fire. If you're running, you're not looking." He broke the rule and sprinted toward Stairwell B.

Five feet from the door, Jay felt a rumble, far worse than when the South Tower fell. The floor heaved. The time was 10:28 a.m., just over two hours after the hijacking of American Flight 11.

Jay lunged for the stairwell doorknob and pulled, but it wouldn't open. He tugged again with all his might. The door swung toward him and Jay dived through, onto the fourth-floor landing of Stairwell B, nearly crashing into Tiller Man Matt Komorowski, who stood wide-eyed as the world rocked and rattled beneath his feet. The rest of Ladder 6 and Josephine were lower, on the steps approaching the third floor and below. Jay didn't know what was happening, but he knew it was bad.

Almost a hundred stories above him, fires had weakened the building's bones: steel floor trusses and core and exterior steel support columns. The south exterior wall became unstable. The wall automatically

tried to shift its load to the core columns and adjacent exterior columns, but they'd been fatally weakened, too, and they couldn't bear
the extra burden. The upper floors sagged, pulling the exterior steel
columns inward. Everything above the impact zone tilted to the
south. Gravity went to work. The North Tower began to collapse on
itself. As it did, bone-rattling vibrations of impending doom bounced
Jay, his men, and Josephine against Stairwell B's walls and floor.

The volume and violence intensified within seconds, as buckling
floors grew closer. Compressed air sent a gale-force wind shooting
down the length of Stairwell B, blasting everyone inside with a storm
of dust and a hail of sheetrock fragments. The squall lifted Matt Komorowski off his feet and blew him down two flights of stairs, injuring his shoulder.

The sounds defied description. In all recorded history, only one
other 110-story building had ever collapsed, and that was twenty-
nine minutes earlier. Everyone who heard it from the inside took the
experience to the grave. Maybe the sound and fury rivaled an Everest avalanche, or a volcanic eruption, or a rocket launch. Or maybe
not. Nothing matched being inside five hundred million fiery, falling
pounds of twisting steel, crumbling concrete, disintegrating office
furnishings, shattering glass, and ending lives.

Curled in a ball, his eyes squeezed shut, his helmet knocked aside,
Jay tumbled head over heels. As the noise grew closer, he braced himself for the crushing pain of a steel beam he anticipated would flatten
him. Time slowed. Jay realized that the ticking time-bomb sensation
in his mind had stopped. He accepted his fate, with thoughts of regret
and disbelief: "I failed my men—I didn't get them out," and "I can't
believe I'm going to die right here in this stairway."

AFTER ESCAPING FROM his elevator holding cell, Chris Young ran
west down Vesey Street toward the Hudson River, never turning to
look back. He reached a bench near the water as a dust storm overtook
him, blotting out the sun and coating him in the powdered remains of
the North Tower and all it contained. His mind blocked out how long

he lay on the ground, sparing him the details of a terrifying memory. Chris still knew nothing about the planes, or about the fallen South Tower, and he didn't know that the rumbling explosion he now heard and felt was the North Tower collapsing.

When the cloud passed, Chris pulled himself onto the bench. He unwrapped his dress shirt from around his head and tugged it over his T-shirt. Functioning but in shock, not sure what else to do, Chris walked north, toward the Marsh & McLennan office four miles uptown, where his day had begun. He wanted to collect a bag of personal items he'd left behind when he grabbed the box of materials for Angela Kyte.

As Chris walked, an EMT asked if he needed help.

"I'm fine," Chris said. "I have to get back to my office."

The EMT gave him a strange look and moved on.

Chris Young, aspiring actor, temp worker, and survivor, blended into the northbound horde of exiles from Lower Manhattan. The last person to escape from a World Trade Center elevator spoke to no one as he walked on.

Hours later, inside a conference room in the shell-shocked Midtown offices of Marsh & McLennan, Chris looked up to a television set as CNN replayed Flight 175 hitting the South Tower. He began to understand all he'd been through and broke down.

AFTER THE SMOKE and dust from the South Tower collapse cleared, EMT Moose Diaz settled his emotions. He exited the bus where he, a television cameraman, and several others had taken refuge. Just as Moose stepped back into the street, the North Tower buckled, following its twin to the ground. As smoke and debris chased him again, Moose ran several blocks north to an open hydrant near an improvised command post. Someone wiped his ash-covered face with a wet cloth. And then he returned to work, helping several asthmatics and a pregnant woman.

Moose had no radio, no helmet, but worst of all, no partners. He turned and walked back toward the World Trade Center to search for Paul Adams, whom he'd last seen when the South Tower began to fall. Moose ran into fellow Battalion 49 EMTs Kevin Barrett and Alwish Moncherry. The three men wept as they held one another for several minutes. Eventually, they made their way to an aid station set up at the recreation complex Chelsea Piers, at Twenty-Third Street and West Side Highway, where they prayed for news of their friends and colleagues Paul Adams, Roberto Abril, and Carlos Lillo.

Paul, meanwhile, had climbed back into his ambulance after dropping off patients at St. Vincent's. While making a U-turn in front of the hospital, to return to the World Trade Center, Paul saw terrified faces looking south. One hundred and two minutes after being struck by American Airlines Flight 11, the North Tower was coming down. Paul drove toward whatever remained of the trade center. He again filled the rear of the ambulance. Paul also invited one man into the passenger seat and allowed another to ride on the hood. Back he went to St. Vincent's.

Before Paul could return for more victims, the ambulance died of smoke inhalation in front of the hospital. Paul noticed a Port Authority police officer riding slowly past on a motorcycle. He asked for a ride, and the cop told him to hop on the back. Paul grabbed his trauma bag, jumped on, and returned to the scene to help anyone he could and to search for Moose, Roberto, Carlos, and the other Battalion 49 EMTs.

TWELVE SECONDS AFTER it began, twelve seconds that felt to Jay Jonas like two minutes, or maybe a lifetime, the din and the heaving stopped. Jay didn't know it yet, but only a few of the lowest floors of Stairwell B had somehow survived relatively intact, entombed within a colossal mound of wreckage, a pile of ruins that began the day as the iconic North Tower of the World Trade Center.

In the crypt-like darkness, Jay took a breath. He gagged on smoke and dust and the faint smell of jet fuel and spit out silt while he rubbed his face and picked out debris from his eyes, nose, and ears. He heard others coughing nearby. Jay couldn't see anyone, so he took roll call for Ladder 6: "Mike? Matt? Billy? Tommy? Sal?"

One after another they yelled back: "I'm here"—"Yep"—"I'm good." All were banged up, but somehow all had survived, and none had serious injuries. Worst off was Mike Meldrum, woozy from a concussion.

"What about the woman?" Jay asked.

"Yeah, she's right here," Billy answered.

Josephine lived through the collapse clinging to the Irons Man's boot. Billy and Tommy lifted mounds of fallen sheet rock and found her relatively unscathed.

Soon they discovered they weren't alone inside the abbreviated remains of Stairwell B. Spread along the stairs of the lowest four floors, covered in dust but with only minor injuries, were six more firefighters: Battalion Chief Rich Picciotto, lieutenants Mickey Kross and Jim McGlynn, and firefighters Bob Bacon, Jeff Coniglio, and James Efthimiades.

A few steps below Jay was a fourteenth survivor inside Stairwell B: Port Authority Police Officer Dave Lim, who'd run toward the North Tower when he heard the plane crash and left his bomb-sniffing dog Sirius in the basement of the South Tower. Dave Lim had made it all the way up to the lower sky lobby, on the 44th floor of the North Tower, when United Flight 175 hit the South Tower. He spent the next hour shepherding North Tower survivors toward exit stairs and checking for stragglers on floors where he didn't see firefighters already working. Dave joined the Stairwell B evacuation when the South Tower fell, only to slow his descent to help more people.

As Jay absorbed the situation, he imagined that the building must have sustained only a partial collapse. He thought they should flee immediately, before the rest crashed down. Jay didn't expect to survive a second fall. He told Billy Butler to fashion his nylon webbing into a full body harness for Josephine. They'd haul her downstairs to whatever remained of the lobby. But Matt Komorowski, two flights down, called out to squelch that plan. Piles of blasted sheetrock and debris clogged the bottom of the stairwell.

As he considered his next move, Jay heard a radio call from Lieutenant Mike Warchola that reinforced his mistaken belief that large parts of the North Tower still stood. Jay saw his longtime friend not ten minutes earlier, helping a man with chest pain.

"Mayday, Mayday, Mayday!" Warchola called. "This is the officer of Ladder Company Five. I'm on the twelfth floor, B stairway. I'm trapped, and I'm hurt bad."

"Cap," called Sal D'Agostino, "you get that?"

"I got it," Jay said. Mike Warchola was saying that he was trapped in the same stairwell, also injured, only eight stories above them. Ladder 6 was on its way.

Jay climbed toward the fifth floor, stepping over holes in the stairs, pushing aside or squeezing through piles of fallen sheetrock, using a twisted handrail to haul himself through debris and beyond gaps in missing steps. It reminded him of a stairwell in a tenement on the verge of collapse. When he reached the fifth-floor landing, Jay heard Mike Warchola call again for help. Halfway to the sixth floor landing, Jay ran into an immovable pile of debris.

"Mayday," Warchola called a third time, his voice weak, his tone distraught.

Jay leaned into the blockage, trying desperately to clear a path, but it wouldn't budge. Jay knew that calling for more men wouldn't be enough; he'd need a crane to lift the load. Jay's heart sank as he pressed the talk button and told his old friend: "I'm sorry, Mike. I can't help you."

He heard no reply.

Jay used his radio to call other firefighters, giving them Warchola's last reported location. Not all his calls went through, but eventually among those outside the North Tower who heard Jay was another friend, Deputy Chief Nick Visconti. Visconti didn't have the heart to tell Jay the truth: Mike Warchola couldn't be inside Stairwell B on the twelfth floor of the North Tower. Anyone outside the wreckage could see that location no longer existed. Warchola and anyone else with him were buried somewhere in the canyons of rubble surrounding the surviving six stories of Stairwell B.

Jay climbed back down, navigating the broken walls and taking care not to fall into holes. Smoke and dust still filled the dark, damaged stairwell. Along the way, Jay pried open the door to the fifth floor and found a custodial bathroom. He knew that the toilet wouldn't work, but it might come in handy if they were trapped more than a short time. He also found Chief Picciotto's bullhorn and passed it down to him.

Jay felt most energized by his discovery of a torn-open freight elevator shaft. When he rejoined his troops on the fourth floor landing, Jay told them: "We've got our ropes. We can rappel down maybe to a subcellar and find the PATH train station." From there, he said, they could walk the tracks in a rail tunnel under the Hudson River and emerge in Hoboken, New Jersey. His men looked at Jay like he'd lost his mind.

"Hey, Cap," Tommy Falco said, "what if we can't get in down below? It's not like we can just run back up the stairs."

"All right, killjoy," Jay said, even as he stored away the idea. He'd consider it again if all else failed and they remained trapped for two or three days. Meanwhile, Matt Komorowski found sprinkler pipes on the third floor. If they got thirsty, they could break into them in the hope of finding water.

For now, there was nothing to do but wait.

As they settled in, Jay heard the radio squawk with a Mayday call from Battalion Chief Richard Prunty, an avuncular figure with a white walrus mustache, somewhere below them in the North Tower's lobby. Chief Prunty said he was hurt, dizzy, pinned under a steel beam. Jay already knew they couldn't reach him, but he and Lieutenant Jim McGlynn tried to keep the chief talking. Unsure whether Rich Prunty could be heard by potential rescuers outside the stairwell, Jay and Jim McGlynn relayed his pleas for help to anyone monitoring their radio channels.

An hour passed. Prunty was slipping away. They heard a final transmission: "Tell my wife and kids that I love them."

The survivors inside Stairwell B had no idea what the world looked like outside their chrysalis. They didn't know that Mayor Giuliani had ordered an evacuation of Lower Manhattan below Canal Street, an evacuation that sent hundreds of thousands of people through dust-covered streets uptown, across bridges, or onto tugs, ferries, fire boats, Coast Guard vessels, and pleasure craft that sailed from the smoldering island like a modern-day Dunkirk.

Jay and his crew coughed and rubbed dust from their eyes. They sat quietly. Most turned off their radios to conserve batteries, while Chief Picciotto, on the stairs a floor below Jay, put out a Mayday call on the command channel. Another hour passed.

Tommy Falco caught Jay's eye. "Hey, Cap. What do we do now?"

"I don't know," Jay replied. "I'm making this up as we go along."

Jay used a different frequency, the primary tactical channel, Channel 1, and put out Maydays of his own. Deputy Chief Tom Haring answered: "Okay, Ladder 6, I got you."

Relief spread. Jay described their location. More captains and chiefs chimed in over the radio, old friends of Jay's whose voices bolstered his spirits. They told Jay that every off-duty member of Ladder 6, along with Jay's old company, Ladder 11, had joined the search. More radio calls, more promises, all with one theme: "I'm coming to get you, brother."

Some of Jay's radio contacts had questions that struck him as odd. They asked for more details of the survivors' location, about how Ladder 6 had entered the North Tower, and where they'd parked their truck. One radio caller flat-out asked Jay how to find the North Tower. With no way to envision the smoking acres of rubble around him, Jay grew annoyed. He thought: "This is not that hard, gentlemen. My five-year-old daughter could follow these directions. It's one of the big buildings on the corner."

Jay urged Josephine and his men to be patient, assuring them that help was on the way. He guessed, incorrectly, that firefighters and other rescuers were overwhelmed by hundreds or even thousands of injured or trapped people. No one on the radio told him that responders could devote unlimited manpower to the fourteen people inside Stairwell B because they comprised the largest single group of known survivors.

Visibility improved enough for Jay to look out a small hole, but he could only see a wall of twisted steel. He and the others heard fires breaking out nearby. Explosions rocked the stairwell, pelting them with a shower of debris. Amid the barrage of noise and falling pieces of sheetrock, Josephine began to cry.

"I'm scared," Josephine said.

"We're all a little scared," Jay told her. "Just hang in there."

Jay was long over the frustration he'd felt earlier about Josephine's slow pace. Josephine was part of Ladder 6, now. During the occasional debris storms, Mike Meldrum draped himself across her body. Sal D'Agostino wrapped her in his coat.

"As long as we're here," Sal told her, "nothing is going to happen to you."

The explosions stopped. The wait continued. K-9 Officer Dave Lim used a radio to ask someone to check on his partner, Sirius, in the basement of the South Tower. He used his cellphone to reach his wife, Diane, who told him that she didn't want to hang up until either she died or he did.

Another hour passed. Someone found a can of soda, so they passed it around. The dust settled more. A beam of light pierced a hole Jay hadn't noticed previously in the stairwell wall above where he sat. Flecks of dust twinkled in the beam, like a sudden spotlight on a darkened stage. Jay stared at it until comprehension dawned.

"Guys," Jay said. "There used to be one hundred and six floors above us, and now I'm seeing sunshine."

Still Jay didn't understand how much, or how little, of the building remained. As far as Jay knew, Mike Warchola remained alive, awaiting rescue eight flights above them.

Smoke and dust cleared further. Within ten minutes they could see through a manhole-sized hole in the stairwell wall to a firefighter from Ladder 43 walking in the rubble, about eighty feet away. They tied a rope around Chief Picciotto and lowered him out the hole to a place where he could stand amid the mound of twisted steel and masonry. Picciotto reached the firefighter and tied off the rope for others to use as a guide. Jay sent out members of his crew, one after another, along with Dave Lim, Mickey Kross, and Bob Bacon. Lieutenant Jim McGlynn stayed behind on a lower level of the stairwell with Jim Efthimiades and Jeff Coniglio, who hoped they could find Chief Prunty alive.

Jay remained inside the stairwell with Tommy, Sal, and Josephine, who still couldn't walk. "Cap," Tommy said, "we can't leave her." Jay told him they'd wait with Josephine until other firefighters arrived with a rescue stretcher to carry her out. Soon Lieutenant Glenn Rohan climbed into the wrecked stairwell with several other members of Ladder 43 to help the men on the lower levels and to evacuate Josephine. One of the newcomers addressed her as "ma'am." Another called her "doll."

Ladder 6 didn't appreciate that. Sal firmly told them: "This is Josephine."

Jay explained her immobility to Glenn Rohan, then reassured Josephine she'd be in good hands. It'd take some time to get a metal-ribbed stretcher called a Stokes basket, but she'd be going home soon.

Jay told Ladder 43 what he knew about Chief Prunty down below. When Jay mentioned that Mike Warchola was hurt somewhere above them, Glenn Rohan gave him a strange look but didn't say anything.

Sal went out through the hole. Tommy followed, then poked his head back inside: "Cap, wait until you get a look at this."

Jay looked out in disbelief. Nothing he'd experienced or imagined prepared him to see the smoking remains of New York's two tallest buildings arrayed around him. The violence of the collapse had pulverized nearly everything inside, leaving no recognizable trace of furniture or computers or phones or carpets. Jay saw no sign of the more than twenty-seven hundred people who'd arrived at the Twin Towers that morning as workers, visitors, or emergency responders, or as airplane passengers and crew, but who'd soon be counted among the departed.

"Oh my god," Jay told himself. "We got bombed. They used airplanes like missiles."

Jay understood that the twelfth floor no longer existed. He realized that if he and the other survivors in Stairwell B had been a little higher or a little lower at 10:28 a.m., they'd almost certainly be dead, too.

Jay followed his men as they traversed the rope line toward West Street, navigating deep crevasses of misshapen steel, fighting to keep their footing on an inch-thick coating of talcum-like dust. At one point, they balanced along a foot-wide beam stretched across a pit of rubble, like explorers crossing a log bridge over a ravine.

Jay looked north and saw fire consuming the forty-seven-story building known as Seven World Trade Center, across Vesey Street from the tower remains. The Secret Service had an office in an adjacent building. As the Stairwell B survivors made their way across the rubble, fire reached an ammunition storage room there. The sound of gunfire convinced Dave Lim that terrorists were launching a ground assault. He reached for his service weapon, knowing that he carried forty-six rounds. Dave resolved to go down fighting, but he soon realized that wouldn't be necessary. The gunshots stopped. (Hours later, at 5:21 p.m., WTC 7 collapsed with no one inside.)

As Stairwell B emptied of survivors, Lieutenant Glenn Rohan and three other firefighters climbed down through the dark. They forced their way into the remains of the lobby to search for Chief Prunty, who'd last been heard from about two hours earlier. They found him fifty feet below where they entered Stairwell B, pinned under a steel beam. They tried CPR but knew it was no use. Unable to lift the beam, they said a prayer and packed up. Before they climbed out, Glenn promised: "We'll come back to get you tomorrow, chief."

Jay kept a close watch on his men as they traversed the rubble. Finally, they faced a steep rise of broken masonry to reach West Street. Fellow firefighters dropped ropes down to them for one hard, final climb. Covered in filth, his eyes burning, Jay emerged on the street feeling spent. He looked back toward the pile and saw Josephine emerging from the stub of the stairwell, carried aloft by firefighters aboard a stretcher. A medic tried to usher Jay toward an ambulance.

"Wait a minute, wait a minute," Jay said. "Where's the command post?"

Jay wanted to check in with the chiefs before leaving. He wasn't driven by protocol or pride. From the radio traffic, Jay knew that scores of firefighters had been searching for him and the others in

Stairwell B. He wouldn't be able to live with himself if someone got hurt or killed while looking for him after the survivors were safe.

Jay walked a block south to Liberty Street. Hundreds of firefighters milled around. They'd removed their helmets to observe a moment of silence, and now they were looking for a fire to fight, someone to help, or a lost comrade to bring home. They understood that many of their number were missing, though they didn't yet know the magnitude of the loss. Deputy Chief Peter Hayden, who hours earlier sent Ladder 6 and so many other firefighters into the North Tower, stood atop a damaged pumper truck.

The chief looked down, Jay looked up. Each man's eyes welled with tears.

"It's good to see you," Chief Hayden said.

"It's good to be here," Jay said.

Endless work remained, the shift wasn't over, and Jay wanted to help. But he knew better. "My guys are shot," Jay said. With a nod, the chief dismissed Captain Jay Jonas and the men of Ladder 6. But first, Hayden offered a small piece of welcome news.

"Now you're going to get promoted to battalion chief."

His eyes moist, his body aching, Jay looked up at Chief Hayden. Jay had devoted countless hours to working, training, and studying, in the hope that someday he'd hear the career-defining words that he'd made chief. But now his mind was elsewhere.

As he walked away, Jay said simply: "It's gonna be good to be around for that."

AFTER HIS FERRY ride, on the train home to Glen Ridge, New Jersey, Ron Clifford thought the worst of his personal ordeal was over, even as he understood that the nation's had only begun. He watched as a woman who worked on Wall Street guzzled vodka straight from a bottle. He called his sister Ruth's cellphone, hoping to tell her he was fine, so she wouldn't worry. No answer, so he left a message.

Fellow passengers read out news updates from their BlackBerry devices about planes crashing into the Pentagon and a Pennsylvania field. Rumors flew wildly around the train car about more planes, more targets, more death.

Ron fell into Brigid's arms at the station. He apologized for ruining his clothes but was too spent to explain all that had happened. At home, Ron stripped out of his soiled blue suit and set aside his yellow tie. As he prepared to shower, Ron's phone rang: David McCourt, his sister Ruth's husband, sounded dazed.

"Hey," David said, "do you know where Ruth is?"

Believing that Ruth had flown west from Boston several days earlier, along with Ruth and David's four-year-old daughter Juliana, Ron said, "Huh? She's in California."

"I think she was on that flight," David answered.

Ron's mind refused to piece it together.

"What flight? What are you talking about?"

"I got a call from Allan Hackel," David said, referring to the husband of Paige Farley-Hackel, Ruth's close friend and Juliana's godmother.

Ron remained fiercely in denial. He told himself that David had recently gone through cancer treatment, and that's why he must be confused. Ron said he'd clear it up by calling Paige's husband. "Give me Allan's number."

Hours earlier, Allan Hackel heard on his car radio that a plane struck the North Tower, but he'd dismissed it. "There's a million planes flying," Allan told himself. He focused on the direct route from Boston to Los Angeles. "I don't think they're going to New York to get to L.A." But when Allan reached the advertising company he owned, his grown son Peter and daughter Jodi, from a previous marriage, crowded into his office.

"What?" Allan had asked. "Was that Paige's flight?" Jodi embraced him. Allan's mind went black. A single thought formed: "My life is gone."

When Ron called, Allan sounded defeated. He assumed that Ron already knew the worst. "They were the two most beautiful, spirited girls that walked the earth," Allan said tearfully of Ron's sister, Ruth, and Allan's wife, Paige. "And that gorgeous little girl."

Ron refused to accept it: "What are you talking about, Allan?"

"They were here, and they were driven to the airport early this morning."

Ron felt the bottom drop out, but still he clung to a whisker of hope. Allan told him that Paige was on a different flight from Ruth and Juliana. Ron went to his computer and joined countless people trying to determine what had happened to their loved ones' flights.

Ron called his half-brother, Spencer, who lived in Boston, and told him to rush to Logan Airport to see what he could learn.

JENNIEANN MAFFEO HAD wanted her sister Andrea to be notified about her burns before their parents, but the request got lost on her way to the hospital. Using the information Ron gave the ambulance driver, someone called Jennieann's boss at UBS PaineWebber, who in turn called the Maffeo home in Brooklyn to deliver the news that Jennieann was alive, but grievously injured.

Just as Jennieann had feared, her elderly mother was too distraught to speak. Frances Maffeo gave the caller Andrea's work number. Eventually, the telephone trail led Andrea back to Ron.

Ron leapt to answer the phone, thinking it might be news about Ruth and Juliana. Andrea heard anguish in Ron's voice when she explained who she was. Ron set aside his fears for his sister and niece and tried to focus on Jennieann.

"Her arms," Ron told Andrea. "My god, her arms are burned." He was too distracted, too distraught, or too kind to tell Andrea the full extent of Jennieann's injuries.

Jennieann's boss didn't know where she'd been taken, so Andrea asked Ron. He didn't know either. "I saw horrendous things," Ron said.

They talked briefly about what happened, about how he had come upon Jennieann in the Marriott lobby. Ron confided in Andrea that they were in the same boat: he was awaiting word about his sister and niece. They heard a click—another call on Ron's line. Ron placed Andrea on hold, then quickly returned.

"I have to go," Ron said. "My sister's dead."

"I'm so sorry," Andrea said. "Can I do anything?"

"Just let me know how your sister is. She has to be okay."

Andrea said she'd stay in touch, and she promised to pray for him.

AT THE COMMAND center at Chelsea Piers, Moose Diaz heard his name read among the missing emergency responders.

"No," he called out. "I'm right here."

He called his wife, Ericka, from a pay phone. Her cries made him weep again.

"Don't go back in!" she pleaded.

"They won't release me," Moose told her. "The city's locked down."

"Come home now!"

Moose and another EMT commandeered an empty ambulance. Bringing along a group of EMT trainees who'd leapt into service, they ignored the citywide lockdown and drove off Manhattan island, across the Fifty-Ninth Street Bridge, back to Battalion 49 in Astoria. He soon learned that his partner Paul Adams was accounted for, as was paramedic Roberto Abril. Carlos Lillo remained among the missing.

Exhausted, covered with bits of building and other remains, coughing and wheezing, Moose drove his car upstate to the little town where his day had begun nearly twenty-four hours earlier. By the time Moose arrived home, President Bush had already delivered his address from the Oval Office. "Missing" posters had already begun spreading around Manhattan. The Pentagon and an old coal mine in Pennsylvania still smoldered. Fighter jets, tankers, and radar planes ruled the skies over the United States. The country and large parts of the world braced for a future where the only certainty seemed to be war.

Ericka was dreaming of water when Moose reached their front stoop. She woke and opened the door to the ash-covered man she loved, a gentle man who loved to help people, a brave man who saved lives that day and nearly lost his own, an American man of Cuban/Palestinian/Haitian descent whose Arabic name meant Moses.

Together they sat on the floor and cried.

ANDREA MAFFEO RUSHED home from work to comfort her parents. She called one hospital after another, but no one had heard of Jennieann. Bridges and tunnels into Manhattan were closed to everyone except emergency responders and public safety workers. Andrea sought help from a family friend, a retired New York City police officer named Gaspere Randazzo.

"C'mon, we're going to the city," Gaspere told her.

Joined by Gaspere's former partner, the two old cops flashed their badges and gained passage across the Verrazzano-Narrows Bridge. Andrea had never seen Manhattan so empty. An occasional emergency vehicle broke the eerie silence. Teams of firefighters walked through the streets, hunched, exhausted, covered in ash. An army in retreat.

They drove to the Beekman Downtown Hospital, but Jennieann wasn't there. A nurse told Andrea to try the city's major burn unit, at Weill Cornell Medical Center on Sixty-Eighth Street. After sorting out confusion about Jennieann's name—she'd been registered as Jeannie—Andrea found her sister. She went upstairs to the burn unit and made what felt like the longest walk in creation. "How am I going to have my mother make this walk?" Andrea wondered.

A nurse named Mike greeted her warmly and asked what she knew.

"Her arms are burned," Andrea said, echoing what Ron Clifford told her.

"It's a little more than that," Mike said.

"What you mean? How bad is she?"

Mike held Andrea's hands. He explained that Jennieann had third-degree burns over more than 80 percent of her body. Fourth-degree might have been more accurate, but that level didn't exist. Jennieann's burns penetrated to her stomach and other internal organs.

Andrea steeled herself and entered the room. Jennieann was mummified in white gauze from neck to feet, with clear bandages on her face and head. Andrea could see that the tops of her sister's ears and part of her nose had burned off. Fingertips as black as coal peeked out from the wraps. A ventilator breathed for her. Doctors had placed Jennieann in a drug-induced coma to spare her from pain. Her eyelids were swollen shut, but her corneas weren't damaged. When the fireball from American Flight 11 raced down the North Tower, Jennieann might have shielded her eyes with her arms. If she recovered she might retain her sight.

Andrea began to whimper, then forced herself to stop, worried that Jennieann might hear her. She moved to the head of her sister's bed.

"You're here, you're safe," Andrea whispered. "You're going to get help."

Jennieann moved her legs.

"Did she hear me?" Andrea asked.

A doctor said it was muscle reflex, but outside, a nurse disagreed. "That was her way of telling you she heard you," the nurse said. Andrea chose to believe it.

When Andrea returned home, Gaspere offered to walk her inside, but Andrea said no—his presence would immediately tip off her mother how awful it was. If Frances Maffeo could be spared for even a minute, Andrea wanted to give her that time.

The two women sat across from each other in the kitchen.

"How is she?" Frances asked.

Andrea put her head down on the table. The silence became too much to bear, so she forced herself to speak: "They don't know if she's going to make it through the night."

Neighbors heard Frances Maffeo's screams.

PART III

RISE

From the Ashes

CHAPTER 22

"YOUR SISTER AND NIECE WILL NEVER BE LONELY"

September 12, 2001, and Beyond

SEPTEMBER 12 DAWNED CLEAR AND WARM ON THE WOUNDED EAST Coast. The attacks had knocked out the two front teeth from the United States' financial center, damaged its military brain, and scarred its rural flesh. During the first hours of the "post-9/11 era," as smoke still rose and tears still flowed, aftershocks reverberated from the epicenters of pain. They rattled stock markets and houses of worship, schools and government offices, airports and stadiums, hearts and minds.

Within weeks of the attacks, the United States went to war in Afghanistan, a conflict that continues at this writing. The war in Iraq followed, starting in 2003 and officially ending in 2011. Before committing to either, President Bush made a statement often overlooked afterward: "The enemy of America is not our many Muslim friends. It is not our many Arab friends. Our enemy is a radical network of terrorists and every government that supports them."

The pursuit of Osama bin Laden consumed a decade, culminating with a May 2011 raid in Abbottabad, Pakistan, during which he was killed by members of SEAL Team Six. Alleged 9/11 mastermind Khalid Sheikh Mohammed was captured in 2003 and tortured for information by the CIA. He and four other men accused of training, financing, and directing the hijackers have spent more than a decade imprisoned at the United States' Guantanamo Naval Base in the Caribbean, with no firm trial date in sight.

Before any of that happened, the nation needed to account for the dead and injured. At first, hope took the form of countless "Missing" posters plastered throughout Manhattan. Soon it became clear that anyone unaccounted for was gone. The death toll was not as high as originally feared, but Mayor Rudy Giuliani proved correct when, hours after the attacks, he said: "The number of casualties will be more than any of us can bear, ultimately."

Not including the hijackers, 2,977 men, women, and children were known to have been killed on the four planes and at the World Trade Center and the Pentagon. Among the dead were 1,462 people in the North Tower, 630 in the South Tower, 421 emergency responders in New York, 246 passengers and crew members on the planes, and 125 men and women in the Pentagon. No one died on the ground at Shanksville.

Roughly six thousand more sustained physical injuries, some of whom would never fully recover. Thousands more, many of them emergency responders and investigators, suffered from respiratory, psychological, and other ailments that revealed themselves later. Nearly seventeen years after 9/11, the FBI announced the death of David LeValley, special agent in charge of the Atlanta bureau, who'd spent several weeks investigating the attacks. "Dave died in the line of duty," the FBI declared, "as a direct result of his work at the World Trade Center." Weeks later, the FDNY chief who led recovery efforts died of cancer traced to toxins from Ground Zero. Ronald Spadafora was the 178th member of the FDNY to die of 9/11-related illnesses. No one expected him to be the last. Authorities estimated that by the twentieth anniversary of 9/11, more people will have died of an illness related to Ground Zero than in the attacks.

Along with counting the dead and injured came the notification of loved ones. Wives lost husbands, husbands lost wives, parents lost children, siblings lost siblings, friends lost friends. The losses touched grandparents and godparents, aunts and uncles, cousins and class-mates, neighbors and colleagues. Every death carved a ragged hole. Some who suffered the most were children who had no previous understanding of the permanency of mortal loss. Roughly three thousand children under age eighteen lost a parent on 9/11, including 108 babies born in the months after their father's death. Each one needed some kind of explanation.

JOHN CREAMER TAUGHT math at an alternative high school in Worcester, Massachusetts, working with students who'd previously dropped out, or were school-age mothers, or who struggled with English. On the morning of September 11, a teacher's aide told John that a plane had struck one of the World Trade Center towers. The news disturbed John, but he didn't worry. He knew that his wife, Tara, was on an American Airlines flight headed from Boston to Los Angeles, not to New York.

Hours later, John left school with a janitor during a free period to run an errand for the school. John didn't have a cellphone, but the janitor took a beeper that would display a phone number in case anyone needed to reach them. Suddenly, the beeper flashed 9-1-1, and the janitor worried that he was in trouble for leaving school grounds without notice.

"We're close to my house," John told him. "We'll call from there."

John called the school secretary, to reassure her that they'd be back soon. But the beeper message wasn't about school rules.

"Wait there," secretary Dorry Lemay told him. "Your dad is coming."

John's mind flashed to his son, Colin, and daughter, Nora, both at daycare.

"What's going on? What's wrong?"

Dorry handed the phone to John's father, Gerry, who ran the school. He'd heard the news about American Flight 11 from Tara's sister Maureen, who'd heard from another sister, Kellie, who worked with Tara at the retailing giant TJX Cos.

"Where are you, John?" Gerry Creamer asked.

"I'm at the house."

"Wait there, I'm coming."

"Why, Dad?"

Gerry wouldn't say.

John met his father in the driveway of the pretty yellow Cape Cod–style cottage that John and Tara had filled with love.

"It was Tara's plane that hit the World Trade Center," Gerry said. "Tara was on that plane." John collapsed into his father's arms.

The next day, September 12, it seemed as though everyone John knew, and many he didn't, converged on the house that would never again be a home. Family members, old and new friends, colleagues from TJX and the Worcester schools, reporters and photographers, neighbors bearing casseroles and condolences.

Everyone except Tara.

John and Gerry drove an hour to Boston to answer routine questions from the FBI, whose work had just begun. Although Osama bin Laden and al-Qaeda were immediately the prime suspects, investigators didn't want to overlook any other possible motive. Before returning home, John sought out a child psychologist for advice on what to tell his children. When John and Gerry reached the house, John saw his mother, Julie, holding one-year-old Nora, who was oblivious to the commotion. Colin, at four, sensed something was amiss. The boy had inherited his mother's oversized smile, but it was nowhere in sight. John led him upstairs.

Father and son lay down on John and Tara's king-sized bed, atop the wedding quilt with interlocking rings made by Tara's aunt. "I need to talk to you about something," John told Colin. John opened a pack of crayons, laid out two sheets of paper, and drew an image of a brown-haired woman with angel wings. Then he helped Colin to sketch his own picture of a "mommy angel."

"There was an accident," John said, fighting sobs. "Your mommy is in heaven and she won't be coming back. She's up in heaven and she's an angel now."

Weeks earlier, Tara had used similar words to help Colin understand why he had only one grandmother. Tara wasn't sure she'd been clear. Now, as John fell silent, Colin erupted in sobs. He'd understood his mother after all.

On the other side of the door, heartbroken, Julie listened to Colin wail for an hour as John tried to comfort him. Finally Colin fell asleep, nestled against his father, on his parents' wedding quilt. Later, John placed Tara's plush robe in Nora's crib so she'd fall asleep to her mother's scent.

Months passed, and John Creamer got a call from the New York medical examiner's office. DNA testing of recovered remains identified a piece of Tara's foot and part of one breast. A few months later, John got a call about more remains. John had what they found of Tara buried at St. John's Cemetery in Worcester, so he and his children could visit whenever they needed solace. They brought flowers every Mother's Day.

Years passed, and the coroner called again and again. John told them to keep whatever else they found entombed with other partial remains at the National September 11 Memorial and Museum in Lower Manhattan, which opened fully to the public in 2014. Victims' remains are kept in a secluded repository, deep within the memorial, with an adjacent Reflection Room open only to 9/11 family members.

Nearby, above ground, are two enormous reflecting pools. The pools are located within the footprints of the missing Twin Towers. The names of all the known fatalities of 9/11, plus the six people killed in the 1993 World Trade Center bombing, are inscribed in bronze on parapets that surround the pools. At night, lights shine through the letters of each name.

Looming over the memorial is a building called One World Trade Center. At a height of 1,776 feet, an elevation chosen for the historical resonance, it opened in October 2014 as the tallest skyscraper in the Western Hemisphere, and sixth-tallest in the world. While celebrating its architectural splendor, the building's owners acknowledged the horrible circumstances that allowed its construction. They advertised the inclusion of "life-safety systems [that] far exceed NYC building code."

John Creamer remarried. His wife, Tina, is the first to say that Tara will always be part of their family. As he entered manhood, Colin often said that his mother is his inspiration to succeed in life. Nora had no memories of her mother, but she grew up basking in stories about Tara. She plans to wear her mother's wedding gown when the time comes.

John and his family bought a new house overlooking a lake. When John sold the yellow Cape, the new owners promised they'd never remove the thick strokes of white paint on the basement wall, the ones that read "Tara ♥s John."

COUNTLESS OTHERS AWOKE on September 12 with their own eternal connections to 9/11. Some, like Lorne Lyles, husband of Flight 93 flight attendant CeeCee Lyles, would hear their lost loved ones calling to them from answering machines. "Hi, baby," CeeCee said in the call, which Lorne found a week later. "I hope to be able to see your face again, baby. I love you. Goodbye."

More than fifteen years after 9/11, after she remarried and became Julie Sweeney Roth, the widow of Flight 175 passenger Brian "Moose" Sweeney could recite every word of his last voicemail from memory: "Jules, this is Brian. Listen, I'm on an airplane that's been hijacked. If things don't go well, and it's not looking good, I just want you to know I absolutely love you. I want you to do good, go have a good time. Same to my parents and everybody. And I just totally love you, and I'll see you when you get there. 'Bye, babe. Hope I'll call you."

Others treasured physical tokens of the missing. Solicitor General Ted Olson, husband of Flight 77 passenger Barbara Olson, fell into bed at 1 a.m. on September 12. There he found the note Barbara left the previous morning on his pillow: "I love you," it said. "When you read this, I will be thinking of you and I will be back on Friday."

In the rubble of the World Trade Center, among thousands of pulverized and partial human remains, searchers found a woman's left hand with the bejeweled wedding ring of Sonia Puopolo, a first-class passenger on American Flight 11. Her family considered it a minor miracle and a message from Sonia about the need to persevere.

For some, the loss proved too much. Before Flight 11 took off, passenger Pendyala "Vamsi" Vamsikrishna had left a voicemail for his wife, Prasanna Kalahasti, a dental student at the University of Southern California, to tell her he'd be home for lunch. A month later, Prasanna hanged herself in their apartment. In a suicide note to her brother, she wrote: "I love Vamsi too much, and the pain is excruciating. . . . If there exists any form after this life, I'll be with him. If not, it will relieve me from this deep pain."

Some accounts of survivors and those left behind would be kept private or be handed down within families like heirlooms. Others would appear in newspapers, magazines, websites, and movies, or in oral histories and aural archives. Some would be memorialized in books.

IN THEIR CONNECTICUT home, Lee and Eunice Hanson watched the televised explosion of the plane carrying their son, Peter, daughter-in-law, Sue Kim, and granddaughter, Christine. The strike into the South Tower ended the Airfone call between Peter and Lee. Later, Eunice realized: "We heard his first cries and his last cries."

They endured the unspeakable, and yet they endured. Lee and Eunice drove to Peter, Sue, and Christine's house to pick through hairbrushes to provide DNA matches. They found Christine's Peter Rabbit doll, tucked into her bed, waiting for her. (They subsequently donated it to the National September 11 Memorial & Museum in New York.)

Eventually, they heard from the medical examiner's office, and Lee asked a friend who ran a funeral home to go to New York. Back at the funeral home, the friend left Lee alone in a room with a little box. Inside rested a single piece of bone that fitted into his hand. Holding it, Lee told himself, "That's all I have of my beautiful red-headed son." Lee and Eunice held out hope that someday remains would be linked to Sue and Christine.

Lee and Eunice walked among the graduates at Boston University in May 2002 to accept Sue's posthumous PhD, which her mentor completed on her behalf. The school established the Annual Sue Kim Hanson Lecture in Immunology. Lee and Eunice created undergraduate writing prizes and an author series in Peter's name at Northeastern University. Peter, Sue, and Christine were remembered in their hometown of Groton, Massachusetts, at a memorial garden with a flagpole, lilac bushes, and three strong trees surrounding a bronze plaque on a boulder.

A young classical composer named Carl Schroeder, who read about the Hansons, wrote a piece for orchestra called "Christine's Lullaby." People who heard it began to donate money, and soon these donations endowed a "pain-free" treatment room in the pediatrics department at the Boston Medical Center, where Sue had studied for her PhD. Butterflies painted on the walls paid tribute to the family, with tie-dyed wings for Peter.

Lee and Eunice spoke to reporters and to groups. They kept their tidy home filled with photos and mementos of their family. Eunice wrote public letters to her lost family. "We miss you so much," she wrote in one. "Peter, I still feel the terrible pain that went through my whole being when Dad, holding the phone, heard your last words. . . . The thought of the three of you in each other's arms in that final moment will never leave me. They tell me that there could not have been any pain, but you knew what was happening. How could those murderers have looked at the innocent people on the plane, at little Christine, and so cruelly kill? How could their leaders, hidden and protected in a far-off land, laugh and joke about their lack of humanity?"

Into their eighties, Lee and Eunice traveled repeatedly to Guantanamo Bay to represent Peter, Sue, and Christine during the run-up to the military prosecutions. When U.S. Navy SEALs killed Osama bin Laden, Eunice told a reporter she felt overcome by complicated feelings. "It was tears and mixed emotions that you shouldn't be happy for a man's death," Eunice said. "Then we came to the reality that this man was responsible for the deaths of thousands of people across the world. I feel so great for the justice—we waited ten years for this."

Then Eunice added: "It's not going to bring my kids back. . . . I don't believe in the word 'closure,' because this is not a book."

During the 2006 trial of al-Qaeda conspirator Zacarias Moussaoui, a prosecutor asked Lee Hanson to express his family's loss. "They took away our dreams," Lee said as jurors wept. "They took away our future. . . . Peter and Sue were just so much, so good, and so alive, and they lived life and they loved people and they loved their families. And we're going to miss all the celebrations. . . . All of those occasions when Christine would have graduated from kindergarten and from the first grade and so forth and so on. And I decided I was going to try to stick around long enough to make sure I got to her wedding someday. All those things are gone. And it is, it is so sad."

PEG OGONOWSKI, WIFE of Flight 11 pilot John Ogonowski, remained with American Airlines as a flight attendant until 2008. She flew less often after 9/11, but she didn't want to give up her job entirely. Partly it was financial, as she raised their three daughters. But she had another reason: "I didn't want terrorists telling me I had to leave my career."

Peg remarried shortly before the tenth anniversary of the attacks, but she kept White Gate Farm. In the years after 9/11, Cambodian farmers continued to grow vegetables there. Seeing them working on the land that John loved, with the skills he had taught them, gave Peg solace.

ON SEPTEMBER 11, as Andrea LeBlanc got ready for work in Rye, New Hampshire, a carpenter friend worked outside, building a new deck. The carpenter heard on the radio that a plane had hit the North Tower, so he called to Andrea and rushed inside to turn on the television. They watched as the second plane hit the South Tower.

"Don't be on that plane," Andrea pleaded. "Don't be on that plane."

As she coped with Bob's death, Andrea LeBlanc grew convinced that the last thing he would have wanted was for anyone else to suffer, anywhere, in his name. She took what she acknowledged to be the unpopular position that the wrong response was more violence, and so became involved with like-minded people affected by the attacks who organized into an antiwar group called September Eleventh Families for Peaceful Tomorrows. The name came from a statement by Dr. Martin Luther King Jr.: "Wars are poor chisels for carving out peaceful tomorrows."

As described in the pages of the group's history, members believed that "the violence that took their loved ones' lives could spin out of control, and fear could be manipulated by politicians and the media to justify foreign and domestic policies that would increase violence while decreasing U.S. citizens' rights and liberties over the years to come."

As Andrea put it: "It was obvious pretty quickly that we would be bombing Afghanistan, and I felt so bad for those people. What happened to those young men, what terrible things had shaped them into people who were willing to kill innocent people? . . . Something bad has to happen, whether it's environment or genetics or society, and I think we bear some responsibility as a nation for our foreign policy. That's one of the scary things right now: Are the decisions being made radicalizing more people?"

REPAIRS TO THE collapsed section of the Pentagon, dubbed the Phoenix Project, proceeded at a breakneck pace. Three thousand construction workers set a one-year completion deadline for themselves, independent of the $700 million contract. Among the hardhats working around the clock was Michael Flocco, a sheet metal worker whose only child, twenty-one-year-old Petty Officer Matthew Flocco, died in the Pentagon attack. New safety features in the rebuilt 400,000-square-foot Wedge One included bright exit signs placed on walls and doors inches above the floor, so they'd be visible to someone crawling through dense smoke.

At a ceremony on September 11, 2002, Defense Secretary Donald H. Rumsfeld declared: "We're here today to honor those who died in this place and to rededicate ourselves to the cause for which they gave their lives, the cause of human liberty."

Later in his speech, Rumsfeld said: "The road ahead is long. But while we have not yet achieved victory, we know in one important sense that the terrorists who attacked us have already been defeated. They were defeated before the first shot was fired in Afghanistan. They were defeated because they failed utterly to achieve their objectives. The terrorists wanted September 11 to be a day when innocence died. Instead it was a day when heroes were born."

ONE HUNDRED TWENTY-FIVE people were killed in the attack on the Pentagon, in addition to the fifty-nine passengers and crew members and five hijackers aboard Flight 77. Seventy of the Pentagon dead were civilians. Ninety-two Pentagon victims died on the first floor, thirty-one on the second floor, and two on the third floor. Among the dead were Lieutenant Colonel Karen Wagner and Chief Warrant Officer William Ruth, apparently of smoke inhalation, in the office where they'd huddled with Major John Thurman, who lived.

Remains of five Pentagon victims were never found, four from inside the building: Retired Army Colonel Ronald Golinski, a civilian Pentagon worker; Navy Petty Officer First Class Ronald Henanway, who left behind a three-year-old son and a one-year-old daughter; James T. Lynch, a civilian worker for the Navy known for passing out candies in the Pentagon hallways; and Rhonda Rasmussen, a civilian worker for the Army and the mother of four. The fifth was the youngest person aboard Flight 77: three-year-old Dana Falkenberg, curly-haired and princess-loving, who died with her parents and her big sister on their way to Australia.

Among the items found in the wreckage was the laminated prayer card Pilot Chic Burlingame carried everywhere, scorched by fire but with his mother's photograph intact beside the poet's words: "I am the soft stars that shine at night. Do not stand at my grave and cry; I am not there, I did not die."

His sister, Debra Burlingame, became an outspoken conservative activist, a board member of the National September 11 Memorial & Museum, and cofounder of 9/11 Families for a Safe & Strong America. Her group's mission: "We support the U.S. military and endorse the doctrine of pre-emption, supported by the 9/11 Commission's statement on terrorist threats: 'Once the danger has fully materialized, evident to all, mobilizing action is easier—but it then may be too late.'"

Debra Burlingame was an outspoken opponent of a plan to build an Islamic cultural center and mosque two blocks from Ground Zero. That proposal was dropped in 2011, replaced by plans for a forty-three-story condominium with fifty apartments, a three-story Islamic museum, and no mosque.

THE BADLY BURNED woman comforted by Father Stephen McGraw was Antoinette "Toni" Sherman, a thirty-five-year-old Army budget analyst. She died six days after the attack, the final Pentagon casualty and the only person to die after reaching a hospital. When he learned her full identity, Father McGraw wrote Antoinette's parents a letter, passing along her love. The man whom Father McGraw granted absolution, a civilian accountant named Juan Cruz-Santiago, suffered terribly from his burns and endured multiple surgeries, but he survived.

FOR RESCUING JERRY Henson at his Pentagon desk, Lieutenant Commander Dave Tarantino, MD, and Captain Dave Thomas each received the Navy and Marine Corps Medal for Heroism, as did SEAL Craig Powell. Lieutenant General Paul Carlton Jr. received an equivalent honor from the Air Force called the Airman's Medal. The awards were among the military's highest honors for noncombat heroism, but that didn't sit right with Jerry. He thought his rescuers deserved the Navy Cross or the Air Force Cross, for extraordinary valor in combat.

Dave Tarantino lent his medal to the Smithsonian Institution, which also collected Dave Thomas's ruined uniform and the TARANTINO nametag he'd snatched from his new friend's shirt. At a Smithsonian press conference, Dave Tarantino deflected praise: "I think hopefully the exhibit will show, and history will show, that it was a setback, but it was not a defeat. Even that day, people started responding. Jerry didn't give up. He could have easily given up in there, and he didn't."

Dave Thomas returned to work on September 12. With his team, he reworked the strategy and policy report they'd spent seventeen months writing, to reflect the post-9/11 reality. They completed the new report within weeks, to meet the original deadline. He called his best friend Bob Dolan's cellphone for days, but Bob never answered. An FBI agent found Bob's Naval Academy ring in the wreckage; a mortuary crew found some remains. Dave Thomas accompanied them for a burial at sea. He thought about Bob every day and never again wore his own academy ring. With Bob gone, it lost its meaning.

Dave Thomas rose to rear admiral and served as commander of Joint Task Force Guantanamo and the Guantanamo Bay detention camp, the military prison that held alleged 9/11 plotters and al-Qaeda combatants. Before he retired from the Navy in 2013, Dave Thomas served as the senior surface warfare officer in the Atlantic Fleet, where two ships, the USS *New York* and the USS *Arlington*, honored the sites of 9/11 attacks. In October 2016, Dave Thomas escorted Bob's widow, Lisa, down the aisle at the wedding of Lisa and Bob's son. At Lisa's request, Dave wore Bob's ceremonial sword.

When the United States went to war after 9/11, Dave Tarantino organized humanitarian assistance in Afghanistan and later Iraq, where he spent ten months as a U.S. adviser to the Iraqi Ministry of Health. During the next decade, he served as the first director of global health in the Office of the Secretary of Defense; spent a year as director of public health for Navy and Marine forces in Okinawa and the Pacific; and became director of medical programs for the Marine Corps, among other jobs devoted to helping others. Throughout, Dave Tarantino gave time and energy to the Yellow Ribbon Fund, a nonprofit devoted to the needs of injured service members and their families.

After retiring from the Navy in 2015, he spent a year traveling the world with a global humanitarian organization then became a senior adviser to the Department of Homeland Security, focused on strategy and health issues at the border. He and his wife, Margie, had twin daughters. Life outside the military gave him more time with them.

WHEN PETTY OFFICER Charles Lewis emerged from the wreckage, he reported that Jack Punches hadn't been there when the plane hit. While Jerry Henson spoke on the phone to cancel his hotel reservation, Jack slipped out to be among the men and women of the Navy Command Center, watching the big TVs to monitor the attack on New York. Jack died as he lived, in the thick of things, shoulder to shoulder with the young sailors he loved almost as much as his own family. Somewhere nearby must have been Bob Dolan, the Navy captain whom Dave Thomas had hoped to save.

Janice Punches had Jack buried in a private cemetery near her house; she wanted access to Jack's grave anytime she liked. She never remarried. Their daughter, Jennifer, married her boyfriend, Mike; they named their son Jack. Jack and Janice's son, Jeremy, named his firstborn Jackson. It hurt to think about how much the original Jack would have loved his grandchildren.

Twenty years old when his father died, Jeremy Punches returned to school and received a master's degree in defense studies from George Mason University. He went to work as a counterterrorism analyst for the CIA, a job he took out of a fierce determination to prevent other families from knowing such loss. Jennifer Punches kept Jack's sense of humor but lost her taste for birthday carrot cake. Even after it grew rancid, she refused to throw away the jug of milk from which she'd poured her father's last glassful. "It never heals," she'd say. "You just get better at putting the Band-Aid on."

JERRY HENSON SPENT four days in the hospital, coughing up clumps of black crud, then recovered at home. Dave Tarantino and Dave Thomas came to the Hensons' house for an emotional reunion and to meet Jerry's family. Weeks later, Jerry returned to work at the Pentagon, where he stayed until his retirement in 2004. He never stopped missing his friend and officemate Jack Punches.

Now and then, at a Virginia pool club where his daughters swam, Dave Tarantino ran into Jerry Henson's daughter Kelly and his granddaughter Kit, who was born seven years after 9/11 and became the light of Jerry's life. In his eighties, Jerry was slowed by Parkinson's disease, and his wife, Kathy, wondered whether the blows he had taken to the head played a role. Jerry woke every day knowing how close he'd come to death on a beautiful September morning. He knew that he might have been burned or buried alive if Dave Thomas hadn't spotted him and Dave Tarantino hadn't crawled through smoke and fire to perform his heroic leg-press. He knew that he might never have met his granddaughter.

When Jerry thought about all that, the former combat aviator, a man who gave a lifetime of service to his country and boundless love to his family, choked up. When words returned, Jerry's voice grew thick as he wondered softly: "How do you thank someone for saving your life?" Jerry Henson lived for more than sixteen years after 9/11. He died at age eighty-one in April 2018 and was buried with full military honors at Arlington National Cemetery.

FOR RESCUING CIVILIAN Lois Stevens and leading survivors from the Army personnel offices on the Pentagon's second floor, Lieutenant Colonel Marilyn Wills was awarded the Soldier's Medal, the Army's highest honor for noncombat heroism. For her burns and other injuries, she received a Purple Heart. Soldier's Medals and Purple Hearts also went to Lieutenant Colonel Robert Grunewald, Colonel Phil McNair, Specialist Michael Petrovich, and Sergeant Major Tony Rose. Major Regina Grant received a Purple Heart. Martha Carden, Lois Stevens, Betty Maxfield, Dalisay Olaes, and John Yates each received the civilian Defense of Freedom Medal.

After several days in the hospital and more recuperation at home, Marilyn returned to work. She learned that her friend and cubiclemate Marian Serva was one of twenty-nine people—military members, civilians, and contractors—killed inside or near the second-floor quarters of the Army personnel office. Also among the dead was the department's commanding officer, General Timothy Maude, the highest-ranking 9/11 casualty at the Pentagon, who'd earned a Bronze Star, a Purple Heart, a Distinguished Service Medal, and other commendations during a thirty-five-year career. Maude died alongside eight officers and civilians in his offices on the outer E Ring, almost directly above where Flight 77 entered the building. One of them was Max Beilke, who had survived two wars and was the last American combat soldier to leave Vietnam. Also among the dead was Odessa Morris, who'd gone to the ladies' room while waiting for Tracy Webb and Dalisay Olaes to get coffee. Another twenty-seven Army personnel department employees were injured.

For Marilyn, the Pentagon was never the same after the loss of her friends and colleagues, above all Marian. She'd also be haunted by the death of Major Dwayne Williams, who was seated in her chair, watching Marian's television, when Flight 77 hit.

After 9/11, Marilyn attended the Industrial College of the Armed Forces at Fort McNair in Washington, D.C., for a master's degree. Promoted to colonel, Marilyn arrived in Afghanistan for a yearlong tour on the day that Osama bin Laden was killed. After thirty years of service, she retired from the military in 2013. Her daughters grown, Marilyn launched a new career as a third-grade teacher.

Parents searched her name on Google and discovered her 9/11 heroism, then sent their children to school with questions. Marilyn decided it was time for show and tell. In early 2017, she led two classes of third-grade students and a surplus of adult chaperones on a field trip to the Pentagon. She showed them the window she had helped to pry open. Marilyn steeled herself and told them about her friend Marian Serva.

ARMY PERSONNEL SECURITY officer John Yates wouldn't remember much until a couple of days later, when his wife, Ellen, woke him in his hospital bed to tell him he had visitors. John was on a ventilator, the pain of his burns dulled by morphine, but he'd never forget the sight of President George W. Bush and First Lady Laura Bush standing at his bedside. As they prepared to leave, George Bush placed a hand on John's arm. John looked up from his bed and saw tears in the president's eyes.

MAJOR KEVIN NASYPANY left the NEADS bunker in Rome, New York, shortly before midnight on September 11. More than thirteen hours had elapsed since the fourth crash. The skies were empty of commercial planes. Yet Nasypany walked off the Ops Floor still expecting "the other shoe to drop" in the form of hijackings from overseas.

Despite his fatigue, he spent the drive home racking his brain, cataloging what he and his team did right and what they didn't. Nasypany believed there were "probably more" than four U.S. hijackings planned but the ground stop by the FAA's Ben Sliney had foiled the other plots. On the night of 9/11, sleep wasn't in the cards. Nasypany knew that the world had become vastly more complicated, which would mean longer workdays for him. He wandered into the half-finished bathroom he'd been remodeling and wondered, "Who's going to do this now?"

For the next year, Nasypany and his crew worked twelve-hour shifts, six days a week, always on edge, watching for the next wave that never came. Later, when he'd win praise and awards for his work juggling information and communications on 9/11, Nasypany would joke that he'd trained by listening to his kids talking over one another at the dinner table.

Despite increased budget and staffing, powerful updated technology to scour the skies, and new lines of communication between the FAA and the military, Nasypany remained on defense. "We're focused on internal and external, always looking for that next threat, because you just never know," he said shortly before he retired in 2014. "I mean, you hear all these talking heads on TV, 'Oh, this will never happen again, never happen again.' . . . Well, look, it *did* happen. So, we're all extremely vigilant, extremely vigilant."

Nasypany grew emotional when he talked about the passengers and crew of Flight 93. "They did my job," he'd say. "They basically did what I was going to have to do in the long run, because I was not going to let another aircraft—I *couldn't* let another aircraft go into D.C."

WORD QUICKLY SPREAD beyond Shanksville about why United Flight 93 crashed short of its target and how the uprising prevented more deaths on the ground. Amid the widespread grief and anger, the rebellion became an inspiration. Coroner Wally Miller put it simply: "The Americans lost some people doing it, but by God, the terrorists lost that day . . . at least in that one."

Wally earned praise as a champion of victims' families and a guardian of the place where they died, which he and many others considered hallowed ground. Yet shortly after the crash, he made an offhand comment about his initial impression of the site that unintentionally fueled conspiracy theorists. "All you saw was debris," he told a reporter for the *New York Times*. "There was no blood. You couldn't see any human remains. You almost thought the passengers had been dropped off somewhere." They hadn't been, of course.

Searchers scrabbling on their knees and climbing high into the hemlocks eventually recovered about 650 pounds of charred body parts. Wally estimated that that represented about 8 percent of the potential total. The remainder was consumed by the explosion or lost in the grass and trees. DNA profiles, dental records, and fingerprints enabled the identification of unique remains from each of the forty passengers and crew. One casket was sent to the family of a victim with a single tooth and a piece of skull the size of a quarter, but it was something.

Forty-eight samples of human remains, weighing less than ten pounds, came from four individuals whose DNA profiles matched none of the passengers or crew. Wally labeled them Terrorist A, B, C, and D, and gave the items to the FBI.

"A lot of people say to me, 'This had to be the worst case you ever had.' Well, it was the most complicated case I ever had," Wally said later. "But to say that this was the worst case I ever had, I can't say that, because every time my phone rings, it's somebody's worst day of their life."

NO MAPMAKER WOULD ever again overlook Shanksville. It took its place alongside another Pennsylvania hamlet forever linked to an epic battle: Gettysburg. Within hours of the crash, the little community was overrun by investigators and international media. Soon it became a destination for people eager to pay tribute to the men and women of Flight 93, tourists who wandered into town and searched for any connection to the heroes. They sought out witnesses and gravitated to makeshift memorials, where they left flags and hats, flowers and ribbons, hearts and teddy bears and stone angels and photos and insignia of first responders. More than a few wore "Let's Roll" T-shirts.

For the most part, the people of Shanksville and Somerset County embraced their place in history. They understood the responsibility that had befallen them. One Sunday months after the crash, Kathie Shaffer's sister, Donna Glessner, stood in church and suggested that locals become volunteer guides, an idea that turned into an enduring Flight 93 Ambassadors program. After weeks of organizing food and donated goods for searchers and investigators, Kathie led a massive oral history project for the National Park Service. Over roughly fifteen years, she conducted more than eight hundred deep and nuanced interviews, preserving the thoughts and experiences of emergency responders and Flight 93 family members, witnesses and investigators, government officials and reporters, and even the flight instructor who taught the terrorist pilot Ziad Jarrah. Donna did many of the transcriptions.

After years of disputes over land values and designs, plans took shape for what would become a bucolic 2,200-acre national park, including an outdoor Memorial Plaza with a quarter-mile walkway adjacent to the crash site, and a concrete-and-glass Visitor Center that movingly tells the story of the courage aboard Flight 93. Among the National Park Service rangers working there was Terry and Kathie

Shaffer's eldest son, Adam. In June 2018, the park service trucked four shipping containers to the site, containing virtually all the recovered wreckage of Flight 93. The contents were buried in a private ceremony, in an area accessible only to loved ones of the victims.

Of all their experiences after the crash, among the Shaffers' most treasured was Shanksville's first and only July Fourth parade, held in 2002. Politicians clamored to speak but were told no. This was a celebration of the country and of the ordinary people who responded that day, willing to do whatever they could. Thousands of cheering, flag-waving paradegoers lined the mile-long route as brass bands, tractors, a costumed Uncle Sam and Betsy Ross, and numerous fire companies rumbled by. Donna organized 273 people—more than the entire population of Shanksville—into a "living flag." Kathie was a star, Donna a stripe. As the "flag" marched down Main Street, crowd members placed hands over hearts and recited the Pledge of Allegiance.

Terry sat proudly on the podium, his tie knotted tightly despite sweltering summer heat. He kept proud watch over his fire company's gleaming new truck, with red, white, and blue hoses, a three-thousand-gallon tank, and a silver plate with the names of all forty heroes of Flight 93, financed partly by donations from people who never knew Shanksville existed before 9/11.

Six years later, in 2008, hundreds of motorcyclists, many of them current or former FDNY members, roared into town escorting a two-ton, fourteen-foot-high cross made from twisted steel from the World Trade Center. They erected it outside the fire station, on a pedestal shaped like the Pentagon, with a plaque that read: "Never Forget. We honor those who saw their untimely fate before them and chose to defeat evil to ensure America's freedom. Flight 93."

Another decade passed. Terry's health deteriorated, and the emotional toll became a burden. He left his job at Pepsi and his calling as Shanksville's fire chief, though he remained the station's president. One spring morning, touring a visitor around the memorial, he squinted into the sunlight as he gazed upon the field where he'd arrived hoping to help on 9/11. "You're seeing grass and I'm seeing plane parts," Terry said. "Every day for me is a Flight 93 day."

ELAINE DUCH, THE severely burned Port Authority senior adminis-trative assistant, was quickly transferred from St. Vincent's Hospital to New York–Presbyterian Hospital's Weill Cornell Burn Center. She arrived the night of September 11 in critical condition, in a medically induced coma, with severe lung injuries and burns over 77 percent of her body. Odds against her survival were high.

Elaine spent weeks on a ventilator, underwent seven skin graft-ing operations between September 18 and December 11, and battled bouts of pneumonia and bacteria in her bloodstream that doctors feared would kill her. Her twin, Janet, and their older sister, Mary-ann, kept vigil.

After three and a half months, on December 29, Elaine regained consciousness. She immediately picked up where she'd left off on 9/11: "You have to call Janet," she told a nurse. "You have to make sure Janet's okay." When Janet heard that Elaine was awake, she told her boyfriend: "I got my Christmas miracle."

Slowly at first, during the next several weeks Elaine learned the full story of what happened. Despite having experienced it firsthand, Elaine Duch might have been the last American adult to know the awful details of 9/11. It took far longer for her to accept that the tow-ers were gone and that so many lives were lost.

A month after she woke, in late January 2002, Elaine became the final 9/11 burn victim released from the Weill Cornell Burn Center. Overwhelmed by obituaries, desperate for any remotely good news related to the attacks, reporters and camera crews swarmed a press conference that marked her departure. Heavily bandaged, seated in a wheelchair, Elaine welcomed the media with a jaunty "Hi, every-one!" She thanked her doctors, nurses, and hospital staffers, then announced that she was ready to visit her favorite casino in Atlantic City. As for her long-term plan, she declared: "I want to get back to the way I was."

Elaine spent the next four and a half months in a rehabilitation cen-ter relearning how to stand, how to walk, how to use her hands, how to live. The last skill she relearned was how to navigate stairs. She finally returned home to Bayonne, New Jersey, on June 5, 2002. When friends from the Port Authority visited, she asked often about the messenger she'd gone to meet on the 88th floor, but no one seemed to know his fate. More than sixteen years passed before she learned that his name was Vaswald George Hall and that he'd been killed by the fireball that maimed her. She underwent more surgeries and extensive physical therapy. "It was hell and back ten times over," she said. Her condition improved, but her body could only heal so much.

As years passed, Elaine accepted that she'd never live without pain and scars; she'd never fully regain her independence; never return to work; never again drive a car. Her personality changed, too. She'd been the more confident, take-charge twin, but after 9/11 she retreated, and those roles fell to Janet. Elaine even lost her taste for coffee, her last drink before the fireball swallowed her. She often said: "The old Elaine died on 9/11."

AFTER 9/11, EMT Moussa "Moose" Diaz took a few days off then refused to sit at a desk on "light duty." He and his partner Paul Adams went back to work, though both felt health effects from their heroism that day. Moose suffered regular bouts of bronchitis, which his doctor suspected resulted from dust he'd inhaled. Over time he fell out of shape, got fitted for hearing aids, and battled difficult memories. But Moose remained on the job, fixing equipment on ambulances as the lead technician with the Medical Equipment Unit. He continued his long commute from upstate Monroe, New York, which had been home to five firefighters who died on 9/11.

As Moose looked ahead toward retirement, newbie EMTs occasionally spotted a photo of him with Elaine Duch, taken as he and Paul strapped her onto their stretcher. "You were there?" they'd ask in awe. Moose would tell his story, often ending with the line "I love my job. I love helping people." Other than that, he rarely talked about 9/11.

AFTER HIS ESCAPE from a North Tower elevator, Chris Young spent several weeks convalescing with family in North Carolina. Upon his return to New York, he worked for six months at the midtown office of Marsh & McLennan, which lost 295 employees and 63 consultants. Among them were Chris's supervisors, Angela Kyte, who dreamed of retiring on Cape Cod with her husband, and Dominique Pandolfo, who had just begun classes for an MBA at New York University.

Chris returned to acting, but his heart wasn't in it. His moods swung from helplessness to grandiosity, from feeling lost to believing that his 9/11 experience empowered him to accomplish anything. Ultimately, he embarked on a successful career in advertising. Chris's role in *Macbeth: The Comedy* remained his only movie credit.

When Chris first returned to New York after 9/11, friends dragged him to a Brooklyn bar to lift his spirits. That night he met a non-profit executive named Christian Lillis. They married in 2015. Long after his ordeal, Chris kept his temporary ID badge to a tower that no longer existed, his eyes shut in the grainy black-and-white photo. He donated it to the permanent collection of the September 11 Memorial & Museum in New York.

ON SEPTEMBER 11, the families of Brian Clark and Stan Praimnath felt certain they were dead. Only twenty-five minutes elapsed between their phone calls from the 31st floor and the South Tower's collapse. Brian's wife, Dianne, fainted when he called her from a ferry terminal at 11:15 a.m., more than an hour after the building fell. Stan found his wife, Jenny, holding their daughters on their front porch. Bloodied and filthy, he rushed to them, only to have his younger daughter recoil, not recognizing him.

Brian called Stan's home late that afternoon and left a message. After visiting a doctor, Stan called back an hour after September 11 passed into history. Each man again credited the other for saving his life. Later, they appeared on television shows and at memorial events, but their private moments meant more. Stan and Jenny sat at the Clark family table at the wedding of Brian and Dianne's eldest daughter, then repaid the honor at their older daughter's wedding. Brian introduced Stan to guests as "my only brother."

After 9/11, it emerged that eighteen people escaped from the 78th floor of the South Tower or above, all by using Stairwell A at least part of the way down. Brian and Stan were two of only four people who survived from above the sky lobby. Another was a telecommunications manager from Euro Brokers, Richard Fern, who reached a barbershop where he listened on his walkie-talkie to the excruciating sound of his friends pleading for help high in the tower. The fourth was Brian's friend, Ron DiFrancesco, who'd initially joined Brian in the search on the 81st floor. After being overcome by smoke, Ron caught up with the men who'd helped the heavyset woman. Ron told a reporter he believed that he climbed as high as the 91st floor before heading down. He escaped moments before the collapse, but suffered a serious head injury and burns to 60 percent of his body.

Stan recovered quickly from his injuries. His relationship with Brian was a salve to his survivor's guilt. Stan took comfort in having sent temp Delis Soriano home, but the collapse killed the colleagues and friends who'd returned to work with him after being assured that their building was safe. Among them were John "Jack" Andreacchio, Manny Gomez, Hideya Kawauchi, Alisha Levin, Joseph Zuccala, and Brian Thompson, with whom Stan had joked in the elevator: "You'd better start thinking relocation."

Stan returned to work but suffered years of nightmares. He obsessed over the question "Of all these good men and women, why me?" He delved deeper into his faith, gained a measure of acceptance, then began to say, "Why not me?" Stan spoke often at churches, relating his survival story through Bible lessons. "Life," he told people, "is about getting up and moving forward."

Still, parts of Stan Praimnath remained in the phantom South Tower. He saved the dust-covered shoes that carried him to safety, with shards of glass embedded in the soles, in a box he labeled DELIVERANCE. Stan also kept Brian's flashlight, which they later donated to the September 11 Memorial & Museum in New York.

Brian shed his first tears when he told his story in church the Sunday after 9/11. Euro Brokers lost sixty-one employees, including Bobby Coll, Dave Vera, Kevin York, and others who had helped the heavyset woman and the frail man up the stairs. Also killed were Susan Pollio and Randy Scott, who dropped the plaintive note from the 84th floor. His wife learned about the message a decade later, in August 2011, after a DNA match of the bloody fingerprint.

A week after 9/11, Brian dreamed that he was visited by José Marrero, the facilities manager he passed in the stairwell at the 68th floor, who also died in the collapse. In the dream, José wore a white shirt and a broad smile. "José, you're alive!" Brian called out. "How did you do that?" Brian heard no response. The dream didn't eliminate his sadness, but it left him at peace.

Before he retired in 2006, Brian helped to rebuild Euro Brokers and ran a relief fund that gave more than $5 million to the families of the company's lost workers.

As they lived with their losses, Brian and Stan cherished one gain. Nearly twenty years after the day they met, they still considered themselves blood brothers.

IT'S UNCLEAR EXACTLY how Alayne Gentul died, but there's evidence she wasn't inside the South Tower when it crumbled. Unlike the fragmented remains of people trapped in the collapse, Alayne's body was found intact, across the street, still dressed in her red blazer. Her husband, Jack, understood the implication: Alayne apparently was one of the scores of men and women who jumped or fell to their deaths. Jack had hoped that she was unconscious from smoke inhalation, inside the building when it fell. But if Alayne chose an ending on her own terms, he accepted that.

"The last thing that she would have wanted to do was die at that time in her life," Jack said years later. "She may have jumped. You can't stop it, but this is something that you can do. To be out of the smoke and the heat, and to be out in the air. It must have felt like flying."

Several days after 9/11, Jack sat weeping on his back porch, where the previous Sunday he and Alayne had shared wine, cheese, gratitude, and a kiss. Alayne, her colleague Ed Emery, and the technology consultants they tried to evacuate were among eighty-seven Fiduciary Trust Company employees and ten contractors who died. Alayne's boss told Jack that she saved forty people by ordering an immediate exit.

As Jack cried, wondering how he'd raise their two sons alone, his pastor called: "You need to come to the church, to see something." There, Jack found hundreds of chrysanthemums lining the sidewalk, a tribute from the children Alayne taught at Sunday school and the families she'd touched. Jack took it as a message.

"I had to recover for those boys, and I did," Jack said. "I didn't want to let her down. I wanted to make sure I would do everything to raise our kids and make sure their lives went on as well as I could, because I know that is what she would want me to do. I didn't want to fall down and become this sad, miserable nothing."

Their son Alex earned a master's degree at Rutgers, where his parents had met, and became a math teacher. Their son Rob went to New York University, where Alayne had earned her MBA, and became an app developer for Major League Baseball. They remained close to Alayne's parents. Jack remarried and became a stepfather to two children.

Jack credited his recovery to keeping Alayne's memory alive. The effort began in earnest several months after 9/11, with a call from a real estate agent. Unaware of Alayne's death, the agent said the beach house the couple liked during their bike survey of Barnegat Light had come onto the market. "I couldn't possibly do this now," Jack said. But later he called back. This was what Alayne had wanted. Jack bought the property and built a house he knew she would have liked.

"It's a way of keeping something special between us," Jack said. "And then handing it down to our children and, hopefully, our grandchildren."

WHEN FDNY PARAMEDIC Carlos Lillo didn't call or come home, his wife, Cecilia, hoped that he was in a hospital. When she learned that he'd died in the line of duty, Cecilia was overcome with grief. She suffered not only from his loss, but also from fear that Carlos had died trying to find her inside the North Tower.

Cecilia gained a measure of relief three days after 9/11, when Carlos's friend and fellow paramedic Edith Torres rushed to her house with a copy of a special edition of *Newsweek* magazine. A photo splashed across two pages showed Carlos tending to a bloody woman in a red dress, an image that confirmed for Cecilia that he'd kept his promise.

It's not entirely clear how Carlos died, but based on reports from colleagues, Cecilia believed that he answered a call to the South Tower lobby for a firefighter with chest pains minutes before the collapse. His partial remains were identified in July 2002, and Cecilia held what she called the "first funeral" that September, to coincide with their second wedding anniversary. The serial discovery of more partial remains, spread over more than a decade, led to more funerals and memorials, including an elaborate shrine to Carlos's memory in their Long Island home. Cecilia grieved, too, for friends, among them the two women with whom she evacuated: Nancy Perez and Arlene Babakitis.

Cecilia canceled her appointment with the fertility specialist. She never had children. She struggled with her unusual, unwanted circumstances as both a 9/11 survivor and the grieving widow of an emergency responder. Yet in a way, that unique status meant that her life and Carlos's death would forever be intertwined.

"My husband was my hero," she'd say. "He will always be my hero. All I learned from him gave me the courage to go on."

Bit by bit, Cecilia rebuilt her life. She left her job with the Port Authority and doted on her nieces and nephews. She kept a special eye on her sister's daughter Casandra, with whom Carlos played on the last day of his life. In time, Cecilia began a relationship with a man who understood that Carlos remains her true love. When she dies, Cecilia wants a container holding her ashes to be placed inside her wedding gown, then sealed inside a double vault in a mausoleum where some of Carlos's remains already rest inside his dress uniform.

Carlos's death was honored in numerous tributes, but one carried special meaning for Cecilia. A few months before 9/11, on one of their daily commutes, they noticed an industrial property across from an elementary school in Astoria, Queens. The trash-strewn lot saddened Carlos, who told Cecilia he thought it would make a fine place for a father to toss a ball or enjoy a milkshake with his child. The City of New York acquired the property in 2007. Today it is an oasis of green space and flowering trees, a haven for city parents and children, named Carlos R. Lillo Park.

RON CLIFFORD WAS shattered by the confirmation that his sister, Ruth Clifford McCourt, and his niece, Juliana, were aboard United Airlines Flight 175. His brothers in Ireland had called earlier in the day, to check on him. Now he called them back. "We're in trouble," Ron said. "It's the end of the world."

Ron told himself that Ruth must have been a beacon of calm at the very end. He thought she must have comforted Juliana, holding her close as their plane veered wildly then dipped low over Manhattan, perhaps whispering an Irish lullaby in her ear. A thought both harrowing and strangely comforting entered Ron's mind. At the instant of Ruth and Juliana's murders, when Flight 175 pierced the South Tower, he'd been nearby. He'd felt the shock wave in every cell of his body.

At home with his family on the night of his daughter's eleventh birthday and the deaths of his sister and niece, Ron could still feel the vibrations of impact, bottomless bass chords resonating in the hollow of his chest. As long as he lived, Ron would remember what he was doing at the moment everything changed. He was kneeling on the floor of the Marriott lobby, tending to Jennieann Maffeo, reciting the Lord's Prayer.

Ron never rescheduled the September 11 meeting he'd considered crucial a day earlier. He learned that the man he was supposed to meet had fled the Marriott after the first explosion. Ron decided that he didn't want to work with someone who might have brushed past without stopping when he and Jennieann huddled on the lobby floor. Ron built a new business that worked out better than he ever imagined.

In the days after 9/11, Ron learned that a young man named Bryan Craig Bennett, a salesman for a division of the investment banking firm Cantor Fitzgerald, had died on the 104th or 105th floor of the North Tower. Bryan grew up in Ron's New Jersey house; Bryan's mother had sold it to Ron and Brigid a year earlier. Adding to the litany of loss, Jennieann Maffeo was right about the colleague she'd stood beside at the shuttle stop. Wai-ching Chung was dead. So was his niece, Maurita Tam, an executive assistant on the 99th floor of the South Tower, whom Wai-ching had amused as a child with floating tissue tricks.

The funeral of Ruth and Juliana drew more than twelve hundred people and received media coverage around the world. Ron organized a celebration of their lives at the Connecticut home of Ruth and her husband, David, complete with an Irish bagpiper. Later, Ruth's wedding ring turned up, undamaged, as workers sifted through the Trade Center wreckage. Its South Sea pearls and diamonds still glistened.

Meanwhile, family and friends sat vigil by Jennieann Maffeo's bedside. She survived the night, then another, and another. The hospital allowed only two visitors in her room at a time, so Jennieann's supporters often filled the waiting area. The two-to-a-room rule didn't apply to Jennieann's mother, Frances, who wouldn't leave anyway.

Andrea Maffeo kept Ron updated about Jennieann's condition, and a few days after the attacks he came to visit. With Ruth and Juliana gone, Ron desperately wanted Jennieann to live, for something positive to emerge from his suffering.

Jennieann's father took Ron's face in both hands and started to cry. "Thank you for giving me my daughter back," Sam Maffeo said. Ron sobbed, too.

Ron moved to the head of Jennieann's bed and whispered: "Please get better." Ron left his yellow silk tie on her pillow.

Weeks went by and Jennieann seemed to defy doctors' dire predictions. She remained in a coma and still hadn't spoken, but she improved enough to be moved to a recovery room. Finally, though, after forty-one days, after more surgeries than Andrea could count, after her percentage of burned skin area was corrected to more than 90, Jennieann's kidneys failed. She died October 23, 2001.

Ron heard the news on the radio after dropping his brothers at New York's John F. Kennedy International Airport, after they had come from Ireland to see the pit that had been dubbed Ground Zero. Ron pulled his car over and wept.

On the same day Jennieann died, the coroner's office called Ron. A DNA match had identified some of Ruth's remains. Ron brought the ashes to Cork, to bury them alongside his and Ruth's grandparents, their brother who died young, and their father, Valentine. A couple of years later, around Christmas, the coroner called again to say that some of Juliana's remains had been identified. Ron returned to Ireland to place them with the remains of her mother.

Ruth and Juliana became symbols of 9/11 in Ireland, and reporters regularly asked Ron to talk about them. On one visit to Ireland with Brigid and their daughter Monica, Ron was cornered by an official at a sports club.

"I know you," she said. "You're the man that was on the television. You're the man whose sister and niece died in 9/11, and they're buried out in St. Finbarr's Cemetery."

"Yeah," Ron said. "That's me."

"Listen to this, now," the woman said. "I drive by there with me mother every week, and we say a Hail Mary to them. As long as me and me mother are alive in Cork, your sister and niece will never be lonely."

Kindness from strangers helped, but for a long time Ron had awful dreams, often featuring Jennieann, but sometimes with Ruth, in which he felt trapped somewhere, trying to escape. Awake, he suffered post-traumatic stress. He'd stand in the shower, scraping and scrubbing his feet raw. Ron's compulsion stemmed from a mental link he'd made between his experiences and the Holocaust movie *Schindler's List*. In a wrenching scene, Otto Schindler steps into a street as the ashes of murdered Jews float down around him. Ron suspected that outside the Marriott, he'd trampled on the ashes of the dead.

Therapy eased his mind, as did sailing in the Long Island Sound, and the love of Brigid and Monica.

For years Ron ached for Ruth's husband, his brother-in-law, David

McCourt. Searchers at Ground Zero found Ruth's battered Hermès wallet, complete with the papal coin from her and David's Vatican wedding. David put the coin on a chain and wore it over his heart. David remarried a decade after 9/11 but died of cancer in 2013.

Allan Hackel paid tribute to his late wife with the Paige Farley-Hackel Memorial Playground in Boston's Dorchester neighborhood. As he approached ninety, Allan still cried for the love of his life.

Brigid and Monica accompanied Ron when he testified at the Moussaoui trial. Andrea Maffeo testified immediately after Ron. Nearly two decades after 9/11, Andrea still cherished the yellow tie Ron left on Jennieann's pillow, a silken bond between them and their lost sisters.

Now and then, Ron pulled out an email that Ruth sent him three months before she died, addressed to "Dearest, Dearest Ron." She wrote: "Congratulations on life. You're doing a great job. It's time to sit back, take a breath, and let the energy flow." It was signed, "Love you always, Ruthie."

As Ron moved ahead with his life, he came to believe that he'd initially misunderstood what happened to him on 9/11. He'd been hailed as a hero, but Ron realized that if not for Jennieann, he might have run through the North Tower lobby and out onto the plaza where fiery wreckage, building debris, and people rained down.

"Do I believe there's a hereafter? There's something. You can't prove it for some reason, but there's something else at work here. The meeting was changed to that place, I happened to be there at the right time, and that got me out of there and got me home to my daughter's eleventh birthday. Jennieann Maffeo was put there for me."

CAPTAIN JAY JONAS and the men of Ladder 6 soon realized that if they hadn't encountered Josephine Harris inside Stairwell B, they might have been among the dead.

If they had stuck to their original double-step evacuation pace, they might have caught up to Chief Richard Prunty in the lobby where he died. If they'd rushed faster, they might have been just outside the

North Tower, where the collapse crushed Chief of Department Peter Ganci Jr., First Deputy Commissioner William Feehan, and numerous others. Or, if they'd hung back like some other firefighters, if Jay hadn't ordered an immediate retreat when the South Tower fell, Ladder 6 might have been higher inside the tower, with no time to rescue themselves or Josephine.

Jay would spend long nights picturing the faces of lost friends and replaying a conversation he had on 9/11 with a fellow captain near the pile of rubble. "Congratulations!" Captain Jim Riches told Jay that day. "I heard all your radio transmissions. That's the most dramatic thing I've ever heard in my career. By the way, did you see Engine 4?" Jim Riches was looking for his eldest son, Firefighter Jimmy Riches. Jay hadn't seen him.

Those thoughts would come later. First, on the afternoon of 9/11, Jay had to get home. The North Tower collapse had destroyed Ladder 6's truck, leaving only his tired legs to carry him. Jay straightened his helmet, gripped his pry bar, and walked north through streets filled with the silt of pulverized buildings, empty of cars and people.

As he entered Chinatown, Jay strode down the center of Canal Street, on any other day a pedestrian death wish. He felt like the cartoon character Pigpen, filth clinging to his skin, trailing a cloud of dust. Jay sensed eyes upon him, so he turned around to see about twenty Chinese men following him at a respectful distance.

One fell in step alongside Jay. "You okay?" the man asked.

"Yeah," Jay said, "I'm okay as long as I keep going. If I stop, I'm not going to want to start again."

"Where you going?"

"The firehouse on Canal and Allen."

"Okay," the man said, "we'll make sure you get there."

With the Chinese men as a solemn honor guard, Jay walked the rest of the way to the Ladder 6 firehouse. There he called his weeping wife, Judy, and told her how much he loved her and their children.

The escort from the Chinese men was among thousands of tributes, official and unofficial, to the heroism of New York firefighters, Port Authority and New York City police officers, and other emergency responders. In the weeks and months that followed, amid funerals and tears, the accolades built into a national outpouring of gratitude.

Among the dead were 343 FDNY firefighters and paramedics. The loss dwarfed the FDNY's previous worst day, in 1966, when a fire in a Manhattan brownstone killed twelve firefighters. Those lost on 9/11 included more than a dozen of the friends and colleagues Jay encountered that day in person or on the radio: Chief Richard Prunty; captains Paddy Brown, Billy Burke, and Terry Hatton, who died without knowing that his wife was pregnant; lieutenants Pete Freund, Dennis Mojica, and Mike Warchola; and firefighters Faustino Apostol, Andy Fredericks, Scott Kopytko, Kenneth Marino, Gerry Nevins, Doug Oelschlager, and Dave Weiss.

Lieutenant William "Billy" McGinn, who worked under Jay years earlier in Ladder 11, and who immediately warned that "it could be a terrorist attack," was also among the dead. Chief Joseph Pfeifer's brother, Lieutenant Kevin Pfeifer, died in the collapse of the North Tower. The FDNY's annual fitness award was renamed for Battalion Chief Orio Palmer, whom Jay saw in the North Tower lobby before Orio led the charge up the South Tower.

Also killed was Firefighter James Riches, whom searchers found in what had been the North Tower lobby. "Big Jimmy" Riches and his three younger sons carried the flag-draped remains of "Little Jimmy" from Ground Zero. Later, all three surviving Riches brothers joined the FDNY, like their father and brother before them.

The death toll included thirty-seven Port Authority Police officers, among them Superintendent Ferdinand "Fred" Morrone, whom Cecilia Lillo passed in the stairwell; twenty-three members of the New York Police Department; eight medics from private ambulance companies; and one member of the New York Fire Patrol. Plus, one bomb-sniffing Labrador retriever, Dave Lim's partner, Sirius.

Amid the grief, word spread about what became known as the Miracle of Stairwell B. Producers for NBC reconnected the men of Ladder 6 with Josephine Harris on the show *Dateline*. With the camera rolling, they credited her with their survival and she lavished them with praise. "They are [the most] strong, brave, caring, kind people I have ever met," Josephine said. "When I was scared, they held my hand. They took off their jackets and gave them to me when I was cold. They told me not to be afraid, they would get me out. And they did. They are magnificent."

Josephine was made an honorary member of Ladder 6, and they bestowed upon her a title even higher than Chief. Her jacket read GUARDIAN ANGEL.

Their mutual affection continued in the years that followed. Sal D'Agostino invited Josephine to his wedding. Jay and the rest included her in media interviews, September 11 memorial events, and parades. At one of them, she waved like royalty from a convertible while they walked alongside, her uniformed footmen.

The bond remained strong despite moves to new firehouses, promotions, and retirements that sent the men of Ladder 6 in different directions. Yet even as she reveled in their connection, Josephine remained intensely private, never letting them know about her financial hardships and significant health issues. Jay expected Josephine to join them during events and media appearances marking the tenth anniversary of 9/11, but in January 2011 she suffered an apparent heart attack in her Brooklyn apartment.

Josephine Harris was pronounced dead at age sixty-nine. Her funeral mass was conducted by Cardinal Edward Egan, archbishop emeritus of New York, with luminaries including former Mayor Rudy Giuliani. She was laid to rest inside a blue steel coffin, its silken interior embroidered with an image of a firefighter walking hand in hand with an angel. Jay and the men of Ladder 6 carried Josephine one last time, as pallbearers.

As the years passed, Jay struggled at times with survivor's guilt and compromised lungs. Still he rose again, from battalion chief to deputy chief, overseeing FDNY units in the Bronx and upper Manhattan. He found satisfaction in raising his family, leading his troops, and publishing a monthly training and safety newsletter that taught lessons and techniques from historic New York fires. He often featured heroic exploits by the men who died around him on 9/11.

As his own retirement neared, Jay sat one autumn afternoon inside a Bronx firehouse whose red door displayed two angels kneeling with a hunched firefighter. Above it, in elaborate script, was inscribed THE DAY THE ANGELS CRIED, SEPTEMBER 11, 2001. Jay poured black coffee, closed his office door, and returned to 9/11. He told how Ladder 6 climbed into Stairwell B of the North Tower, how they saved Josephine and she saved them, and how so many innocent, brave people died at the hands of murderous zealots.

Scene by scene, stair by stair, Jay told the story of heroes named Irons Man Billy Butler, Roof Man Sal D'Agostino, Can Man Tommy Falco, Tiller Man Matt Komorowski, Chauffeur Mike Meldrum, and himself, Captain Jay Jonas. Hours passed, and night fell. At the story's end, Jay leaned back in his chair and said softly: "I was never so proud to be a fireman as I was that day."

APPENDIX 1

The Fallen

FOR EVERY STORY TOLD IN THIS BOOK, A THOUSAND OTHERS ALSO deserve to be remembered. Stories of close calls and selfless acts, of waiting and deliverance, of loss and pain. Although it was impossible to include them all here, this book would feel incomplete without the names of all the men, women, and children known to have been killed in the attacks and the immediate aftermath.

Below is an alphabetical list of the nearly three thousand names as they appear inscribed in bronze on the 9/11 Memorial in New York. The "address" following each indicates where the name is located—at the north or the south reflecting pool—and the panel number on which it appears.

Gordon M. Aamoth, Jr. S-49
Edelmiro Abad S-40
Marie Rose Abad S-34
Andrew Anthony Abate N-57
Vincent Paul Abate N-57
Laurence Christopher Abel N-32
Alona Abraham S-4
William F. Abrahamson N-7
Richard Anthony Aceto N-4
Heinrich Bernhard Ackermann S-55
Paul Acquaviva N-37
Christian Adams S-68
Donald LaRoy Adams N-55
Patrick Adams S-45

Shannon Lewis Adams N-49
Stephen George Adams N-70
Ignatius Udo Adanga N-71
Christy A. Addamo N-8
Terence Edward Adderley, Jr. N-58
Sophia B. Addo N-68
Lee Adler N-37
Daniel Thomas Afflitto N-25
Emmanuel Akwasi Afuakwah N-71
Alok Agarwal N-36
Mukul Kumar Agarwala S-43
Joseph Agnello S-11
David Scott Agnes N-47
Joao Alberto da Fonseca Aguiar, Jr. S-34

Brian G. Ahearn S-13

Jeremiah Joseph Ahern S-47

Joanne Marie Ahladiotis N-37

Shabbir Ahmed N-70

Terrance Andre Aiken N-17

Godwin O. Ajala S-65

Trudi M. Alagero N-5

Andrew Alameno N-52

Margaret Ann Alario S-63

Gary M. Albero S-63

Jon Leslie Albert N-7

Peter Craig Alderman N-21

Jacquelyn Delaine Aldridge-Frederick N-10

David D. Alger N-59

Ernest Alikakos S-47

Edward L. Allegretto N-40

Eric Allen S-21

Joseph Ryan Allen N-41

Richard Dennis Allen S-21

Richard L. Allen N-19

Christopher E. Allingham N-42

Anna S. W. Allison N-2

Janet Marie Alonso N-5

Anthony Alvarado N-23

Antonio Javier Alvarez N-70

Victoria Alvarez-Brito N-8

Telmo E. Alvear N-71

Cesar Amoranto Alviar N-16

Tariq Amanullah S-42

Angelo Amaranto N-64

James M. Amato S-7

Joseph Amatuccio S-24

Paul W. Ambrose S-70

Christopher Charles Amoroso S-28

Craig Scott Amundson S-74

Kazuhiro Anai N-63

Calixto Anaya, Jr. S-21

Joseph P. Anchundia S-52

Kermit Charles Anderson N-9

Yvette Constance Anderson S-48

John Jack Andreacchio S-44

Michael Rourke Andrews N-53

Jean Ann Andrucki N-66

Siew-Nya Ang N-5

Joseph Angelini, Sr. S-9

Joseph John Angelini, Jr. S-9

David Lawrence Angell N-1

Mary Lynn Edwards Angell N-1

Laura Angilletta N-32

Doreen J. Angrisani N-15

Lorraine Antigua N-53

Seima David Aoyama N-2

Peter Paul Apollo N-26

Faustino Apostol, Jr. S-6

Frank Thomas Aquilino N-39

Patrick Michael Aranyos S-30

David Gregory Arce S-13

Michael George Arczynski S-54

Louis Arena S-5

Barbara Jean Arestegui N-74

Adam P. Arias S-31

Michael J. Armstrong N-43

Jack Charles Aron N-4

Joshua Todd Aron N-42

Richard Avery Aronow N-66

Myra Joy Aronson N-74

Japhet Jesse Aryee S-48

Carl Francis Asaro S-10

Michael A. Asciak N-63

Michael Edward Asher N-36

Janice Marie Ashley N-58

Thomas J. Ashton N-19

Manuel O. Asitimbay N-68

Gregg A. Atlas S-5

Gerald Thomas Atwood S-11

James Audiffred N-64

Louis F. Aversano, Jr. S-58

Ezra Aviles N-65

Sandy Ayala N-70

Arlene T. Babakitis N-66

Eustace R. Bacchus N-71

John J. Badagliacca N-52

Jane Ellen Baeszler N-43

Robert J. Baierwalter S-63

Andrew J. Bailey N-12

Brett T. Bailey S-31

Garnet Ace Bailey S-3
Tatyana Bakalinskaya N-17
Michael S. Baksh N-16
Sharon M. Balkcom N-7
Michael Andrew Bane N-14
Katherine Bantis N-12
Gerard Baptiste S-14
Walter Baran S-40
Gerard A. Barbara S-18
Paul Vincent Barbaro N-36
James William Barbella S-26
Victor Daniel Barbosa S-37
Christine Johnna Barbuto N-1
Colleen Ann Barkow N-32
David Michael Barkway N-42
Matthew Barnes S-21
Melissa Rose Barnes S-72
Sheila Patricia Barnes S-58
Evan Jay Baron N-60
Renee Barrett-Arjune N-48
Arthur Thaddeus Barry S-20
Diane G. Barry S-56
Maurice Vincent Barry S-28
Scott D. Bart N-9
Carlton W. Bartels N-50
Guy Barzvi N-48
Inna B. Basina N-48
Alysia Christine Burton Basmajian N-47
Kenneth William Basnicki N-21
Steven Joseph Bates S-6
Paul James Battaglia N-4
W. David Bauer N-37
Ivhan Luis Carpio Bautista N-69
Marlyn Capito Bautista N-6
Mark Lawrence Bavis S-3
Jasper Baxter S-45
Lorraine G. Bay S-67
Michele Beale N-20
Todd M. Beamer S-68
Paul Frederick Beatini S-63
Jane S. Beatty N-9
Alan Anthony Beaven S-67
Lawrence Ira Beck N-31
Manette Marie Beckles S-42

Carl John Bedigian S-21
Michael Ernest Beekman S-48
Maria A. Behr N-27
Max J. Beilke S-1
Yelena Belilovsky N-61
Nina Patrice Bell N-8
Debbie S. Bellows N-37
Stephen Elliot Belson S-17
Paul M. Benedetti S-62
Denise Lenore Benedetto S-60
Bryan Craig Bennett N-55
Eric L. Bennett N-65
Oliver Bennett N-20
Margaret L. Benson N-66
Dominick J. Berardi N-31
James Patrick Berger S-56
Steven Howard Berger S-48
John P. Bergin S-6
Alvin Bergsohn N-25
Daniel David Bergstein N-66
Graham Andrew Berkeley S-3
Michael J. Berkeley N-67
Donna M. Bernaerts N-16
David W. Bernard S-66
William H. Bernstein N-56
David M. Berray N-20
David Shelby Berry S-36
Joseph John Berry S-36
William Reed Bethke N-10
Yeneneh Betru S-69
Timothy D. Betterly N-41
Carolyn Mayer Beug N-1
Edward Frank Beyea N-65
Paul Michael Beyer S-14
Anil Tahilram Bharvaney N-22
Bella J. Bhukhan N-49
Shimmy D. Biegeleisen S-42
Peter Alexander Bielfeld S-18
William G. Biggart S-66
Brian Eugene Bilcher S-14
Mark Bingham S-67
Carl Vincent Bini S-6
Gary Eugene Bird N-13
Joshua David Birnbaum N-42

George John Bishop S-59

Kris Romeo Bishundat S-72

Jeffrey Donald Bittner S-35

Albert Balewa Blackman, Jr. N-48

Christopher Joseph Blackwell S-15

Carrie Rosetta Blagburn S-1

Susan Leigh Blair S-56

Harry Blanding, Jr. S-62

Janice Lee Blaney N-16

Craig Michael Blass N-28

Rita Blau S-41

Richard Middleton Blood, Jr. S-62

Michael Andrew Boccardi N-59

John Paul Bocchi N-46

Michael L. Bocchino S-19

Susan M. Bochino S-62

Deora Frances Bodley S-68

Bruce Douglas Boehm N-41

Mary Catherine Murphy Boffa N-3

Nicholas Andrew Bogdan N-13

Darren Christopher Bohan S-56

Lawrence Francis Boisseau S-23

Vincent M. Boland, Jr. N-10

Touri Hamzavi Bolourchi S-4

Alan Bondarenko S-65

Andre Bonheur, Jr. N-58

Colin Arthur Bonnett N-14

Frank J. Bonomo S-12

Yvonne Lucia Bonomo N-18

Sean Booker, Sr. N-19

Kelly Ann Booms N-1

Canfield D. Boone S-74

Mary Jane Booth S-69

Sherry Ann Bordeaux S-42

Krystine Bordenabe S-34

Jerry J. Borg S-66

Martin Michael Boryczewski N-26

Richard Edward Bosco N-58

Klaus Bothe S-3

Carol Marie Bouchard N-75

J. Howard Boulton S-31

Francisco Eligio Bourdier S-38

Thomas Harold Bowden, Jr. N-26

Donna M. Bowen S-75

Kimberly S. Bowers N-36

Veronique Nicole Bowers N-70

Larry Bowman S-65

Shawn Edward Bowman, Jr. N-49

Kevin L. Bowser N-16

Gary R. Box S-6

Gennady Boyarsky N-18

Pamela Boyce N-58

Allen P. Boyle S-73

Michael Boyle S-13

Alfred J. Braca N-41

Sandra Conaty Brace N-18

Kevin Hugh Bracken S-15

Sandy Waugh Bradshaw S-67

David Brian Brady N-22

Alexander Braginsky N-22

Nicholas W. Brandemarti S-33

Daniel Raymond Brandhorst S-4

David Reed Gamboa Brandhorst S-4

Michelle Renee Bratton N-34

Patrice Braut N-10

Lydia Estelle Bravo N-11

Ronald Michael Breitweiser S-42

Edward A. Brennan III N-53

Frank H. Brennan N-55

Michael E. Brennan S-10

Peter Brennan S-8

Thomas More Brennan S-52

Daniel J. Brethel S-17

Gary Lee Bright S-64

Jonathan Eric Briley N-68

Mark A. Brisman S-45

Paul Gary Bristow N-20

Marion R. Britton S-67

Mark Francis Broderick N-28

Herman Charles Broghammer S-58

Keith A. Broomfield N-64

Bernard C. Brown II S-70

Janice Juloise Brown N-11

Lloyd Stanford Brown N-29

Patrick John Brown S-8

Bettina B. Browne-Radburn S-61

Mark Bruce S-52

Richard George Bruehert N-5

Andrew Brunn S-6

Vincent Edward Brunton S-20

Ronald Bucca S-14

Brandon J. Buchanan N-29

Greg J. Buck S-12

Dennis Buckley N-43

Nancy Clare Bueche S-61

Patrick Joseph Buhse N-53

John Edward Bulaga, Jr. N-34

Stephen Bruce Bunin N-37

Christopher L. Burford S-71

Matthew J. Burke N-29

Thomas Daniel Burke N-54

William Francis Burke, Jr. S-18

Charles F. Burlingame III S-69

Thomas E. Burnett, Jr. S-68

Donald J. Burns S-18

Kathleen Anne Burns S-43

Keith James Burns N-28

John Patrick Burnside S-12

Irina Buslo S-44

Milton G. Bustillo N-34

Thomas M. Butler S-7

Patrick Dennis Byrne S-8

Timothy G. Byrne S-50

Daniel M. Caballero S-72

Jesus Neptali Cabezas N-68

Lillian Caceres N-4

Brian Joseph Cachia N-34

Steven Dennis Cafiero, Jr. S-55

Richard Michael Caggiano N-26

Cecile Marella Caguicla N-7

John Brett Cahill S-3

Michael John Cahill N-11

Scott Walter Cahill N-42

Thomas Joseph Cahill N-40

George C. Cain S-20

Salvatore B. Calabro S-8

Joseph M. Calandrillo N-18

Philip V. Calcagno N-15

Edward Calderon S-26

Jose O. Calderon-Olmedo S-74

Kenneth Marcus Caldwell N-65

Dominick E. Calia N-43

Felix Bobby Calixte N-73

Francis Joseph Callahan S-17

Liam Callahan S-29

Suzanne M. Calley S-71

Gino Luigi Calvi N-51

Roko Camaj S-37

Michael F. Cammarata S-15

David Otey Campbell S-34

Geoffrey Thomas Campbell N-22

Robert Arthur Campbell S-44

Sandra Patricia Campbell N-37

Sean Thomas Canavan S-64

John A. Candela N-26

Vincent A. Cangelosi N-41

Stephen J. Cangialosi N-43

Lisa Bella Cannava N-58

Brian Cannizzaro S-8

Michael R. Canty N-61

Louis Anthony Caporicci N-53

Jonathan Neff Cappello N-52

James Christopher Cappers N-15

Richard Michael Caproni N-10

Jose Manuel Cardona N-62

Dennis M. Carey, Sr. S-7

Edward Carlino N-11

Michael Scott Carlo S-12

David G. Carlone S-63

Rosemarie C. Carlson N-67

Mark Stephen Carney N-65

Joyce Ann Carpeneto N-72

Jeremy Caz Carrington N-45

Michael T. Carroll S-8

Peter J. Carroll S-6

James Joseph Carson, Jr. N-35

Christoffer Mikael Carstanjen S-3

Angelene C. Carter S-76

James Marcel Cartier S-64

Sharon Ann Carver S-1

Vivian Casalduc N-65

John Francis Casazza N-52

Paul Regan Cascio S-30

Neilie Anne Heffernan Casey N-75

William Joseph Cashman S-68

Thomas Anthony Casoria S-18

William Otto Caspar N-13

Alejandro Castaño S-38

Arcelia Castillo N-5

Leonard M. Castrianno N-44

Jose Ramon Castro N-23

William E. Caswell S-70

Richard G. Catarelli N-9

Christopher Sean Caton N-54

Robert John Caufield N-19

Mary Teresa Caulfield N-9

Judson Cavalier S-52

Michael Joseph Cawley S-11

Jason David Cayne N-43

Juan Armando Ceballos S-37

Marcia G. Cecil-Carter N-63

Jason Michael Cefalu N-56

Thomas Joseph Celic N-12

Ana Mercedes Centeno N-14

Joni Cesta S-38

John J. Chada S-1

Jeffrey Marc Chairnoff S-51

Swarna Chalasani S-42

William A. Chalcoff N-16

Eli Chalouh S-48

Charles Lawrence Chan N-44

Mandy Chang S-44

Rosa Maria Chapa S-71

Mark Lawrence Charette N-4

David M. Charlebois S-69

Gregorio Manuel Chavez N-70

Pedro Francisco Checo S-39

Douglas MacMillan Cherry S-60

Stephen Patrick Cherry N-26

Vernon Paul Cherry S-11

Nestor Julio Chevalier, Jr. N-33

Swede Joseph Chevalier N-28

Alexander H. Chiang N-10

Dorothy J. Chiarchiaro N-58

Luis Alfonso Chimbo N-70

Robert Chin S-39

Eddie Wing-Wai Ching N-23

Nicholas Paul Chiofalo S-7

John G. Chipura S-21

Peter A. Chirchirillo N-5

Catherine Ellen Chirls N-55

Kyung Hee Casey Cho N-14

Abul K. Chowdhury N-36

Mohammad Salahuddin Chowdhury N-67

Kirsten Lail Christophe S-54

Pamela Chu N-29

Steven Paul Chucknick S-31

Wai Ching Chung S-53

Christopher Ciafardini N-60

Alex F. Ciccone N-8

Frances Ann Cilente N-37

Elaine Cillo N-6

Patricia Ann Cimaroli Massari and her unborn child N-11

Edna Cintron N-12

Nestor Andre Cintron III N-44

Robert D. Cirri, Sr. S-29

Juan Pablo Cisneros N-52

Benjamin Keefe Clark S-39

Eugene Clark S-56

Gregory Alan Clark N-31

Mannie Leroy Clark N-10

Sara M. Clark S-70

Thomas R. Clark S-51

Christopher Robert Clarke S-50

Donna Marie Clarke N-14

Michael J. Clarke S-16

Suria Rachel Emma Clarke N-34

Kevin Francis Cleary S-32

James D. Cleere N-5

Geoffrey W. Cloud N-47

Susan Marie Clyne N-8

Steven Coakley S-13

Jeffrey Alan Coale N-69

Patricia A. Cody N-8

Daniel Michael Coffey N-5

Jason Matthew Coffey N-5

Florence G. Cohen S-47

Kevin S. Cohen N-33

Anthony Joseph Coladonato N-36

Mark Joseph Colaio N-42

Stephen J. Colaio N-42

Christopher Michael Colasanti N-53

Kevin Nathaniel Colbert S-35

Michel P. Colbert N-52

Keith E. Coleman N-30

Scott Thomas Coleman N-30

Tarel Coleman S-23

Liam Joseph Colhoun N-73

Robert D. Colin S-61

Robert J. Coll S-31

Jean Marie Collin S-63

John Michael Collins S-22

Michael L. Collins N-36

Thomas Joseph Collins S-50

Joseph Kent Collison N-72

Jeffrey Dwayne Collman N-74

Patricia Malia Colodner N-6

Linda M. Colon N-3

Sol E. Colon S-58

Ronald Edward Comer N-11

Jaime Concepcion N-70

Albert Conde S-63

Denease Conley S-65

Susan P. Conlon N-73

Margaret Mary Conner N-31

Cynthia Marie Lise Connolly S-56

John E. Connolly, Jr. S-32

James Lee Connor S-50

Jonathan M. Connors N-25

Kevin Patrick Connors S-30

Kevin F. Conroy N-4

Brenda E. Conway N-12

Dennis Michael Cook N-40

Helen D. Cook N-72

Jeffrey W. Coombs N-2

John A. Cooper S-49

Julian T. Cooper S-73

Joseph John Coppo, Jr. N-43

Gerard J. Coppola N-63

Joseph Albert Corbett N-53

John J. Corcoran III S-4

Alejandro Cordero N-6

Robert Joseph Cordice S-7

Ruben D. Correa S-9

Danny A. Correa-Gutierrez N-7

Georgine Rose Corrigan S-68

James J. Corrigan, Ret. S-5

Carlos Cortés-Rodriguez S-65

Kevin Michael Cosgrove S-60

Dolores Marie Costa N-58

Digna Alexandra Costanza N-13

Charles Gregory Costello, Jr. N-64

Michael S. Costello N-26

Asia S. Cottom S-70

Conrod Kofi Cottoy, Sr. N-62

Martin John Coughlan S-64

John G. Coughlin S-23

Timothy J. Coughlin N-54

James E. Cove S-59

Andre Colin Cox N-23

Frederick John Cox S-50

James Raymond Coyle S-7

Michele Coyle-Eulau N-11

Christopher Seton Cramer S-42

Eric A. Cranford S-72

Denise Elizabeth Crant N-10

James Leslie Crawford, Jr. N-27

Robert James Crawford S-18

Tara Kathleen Creamer N-75

Joanne Mary Cregan N-37

Lucia Crifasi N-18

John A. Crisci S-8

Daniel Hal Crisman N-15

Dennis A. Cross S-6

Kevin R. Crotty S-52

Thomas G. Crotty S-53

John R. Crowe S-55

Welles Remy Crowther S-50

Robert L. Cruikshank N-58

John Robert Cruz N-49

Grace Alegre Cua S-39

Kenneth John Cubas S-43

Francisco Cruz Cubero S-65

Thelma Cuccinello N-1

Richard Joseph Cudina N-51

Neil James Cudmore N-20

Thomas Patrick Cullen III S-13

Joan Cullinan N-31

Joyce Rose Cummings S-39

Brian Thomas Cummins N-27

Michael Joseph Cunningham S-31

Robert Curatolo S-19

Laurence Damian Curia N-41

Paul Dario Curioli S-63

Patrick Joseph Currivan N-74

Beverly L. Curry N-35

Andrew Peter Charles Curry Green N-1

Michael Sean Curtin S-24

Patricia Cushing S-67

Gavin Cushny N-31

Caleb Arron Dack N-21

Carlos S. da Costa S-25

Jason M. Dahl S-67

Brian Paul Dale N-76

John D'Allara S-24

Vincent Gerard D'Amadeo N-32

Thomas A. Damaskinos N-32

Jack L. D'Ambrosi, Jr. N-45

Jeannine Damiani-Jones N-42

Manuel João DaMota N-71

Patrick W. Danahy S-40

Mary D'Antonio N-6

Vincent G. Danz S-24

Dwight Donald Darcy N-66

Elizabeth Ann Darling N-12

Annette Andrea Dataram N-69

Edward A. D'Atri S-6

Michael D. D'Auria S-16

Lawrence Davidson S-62

Michael Allen Davidson N-30

Scott Matthew Davidson S-10

Titus Davidson S-46

Niurka Davila N-66

Ada M. Davis S-75

Clinton Davis, Sr. S-28

Wayne Terrial Davis N-21

Anthony Richard Dawson N-22

Calvin Dawson S-32

Edward James Day S-15

William Thomas Dean N-11

Robert J. DeAngelis, Jr. S-64

Thomas Patrick DeAngelis S-16

Dorothy Alma de Araujo S-4

Ana Gloria Pocasangre Debarrera S-2

Tara E. Debek N-9

James D. Debeuneure S-70

Anna M. DeBin N-47

James V. DeBlase, Jr. N-51

Jayceryll Malabuyoc de Chavez S-40

Paul DeCola N-36

Gerald F. DeConto S-72

Simon Marash Dedvukaj N-64

Jason Christopher DeFazio N-40

David A. DeFeo S-49

Jennifer De Jesus S-46

Monique Effie DeJesus N-29

Nereida De Jesus S-60

Emy De La Peña S-40

Donald Arthur Delapenha S-36

Azucena Maria de la Torre N-47

Vito Joseph DeLeo N-63

Danielle Anne Delie N-3

Joseph A. Della Pietra N-40

Andrea DellaBella S-58

Palmina DelliGatti N-4

Colleen Ann Deloughery S-59

Joseph DeLuca S-68

Manuel Del Valle, Jr. S-16

Francis Albert De Martini S-27

Anthony Demas S-55

Martin N. DeMeo S-9

Francis Deming N-17

Carol Keyes Demitz S-42

Kevin Dennis N-44

Thomas Francis Dennis, Sr. N-56

Jean C. DePalma N-12

Jose Nicolas De Pena N-69

Robert John Deraney N-21

Michael DeRienzo N-53

David Paul DeRubbio S-14

Jemal Legesse DeSantis N-58

Christian Louis DeSimone N-4

Edward DeSimone III N-53

Andrew J. Desperito S-18

Michael Jude D'Esposito N-6

Cindy Ann Deuel N-59

Melanie Louise de Vere N-20

Jerry DeVito N-60
Robert P. Devitt, Jr. N-32
Dennis Lawrence Devlin S-15
Gerard P. Dewan S-8
Sulemanali Kassamali Dhanani S-53
Michael Louis DiAgostino N-49
Matthew Diaz N-24
Nancy Diaz N-70
Obdulio Ruiz Diaz N-71
Michael A. Diaz-Piedra III N-72
Judith Berquis Diaz-Sierra S-40
Patricia Florence Di Chiaro N-8
Rodney Dickens S-70
Jerry D. Dickerson S-74
Joseph Dermot Dickey, Jr. N-46
Lawrence Patrick Dickinson N-67
Michael D. Diehl S-40
John Difato N-58
Vincent Francis DiFazio N-55
Carl Anthony DiFranco N-4
Donald Joseph DiFranco N-64
John DiGiovanni N-73
Eddie A. Dillard S-70
Debra Ann Di Martino S-36
David DiMeglio N-2
Stephen Patrick Dimino N-53
William John Dimmling N-12
Christopher More Dincuff N-60
Jeffrey Mark Dingle N-21
Rena Sam Dinnoo N-12
Anthony Dionisio N-33
George DiPasquale S-17
Joseph Di Pilato S-46
Douglas Frank DiStefano N-49
Donald Americo DiTullio N-75
Ramzi A. Doany N-14
Johnnie Doctor, Jr. S-72
John Joseph Doherty S-60
Melissa Cándida Doi S-46
Brendan Dolan N-61
Robert E. Dolan, Jr. S-73
Neil Matthew Dollard N-40
James Domanico S-48
Benilda Pascua Domingo S-37

Alberto Dominguez N-2
Carlos Dominguez N-3
Jerome Mark Patrick Dominguez S-25
Kevin W. Donnelly S-6
Jacqueline Donovan S-33
William H. Donovan S-73
Stephen Scott Dorf S-32
Thomas Dowd N-55
Kevin Christopher Dowdell S-11
Mary Yolanda Dowling S-59
Raymond Matthew Downey, Sr. S-9
Frank Joseph Doyle S-34
Joseph Michael Doyle N-33
Randall L. Drake S-38
Patrick Joseph Driscoll S-68
Stephen Patrick Driscoll S-24
Charles A. Droz III S-70
Mirna A. Duarte N-16
Luke A. Dudek N-70
Christopher Michael Duffy S-35
Gerard J. Duffy S-10
Michael Joseph Duffy S-35
Thomas W. Duffy N-4
Antoinette Duger N-72
Jackie Sayegh Duggan N-69
Sareve Dukat S-48
Patrick Dunn S-72
Felicia Gail Dunn-Jones S-66
Christopher Joseph Dunne N-13
Richard Anthony Dunstan S-59
Patrick Thomas Dwyer N-25

Joseph Anthony Eacobacci N-50
John Bruce Eagleson S-66
Edward T. Earhart S-72
Robert Douglas Eaton N-46
Dean Phillip Eberling S-33
Margaret Ruth Echtermann S-48
Paul Robert Eckna N-28
Constantine Economos S-51
Barbara G. Edwards S-70
Dennis Michael Edwards N-54
Michael Hardy Edwards S-50
Christine Egan S-53

Lisa Erin Egan N-49
Martin J. Egan, Jr. S-11
Michael Egan S-53
Samantha Martin Egan N-49
Carole Eggert N-6
Lisa Caren Ehrlich S-62
John Ernst Eichler N-71
Eric Adam Eisenberg S-58
Daphne Ferlinda Elder N-8
Michael J. Elferis S-18
Mark Joseph Ellis S-25
Valerie Silver Ellis N-25
Albert Alfy William Elmarry N-36
Robert R. Elseth S-73
Edgar Hendricks Emery, Jr. S-41
Doris Suk-Yuen Eng N-70
Christopher Epps N-6
Ulf Ramm Ericson S-65
Erwin L. Erker N-5
William John Erwin N-46
Sarah Ali Escarcega N-20
Jose Espinal S-66
Fanny Espinoza N-47
Billy Scoop Esposito N-40
Bridget Ann Esposito N-18
Francis Esposito S-7
Michael A. Esposito S-7
Ruben Esquilin, Jr. S-39
Sadie Ette N-69
Barbara G. Etzold N-59
Eric Brian Evans S-59
Robert Edward Evans S-15
Meredith Emily June Ewart S-54

Catherine K. Fagan N-13
Patricia Mary Fagan S-55
Ivan Kyrillos Fairbanks-Barbosa N-43
Keith George Fairben S-26
Sandra Fajardo-Smith N-7
Charles S. Falkenberg S-69
Dana Falkenberg S-69
Zoe Falkenberg S-69
Jamie L. Fallon S-72
William F. Fallon N-65

William Lawrence Fallon, Jr. N-37
Anthony J. Fallone, Jr. N-51
Dolores Brigitte Fanelli N-5
Robert John Fangman S-2
John Joseph Fanning S-11
Kathleen Anne Faragher N-22
Thomas James Farino S-19
Nancy C. Doloszycki Farley N-18
Paige Marie Farley-Hackel N-75
Elizabeth Ann Farmer N-47
Douglas Jon Farnum N-10
John Gerard Farrell N-53
John W. Farrell S-51
Terrence Patrick Farrell S-11
Joseph D. Farrelly S-22
Thomas Patrick Farrelly N-17
Syed Abdul Fatha S-49
Christopher Edward Faughnan N-54
Wendy R. Faulkner S-61
Shannon Marie Fava N-35
Bernard D. Favuzza N-42
Robert Fazio, Jr. S-24
Ronald Carl Fazio, Sr. S-60
William M. Feehan S-18
Francis Jude Feely N-7
Garth Erin Feeney N-21
Sean Bernard Fegan N-60
Lee S. Fehling S-7
Peter Adam Feidelberg S-54
Alan D. Feinberg S-10
Rosa Maria Feliciano N-15
Edward P. Felt S-68
Edward Thomas Fergus, Jr. N-41
George J. Ferguson III S-37
J. Joseph Ferguson S-69
Henry Fernandez N-70
Judy Hazel Santillan Fernandez N-36
Julio Fernandez S-45
Elisa Giselle Ferraina N-20
Anne Marie Sallerin Ferreira N-44
Robert John Ferris S-60
David Francis Ferrugio N-56
Louis V. Fersini, Jr. N-43
Michael David Ferugio S-63

Bradley James Fetchet S-35
Jennifer Louise Fialko S-59
Kristen Nicole Fiedel N-6
Amelia V. Fields S-75
Samuel Fields S-65
Alexander Milan Filipov N-2
Michael Bradley Finnegan N-45
Timothy J. Finnerty N-52
Michael C. Fiore S-5
Stephen J. Fiorelli N-66
Paul M. Fiori N-24
John B. Fiorito N-41
John R. Fischer S-13
Andrew Fisher N-22
Bennett Lawson Fisher S-40
Gerald P. Fisher S-75
John Roger Fisher N-66
Thomas J. Fisher S-41
Lucy A. Fishman S-61
Ryan D. Fitzgerald S-40
Thomas James Fitzpatrick S-52
Richard P. Fitzsimons S-23
Salvatore Fiumefreddo N-24
Darlene E. Flagg S-70
Wilson F. Flagg S-70
Christina Donovan Flannery S-50
Eileen Flecha S-41
Andre G. Fletcher S-7
Carl M. Flickinger N-40
Matthew M. Flocco S-72
John Joseph Florio S-22
Joseph Walkden Flounders S-32
Carol Ann Flyzik N-1
David Fodor S-41
Michael N. Fodor S-11
Stephen Mark Fogel N-47
Thomas J. Foley S-16
Jane C. Folger S-67
David J. Fontana S-6
Chih Min Foo S-44
Delrose E. Forbes Cheatham N-48
Godwin Forde S-46
Donald A. Foreman S-27
Christopher Hugh Forsythe N-44

Claudia Alicia Foster N-56
Noel John Foster S-62
Sandra N. Foster S-71
Ana Fosteris S-61
Robert Joseph Foti S-20
Jeffrey Fox S-35
Virginia Elizabeth Fox N-10
Pauline Francis N-24
Virgin Lucy Francis N-69
Gary Jay Frank S-58
Morton H. Frank N-26
Peter Christopher Frank N-59
Colleen L. Fraser S-68
Richard K. Fraser S-59
Kevin J. Frawley S-33
Clyde Frazier, Jr. S-27
Lillian Inez Frederick S-58
Andrew Fredericks S-21
Tamitha Freeman S-58
Brett Owen Freiman S-46
Peter L. Freund S-7
Arlene Eva Fried N-46
Alan W. Friedlander S-58
Andrew Keith Friedman N-59
Paul J. Friedman N-75
Gregg J. Froehner S-29
Lisa Anne Frost S-3
Peter Christian Fry S-32
Clement A. Fumando N-33
Steven Elliot Furman N-50
Paul James Furmato N-26
Karleton Douglas Beye Fyfe N-1

Fredric Neal Gabler N-26
Richard Peter Gabriel S-70
Richard S. Gabrielle S-55
James Andrew Gadiel N-31
Pamela Lee Gaff S-55
Ervin Vincent Gailliard S-66
Deanna Lynn Galante and her unborn
    child N-37
Grace Catherine Galante N-37
Anthony Edward Gallagher N-50
Daniel James Gallagher N-28

John Patrick Gallagher N-49

Lourdes J. Galletti N-47

Cono E. Gallo N-61

Vincent Gallucci N-5

Thomas E. Galvin N-39

Giovanna Galletta Gambale N-34

Thomas Gambino, Jr. S-15

Giann F. Gamboa S-37

Ronald L. Gamboa S-4

Peter James Ganci, Jr. S-17

Michael Gann N-20

Charles William Garbarini S-12

Andrew Sonny Garcia S-68

Cesar R. Garcia N-5

David Garcia N-17

Jorge Luis Morron Garcia S-65

Juan Garcia N-23

Marlyn Del Carmen Garcia N-3

Christopher Samuel Gardner S-57

Douglas Benjamin Gardner N-38

Harvey Joseph Gardner III N-72

Jeffrey Brian Gardner N-4

Thomas A. Gardner S-8

William Arthur Gardner N-37

Frank Garfi N-25

Rocco Nino Gargano N-28

James M. Gartenberg N-64

Matthew David Garvey S-6

Bruce Gary S-15

Boyd Alan Gatton S-43

Donald Richard Gavagan, Jr. N-42

Peter Alan Gay N-2

Terence D. Gazzani N-51

Gary Paul Geidel S-10

Paul Hamilton Geier N-51

Julie M. Geis S-57

Peter Gerard Gelinas N-56

Steven Paul Geller N-29

Howard G. Gelling, Jr. S-51

Peter Victor Genco, Jr. N-41

Steven Gregory Genovese N-26

Alayne Gentul S-42

Linda M. George N-75

Edward F. Geraghty S-9

Suzanne Geraty N-35

Ralph Gerhardt N-45

Robert Gerlich N-18

Denis P. Germain S-16

Marina Romanovna Gertsberg N-48

Susan M. Getzendanner S-40

Lawrence D. Getzfred S-72

James G. Geyer N-55

Cortez Ghee S-75

Joseph M. Giaccone N-36

Vincent Francis Giammona S-6

Debra Lynn Gibbon S-54

James Andrew Giberson S-16

Brenda C. Gibson S-1

Craig Neil Gibson N-16

Ronnie E. Gies S-8

Andrew Clive Gilbert N-45

Timothy Paul Gilbert N-45

Paul Stuart Gilbey S-32

Paul John Gill S-9

Mark Y. Gilles N-50

Evan Hunter Gillette S-50

Ronald Lawrence Gilligan N-33

Rodney C. Gillis S-24

Laura Gilly N-35

John F. Ginley S-16

Donna Marie Giordano S-55

Jeffrey John Giordano S-8

John Giordano S-18

Steven A. Giorgetti N-13

Martin Giovinazzo N-3

Kum-Kum Girolamo S-54

Salvatore Gitto N-10

Cynthia Giugliano N-64

Mon Gjonbalaj S-37

Dianne Gladstone S-47

Keith Alexander Glascoe S-11

Thomas Irwin Glasser S-49

Edmund Glazer N-75

Harry Glenn N-16

Barry H. Glick N-66

Jeremy Logan Glick S-67

Steven Glick N-21

John T. Gnazzo N-32

William Robert Godshalk S-35

Michael Gogliormella N-35

Brian F. Goldberg S-42

Jeffrey G. Goldflam N-38

Michelle Goldstein S-62

Monica Goldstein N-48

Steven Ian Goldstein N-50

Ronald F. Golinski S-75

Andrew H. Golkin N-46

Dennis James Gomes S-43

Enrique Antonio Gomez N-68

Jose Bienvenido Gomez N-68

Manuel Gomez, Jr. S-44

Wilder Alfredo Gomez N-69

Jenine Nicole Gonzalez S-53

Mauricio Gonzalez S-64

Rosa J. Gonzalez N-66

Lynn Catherine Goodchild S-3

Calvin Joseph Gooding N-39

Peter Morgan Goodrich S-3

Harry Goody S-48

Kiran Kumar Reddy Gopu N-8

Catherine C. Gorayeb N-22

Lisa Fenn Gordenstein N-75

Kerene Gordon N-24

Sebastian Gorki S-38

Kieran Joseph Gorman S-36

Thomas Edward Gorman S-28

Michael Edward Gould N-25

O. Kristin Osterholm White Gould S-68

Douglas Alan Gowell S-4

Yuji Goya S-45

Jon Richard Grabowski N-15

Christopher Michael Grady N-46

Edwin J. Graf III N-41

David Martin Graifman S-34

Gilbert Franco Granados S-58

Lauren Catuzzi Grandcolas and her
  unborn child S-68

Elvira Granitto N-64

Winston Arthur Grant N-65

Christopher S. Gray N-44

Ian J. Gray S-71

James Michael Gray S-13

Tara McCloud Gray N-72

John M. Grazioso N-25

Timothy George Grazioso N-25

Derrick Auther Green S-42

Wade B. Green N-23

Wanda Anita Green S-67

Elaine Myra Greenberg N-20

Donald Freeman Greene S-67

Gayle R. Greene N-9

James Arthur Greenleaf, Jr. N-62

Eileen Marsha Greenstein S-56

Elizabeth Martin Gregg N-59

Denise Marie Gregory N-63

Donald H. Gregory N-39

Florence Moran Gregory S-58

Pedro Grehan N-51

John Michael Griffin N-63

Tawanna Sherry Griffin N-23

Joan Donna Griffith S-39

Warren Grifka N-15

Ramon B. Grijalvo N-65

Joseph F. Grillo N-66

David Joseph Grimner N-12

Francis Edward Grogan S-4

Linda Gronlund S-68

Kenneth George Grouzalis S-25

Joseph Grzelak S-19

Matthew James Grzymalski N-54

Robert Joseph Gschaar S-53

Liming Gu N-3

Richard J. Guadagno S-67

Jose A. Guadalupe S-10

Cindy Yan Zhu Guan S-48

Geoffrey E. Guja S-12

Joseph P. Gullickson S-9

Babita Girjamatie Guman S-39

Douglas Brian Gurian N-39

Janet Ruth Gustafson S-61

Philip T. Guza S-53

Barbara Guzzardo S-55

Peter Mark Gyulavary S-65

Gary Robert Haag N-5

Andrea Lyn Haberman N-61

Barbara Mary Habib N-9

Philip Haentzler N-73

Nezam A. Hafiz N-6

Karen Elizabeth Hagerty S-54

Steven Michael Hagis N-55

Mary Lou Hague S-35

David Halderman S-21

Maile Rachel Hale N-21

Diane Hale-McKinzy S-1

Richard B. Hall S-54

Stanley R. Hall S-70

Vaswald George Hall N-67

Robert J. Halligan S-54

Vincent Gerard Halloran S-13

Carolyn B. Halmon S-75

James Douglas Halvorson N-0

Mohammad Salman Hamdani S-66

Felicia Hamilton S-41

Robert W. Hamilton S-12

Carl Max Hammond, Jr. S-3

Frederic K. Han N-46

Christopher James Hanley N-22

Sean S. Hanley S-12

Valerie Joan Hanna N-9

Thomas Paul Hannafin S-5

Kevin James Hannaford, Sr. N-50

Michael Lawrence Hannan N-10

Dana Rey Hannon S-19

Christine Lee Hanson S-4

Peter Burton Hanson S-4

Sue Kim Hanson S-4

Vassilios G. Haramis S-65

James A. Haran N-51

Gerald Francis Hardacre S-4

Jeffrey Pike Hardy N-24

T.J. Hargrave N-55

Daniel Edward Harlin S-16

Frances Haros S-35

Harvey L. Harrell S-5

Stephen G. Harrell S-5

Melissa Harrington-Hughes N-22

Aisha Ann Harris N-72

Stewart D. Harris N-47

John Patrick Hart S-39

Eric Hartono S-4

John Clinton Hartz S-43

Emeric Harvey N-67

Peter Paul Hashem N-2

Thomas Theodore Haskell, Jr. S-22

Timothy Shawn Haskell S-22

Joseph John Hasson III N-55

Leonard W. Hatton, Jr. S-26

Terence S. Hatton S-9

Michael Helmut Haub S-10

Timothy Aaron Haviland N-14

Donald G. Havlish, Jr. S-56

Anthony Maurice Hawkins N-31

Nobuhiro Hayatsu S-39

James Edward Hayden S-4

Robert Jay Hayes N-76

Philip T. Hayes, Ret. S-13

W. Ward Haynes N-49

Scott Jordan Hazelcorn N-54

Michael K. Healey S-12

Roberta B. Heber N-7

Charles Francis Xavier Heeran N-29

John F. Heffernan S-15

Michele M. Heidenberger S-69

Sheila M.S. Hein S-75

H. Joseph Heller, Jr. N-62

JoAnn L. Heltibridle N-14

Ronald John Hemenway S-71

Mark F. Hemschoot S-62

Ronnie Lee Henderson S-23

Brian Hennessey N-35

Edward R. Hennessy, Jr. N-76

Michelle Marie Henrique S-41

Joseph Patrick Henry S-10

William L. Henry, Jr. S-10

Catherina Henry-Robinson N-72

John Christopher Henwood N-52

Robert Allan Hepburn N-14

Mary Herencia S-55

Lindsay C. Herkness III S-46

Harvey Robert Hermer N-24

Norberto Hernandez N-68

Raul Hernandez N-31

Gary Herold S-58

Jeffrey Alan Hersch N-47
Thomas J. Hetzel S-17
Leon Bernard Heyward MC
Sundance S-36
Brian Christopher Hickey S-12
Enemencio Dario Hidalgo Cedeño N-69
Timothy Brian Higgins S-22
Robert D. W. Higley II S-59
Todd Russell Hill S-46
Clara Victorine Hinds N-69
Neal O. Hinds S-37
Mark Hindy N-25
Katsuyuki Hirai S-39
Heather Malia Ho N-70
Tara Yvette Hobbs S-59
Thomas Anderson Hobbs N-50
James J. Hobin N-9
Robert Wayne Hobson III N-49
DaJuan Hodges N-8
Ronald G. Hoerner S-65
Patrick A. Hoey N-66
John A. Hofer N-2
Marcia Hoffman N-36
Stephen Gerard Hoffman N-42
Frederick Joseph Hoffmann N-39
Michele L. Hoffmann N-39
Judith Florence Hofmiller N-16
Wallace Cole Hogan, Jr. S-74
Thomas Warren Hohlweck, Jr. S-60
Jonathan R. Hohmann S-8
Cora Hidalgo Holland N-2
John Holland N-70
Joseph F. Holland N-61
Jimmie I. Holley S-75
Elizabeth Holmes S-32
Thomas P. Holohan S-14
Herbert Wilson Homer S-2
LeRoy W. Homer, Jr. S-67
Bradley V. Hoorn N-58
James P. Hopper N-30
Montgomery McCullough Hord N-29
Michael Joseph Horn N-27
Matthew Douglas Horning N-16
Robert L. Horohoe, Jr. N-39

Michael Robert Horrocks S-2
Aaron Horwitz N-42
Charles J. Houston S-32
Uhuru G. Houston S-28
Angela M. Houtz S-73
George Gerard Howard S-28
Brady Kay Howell S-73
Michael C. Howell N-60
Steven Leon Howell N-3
Jennifer L. Howley and her unborn
    child S-56
Milagros Hromada S-55
Marian R. Hrycak S-48
Stephen Huczko, Jr. S-30
Kris Robert Hughes S-34
Paul Rexford Hughes N-16
Robert T. Hughes, Jr. N-73
Thomas F. Hughes, Jr. N-71
Timothy Robert Hughes N-44
Susan Huie N-20
Lamar Demetrius Hulse N-17
John Nicholas Humber, Jr. N-1
William Christopher Hunt S-33
Kathleen Anne Hunt-Casey S-50
Joseph Gerard Hunter S-8
Peggie M. Hurt S-75
Robert R. Hussa N-62
Stephen N. Hyland, Jr. S-74
Robert J. Hymel S-71
Thomas Edward Hynes S-37
Walter G. Hynes S-17

Joseph Anthony Ianelli N-9
Zuhtu Ibis N-36
Jonathan Lee Ielpi S-7
Michael Patrick Iken S-33
Daniel Ilkanayev N-48
Frederick J. Ill, Jr. S-16
Abraham Nethanel Ilowitz N-64
Anthony P. Infante, Jr. S-27
Louis S. Inghilterra S-43
Christopher Noble Ingrassia N-30
Paul Innella N-36
Stephanie Veronica Irby N-7

Douglas Jason Irgang S-50
Kristin Irvine-Ryan S-51
Todd Antione Isaac N-56
Erik Hans Isbrandtsen N-25
Taizo Ishikawa S-45
Waleed Joseph Iskandar N-1
Aram Iskenderian, Jr. N-47
John F. Iskyan N-52
Kazushige Ito S-45
Aleksandr Valeryevich Ivantsov N-27
Lacey Bernard Ivory S-74

Virginia May Jablonski N-5
Bryan C. Jack S-70
Brooke Alexandra Jackman N-41
Aaron Jeremy Jacobs N-29
Ariel Louis Jacobs N-21
Jason Kyle Jacobs S-40
Michael G. Jacobs S-42
Steven A. Jacobson N-71
Steven D. Jacoby S-70
Ricknauth Jaggernauth N-71
Jake Denis Jagoda N-34
Yudhvir S. Jain N-37
Maria Jakubiak N-11
Robert Adrien Jalbert S-2
Ernest James N-5
Gricelda E. James N-67
Mark Steven Jardim N-23
Amy Nicole Jarret S-2
Muhammadou Jawara N-70
Francois Jean-Pierre N-71
Maxima Jean-Pierre N-24
Paul Edward Jeffers N-52
John Charles Jenkins N-76
Joseph Jenkins, Jr. S-64
Alan Keith Jensen S-43
Prem Nath Jerath N-67
Farah Jeudy S-60
Hweidar Jian N-27
Eliezer Jimenez, Jr. N-69
Luis Jimenez, Jr. N-13
Charles Gregory John S-45
Nicholas John N-23

Dennis M. Johnson S-74
LaShawna Johnson N-72
Scott Michael Johnson S-33
William R. Johnston S-14
Allison Horstmann Jones S-51
Arthur Joseph Jones III N-59
Brian Leander Jones S-39
Charles Edward Jones N-74
Christopher D. Jones N-41
Donald T. Jones II N-43
Donald W. Jones N-55
Judith Lawter Jones S-73
Linda Jones S-56
Mary S. Jones N-65
Andrew Brian Jordan, Sr. S-22
Robert Thomas Jordan N-42
Albert Gunnis Joseph S-46
Ingeborg Joseph S-46
Karl Henry Joseph S-20
Stephen Joseph S-44
Jane Eileen Josiah S-43
Anthony Jovic S-10
Angel L. Juarbe, Jr. S-16
Karen Sue Juday N-31
Ann C. Judge S-70
Mychal F. Judge S-18
Paul William Jurgens S-30
Thomas Edward Jurgens S-26

Shashikiran Lakshmikantha
  Kadaba N-18
Gavkharoy Kamardinova S-64
Shari Kandell N-32
Howard Lee Kane N-69
Jennifer Lynn Kane N-4
Vincent D. Kane S-18
Joon Koo Kang N-29
Sheldon Robert Kanter N-36
Deborah H. Kaplan N-66
Robin Lynne Kaplan N-1
Alvin Peter Kappelmann, Jr. S-63
Charles H. Karczewski S-56
William A. Karnes N-9
Douglas Gene Karpiloff S-26

Charles L. Kasper S-11
Andrew K. Kates N-38
John A. Katsimatides N-39
Robert Michael Kaulfers S-28
Don Jerome Kauth, Jr. S-36
Hideya Kawauchi S-44
Edward T. Keane N-66
Richard M. Keane N-15
Lisa Yvonne Kearney-Griffin N-18
Karol Ann Keasler S-34
Barbara A. Keating N-76
Paul Hanlon Keating S-5
Leo Russell Keene III S-33
Brenda Kegler S-1
Chandler Raymond Keller S-69
Joseph John Keller S-46
Peter R. Kellerman N-28
Joseph P. Kellett N-61
Frederick H. Kelley III N-43
James Joseph Kelly N-56
Joseph A. Kelly N-51
Maurice P. Kelly N-24
Richard John Kelly, Jr. S-15
Thomas Michael Kelly S-30
Thomas Richard Kelly S-20
Thomas W. Kelly S-20
Timothy Colin Kelly N-43
William Hill Kelly, Jr. N-21
Robert Clinton Kennedy N-12
Thomas J. Kennedy S-8
Yvonne E. Kennedy S-69
John Richard Keohane S-63
Ralph Francis Kershaw S-3
Ronald T. Kerwin S-8
Howard L. Kestenbaum S-53
Douglas D. Ketcham N-29
Ruth Ellen Ketler S-40
Boris Khalif N-17
Norma Cruz Khan S-71
Sarah Khan N-24
Taimour Firaz Khan N-62
Rajesh Khandelwal N-12
SeiLai Khoo N-59
Michael Vernon Kiefer S-22

Satoshi Kikuchihara S-39
Andrew Jay-Hoon Kim N-60
Lawrence Don Kim N-10
Mary Jo Kimelman N-54
Heinrich Kimmig S-3
Karen Ann Kincaid S-70
Amy R. King S-2
Andrew M. King N-44
Lucille Teresa King S-61
Robert King, Jr. S-14
Lisa King-Johnson S-36
Brian K. Kinney S-3
Takashi Kinoshita S-44
Chris Michael Kirby S-64
Robert Kirkpatrick N-73
Howard Barry Kirschbaum N-8
Glenn Davis Kirwin N-38
Helen Crossin Kittle and her unborn
    child N-35
Richard Joseph Klares S-63
Peter Anton Klein N-17
Alan David Kleinberg N-52
Karen Joyce Klitzman N-45
Ronald Philip Kloepfer S-25
Stephen A. Knapp N-73
Eugueni Kniazev N-69
Andrew James Knox N-24
Thomas Patrick Knox N-50
Rebecca Lee Koborie N-4
Deborah A. Kobus S-39
Gary Edward Koecheler S-32
Frank J. Koestner N-28
Ryan Kohart N-27
Vanessa Lynn Przybylo Kolpak S-36
Irina Kolpakova S-45
Suzanne Rose Kondratenko S-63
Abdoulaye Koné N-68
Bon Seok Koo N-73
Dorota Kopiczko N-15
Scott Michael Kopytko S-21
Bojan George Kostic N-27
Danielle Kousoulis N-40
David P. Kovalcin N-2
John J. Kren S-32

William Edward Krukowski S-11
Lyudmila Ksido N-17
Toshiya Kuge S-68
Shekhar Kumar N-35
Kenneth Bruce Kumpel S-22
Frederick Kuo, Jr. S-65
Patricia A. Kuras N-3
Nauka Kushitani S-41
Thomas Joseph Kuveikis S-22
Victor Kwarkye N-68
Raymond Kui Fai Kwok N-33
Angela Reed Kyte N-11

Andrew La Corte N-62
Carol Ann La Plante N-15
Jeffrey G. La Touche N-70
Kathryn L. LaBorie S-2
Amarnauth Lachhman N-24
Ganesh K. Ladkat N-34
James Patrick Ladley N-40
Joseph A. Lafalce N-32
Jeanette Louise Lafond-Menichino N-10
David James LaForge S-12
Michael Patrick LaForte N-53
Alan Charles LaFrance N-69
Juan Mendez Lafuente N-71
Neil Kwong-Wah Lai S-47
Vincent Anthony Laieta S-53
William David Lake S-16
Franco Lalama N-66
Chow Kwan Lam S-48
Michael S. Lamana S-72
Stephen LaMantia N-56
Amy Hope Lamonsoff N-20
Robert T. Lane S-7
Brendan Mark Lang N-26
Rosanne P. Lang N-26
Vanessa Lang Langer and her unborn
    child S-49
Mary Lou Langley S-53
Peter J. Langone S-23
Thomas Michael Langone S-23
Michele Bernadette Lanza S-40
Ruth Sheila Lapin S-37

Ingeborg A.D. Lariby S-49
Robin Blair Larkey N-44
Judith Camilla Larocque N-2
Christopher Randall Larrabee N-25
Hamidou S. Larry N-9
Scott Larsen S-21
John Adam Larson S-57
Natalie Janis Lasden N-75
Gary Edward Lasko N-7
Nicholas Craig Lassman N-36
Paul Laszczynski S-29
Charles A. Laurencin S-46
Stephen James Lauria N-7
Maria LaVache N-6
Denis Francis Lavelle N-16
Jeannine Mary LaVerde S-36
Anna A. Laverty S-39
Steven Lawn S-54
Robert A. Lawrence, Jr. S-49
Nathaniel Lawson N-23
David W. Laychak S-1
Eugen Gabriel Lazar N-33
James Patrick Leahy S-25
Joseph Gerard Leavey S-21
Neil J. Leavy S-13
Robert G. LeBlanc S-3
Leon Lebor N-64
Kenneth Charles Ledee N-14
Alan J. Lederman S-60
Elena F. Ledesma N-9
Alexis Leduc S-43
Daniel John Lee N-2
David S. Lee S-42
Dong Chul Lee S-70
Gary H. Lee N-35
Hyun Joon Lee S-48
Juanita Lee S-54
Kathryn Blair Lee N-9
Linda C. Lee N-22
Lorraine Mary Greene Lee S-56
Myoung Woo Lee S-47
Richard Y.C. Lee N-29
Stuart Soo-Jin Lee N-21
Yang Der Lee N-70

Stephen Paul Lefkowitz S-48
Adriana Legro N-61
Edward Joseph Lehman S-54
Eric Lehrfeld N-22
David R. Leistman N-39
David Prudencio Lemagne S-29
Joseph Anthony Lenihan S-34
John Joseph Lennon, Jr. S-28
John Robinson Lenoir S-52
Jorge Luis León, Sr. N-35
Matthew G. Leonard N-46
Michael Lepore N-13
Charles A. Lesperance N-71
Jeff LeVeen N-26
John Dennis Levi S-29
Alisha Caren Levin S-44
Neil David Levin N-65
Robert Levine N-39
Robert Michael Levine S-37
Shai Levinhar N-29
Daniel M. Lewin N-75
Adam Jay Lewis S-35
Jennifer Lewis S-69
Kenneth E. Lewis S-69
Margaret Susan Lewis N-66
Ye Wei Liang N-8
Orasri Liangthanasarn N-69
Daniel F. Libretti S-17
Ralph Michael Licciardi S-64
Edward Lichtschein N-36
Samantha L. Lightbourn-Allen S-76
Steven Barry Lillianthal N-56
Carlos R. Lillo S-11
Craig Damian Lilore N-25
Arnold Arboleda Lim S-41
Darya Lin S-63
Wei Rong Lin N-67
Nickie L. Lindo N-58
Thomas V. Linehan, Jr. N-12
Robert Thomas Linnane S-12
Alan Patrick Linton, Jr. S-52
Diane Theresa Lipari N-61
Kenneth P. Lira Arévalo S-45
Francisco Alberto Liriano N-58

Lorraine Lisi S-40
Paul Lisson S-49
Vincent M. Litto N-25
Ming-Hao Liu S-64
Nancy Liz S-56
Harold Lizcano N-59
Martin Lizzul N-36
George A. Llanes N-63
Elizabeth C. Logler N-34
Catherine Lisa Loguidice N-55
Jérôme Robert Lohez N-65
Michael William Lomax S-57
Stephen V. Long S-73
Laura Maria Longing N-8
Salvatore P. Lopes S-53
Daniel Lopez N-62
George Lopez S-41
Luis Manuel Lopez S-37
Maclovio Lopez, Jr. S-3
Manuel L. Lopez N-14
Joseph Lostrangio N-17
Chet Dek Louie N-46
Stuart Seid Louis S-50
Joseph Lovero S-29
Sara Elizabeth Low N-74
Jenny Seu Kueng Low Wong N-14
Michael W. Lowe S-46
Garry W. Lozier S-52
John P. Lozowsky N-17
Charles Peter Lucania S-64
Edward Hobbs Luckett N-55
Mark Gavin Ludvigsen S-36
Lee Charles Ludwig S-42
Sean Thomas Lugano S-35
Daniel Lugo S-65
Marie Lukas N-35
William Lum, Jr. N-18
Michael P. Lunden N-53
Christopher E. Lunder N-42
Anthony Luparello S-37
Gary Frederick Lutnick N-38
Linda Anne Luzzicone N-45
Alexander Lygin N-48
CeeCee Lyles S-67

Farrell Peter Lynch N-57
James Francis Lynch S-28
James T. Lynch, Jr. S-73
Louise A. Lynch N-15
Michael Cameron Lynch N-41
Michael Francis Lynch S-15
Michael Francis Lynch S-9
Richard D. Lynch, Jr. S-31
Robert Henry Lynch, Jr. S-26
Sean P. Lynch N-26
Sean Patrick Lynch N-57
Terence M. Lynch S-75
Michael J. Lyons S-13
Monica Anne Lyons N-0
Nehamon Lyons IV S-72
Patrick John Lyons S-23

Robert Francis Mace N-47
Marianne MacFarlane S-2
Jan Maciejewski N-69
Susan A. Mackay N-1
William Macko N-73
Catherine Fairfax MacRae N-59
Richard Blaine Madden S-58
Simon Maddison N-31
Noell C. Maerz S-30
Jennieann Maffeo N-73
Joseph Maffeo S-9
Jay Robert Magazine N-71
Brian Magee N-20
Charles W. Magee N-63
Joseph V. Maggitti N-4
Ronald Magnuson N-48
Daniel L. Maher N-13
Thomas A. Mahon N-51
William J. Mahoney S-11
Joseph Daniel Maio N-30
Linda C. Mair-Grayling N-8
Takashi Makimoto S-44
Abdu Ali Malahi S-45
Debora I. Maldonado N-0
Myrna T. Maldonado-Agosto N-66
Alfred Russell Maler N-54
Gregory James Malone S-32

Edward Francis Maloney III N-50
Joseph E. Maloney S-7
Gene Edward Maloy N-3
Christian H. Maltby N-44
Francisco Miguel Mancini N-71
Joseph Mangano N-3
Sara Elizabeth Manley N-59
Debra M. Mannetta N-61
Marion Victoria Manning N-13
Terence John Manning N-21
James Maounis S-40
Alfred Gilles Padre Joseph
Marchand S-2
Joseph Ross Marchbanks, Jr. S-5
Laura A. Marchese N-65
Hilda Marcin S-67
Peter Edward Mardikian N-21
Edward Joseph Mardovich S-33
Charles Joseph Margiotta S-16
Louis Neil Mariani S-4
Kenneth Joseph Marino S-9
Lester V. Marino N-24
Vita Marino S-51
Kevin D. Marlo S-50
Jose Juan Marrero S-32
John Daniel Marshall S-15
Shelley A. Marshall S-71
James Martello N-26
Michael A. Marti N-51
Karen Ann Martin N-74
Peter C. Martin S-18
Teresa M. Martin S-75
William J. Martin, Jr. N-51
Brian E. Martineau S-62
Betsy Martinez N-32
Edward J. Martinez N-35
Jose Angel Martinez, Jr. N-24
Robert Gabriel Martinez S-65
Waleska Martinez S-67
Lizie D. Martinez-Calderon S-55
Paul Richard Martini S-12
Anne Marie Martino-Cramer S-42
Joseph A. Mascali S-6
Bernard Mascarenhas N-7

Stephen Frank Masi N-35
Ada L. Mason-Acker S-1
Nicholas George Massa S-53
Michael Massaroli N-32
Philip William Mastrandrea, Jr. N-30
Rudy Mastrocinque N-5
Joseph Mathai N-21
Charles William Mathers N-4
William A. Mathesen S-32
Marcello Matricciano N-36
Margaret Elaine Mattic N-72
Dean E. Mattson S-74
Robert D. Mattson S-40
Walter A. Matuza, Jr. N-63
Timothy J. Maude S-74
Jill Maurer-Campbell S-37
Charles A. Mauro, Jr. S-56
Charles J. Mauro N-68
Dorothy Mauro N-9
Nancy T. Mauro N-8
Robert J. Maxwell S-1
Renée A. May and her unborn child S-69
Tyrone May S-48
Keithroy Marcellus Maynard S-14
Robert J. Mayo S-23
Kathy N. Mazza S-29
Edward Mazzella, Jr. N-28
Jennifer Lynn Mazzotta N-33
Kaaria Mbaya N-37
James Joseph McAlary, Jr. N-61
Brian Gerard McAleese S-15
Patricia Ann McAneney N-8
Colin R. McArthur S-58
John Kevin McAvoy S-6
Kenneth M. McBrayer S-52
Brendan F. McCabe S-43
Michael McCabe N-28
Thomas Joseph McCann S-14
Justin McCarthy N-30
Kevin M. McCarthy N-40
Michael Desmond McCarthy N-60
Robert G. McCarthy N-27
Stanley McCaskill N-16
Katie Marie McCloskey N-17

Juliana Valentine McCourt S-3
Ruth Magdaline McCourt S-3
Charles Austin McCrann N-12
Tonyell F. McDay N-13
Matthew T. McDermott N-30
Joseph P. McDonald N-45
Brian Grady McDonnell S-24
Michael P. McDonnell S-36
John F. McDowell, Jr. S-51
Eamon J. McEneaney N-57
John Thomas McErlean, Jr. N-39
Daniel Francis McGinley S-35
Mark Ryan McGinly N-60
William E. McGinn S-21
Thomas Henry McGinnis N-61
Michael Gregory McGinty N-4
Ann Walsh McGovern S-55
Scott Martin McGovern S-31
William J. McGovern S-6
Stacey Sennas McGowan S-51
Francis Noel McGuinn N-51
Thomas F. McGuinness, Jr. N-74
Patrick J. McGuire S-30
Thomas M. McHale N-56
Keith David McHeffey N-28
Ann M. McHugh S-30
Denis J. McHugh III S-33
Dennis P. McHugh S-18
Michael Edward McHugh, Jr. N-34
Robert G. McIlvaine N-22
Donald James McIntyre S-30
Stephanie Marie McKenna N-18
Molly L. McKenzie S-75
Barry J. McKeon S-40
Evelyn C. McKinnedy S-37
Darryl Leron McKinney N-29
George Patrick McLaughlin, Jr. N-59
Robert C. McLaughlin, Jr. N-52
Gavin McMahon S-59
Robert D. McMahon S-13
Edmund M. McNally S-43
Daniel Walker McNeal S-51
Walter Arthur McNeil S-28
Christine Sheila McNulty N-19

Sean Peter McNulty N-28
Robert William McPadden S-15
Terence A. McShane S-9
Timothy Patrick McSweeney S-7
Martin E. McWilliams S-17
Rocco A. Medaglia N-71
Abigail Medina N-16
Ana Iris Medina S-54
Damian Meehan N-61
William J. Meehan, Jr. N-27
Alok Kumar Mehta N-34
Raymond Meisenheimer S-14
Manuel Emilio Mejia N-69
Eskedar Melaku N-14
Antonio Melendez N-70
Mary P. Melendez S-43
Christopher D. Mello N-75
Yelena Melnichenko N-10
Stuart Todd Meltzer N-50
Diarelia Jovanah Mena N-27
Dora Marie Menchaca S-69
Charles R. Mendez S-20
Lizette Mendoza S-60
Shevonne Olicia Mentis N-7
Wolfgang Peter Menzel S-3
Steve John Mercado S-16
Wilfredo Mercado N-73
Wesley Mercer S-47
Ralph Joseph Mercurio N-50
Alan Harvey Merdinger N-0
George L. Merino S-42
Yamel Josefina Merino S-26
George Merkouris N-60
Deborah Merrick N-66
Raymond Joseph Metz III S-32
Jill Ann Metzler S-62
David Robert Meyer N-41
Nurul H. Miah N-15
William Edward Micciulli N-29
Martin Paul Michelstein S-63
Patricia E. Mickley S-71
Ronald D. Milam S-73
Peter Teague Milano N-40

Gregory Milanowycz S-58
Lukasz Tomasz Milewski N-23
Sharon Christina Millan S-45
Corey Peter Miller N-31
Craig J. Miller S-27
Douglas C. Miller S-6
Henry Alfred Miller, Jr. S-20
Joel Miller N-16
Michael Matthew Miller N-55
Nicole Carol Miller S-67
Philip D. Miller S-58
Robert Alan Miller S-48
Robert Cromwell Miller, Jr. S-61
Benny Millman S-64
Charles M. Mills, Jr. S-26
Ronald Keith Milstein S-39
Robert J. Minara S-22
William George Minardi N-54
Louis Joseph Minervino N-15
Thomas Mingione S-22
Wilbert Miraille N-31
Domenick N. Mircovich S-31
Rajesh Arjan Mirpuri N-21
Joseph D. Mistrulli N-71
Susan J. Miszkowicz N-66
Paul Thomas Mitchell S-20
Richard P. Miuccio S-47
Jeffrey Peter Mladenik N-1
Frank V. Moccia, Sr. S-65
Louis Joseph Modafferi S-6
Boyie Mohammed N-62
Dennis Mojica S-8
Manuel D. Mojica, Jr. S-21
Kleber Rolando Molina S-43
Manuel De Jesus Molina N-64
Carl Molinaro S-17
Justin John Molisani, Jr. S-30
Brian Patrick Monaghan S-64
Franklyn Monahan N-32
John Gerard Monahan N-33
Kristen Leigh Montanaro N-3
Craig Montano N-42
Michael G. Montesi S-9

Carlos Alberto Montoya N-75

Antonio De Jesus Montoya Valdes N-74

Cheryl Ann Monyak N-9

Thomas Carlo Moody S-18

Sharon Moore S-52

Krishna V. Moorthy S-43

Laura Lee Defazio Morabito N-75

Abner Morales S-41

Carlos Manuel Morales N-31

Paula E. Morales S-59

Sonia Mercedes Morales Puopolo N-76

Gerard P. Moran, Jr. S-73

John Christopher Moran N-20

John Michael Moran S-11

Kathleen Moran S-63

Lindsay Stapleton Morehouse S-36

George William Morell N-54

Steven P. Morello N-3

Vincent S. Morello S-16

Yvette Nicole Moreno N-59

Dorothy Morgan N-15

Richard J. Morgan S-30

Nancy Morgenstern N-31

Sanae Mori N-22

Blanca Robertina Morocho
    Morocho N-68

Leonel Geronimo Morocho
    Morocho N-68

Dennis Gerard Moroney N-47

Lynne Irene Morris N-33

Odessa V. Morris S-76

Seth Allan Morris N-54

Steve Morris N-19

Christopher Martel Morrison N-23

Ferdinand V. Morrone S-27

William David Moskal N-3

Brian A. Moss S-71

Marco Motroni N-62

Cynthia Motus-Wilson N-67

Iouri A. Mouchinski N-71

Jude Joseph Moussa N-50

Peter Moutos N-9

Damion O'Neil Mowatt N-23

Teddington H. Moy S-1

Christopher Michael Mozzillo S-7

Stephen Vincent Mulderry S-33

Richard T. Muldowney, Jr. S-21

Michael D. Mullan S-17

Dennis Michael Mulligan S-17

Peter James Mulligan N-29

Michael Joseph Mullin N-26

James Donald Munhall S-52

Nancy Muñiz N-65

Francisco Heladio Munoz N-4

Carlos Mario Muñoz N-70

Theresa Munson S-57

Robert Michael Murach N-47

Cesar Augusto Murillo N-29

Marc A. Murolo N-53

Brian Joseph Murphy N-55

Charles Anthony Murphy N-56

Christopher W. Murphy S-33

Edward Charles Murphy N-50

James F. Murphy IV N-23

James Thomas Murphy N-54

Kevin James Murphy N-5

Patrick Jude Murphy S-73

Patrick Sean Murphy N-5

Raymond E. Murphy S-19

Robert Eddie Murphy, Jr. S-45

John Joseph Murray N-45

John Joseph Murray S-45

Susan D. Murray N-14

Valerie Victoria Murray N-64

Richard Todd Myhre N-33

Louis J. Nacke II S-68

Robert B. Nagel S-10

Mildred Rose Naiman N-75

Takuya Nakamura N-63

Alexander John Robert Napier S-54

Frank Joseph Naples III N-45

John Philip Napolitano S-17

Catherine Ann Nardella S-61

Mario Nardone, Jr. S-30

Manika K. Narula N-33

Shawn M. Nassaney S-3
Narender Nath N-11
Karen Susan Navarro N-62
Joseph M. Navas S-28
Francis Joseph Nazario N-32
Glenroy I. Neblett N-18
Rayman Marcus Neblett S-60
Jerome O. Nedd N-71
Laurence F. Nedell S-58
Luke G. Nee N-43
Pete Negron S-0
Laurie Ann Neira N-76
Ann N. Nelson N-42
David William Nelson N-61
Ginger Risco Nelson N-60
James A. Nelson S-30
Michele Ann Nelson N-49
Peter Allen Nelson S-12
Oscar Francis Nesbitt S-47
Gerard Terence Nevins S-8
Renee Tetreault Newell N-74
Christopher C. Newton S-71
Christopher Newton-Carter S-51
Nancy Yuen Ngo N-17
Khang Ngoc Nguyen S-73
Jody Tepedino Nichilo N-47
Kathleen Ann Nicosia N-74
Martin Stewart Niederer N-26
Alfonse Joseph Niedermeyer S-28
Frank John Niestadt, Jr. S-62
Gloria Nieves S-40
Juan Nieves, Jr. N-70
Troy Edward Nilsen N-35
Paul Nimbley N-30
John Ballantine Niven S-61
Katherine McGarry Noack N-22
Curtis Terrance Noel N-72
Michael A. Noeth S-72
Daniel R. Nolan N-3
Robert Walter Noonan N-49
Jacqueline June Norton N-2
Robert Grant Norton N-2
Daniela Rosalia Notaro N-58
Brian Christopher Novotny N-45

Soichi Numata S-44
Brian Nunez N-45
Jose Nunez N-71
Jeffrey Roger Nussbaum N-62

Dennis Patrick O'Berg S-20
James P. O'Brien, Jr. N-55
Michael P. O'Brien N-43
Scott J. O'Brien N-22
Timothy Michael O'Brien N-57
Daniel O'Callaghan S-10
Dennis James O'Connor, Jr. N-30
Diana J. O'Connor S-49
Keith Kevin O'Connor S-34
Richard J. O'Connor N-12
Amy O'Doherty N-39
Marni Pont O'Doherty S-36
James Andrew O'Grady S-50
Thomas G. O'Hagan S-13
Patrick J. O'Keefe S-10
William O'Keefe S-11
Gerald Thomas O'Leary N-27
Matthew Timothy O'Mahony N-57
John P. O'Neill N-63
Peter J. O'Neill, Jr. S-52
Sean Gordon Corbett O'Neill N-25
Kevin M. O'Rourke S-17
Patrick J. O'Shea N-61
Robert William O'Shea N-60
Timothy Franklin O'Sullivan N-73
James A. Oakley N-9
Douglas E. Oelschlager S-20
Takashi Ogawa N-22
Albert Ogletree N-24
Philip Paul Ognibene S-36
John A. Ogonowski N-74
Joseph J. Ogren S-7
Samuel Oitice S-9
Gerald Michael Olcott N-11
Christine Anne Olender N-68
Linda Mary Oliva N-59
Edward K. Oliver N-61
Leah Elizabeth Oliver N-12
Eric Taube Olsen S-20

Jeffrey James Olsen S-5
Barbara K. Olson S-70
Maureen Lyons Olson N-7
Steven John Olson S-8
Toshihiro Onda S-44
Seamus L. Oneal N-37
Betty Ann Ong N-74
Michael C. Opperman S-61
Christopher T. Orgielewicz S-49
Margaret Quinn Orloske N-8
Virginia Anne Ormiston N-5
Ruben S. Ornedo S-70
Ronald Orsini N-56
Peter Keith Ortale S-33
Juan Ortega-Campos S-38
Jane Marie Orth N-75
Alexander Ortiz N-65
David Ortiz S-27
Emilio Pete Ortiz N-62
Pablo Ortiz N-67
Paul Ortiz, Jr. N-21
Sonia Ortiz N-64
Masaru Ose S-44
Elsy Carolina Osorio Oliva N-72
James R. Ostrowski N-46
Jason Douglas Oswald N-48
Michael John Otten S-16
Isidro D. Ottenwalder N-68
Michael Chung Ou S-48
Todd Joseph Ouida N-44
Jesus Ovalles N-69
Peter J. Owens, Jr. N-42
Adianes Oyola S-45

Angel M. Pabon, Jr. N-28
Israel Pabon, Jr. N-24
Roland Pacheco N-65
Michael Benjamin Packer N-22
Diana B. Padro S-76
Deepa Pakkala N-17
Jeffrey Matthew Palazzo S-5
Thomas Palazzo N-54
Richard A. Palazzolo N-54
Orio Joseph Palmer S-17

Frank Anthony Palombo S-21
Alan N. Palumbo N-51
Christopher Matthew Panatier N-45
Dominique Lisa Pandolfo N-14
Jonas Martin Panik S-73
Paul J. Pansini S-5
John M. Paolillo S-11
Edward Joseph Papa N-54
Salvatore T. Papasso S-26
James Nicholas Pappageorge S-14
Marie Pappalardo S-2
Vinod Kumar Parakat N-29
Vijayashanker Paramsothy S-57
Nitin Ramesh Parandkar N-19
Hardai Parbhu S-56
James Wendell Parham S-29
Debra Marie Paris S-53
George Paris N-33
Gye Hyong Park N-64
Philip Lacey Parker S-61
Michael Alaine Parkes N-12
Robert E. Parks, Jr. N-46
Hashmukh C. Parmar N-37
Robert Parro S-16
Diane Marie Parsons S-47
Leobardo Lopez Pascual N-70
Michael J. Pascuma, Jr. N-67
Jerrold Hughes Paskins N-17
Horace Robert Passananti N-11
Suzanne H. Passaro S-53
Avnish Ramanbhai Patel N-59
Dipti Patel N-33
Manish Patel S-30
Steven Bennett Paterson N-51
James Matthew Patrick N-51
Manuel D. Patrocino N-70
Bernard E. Patterson N-43
Clifford L. Patterson, Jr. S-74
Cira Marie Patti S-34
Robert E. Pattison N-63
James Robert Paul N-60
Patrice Paz S-61
Victor Hugo Paz N-69
Stacey Lynn Peak N-50

Richard Allen Pearlman S-27

Durrell V. Pearsall, Jr. S-11

Thomas Nicholas Pecorelli N-74

Thomas Pedicini N-42

Todd Douglas Pelino N-54

Mike Adrian Pelletier N-49

Anthony G. Peluso S-36

Angel R. Pena S-56

Robert Penninger S-69

Richard Al Penny S-49

Salvatore F. Pepe N-3

Carl Allen B. Peralta N-30

Robert David Peraza N-32

Jon A. Perconti, Jr. N-27

Alejo Perez N-67

Angel Perez, Jr. N-33

Angela Susan Perez N-32

Anthony Perez N-37

Ivan Antonio Perez S-41

Nancy E. Perez N-66

Berry Berenson Perkins N-76

Joseph John Perroncino N-32

Edward J. Perrotta N-50

Emelda H. Perry S-64

Glenn C. Perry, Sr. S-22

John William Perry S-24

Franklin Allan Pershep S-56

Danny Pesce N-55

Michael John Pescherine S-34

Davin N. Peterson N-28

Donald Arthur Peterson S-67

Jean Hoadley Peterson S-67

William Russell Peterson N-15

Mark James Petrocelli N-61

Philip Scott Petti S-16

Glen Kerrin Pettit S-25

Dominick A. Pezzulo S-29

Kaleen Elizabeth Pezzuti N-54

Kevin J. Pfeifer S-14

Tu-Anh Pham N-60

Kenneth John Phelan, Sr. S-13

Sneha Anne Philip S-66

Eugenia McCann Piantieri N-13

Ludwig John Picarro S-63

Matthew Picerno N-43

Joseph O. Pick S-41

Christopher J. Pickford S-12

Dennis J. Pierce S-47

Bernard Pietronico N-41

Nicholas P. Pietrunti N-30

Theodoros Pigis S-49

Susan Elizabeth Pinto N-35

Joseph Piskadlo N-63

Christopher Todd Pitman N-45

Joshua Michael Piver N-33

Robert R. Ploger III S-71

Zandra F. Ploger S-71

Joseph Plumitallo N-41

John M. Pocher N-41

William Howard Pohlmann S-47

Laurence Michael Polatsch N-27

Thomas H. Polhemus N-17

Steve Pollicino N-39

Susan M. Pollio S-33

Darin H. Pontell S-73

Joshua Iosua Poptean N-71

Giovanna Porras N-72

Anthony Portillo S-49

James Edward Potorti N-11

Daphne Pouletsos S-55

Richard N. Poulos N-30

Stephen Emanual Poulos S-60

Brandon Jerome Powell N-23

Scott Alan Powell S-75

Shawn Edward Powell S-20

Antonio Dorsey Pratt N-23

Gregory M. Preziose N-53

Wanda Ivelisse Prince S-42

Vincent A. Princiotta S-20

Kevin M. Prior S-22

Everett Martin Proctor III N-48

Carrie Beth Progen S-59

David Lee Pruim S-62

Richard A. Prunty S-5

John Foster Puckett N-68

Robert David Pugliese N-10

Edward F. Pullis S-62

Patricia Ann Puma N-64

Jack D. Punches S-73
Hemanth Kumar Puttur N-17
Joseph J. Pycior, Jr. S-72
Edward R. Pykon N-61

Christopher Quackenbush S-52
Lars Peter Qualben N-15
Lincoln Quappé S-16
Beth Ann Quigley N-27
Patrick J. Quigley IV S-4
Michael T. Quilty S-15
James Francis Quinn N-30
Ricardo J. Quinn S-18

Carol Millicent Rabalais S-61
Christopher Peter Anthony
    Racaniello N-32
Leonard J. Ragaglia S-10
Eugene J. Raggio S-24
Laura Marie Ragonese-Snik S-54
Michael Paul Ragusa S-23
Peter Frank Raimondi N-59
Harry A. Raines N-36
Lisa J. Raines S-71
Ehtesham Raja S-39
Valsa Raju N-63
Edward J. Rall S-17
Lukas Rambousek N-58
Maria Ramirez S-45
Harry Ramos N-63
Vishnoo Ramsaroop N-64
Deborah A. Ramsaur S-1
Lorenzo E. Ramzey S-55
Alfred Todd Rancke S-50
Adam David Rand S-8
Jonathan C. Randall N-6
Shreyas S. Ranganath N-7
Anne T. Ransom N-18
Faina Rapoport N-17
Rhonda Sue Rasmussen S-76
Robert A. Rasmussen S-37
Amenia Rasool N-11
R. Mark Rasweiler N-9
Marsha D. Ratchford S-72

David Alan James Rathkey S-46
William Ralph Raub N-25
Gerard F. Rauzi S-47
Alexey Razuvaev S-32
Gregory Reda N-6
Sarah Anne Redheffer N-20
Michele Marie Reed S-62
Judith Ann Reese N-67
Donald J. Regan S-14
Robert M. Regan S-10
Thomas Michael Regan S-54
Christian Michael Otto Regenhard S-23
Howard Reich S-49
Gregg Reidy N-28
James Brian Reilly S-34
Kevin O. Reilly S-20
Timothy E. Reilly N-11
Joseph Reina, Jr. N-33
Thomas Barnes Reinig N-55
Frank Bennett Reisman N-28
Joshua Scott Reiss N-51
Karen Renda N-18
John Armand Reo N-40
Richard Cyril Rescorla S-46
John Thomas Resta N-62
Sylvia San Pio Resta and her unborn
    child N-62
Martha M. Reszke S-1
David E. Retik N-1
Todd H. Reuben S-69
Luis Clodoaldo Revilla Mier S-65
Eduvigis Reyes, Jr. N-72
Bruce Albert Reynolds S-28
John Frederick Rhodes S-55
Francis Saverio Riccardelli S-25
Rudolph N. Riccio N-34
Ann Marie Riccoboni N-64
David Harlow Rice S-52
Eileen Mary Rice N-6
Kenneth Frederick Rice III N-13
CeCelia E. Richard S-76
Vernon Allan Richard S-20
Claude Daniel Richards S-25
Gregory David Richards N-39

Michael Richards N-63
Venesha Orintia Richards N-6
Jimmy Riches S-21
Alan Jay Richman N-11
John M. Rigo N-10
Frederick Charles Rimmele III S-2
Rose Mary Riso S-47
Moises N. Rivas N-67
Joseph R. Rivelli, Jr. S-22
Carmen Alicia Rivera S-42
Isaias Rivera N-63
Juan William Rivera N-72
Linda Ivelisse Rivera N-15
David E. Rivers N-20
Joseph R. Riverso N-51
Paul V. Rizza S-40
John Frank Rizzo S-64
Stephen Louis Roach N-54
Joseph Roberto S-35
Leo Arthur Roberts N-43
Michael E. Roberts S-21
Michael Edward Roberts S-16
Donald Walter Robertson, Jr. N-45
Jeffrey Robinson N-16
Michell Lee Jean Robotham S-56
Donald Arthur Robson N-39
Antonio A. Rocha N-51
Raymond James Rocha N-44
Laura Rockefeller N-20
John Michael Rodak S-51
Antonio José Rodrigues S-29
Anthony Rodriguez S-22
Carmen Milagros Rodriguez S-58
Gregory E. Rodriguez N-48
Marsha A. Rodriguez N-6
Mayra Valdes Rodriguez S-59
Richard Rodriguez S-29
David Bartolo Rodriguez-Vargas N-69
Matthew Rogan S-14
Jean Destrehan Rogér N-74
Karlie Rogers N-20
Scott William Rohner N-44
Keith Michael Roma S-25
Joseph M. Romagnolo N-24

Efrain Romero, Sr. S-44
Elvin Romero N-28
James A. Romito S-27
Sean Paul Rooney S-57
Eric Thomas Ropiteau N-33
Aida Rosario N-18
Angela Rosario N-29
Mark H. Rosen S-52
Brooke David Rosenbaum N-33
Linda Rosenbaum N-12
Sheryl Lynn Rosenbaum N-47
Lloyd Daniel Rosenberg N-40
Mark Louis Rosenberg N-7
Andrew Ira Rosenblum N-40
Joshua M. Rosenblum N-27
Joshua Alan Rosenthal S-41
Richard David Rosenthal N-48
Philip Martin Rosenzweig N-2
Daniel Rosetti S-64
Richard Barry Ross N-2
Norman S. Rossinow S-61
Nicholas P. Rossomando S-5
Michael Craig Rothberg N-29
Donna Marie Rothenberg S-60
Mark David Rothenberg S-68
James Michael Roux S-2
Nicholas Charles Alexander Rowe N-23
Edward V. Rowenhorst S-76
Judy Rowlett S-1
Timothy Alan Roy, Sr. S-24
Paul G. Ruback S-21
Ronald J. Ruben S-34
Joanne Rubino N-14
David M. Ruddle S-66
Bart Joseph Ruggiere N-49
Susan A. Ruggiero N-13
Adam Keith Ruhalter N-47
Gilbert Ruiz N-69
Robert E. Russell S-1
Stephen P. Russell S-7
Steven Harris Russin N-52
Michael Thomas Russo, Sr. S-7
Wayne Alan Russo N-6
William R. Ruth S-74

Edward Ryan N-61
John Joseph Ryan S-34
Jonathan Stephan Ryan S-30
Matthew L. Ryan S-9
Tatiana Ryjova S-48
Christina Sunga Ryook N-49

Thierry Saada N-41
Jason Elazar Sabbag S-42
Thomas E. Sabella S-17
Scott H. Saber N-23
Charles E. Sabin, Sr. S-71
Joseph Francis Sacerdote N-44
Jessica Leigh Sachs N-74
Francis John Sadocha N-24
Jude Elias Safi N-26
Brock Joel Safronoff N-7
Edward Saiya S-45
John Patrick Salamone N-40
Marjorie C. Salamone S-75
Hernando Rafael Salas S-38
Juan G. Salas N-70
Esmerlin Antonio Salcedo S-65
John Pepe Salerno N-30
Rahma Salie and her unborn child N-1
Richard L. Salinardi, Jr. S-37
Wayne John Saloman N-35
Nolbert Salomon S-46
Catherine Patricia Salter S-60
Frank G. Salvaterra S-51
Paul Richard Salvio N-62
Samuel Robert Salvo, Jr. S-59
Carlos Alberto Samaniego N-42
John P. Sammartino S-71
James Kenneth Samuel, Jr. N-60
Michael San Phillip S-51
Hugo M. Sanay S-31
Alva Cynthia Jeffries Sanchez N-16
Jacquelyn Patrice Sanchez N-47
Jesus Sanchez S-2
Raymond Sanchez S-66
Eric M. Sand N-28
Stacey Leigh Sanders N-3
Herman S. Sandler S-52

Jim Sands, Jr. N-36
Ayleen J. Santiago N-65
Kirsten Reese Santiago N-67
Maria Theresa Concepcion
    Santillan N-36
Susan Gayle Santo N-9
Christopher A. Santora S-10
John August Santore S-5
Mario L. Santoro S-26
Rafael Humberto Santos N-34
Rufino C.F. Santos III N-17
Victor J. Saracini S-2
Kalyan K. Sarkar N-66
Chapelle Renee Stewart Sarker N-14
Paul F. Sarle N-56
Deepika Kumar Sattaluri N-18
Gregory Thomas Saucedo S-6
Susan M. Sauer N-11
Anthony Savas N-67
Vladimir Savinkin N-48
John Michael Sbarbaro N-56
David M. Scales S-74
Robert Louis Scandole N-52
Michelle Scarpitta S-31
Dennis Scauso S-8
John Albert Schardt S-12
John G. Scharf S-63
Fred C. Scheffold, Jr. S-6
Angela Susan Scheinberg N-64
Scott Mitchell Schertzer N-33
Sean Schielke N-44
Steven Francis Schlag N-51
Robert A. Schlegel S-72
Jon Schlissel S-48
Karen Helene Schmidt S-46
Ian Schneider N-52
Thomas G. Schoales S-21
Marisa Dinardo Schorpp N-49
Frank G. Schott, Jr. N-13
Gerard Patrick Schrang S-14
Jeffrey H. Schreier N-31
John T. Schroeder N-59
Susan Lee Schuler S-53
Edward W. Schunk N-55

Mark Evan Schurmeier N-22
John Burkhart Schwartz N-40
Mark Schwartz S-26
Adriane Victoria Scibetta N-48
Raphael Scorca N-3
Janice M. Scott S-1
Randolph Scott S-31
Christopher Jay Scudder S-37
Arthur Warren Scullin N-14
Michael H. Seaman N-46
Margaret M. Seeliger S-53
Anthony Segarra N-64
Carlos Segarra N-72
Jason M. Sekzer N-31
Matthew Carmen Sellitto N-46
Michael L. Selves S-75
Howard Selwyn S-31
Larry John Senko N-65
Arturo Angelo Sereno N-58
Frankie Serrano S-45
Marian H. Serva S-75
Alena Sesinova N-3
Adele Christine Sessa N-27
Sita Nermalla Sewnarine S-43
Karen Lynn Seymour N-73
Davis Grier Sezna, Jr. S-52
Thomas Joseph Sgroi N-8
Jayesh Shantilal Shah N-37
Khalid M. Shahid N-33
Mohammed Shajahan N-14
Gary Shamay N-31
Earl Richard Shanahan N-5
Dan F. Shanower S-72
Neil G. Shastri N-58
Kathryn Anne Shatzoff N-10
Barbara A. Shaw N-20
Jeffrey James Shaw N-24
Robert John Shay, Jr. N-53
Daniel James Shea N-38
Joseph Patrick Shea N-38
Kathleen Shearer S-3
Robert M. Shearer S-3
Linda June Sheehan S-50

Hagay Shefi N-21
Antionette M. Sherman S-75
John Anthony Sherry S-30
Atsushi Shiratori N-44
Thomas Joseph Shubert N-29
Mark Shulman N-10
See Wong Shum N-71
Allan Abraham Shwartzstein N-30
Clarin Shellie Siegel-Schwartz S-53
Johanna Sigmund N-60
Dianne T. Signer and her unborn
    child N-60
Gregory Sikorsky S-12
Stephen Gerard Siller S-5
David Silver N-29
Craig A. Silverstein S-50
Nasima H. Simjee S-41
Bruce Edward Simmons S-51
Diane M. Simmons S-69
Donald D. Simmons S-76
George W. Simmons S-69
Arthur Simon N-58
Kenneth Alan Simon N-58
Michael J. Simon N-49
Paul Joseph Simon N-17
Marianne Liquori Simone N-35
Barry Simowitz S-48
Jane Louise Simpkin S-2
Jeff Lyal Simpson S-27
Cheryle D. Sincock S-75
Khamladai Khami Singh N-68
Roshan Ramesh Singh N-68
Thomas E. Sinton III N-55
Peter A. Siracuse N-39
Muriel F. Siskopoulos S-33
Joseph Michael Sisolak N-6
John P. Skala S-27
Francis Joseph Skidmore, Jr. S-32
Toyena Corliss Skinner N-72
Paul Albert Skrzypek N-50
Christopher Paul Slattery N-30
Vincent Robert Slavin N-27
Robert F. Sliwak N-56

Paul Kenneth Sloan S-33

Stanley S. Smagala, Jr. S-15

Wendy L. Small N-54

Gregg H. Smallwood S-72

Catherine T. Smith N-16

Daniel Laurence Smith S-31

Gary F. Smith S-1

George Eric Smith S-39

Heather Lee Smith N-75

James Gregory Smith N-40

Jeffrey R. Smith S-52

Joyce Patricia Smith N-24

Karl T. Smith, Sr. N-43

Kevin Joseph Smith S-9

Leon Smith, Jr. S-11

Moira Ann Smith S-24

Monica Rodriguez Smith and her
unborn child N-73

Rosemary A. Smith N-73

Bonnie Shihadeh Smithwick N-61

Rochelle Monique Snell S-49

Christine Ann Snyder S-67

Dianne Bullis Snyder N-74

Leonard J. Snyder, Jr. S-54

Astrid Elizabeth Sohan N-6

Sushil S. Solanki N-34

Rubén Solares N-31

Naomi Leah Solomon N-21

Daniel W. Song N-56

Mari-Rae Sopper S-69

Michael Charles Sorresse N-5

Fabian Soto N-63

Timothy Patrick Soulas N-44

Gregory Thomas Spagnoletti S-35

Donald F. Spampinato, Jr. N-39

Thomas Sparacio S-32

John Anthony Spataro N-10

Robert W. Spear, Jr. S-19

Robert Speisman S-70

Maynard S. Spence, Jr. N-6

George Edward Spencer III S-31

Robert Andrew Spencer N-45

Mary Rubina Sperando N-21

Frank Spinelli N-44

William E. Spitz N-42

Joseph Patrick Spor, Jr. S-15

Klaus Johannes Sprockamp S-47

Saranya Srinuan N-52

Fitzroy St. Rose N-72

Michael F. Stabile S-32

Lawrence T. Stack S-18

Timothy M. Stackpole S-20

Richard James Stadelberger S-40

Eric Adam Stahlman N-46

Gregory Stajk S-17

Alexandru Liviu Stan N-34

Corina Stan N-34

Mary Domenica Stanley N-14

Anthony Starita N-42

Jeffrey Stark S-13

Derek James Statkevicus S-34

Patricia J. Statz S-75

Craig William Staub S-34

William V. Steckman N-67

Eric Thomas Steen S-30

William R. Steiner N-12

Alexander Robbins Steinman N-25

Edna L. Stephens S-1

Andrew Stergiopoulos N-45

Andrew J. Stern N-43

Norma Lang Steuerle S-69

Martha Jane Stevens S-62

Michael James Stewart N-61

Richard H. Stewart, Jr. N-41

Sanford M. Stoller N-17

Douglas Joel Stone N-74

Lonny Jay Stone N-63

Jimmy Nevill Storey N-12

Timothy Stout N-35

Thomas Strada N-40

James J. Straine, Jr. N-52

Edward W. Straub S-55

George J. Strauch, Jr. S-60

Edward Thomas Strauss S-24

Steven R. Strauss S-46

Larry L. Strickland S-74

Steven F. Strobert N-55
Walwyn Wellington Stuart, Jr. S-29
Benjamin Suarez S-11
David Scott Suarez N-17
Ramon Suarez S-25
Dino Xavier Suarez Ramirez N-75
Yoichi Sumiyama Sugiyama S-44
William Christopher Sugra N-34
Daniel Thomas Suhr S-14
David Marc Sullins S-25
Christopher P. Sullivan S-22
Patrick Sullivan N-40
Thomas G. Sullivan N-67
Hilario Soriano Sumaya, Jr. N-8
James Joseph Suozzo N-41
Colleen M. Supinski S-51
Robert Sutcliffe N-67
Seline Sutter N-65
Claudia Suzette Sutton N-48
John Francis Swaine N-39
Kristine M. Swearson N-34
Brian David Sweeney S-2
Brian Edward Sweeney S-9
Madeline Amy Sweeney N-74
Kenneth J. Swenson N-48
Thomas F. Swift S-46
Derek Ogilvie Sword S-35
Kevin Thomas Szocik S-35
Gina Sztejnberg N-15
Norbert P. Szurkowski N-50

Harry Taback N-4
Joann C. Tabeek N-35
Norma C. Taddei N-13
Michael Taddonio S-31
Keiichiro Takahashi S-32
Keiji Takahashi S-44
Phyllis Gail Talbot N-11
Robert R. Talhami N-27
John Talignani S-68
Sean Patrick Tallon S-5
Paul Talty S-24
Maurita Tam S-53
Rachel Tamares S-61

Hector Rogan Tamayo S-45
Michael Andrew Tamuccio N-59
Kenichiro Tanaka S-44
Rhondelle Cherie Tankard S-59
Michael Anthony Tanner N-25
Dennis Gerard Taormina, Jr. N-12
Kenneth Joseph Tarantino N-46
Allan Tarasiewicz S-7
Michael C. Tarrou S-2
Ronald Tartaro N-60
Deborah Tavolarella S-2
Darryl Anthony Taylor N-72
Donnie Brooks Taylor S-59
Hilda E. Taylor S-70
Kip P. Taylor S-74
Leonard E. Taylor S-71
Lorisa Ceylon Taylor N-15
Michael Morgan Taylor N-40
Sandra C. Taylor S-1
Sandra Dawn Teague S-69
Karl W. Teepe S-71
Paul A. Tegtmeier S-21
Yeshavant Moreshwar Tembe S-47
Anthony Tempesta N-53
Dorothy Pearl Temple S-47
Stanley L. Temple N-31
David Gustaf Peter Tengelin N-4
Brian John Terrenzi N-47
Lisa Marie Terry N-11
Goumatie Thackurdeen S-41
Harshad Sham Thatte N-17
Michael Theodoridis N-1
Thomas F. Theurkauf, Jr. S-36
Lesley Anne Thomas N-49
Brian Thomas Thompson S-44
Clive Ian Thompson S-32
Glenn Thompson N-43
Nigel Bruce Thompson N-44
Perry A. Thompson S-60
Vanavah Alexei Thompson N-64
William H. Thompson S-26
Eric Raymond Thorpe S-35
Nichola Angela Thorpe S-33
Tamara C. Thurman S-74

Sal Edward Tieri, Jr. N-10
John Patrick Tierney S-13
Mary Ellen Tiesi S-62
William Randolph Tieste N-25
Kenneth Tietjen S-29
Stephen Edward Tighe N-56
Scott Charles Timmes N-62
Michael E. Tinley N-15
Jennifer M. Tino N-11
Robert Frank Tipaldi N-26
John James Tipping II S-10
David Tirado N-23
Hector Luis Tirado, Jr. S-15
Michelle Lee Titolo N-48
Alicia Nicole Titus S-2
John J. Tobin N-8
Richard J. Todisco S-51
Otis V. Tolbert S-73
Vladimir Tomasevic N-22
Stephen Kevin Tompsett N-22
Thomas Tong S-39
Doris Torres S-42
Luis Eduardo Torres N-51
Amy Elizabeth Toyen N-23
Christopher Michael Traina N-63
Daniel Patrick Trant N-43
Abdoul Karim Traore N-68
Glenn J. Travers, Sr. N-32
Walter Philip Travers N-56
Felicia Yvette Traylor-Bass N-65
James Anthony Trentini N-2
Mary Barbara Trentini N-2
Lisa L. Trerotola N-67
Karamo Baba Trerra S-39
Michael Angel Trinidad N-31
Francis Joseph Trombino S-38
Gregory James Trost S-33
Willie Q. Troy S-1
William P. Tselepis, Jr. N-45
Zhanetta Valentinovna Tsoy N-13
Michael Patrick Tucker N-28
Lance Richard Tumulty S-31
Ching Ping Tung S-44
Simon James Turner N-20

Donald Joseph Tuzio S-39
Robert T. Twomey N-67
Jennifer Lynn Tzemis N-58

John G. Ueltzhoeffer N-15
Tyler Victor Ugolyn N-59
Michael A. Uliano N-56
Jonathan J. Uman N-38
Anil Shivhari Umarkar N-34
Allen V. Upton N-39
Diane Marie Urban S-47

John Damien Vaccacio N-43
Bradley Hodges Vadas S-35
William Valcarcel S-48
Felix Antonio Vale N-32
Ivan Vale N-32
Benito Valentin N-18
Santos Valentin, Jr. S-25
Carlton Francis Valvo II N-46
Pendyala Vamsikrishna N-74
Erica H. Van Acker S-55
Kenneth W. Van Auken N-52
R. Bruce Van Hine S-13
Daniel M. Van Laere S-62
Edward Raymond Vanacore S-41
Jon Charles Vandevander N-62
Frederick T. Varacchi N-38
Gopalakrishnan Varadhan N-46
David Vargas S-49
Scott C. Vasel N-16
Azael Ismael Vasquez N-24
Ronald J. Vauk S-73
Arcangel Vazquez S-41
Santos Vazquez N-31
Peter Vega S-11
Sankara Sastry Velamuri S-47
Jorge Velazquez S-47
Lawrence G. Veling S-7
Anthony Mark Ventura S-41
David Vera S-31
Loretta Ann Vero N-18
Christopher James Vialonga N-62
Matthew Gilbert Vianna N-34

Robert Anthony Vicario N-24
Celeste Torres Victoria N-20
Joanna Vidal N-20
John T. Vigiano II S-23
Joseph Vincent Vigiano S-23
Frank J. Vignola, Jr. N-48
Joseph Barry Vilardo N-28
Claribel Villalobos Hernandez N-23
Sergio Gabriel Villanueva S-23
Chantal Vincelli N-21
Melissa Renée Vincent N-65
Francine Ann Virgilio S-61
Lawrence Virgilio S-20
Joseph Gerard Visciano S-34
Joshua S. Vitale N-26
Maria Percoco Vola S-62
Lynette D. Vosges S-59
Garo H. Voskerijian N-13
Alfred Anton Vukosa N-35

Gregory Kamal Bruno Wachtler N-60
Karen J. Wagner S-74
Mary Alice Wahlstrom N-1
Honor Elizabeth Wainio S-67
Gabriela Silvina Waisman N-23
Wendy Alice Rosario Wakeford N-53
Courtney Wainsworth Walcott S-46
Victor Wald N-63
Kenneth E. Waldie N-2
Benjamin James Walker N-16
Glen Wall N-57
Mitchel Scott Wallace S-26
Peter Guyder Wallace N-6
Robert Francis Wallace S-12
Roy Michael Wallace N-44
Jeanmarie Wallendorf S-36
Matthew Blake Wallens N-39
Meta L. Waller S-1
John Wallice, Jr. N-30
Barbara P. Walsh N-9
Jim Walsh N-34
Jeffrey P. Walz S-14
Ching Wang S-44
Weibin Wang N-36

Michael Warchola S-6
Stephen Gordon Ward N-48
Timothy Ray Ward S-2
James A. Waring N-31
Brian G. Warner N-37
Derrick Christopher Washington S-66
Charles Waters N-32
James Thomas Waters, Jr. S-34
Patrick J. Waters S-8
Kenneth Thomas Watson S-21
Michael Henry Waye N-8
Todd Christopher Weaver S-43
Walter Edward Weaver S-25
Nathaniel Webb S-28
Dinah Webster N-20
William Michael Weems S-4
Joanne Flora Weil S-45
Michael T. Weinberg S-17
Steven Weinberg S-37
Scott Jeffrey Weingard N-27
Steven George Weinstein N-13
Simon Weiser N-65
David M. Weiss S-8
David Thomas Weiss N-46
Chin Sun Pak Wells S-74
Vincent Michael Wells N-44
Deborah Jacobs Welsh S-67
Timothy Matthew Welty S-7
Christian Hans Rudolf Wemmers N-21
Ssu-Hui Wen N-34
John Joseph Wenckus N-2
Oleh D. Wengerchuk S-65
Peter M. West N-43
Whitfield West, Jr. N-35
Meredith Lynn Whalen N-60
Eugene Michael Whelan S-12
Adam S. White N-50
Edward James White III S-13
James Patrick White N-39
John Sylvester White N-63
Kenneth Wilburn White, Jr. N-24
Leonard Anthony White S-66
Malissa Y. White N-15
Maudlyn A. White S-74

Sandra L. White S-75
Wayne White N-9
Leanne Marie Whiteside S-59
Mark P. Whitford S-15
Leslie A. Whittington S-69
Michael T. Wholey S-29
Mary Lenz Wieman S-59
Jeffrey David Wiener N-12
William J. Wik S-60
Alison Marie Wildman N-61
Glenn E. Wilkinson S-14
Ernest M. Willcher S-75
John Charles Willett N-50
Brian Patrick Williams N-41
Candace Lee Williams N-75
Crossley Richard Williams, Jr. S-41
David J. Williams N-64
David Lucian Williams S-73
Debbie L. Williams S-54
Dwayne Williams S-74
Kevin Michael Williams S-50
Louie Anthony Williams N-66
Louis Calvin Williams III S-37
John P. Williamson S-8
Donna Ann Wilson S-56
William Eben Wilson S-61
David Harold Winton S-35
Glenn J. Winuk S-27
Thomas Francis Wise N-9
Alan L. Wisniewski S-52
Frank Paul Wisniewski N-53
David Wiswall S-55
Sigrid Charlotte Wiswe N-18
Michael R. Wittenstein N-52
Christopher W. Wodenshek N-49
Martin Phillips Wohlforth S-52
Katherine Susan Wolf N-3
Jennifer Yen Wong N-20
Siucheung Steve Wong N-4
Yin Ping Wong S-60
Yuk Ping Wong S-48
Brent James Woodall S-33
James John Woods N-26
Marvin Roger Woods S-73

Patrick J. Woods S-64
Richard Herron Woodwell S-35
David Terence Wooley S-9
John Bentley Works S-34
Martin Michael Wortley N-46
Rodney James Wotton S-43
William Wren, Ret. S-22
John W. Wright, Jr. S-50
Neil Robin Wright N-46
Sandra Lee Wright S-57

Jupiter Yambem N-69
John D. Yamnicky, Sr. S-71
Suresh Yanamadala N-16
Vicki Yancey S-70
Shuyin Yang S-70
Matthew David Yarnell S-41
Myrna Yaskulka N-60
Shakila Yasmin N-15
Olabisi Shadie Layeni Yee N-67
Kevin W. Yokum S-72
Edward P. York N-49
Kevin Patrick York S-31
Raymond R. York S-20
Suzanne Martha Youmans S-54
Barrington Leroy Young, Jr. S-31
Donald McArthur Young S-72
Edmond G. Young, Jr. S-74
Jacqueline Young N-3
Lisa L. Young S-1
Elkin Yuen N-61

Joseph C. Zaccoli N-43
Adel Agayby Zakhary N-63
Arkady Zaltsman S-63
Edwin J. Zambrana, Jr. S-49
Robert Alan Zampieri N-62
Mark Zangrilli S-63
Christopher R. Zarba, Jr. N-1
Ira Zaslow S-46
Kenneth Albert Zelman N-19
Abraham J. Zelmanowitz N-65
Martin Morales Zempoaltecatl N-68
Zhe Zeng S-37

Marc Scott Zeplin N-27
Jie Yao Justin Zhao S-39
Yuguang Zheng S-70
Ivelin Ziminski N-5
Michael Joseph Zinzi N-14
Charles Alan Zion N-25

Julie Lynne Zipper S-49
Salvatore J. Zisa N-5
Prokopios Paul Zois N-18
Joseph J. Zuccala S-44
Andrew Steven Zucker S-45
Igor Zukelman S-43

APPENDIX 2

Timeline of Key Events on September 11, 2001

7:59 A.M. American Airlines Flight 11 takes off from Boston's Logan International Airport. Bound for Los Angeles, the Boeing 767 carries eleven crew members, seventy-six passengers, and five hijackers.

8:14 A.M. (APPROX.) American Flight 11 is hijacked.

8:15 A.M. United Airlines Flight 175 takes off from Boston's Logan International Airport. Bound for Los Angeles, the Boeing 767 carries nine crew members, fifty-one passengers, and five hijackers.

8:19 A.M. Flight attendant Betty Ong alerts American Airlines that Flight 11 has been hijacked.

8:20 A.M. American Airlines Flight 77 takes off from Washington Dulles International Airport. Bound for Los Angeles, the Boeing 757 carries six crew members, fifty-three passengers, and five hijackers.

8:37 A.M. The FAA's Boston Air Traffic Control Center alerts the military to a problem on board Flight 11.

8:42 A.M. United Airlines Flight 93 takes off from Newark International Airport. Bound for San Francisco, the Boeing 757 carries seven crew members, thirty-three passengers, and four hijackers.

8:42 A.M. TO 8:46 A.M. (APPROX.) United Flight 175 is hijacked.

8:46 A.M. American Flight 11 crashes into the 93rd through 99th floors of the North Tower of the World Trade Center in Manhattan, killing everyone aboard.

8:50 A.M. TO 8:54 A.M. (APPROX.) American Flight 77 is hijacked.

9:03 A.M. United Flight 175 crashes into the 77th through 85th floors of the South Tower of the World Trade Center in Manhattan, killing everyone aboard.

9:05 A.M. While visiting an elementary school in Sarasota, Florida, President George W. Bush learns that a second plane has hit the World Trade Center.

9:28 A.M. United Flight 93 is hijacked.

9:37 A.M. American Flight 77 crashes into the first and second floors of the Pentagon in Arlington, Virginia, killing everyone aboard.

9:42 A.M. The Federal Aviation Administration grounds all flights awaiting takeoff at U.S. airports and orders all civilian flights to land at the nearest available airport.

9:59 A.M. Although it is the second tower struck, the South Tower of the World Trade Center is the first to collapse. More than 800 people die as a result of the crash and the collapse.

10:03 A.M. United Flight 93 crashes in an empty field that formerly was the site of a coal mine near the village of Shanksville, Pennsylvania, killing everyone aboard.

10:15 A.M. The damaged west side of Pentagon's outer "E Ring" collapses. 125 people are killed in the attack on the Pentagon.

10:28 A.M. 102 minutes after being struck, the North Tower of the World Trade Center collapses. More than 1,600 people die as a result of the crash and the collapse.

11:02 A.M. New York City mayor Rudy Giuliani orders the evacuation of Lower Manhattan.

5:20 P.M. The empty 47-story building known as 7 World Trade Center collapses as a result of damage and fire caused by the collapse of the North Tower.

8:30 P.M. President Bush delivers an address from the Oval Office in which he expresses sympathy for everyone mourning the deaths of innocent victims, pledges support for the injured, and vows to track down all those responsible.

Sources: 9/11 Commission Report, September 11 Memorial & Museum.

ACKNOWLEDGMENTS

ONE COLD WINTER MORNING, I SAT IN THE CONNECTICUT HOME OF two gracious, heartbroken people. Lee and Eunice Hanson tore open the scars of loss and recounted the agony of watching on television as United Flight 175 crashed into the South Tower bearing their son, Peter, his wife, Sue Kim, and their granddaughter, Christine. I am filled with gratitude to the Hansons and everyone who shared their stories, all of whom asked only in return that I write to the best of my ability.

Special thanks also to Paul Adams, Brynn Bender, Bernard Brown Sr., Sinita Brown, Debra Burlingame, Janet Cardwell, Brian Clark, Ron Clifford, John Creamer, Colin Creamer, Nora Creamer, Gerry Creamer, Julie Creamer, Tina Creamer, Moussa Diaz, Elaine Duch, Rob "Boomer" Elliott, Harry Friedenreich, Gerry Gaeta, Jack Gentul, Robert Grunewald, Allan Hackel, Peg Ogonowski Hatch, Jerry and Kathy Henson, Jay Jonas, Jack Keane, Andrea LeBlanc, Christian John Lillis, Cecilia Lillo, Mike Low, Andrea Maffeo, Father Stephen McGraw, Rasha McMillon, Wally Massenburg, Ray Murray, Kevin Nasypany, Cathie Ong-Herrera, Heather "Lucky" Penney, Stanley Praimnath, Janice Punches, Jennifer Punches-Botta, Jeremy Punches, Sonia Tita Puopolo, Julie Sweeney Roth, Bruce Russell, Terry Shaffer, Dr. David Tarantino, Dave Thomas, Millie Wears, Marilyn Wills, Michael Woodward, and Chris Briggs Young.

At the National September 11 Memorial & Museum, a magnificent place and a tremendous educational resource, I received help and encouragement from Alice Greenwald, Jan Ramirez, and Cliff Chanin. Among many other benefits, they connected me to Martha

Feltenstein and Jim Connors, who helped solve a longstanding mystery about the identity of the messenger killed alongside Elaine Duch.

In addition to participating in the events described here, Donna Glessner and Kathie Shaffer provided insights into the events of Flight 93 and valuable thoughts on the entire manuscript. One huge benefit of this work is making friends with people like them.

This book would never have been written without Richard Abate, a true friend and helicopter parent to my career. Claire Wachtel, who edited two of my previous books, embraced this idea with her trademark passion. Luke Dempsey battled my bad habits and improved this work in countless ways. Jonathan Burnham advocated for this book from start to finish. I'm grateful to the entire Harper team, including Doug Jones, Leah Wasielewski, Katie O'Callaghan, Tina Andreadis, Kate D'Esmond, Haley Swanson, Nate Knaebel, Emily DeHuff, Robin Bilardello, and Leah Carlson-Stanisic.

Researching this book was like planning an assault on Everest. Sarah Kess created order out of chaos by building elaborate timelines that guided every step of the way, then offered wise comments on the end result. Mariya Manzhos proved her gifts as an investigative reporter by helping to track down hundreds of people with 9/11–related stories, then spotting glitches in the manuscript. Thanks also to graduate research assistants Ana Goni-Lessan, Geoff Line, and Meggie Quackenbush. Steve Wylie of Pat Casteel Transcripts is a lifesaver. Thanks to Rachel Kim of 3Arts for help in a million ways.

Anneliese J. Thomson, official court reporter for the Honorable Leonie M. Brinkema, enabled me to obtain thousands of pages of the Moussaoui trial transcript. Thanks to Stephen Lofgren and Kate Richards at the U.S. Army Center for Military History; LCDR Lauren Cole, director of the Navy Office of Information East; and Adam Berenbak, archivist in the Center for Legislative Archives at the National Archives and Records Administration.

Retired Army colonel Miles Kara, an investigator on the 9/11 Commission and a scholar of 9/11 events in the air, provided invaluable help and advice. Miles introduced me to John Farmer, senior counsel to the 9/11 Commission and author of the excellent book *The Ground*

Truth, whose encouragement and suggestions enriched this work immeasurably. I'm beyond grateful for the expertise of Priscilla Jones, PhD, chief of Histories and Studies, Air Force Historical Support Division.

My professional home is Boston University, where I'm privileged to teach journalism and hold the Sumner M. Redstone Chair in Narrative Studies. Dean Tom Fiedler has been a true friend and an outstanding leader. I'm buoyed by the encouragement of my Department of Journalism comrades: Chris Daly, Anne Donohue, Noelle Graves, Michael Holley, Michelle Johnson, Greg Marinovich, Bill McKeen, Maggie Mulvihill, Safoura Rafeizadeh, Jane Regan, Caryl Rivers, Peter Smith, Susan Walker, and Brooke Williams. I enjoy the great fortune of a thirty-plus-year friendship with my BU colleague Dick Lehr, a gifted author and journalist. He's a superb editor, too.

At the *Boston Globe,* I shared the byline on the lead news story on 9/11 with Matthew Brelis, a first-class aviation reporter and a great guy. We received help from Globies including Anne Barnard, Sandy Coleman, Bud Collins, John Ellement, Mary Leonard, Raja Mishra, Brian Mooney, Shelley Murphy, Michael Paulson, Ralph Ranalli, and Robert Schlesinger. Our all-star editors included Marty Baron, Helen Donovan, Greg Moore, Ben Bradlee Jr., Tom Mulvoy, Mark Morrow, Ken Cooper, and Charlie Mansbach.

Globe editor Brian McGrory is the best newsman I know and my closest friend. He contributed to that first 9/11 story and to my life in more ways than I can count. Also part of that original team was the peerless Steve Kurkjian, who magically appeared at the beginning of my career and has been a guiding light ever since. Before Mike Rezendes was immortalized in the movie *Spotlight,* we worked together on early stories about the hijackers. He is every bit as dogged as Hollywood portrayed him.

I owe extra thanks to four exceptional journalists who contributed to "Six Lives," the narrative model for this book, published the Sunday after 9/11: Tina Cassidy, Caroline Louise Cole, Bella English, and Tatsha Robertson. Earlier in my career, I learned the craft from Marty Nolan, Gerry O'Neill, Matt Storin, Steve Bailey, Michael Larkin, Al

Larkin, Ellen Clegg, the late John C. Burke, and other great journalists who helped me every day.

I'm blessed with the support of longtime friends, among them Naftali Bendavid, Chris Callahan, Dan Field, Colleen Granahan, Joann Muller, Ruth and Bill Weinstein, and my oldest buddy, Jeff Feigelson. I wouldn't be here without my first mentor, the late, great Wilbur Doctor.

This is the only place where my family comes last. The small but fierce Zuckoff and Kreiter clans have encouraged my professional pursuits in every way, starting with my brother, Allan Zuckoff, whose intellect and integrity remain models for me.

Our late mother, Gerry Zuckoff, is a blessed memory and a cherished presence. Her empathy is my greatest influence. My father, Sid Zuckoff, a retired New York City high school history teacher, was my first editor in life and on this book. As his granddaughters say, we should all be more like Sid.

Speaking of my daughters, Isabel and Eve, they are constant sources of inspiration, joy, and *naches*. They are my heart. My wife, Suzanne, got me through this. She makes everything that matters possible. Thank you for my life.

NOTES

The work of the 9/11 Commission is an invaluable research resource to anyone interested in these events. In addition to the commission's report and hearings, commission staff members conducted more than twelve hundred fact-finding interviews, which I drew on extensively and used as background information for follow-up interviews. Each commission interview was summarized in a Memorandum for the Record (MFR); these are available online at the National Archives, at www.archives.gov/research/9-11/commission-memoranda.html.

The National Archives also contain voluminous other commission records, known as the 9/11 Commission Series, which can be found at www.archives.gov/research/9-11/commission-series.html. Also see NARA's 9/11 Federal Aviation Administration Finding Aid, at https://www.archives.gov/research/foia/faa-finding-aid. FBI interviews and other essential primary documents, collected by the 9/11 Document Archive, can be found at www.scribd.com.

I chose to follow the 9/11 Commission's approach to Islamic surnames, referring to individuals on second and subsequent references by the last word in the names by which they are known, i.e., Marwan al-Shehhi on first reference, then Shehhi afterward. One exception is Osama bin Laden, primarily to avoid confusing readers familiar with seeing him referred to as "bin Laden."

EPIGRAPH

ix   "The ravages of many a forest fire": Dr. J. S. Boyce, "Fire Scars and Decay," *The Timberman*, vol. 22, p. 7, May 1921.

INTRODUCTION: "THE DARKNESS OF IGNORANCE"

xvii   "A stream of light": Unbylined story, "The Statue Unveiled," *New York Times*, October 29, 1886, p. 2. Also see "France's Gift Accepted," *New York Times*, October 29, 1886, p. 1.

xvii   "In a moment the air": Unbylined story, "The Sights and Sightseers," *New York Times*, October 29, 1886, p. 2.

xviii   "84th floor / west office": John Breunig, "Father's Note Changes Family's 9/11 Account," *The Stamford Advocate*, December 28, 2012, p. 1.

xviii   couldn't stomach a ticker-tape parade: Although there was a hiatus of ticker-tape parades in New York after 9/11, the city did hold other parades, including the annual Macy's Thanksgiving Day Parade in 2001, with floats that commemorated September 11.

xix   what happened to other people: The phrase has been used by many writers. In this instance, it bears noting that the reporter Sam Roberts applied it to 9/11 less than a year after the attacks: "When History Isn't Something That Happens to Other People," *New York Times*, April 24, 2002, p. G15.

xix   mention of the "homeland": Elizabeth Becker, "Washington Talk: Prickly Roots of 'Homeland Security,'" *New York Times*, August 31, 2002.

xix   "The passage of time": Ian W. Toll, "The Paradox of Pearl Harbor," *Boston Globe*, Dec. 4, 2016, p. K1. (Hereafter, Toll)

xx   Falling Man: See Tom Junod, "The Falling Man: An Unforgettable Story," *Esquire*, September 9, 2016.

xx   lead news story: Mitchell Zuckoff and Matthew Brelis, "New Day of Infamy," *Boston Globe*, September 12, 2001, p. 1. Also see Mitchell Zuckoff, "Six Lives: Reliving the Morning of Death," *Boston Globe*, September 16, 2001, p. 1.

xx   twenty-eight 9/11 victims earned degrees: https://www.bu.edu/remember /in-memoriam/. Accessed December 4, 2015.

xxii   "Christianity—especially the evangelizing . . .": Lawrence Wright, *The Looming Tower: Al-Qaeda and the Road to 9/11* (New York: First Vintage Books Edition, September 2007), p. 194.

xxiii   "Anger, resentment, and humiliation . . .": Wright, p. 123.

PROLOGUE: "A CLEAR DECLARATION OF WAR"

1   1918, with the defeat of the last great Muslim empire: The conflict between Western and Islamic societies is covered powerfully, accessibly, and succinctly in Bernard Lewis, *What Went Wrong: Western Impact and Middle Eastern Response* (New York: Oxford University Press, 2002). Also see Lewis, "The Revolt of Islam: When did the conflict with the West begin, and how could it end?," *The New Yorker*, November 19, 2001.

2   "All of these crimes and sins": Bernard Lewis, "License to Kill: Usama bin Ladin's Declaration of Jihad," *Foreign Affairs*, November/December 1998. Also translated version of World Islamic Front statement, titled "Jihad Against Jews and Crusaders," February 23, 1998, http://fas.org/irp/world/para/docs/980223-fatwa. htm, accessed May 3, 2016 (complete translation is contained in Appendix III.)

2   headed his own terrorist group: 9/11 Commission Report, p. 341.

3   apparently tipped off: 9/11 Commission report, pp. 117–18.

3   indicted him in absentia: Benjamin Weiser, "Saudi Is Indicted in Bomb Attacks on U.S. Embassies," *New York Times*, Nov. 5, 1998.

3   formally described his terror group: 9/11 Commission Report, p. 341.

3   the USS *Cole*: 9/11 Commission report, p. 190.

3   "It would be a mistake": 9/11 Commission Report, p. 343. See: Paul R. Pillar, *Terrorism and U.S. Foreign Policy* (Brookings Institution Press, 2001), p. 23.

4   "multimillionaire Saudi dissident": Douglas Jehl, "In a Bahrain Port, No More Sailors on the Town," *New York Times*, May 4, 1997.

4   "recent reports": Benjamin Weiser, "Suspected Chief Plotter in Trade Center Blast Goes on Trial Today," *New York Times*, August 4, 1997.

4   nearly six months later: Philip Shenon, "Bombings in East Africa: In Washington, Focus on Suspects in Past Attacks," *New York Times*, August 8, 1998.

4   downplaying the apparent threat: Tim Weiner, "U.S. Hard Put to Find Proof that Bin Laden Directed Attacks," *New York Times*, April 13, 1999.

5   a pointed story: Walter Pincus, "Anti-U.S. Calls for Attacks Are Seen as Serious," *Washington Post*, February 25, 1998.

5   "[W]e had our little story": John Miller, "Greetings America, My Name Is Osama bin Laden," *Esquire*, February 1999.

5   "To most Americans": Bernard Lewis, "License to Kill: Usama bin Ladin's Declaration of Jihad," *Foreign Affairs*, November/December 1998.

6   state of the nation: A Gallup poll taken on September 10, 2001, found that 55 percent of Americans were "dissatisfied with the way things are going in the United States." Rick Hampson, "The Day Before," *USA Today*, September 9, 2002.

6   A Gallup poll: Jeffrey M. Jones, Gallup Poll Senior Managing Editor, "Sept. 11 Effects, Though Largely Faded, Persist," September 9, 2003. http://www.gallup.com/poll/9208/sept-effects-though-largely-faded-persist.aspx.

CHAPTER 1: "QUIET'S A GOOD THING"

9   "Dad, I need help": This account of the Ogonowski family on September 10, 2001, comes primarily from an interview with Margaret "Peg" Ogonowski Hatch on January 27, 2017, with some details and phrases taken from Mitchell Zuckoff, "Six Lives: Reliving The Morning of Death," *The Boston Globe*, September 16, 2001, p. A1.

11   a program John felt passionate about: Interview with Peg Ogonowski Hatch. Also see Caroline Louise Cole, "Immigrant Farmers Return to the Land, Growing Asian Foods," *The Boston Globe*, July 25, 1999, South Weekly Section, p. 1.

11   "These guys are putting more care": Caroline Louise Cole, "Immigrant Farmers Return to the Land, Growing Asian Foods," *Boston Globe*, July 25, 1999, South Weekly Section, p. 1.

12   vibrant young woman slalomed: The story of the Hanson family comes from interviews with C. Lee and Eunice Hanson, February 22, 2017 and follow-up interview phone calls; transcript of C. Lee Hanson's testimony at Moussaoui trial; FBI interviews with C. Lee Hanson, September 11, 2001 and June 20, 2002. Also Brian McGrory, "Up From the Ashes," *Boston Globe*, September 11, 2011.

13   joining the faculty: Albert Lin, "Young Promise Turned to Tragedy in Sept. 11 Hijackings," *The Korea Herald*, October 6, 2001.

14   *Washington Journal*: Video of September 9, 2001, edition of CSPAN show, https://www.c-span.org/video/?165914-2/news-review&start=2217.

15   at forty-five: Neil A. Lewis, "Barbara Olson, 45, Advocate and Conservative Commentator," *New York Times*, September 13, 2001; video interview of Theodore Olson at the Hudson Union Society, www.youtube.com/watch?v=5ppFvUc10nc; Amy Argetsinger and Roxanne Roberts, "Napa Nuptials For Olson and His Lady," *Washington Post*, October 22, 2006.

15   at their home in Virginia: Toby Harnden, "She Asked Me How to Stop the Plane," *The Telegraph*, March 5, 2002.

16   CeeCee Lyles: Linda Shrieves, "CeeCee Lyles Was Soaring Through Life, Then Destiny Came Calling," *Orlando Sentinel*, September 29, 2001. Testimony of Lorne Lyles at Moussaoui trial, transcript pp. 3503–16. Also FBI interviews with Lorne Lyles, September 12, 2001; Carrie Louise Ross, October 5, 2001; and "Flight Crew: CeeCee Lyles," *(Pittsburgh) Post-Gazette*, October 28, 2001.

17   "Man! She is beautiful": Flight 93 Memorial Oral History Transcript, interview with Lorne Lyles, October 25, 2006, conducted by Kathie Shaffer. The Flight 93 Memorial Oral History Transcripts are held by the National Park Service at the Flight 93 Memorial in Somerset County, PA.

17   "everything and nothing": Dennis B. Roddy et al., "Flight 93: Forty Lives, One Destiny," *(Pittsburgh) Post-Gazette*, October 28, 2001.

19   a full plate: Interview with Kevin Nasypany, March 20, 2017.

19   leading defenseman: Rich Thompson, "Hockey's Dismal Season Raises Questions," *Michigan Journal*, March 29, 1978, p. 7.

20   potential national security threats: 9/11 Commission Report, pp. 14–15.

21   a sharp drop: 9/11 Commission Staff Interview with General Richard Myers, NORAD Commander in Chief, February 17, 2004.

21   on-alert fighter jets: Ibid.

21   a plot by terrorists: Interview with Kevin Nasypany, March 20, 2017.

22   "really quiet in here": Ibid.

22   polo shirt: Video cameras at a Walmart in Portland, Maine, captured Atta's image and clothing on September 10, 2001. The images were later distributed by the FBI.

22   flimsy vinyl Travelpro suitcase: FBI interview with Lynn Howland, October 13, 2001. Howland, a first officer with American Airlines, encountered a man she identified as Atta at Logan Airport the morning of September 11, 2001. Based on his dress and his question about whether she would be flying Flight 11 to Los Angeles, she thought he might be a pilot. She recalled his bag, according to the FBI, as "black vinyl, round top, and incredibly cheap."

22   room 308: 9/11 Commission Memorandum for the Record, titled "Boston, Massachusetts, Summary," prepared by Quinn John Tamm Jr., February 2, 2004, p. 6.

23   early childhood: Terry McDermott, *Perfect Soldiers* (New York: HarperCollins, 2005), pp. 12–14.

23   raising him like a girl: Neil MacFarquhar, "A Nation Challenged: The Mastermind; A Portrait of the Terrorist: From Shy Child to Single-Minded Killer," *New York Times*, October 10, 2001.

23   doctor and a professor: FBI document, "American Airlines Flight #11," FBI PENTTBOM Investigation, April 19, 2002, p. 02998.

23   joined a trade group: Peter Finn, "A Fanatic's Quiet Path to Terror," *Washington Post*, September 22, 2001.

23   high enough grades: McDermott, p. 19.

23   became a fixture: McDermott, p. 34; 9/11 Commission Staff Statement No. 16, "Outline of the 9/11 Plot," pp. 3–4.

24   "throats slit": Douglas Frantz and Desmond Butler, "Imam at German Mosque Preached Hate to 9/11 Pilots," *New York Times*, July 16, 2002.

24   Islamic scriptures: McDermott, p. 54.

24   private Christian schools: McDermott, p. 50.

24   Beirut discos: 9/11 Commission Report, p. 163.

24   pledged *bayat:* 9/11 Commission Staff Statement No. 16, "Outline of the 9/11 Plot," p. 4.

25   suicide hijacking plot: 9/11 Commission Report, pp. 145–50.

25   met bin Laden: 9/11 Commission Staff Statement No. 16, "Outline of the 9/11 Plot," p. 2.

25   "Look down there": Benjamin Weiser, "Suspected Chief Plotter in Trade Center Blast Goes on Trial Today," *New York Times,* August 4, 1997.

25   ten planes: 9/11 Commission Staff Statement No. 16, "Outline of the 9/11 Plot," p. 13.

25   tactical commander: 9/11 Commission Report, p. 167.

25   thirty-one flight schools: 9/11 Commission Report, p. 168.

25   Before they applied for visas: 9/11 Commission Report, p. 168.

26   Hani Hanjour: 9/11 Commission Staff Statement No. 16, "Outline of the 9/11 Plot," pp. 6–7.

26   handpicked them: 9/11 Commission Staff Statement No. 16, "Outline of the 9/11 Plot," pp. 8–9.

26   weren't especially imposing: 9/11 Commission Staff Statement No. 16, "Outline of the 9/11 Plot," p. 8.

26   generally avoiding trouble: One member of al-Qaeda to whom this didn't apply was Zacarias Moussaoui, who was arrested in Minnesota in August 2001 on immigration charges by members of the FBI and the Immigration and Naturalization Service. At the time of his arrest, according to court filings, he was in possession of two knives, flight manuals for the Boeing 747 model 400, a flight simulator computer program, shin guards, fighting gloves, and a handheld aviation radio, among other items.

26   "muscle" group members: 9/11 Commission Staff Statement No. 16, "Outline of the 9/11 Plot," pp. 5–6.

26   Experienced jihadists: 9/11 Commission Report, p. 155.

27   identified Mihdhar: 9/11 and Terrorist Travel Monograph, 9/11 Commission, Staff Report, August 21, 2004, p. 10.

27   terrorist watchlist: The intelligence failures surrounding Mihdhar and Hazmi remain significant and troubling. They are discussed at length in the 9/11 Commission Report, pp. 269–72, and examined in detail by Summers and Swan in *The Eleventh Day,* pp. 375–88. Also see the Inspector General's review of FBI handling of information prior to the 9/11 attacks, https://oig.justice.gov/special/0506/chapter5.htm.

27   other countries: Office of the Inspector General, "A Review of the FBI's Handling of Intelligence Information Prior to the September 11 Attacks," Chapter 5, Footnote 115, released publicly June 2005. https://oig.justice.gov/special/0506/chapter5.htm#115.

27   "individual and systemic failings": See Chapter 5 of "A Review of the FBI's Handling of Intelligence Information Related to the September 11 Attacks," Office

of the Inspector General, released publicly June 2006. https://oig.justice.gov /special/s0606/chapter5.htm#IIIF.

27   no interest in takeoffs or landings: 9/11 Commission Staff Statement No. 16, "Outline of the 9/11 Plot," p. 14.

27   cross-country flights: 9/11 Commission Staff Statement No. 16, "Outline of the 9/11 Plot," pp. 9–11.

27   became frustrated: 9/11 Commission Staff Statement No. 16, "Outline of the 9/11 Plot," p. 18. This information is based on interrogations of Khalid Sheikh Mohammed and alleged plot facilitator Ramzi Binalshibh, so should be considered accordingly.

28   Battle of Vienna: Wright, p. 194; also see Christopher Hitchens, "Why the Suicide Killers Chose September 11," *The Guardian,* October 3, 2001.

28   computers in public libraries: This fact led to a controversial provision of Public Law 107-56, 115 Stat. 272 (2001), the so-called PATRIOT Act, that required librarians to hand over patron reading and computer records to law enforcement.

28   less than half a million dollars: The 9/11 Commission report concluded that the attacks cost between $400,000 and $500,000, including $270,000 spent inside the United States. Other costs included "travel to obtain passports and visas, travel to the United States, expenses incurred by the plot leader and facilitators outside the United States, and expenses incurred by the people selected to be hijackers who ultimately did not participate." That total did not include the cost of running camps in Afghanistan where the plotters trained. See pp. 14, 169, 172, and 499.

28   Sweet Temptations: 9/11 Commission Memorandum for the Record, titled "Boston, Massachusetts, Summary," prepared by Quinn John Tamm Jr., February 2, 2004, p. 3.

28   private dance: Laura Mansnerus and David Kocieniewski, "A Hub for Hijackers Found in New Jersey," *New York Times,* September 27, 2001.

28   single-engine plane: FBI document, "Translation of the interview conducted by German authorities of the girlfriend of Ziad Jarrah, September 18, 2001."

29   threw Jarrah out: Ibid.

29   mail a letter: A mistake on the address caused the package containing the letter to be returned, after which the FBI seized it and had it translated. Dirk Laabs and Terry McDermott, "Prelude to 9/11: A Hijacker's Love, Lies," *Los Angeles Times,* January 27, 2003. Also see FBI document, "Translation of the interview conducted by German authorities of the girlfriend of Ziad Jarrah, September 18, 2001."

29   private pilot's license: FBI "Final Movements" document, Exhibit ST0001 01-455-A, Part B, introduced at Moussaoui trial, p. 82; Summers and Swan, p. 149.

29   received a commercial pilot's license: FBI document, "Summary of Penttbom Investigation," February 29, 2004, p. 20. https://vault.fbi.gov/9-11%20Commission %20Report/9-11-fbi-report-2004-02(feb). FBI Summary of Penttbom Investigation, February 29, 2004, p. 20.

29   videotaped lessons: FBI memorandum on Atta's suitcases, February 10, 2004.

29   a four-page letter: FBI document, "Summary of Penttbom Investigation," February 29, 2004, p. 69. https://vault.fbi.gov/9-11%20Commission%20Report

/9-11-fbi-report-2004-02(feb). This document describes it as a four-page letter, although elsewhere it is said to be five pages.

29   detailed instructions: Translation of "The Last Night" letter, introduced as Government Exhibit BS01101T, 01-455-A, at Moussaoui trial.

30   copies of the same letter: On September 28, 2001, Attorney General John Ashcroft held a news conference where he described the letter and reported that copies were found in Atta's suitcase, which never made Flight 11; amid the wreckage of Flight 93; and in a car left at Dulles Airport by Flight 77 hijacker Nawaf al-Haznawi.

30   believed to have written: McDermott, p. 232.

CHAPTER 2: "HE'S NORDO"

31   1.8 million passengers: 9/11 Commission Staff Monograph, "Four Flights and Civil Aviation Security," September 12, 2005, p. 60. (Hereafter: Four Flights Monograph.) Link at: www.archives.gov/research/9-11/staff-monographs.html.

32   early train: Interview with Michael Woodward, February 2, 2017. Elements of Woodward's account also come from his FBI interview on September 12, 2001, and his 9/11 Commission interview on January 25, 2004.

32   Betty Ong: Interview with Cathie Ong-Herrera, January 16, 2017; interview with Michael Woodward, February 2, 2017; and multiple FBI 302s of Interest re Flight 11.

32   Chinese opera: Interview with Cathie Ong-Herrera, January 16, 2017.

33   Kathleen "Kathy" Nicosia: FBI 302s, Homer file re Kathleen Nicosia. FBI interview with Nancy Wyatt, September 15, 2001.

33   Around 7:15 a.m.: Some preflight times are approximate, based on recollections of individuals. Not all activities during preflight preparation are logged by time.

33   Shawn Trotman: FBI interview, September 16, 2001.

33   an hour earlier: Flight 11 had previously been Flight 198, which arrived at approximately 6:06 a.m. FBI interview with ramp service manager Salvatore Misuraca, September 17, 2001.

34   76,400 pounds: Trotman told the FBI he filled the wings with 76,000 pounds of fuel, but the 9/11 Commission "Four Flights Monograph" put the amount at 76,400, p. 7.

34   forty-foot fire truck: FAMA Emergency Vehicle Size and Weight Regulation Guideline, p. 9. https://fama.org/wp-content/uploads/2015/09/1441593313_55ecf7e 17d32d.pdf, accessed February 2, 2017.

34   landing gear: FBI interview with cleaning crew member Wayne Kirk, September 12, 2001. Kirk said he left the plane between 7 a.m. and 7:30 a.m., as the captain was outside checking the landing gear.

34   Madeline "Amy" Sweeney: FBI interview with Michael Sweeney, September 20, 2001. Details of the call regarding her daughter come from multiple sources, including Sally Heaney, "Separate Lives Joined By Shared Loss," *Boston Globe*, September 11, 2003.

34   Candace Lee Williams: "Portraits of Grief: Candace Lee Williams," *New York Times*, November 30, 2001.

34	Robert Norton: Rachael Rees, "Area Man's Sister Was on Fatal Flight," *(Oregon) Bulletin*, September 10, 2001.

34	Daniel Lee: "Allison Lee, Whose Dad Died on 9/11 on His Way to See Her Being Born," *People*, September 12, 2016.

34	Cora Hidalgo Holland: Marcella Bombardieri, "Four Families' Enduring Grief," *Boston Globe*, September 11, 2002.

35	Berry Berenson: Amanda Hopkinson, "Berry Berenson," *The Guardian*, September 14, 2001.

35	Alexander Filipov: "Portraits of Grief: Alexander Filipov," *New York Times*, October 8, 2001.

35	Pendyala "Vamsi" Vamsikrishna: Steve Lopez, "When Love Stands Bravely Against Unbearable Grief," *Los Angeles Times*, October 24, 2001.

35	didn't rattle easily: Background on Tara Creamer comes from interviews with John Creamer, Colin Creamer, Nora Creamer, Gerry Creamer, Julie Creamer, and Tina Creamer, January 27, 2017. Also from several follow-up telephone interviews with John Creamer and information from Mitchell Zuckoff, "Six Lives: Reliving the Morning of Death," *Boston Globe*, September 16, 2001, p. A1.

37	run a 5K race: Richard Chacon and Charles M. Sennott, "Neilie Heffernan Casey, 32," *Boston Globe*, September 21, 2001.

37	saying hi for him: Holly Ramer, "Survivors of 9/11 Victims Look for Meaning and Find Ways to Move Forward," The Associated Press, September 11, 2002.

37	David Retik: "Portraits of Grief: David Retik," *New York Times*, November 17, 2001.

37	Richard Ross: "Portraits of Grief: Richard Ross," *New York Times*, September 8, 2002.

37	worst day of his life: FBI interview with unidentified American Airlines gate agent, September 14, 2001.

37	Sonia Puopolo: Interview with Sonia Tita Puopolo, January 20, 2017.

38	"gratitude list": Interview with Allan Hackel, January 28, 2017.

38	David Angell: Unbylined story, "David Angell, 54; A Creator and Writer for "Frasier" Sitcom," *New York Times*, September 14, 2001.

38	"American Flight 11": *Frasier* episode "Odd Man Out," aired May 21, 1997. Script found at www.kacl780.net/frasier/transcripts/season_4/episode_24/odd_man_out.html.

39	$400 million deal: FBI interview with Anne Lewin, September 21, 2001.

39	hundred times that price: Hiawatha Bray, "A Lost Spirit Still Inspires," *Boston Globe*, September 4, 2011. For a full account of Lewin's remarkable life, see Molly Knight Raskin, *No Better Time* (Boston: Da Capo Press, 2013).

39	roundabout route: 9/11 Commission Staff Monograph, "Four Flights and Civilian Aviation Security," September 12, 2005, http://www.archives.gov/legislative/research/9-11/staff-report-sept2005.pdf (hereafter: Four Flights Monograph), p. 2.

39	One possibility: Four Flights Monograph, pp. 3–4.

39	innocuous items: FBI Memo on Atta's suitcases, February 10, 2004.

40	clenched his jaw: Four Flights Monograph, p. 2.

40   5:45 a.m.: Four Flights Monograph, p. 3.

40   stern expression: A security image of Mohamed Atta and Abdulaziz al-Omari passing through security at the Portland jetport was introduced as evidence at the Moussaoui trial.

40   CAPPS: Four Flights Monograph, p. 70.

40   threat to air travel: Four Flights Monograph, pp. 70–79.

40   sixty-four hijackings: Four Flights Monograph, p. 58.

41   stricter airport security rules: Four Flights Monograph, p. 71.

41   Swiss Army knives: Four Flights Monograph, p. 5.

41   carry box cutters: 9/11 Commission Staff Statement No. 16, "Outline of the 9/11 Plot," p. 9.

41   "common sense": Four Flights Monograph, pp. 73–74.

42   thirty-three such marshals: Four Flights Monograph, p. 80.

42   "Terrorism can occur": Four Flights Monograph, p. 63.

42   a strange interaction: FBI interview with Lynn Howland, October 13, 2001.

42   "No, I just brought": Ibid.

42   asked a gate agent: FBI interview with Salvatore Misuraca, September 11, 2001; FBI interview with Philip Depasquale, September 17, 2001; FBI memorandum on Atta's suitcases, February 10, 2004. Also, Commission staff memorandum for the record, Boston, MA, Summary, February 2, 2004, p. 10.

43   handwritten instruction letter: FBI document, "Summary of Penttbom Investigation," February 29, 2004, p. 69. https://vault.fbi.gov/9-11%20Commission%20Report/9-11-fbi-report-2004-02(feb). This document describes it as a four-page letter.

43   videotaped lessons: FBI memorandum on Atta's suitcases, February 10, 2004.

43   Already seated: Four Flights Monograph, p. 6.

43   Shortly after: This time, from the Four Flights Monograph, p. 6, is considered approximate. It is the same time, 7:40 a.m., given for when Flight 11 pushed back from the gate, p. 7.

44   158 passengers: Four Flights Monograph, p. 6.

44   walked aboard: Interview with Michael Woodward, February 2, 2017.

44   "Type A-plus": "Portraits of Grief: They liked her style," New York Times, December 9, 2001; Sam Trapani, "Seasons in the Sun: Remember Karen Martin of Danvers," Danvers Herald, September 8, 2011.

44   petite and patient: "Portraits of Grief: Taking Time to Relax," New York Times, September 9, 2002.

44   Runway 4R: National Transportation Safety Board Flight Path Study, American Airlines Flight 11, February 19, 2002, p/ 1. http://nsarchive.gwu.edu/NSAEBB/NSAEBB196/doc01.pdf. Hereafter: NTSB Flight 11.

44   7:59 a.m.: Four Flights Monograph, p. 7.

44   forty-five hundred: 9/11 Commission Report, p. 29.

45   nineteen times: NTSB Flight 11, pp. 5–6. Also see Four Flights Monograph, p. 7.

45   smell of fresh-brewed coffee: Interview with Peg Ogonowski Hatch, January 24, 2017, who by her estimates worked as a flight attendant on Flight 11

hundreds of times. She said coffee would have been brewing during the first fifteen minutes of the flight, even if service had not yet begun.

45 "silver service": Interview with Peg Ogonowski, January 24, 2017.

45 Sixteen seconds later: NTSB Flight 11, p. 6.

45 nineteen years: Peter Zalewski interview with 9/11 Commission staff, September 23, 2003, p. 3. Hereafter: Zalewski, 9/11 Commission.

46 260 other controllers: Lynn Spencer, *Touching History: The Untold Story of the Drama that Unfolded in the Skies over America on 9/11* (New York: Free Press, 2011), p. 6. Hereafter: Spencer, *Touching History*. The full name of Boston Center is Boston Air Route Traffic Control Center.

46 he wondered: Tom Brokaw, "The Skies Over America," NBC News, September 9, 2006. www.nbcnews.com/id/14754701/ns/dateline_nbc/t/skies-over-america/#.WKI-jmQrJPM. Hereafter: NBC Skies Report.

46 collision course: Zalewski, 9/11 Commission.

46 "He's NORDO": NBC Skies Report.

46 happened often enough: Zalewski, 9/11 Commission.

46 "American Eleven, Boston": NTSB Flight 11, p. 6. Radio transmissions and time stamps between Zalewski and Flight 11 come from this source.

47 brief, unknown sound: "FAA Summary of Air Traffic Hijack Events, September 11, 2001." Radio communication timeline dated September 17, 2001. http://ns archive.gwu.edu/NSAEBB/NSAEBB165/faa7.pdf.

47 at 8:21 a.m.: 9/11 Commission Report, p. 18.

47 said quietly: Zalewski 9/11 Commission interview and NBC Skies Report.

48 "Absolutely not": Spencer, *Touching History*, p. 13.

48 declaring a hijacking: 9/11 Commission interview with Terry Biggio, September 23, 2003.

CHAPTER 3: "A BEAUTIFUL DAY TO FLY"

49 As they ate: Interview with Lee and Eunice Hanson, February 22, 2017.

50 flight path: See NTSB Flight Path Study, United Flight 175, p. 3. The plane flew roughly over Sharon, Connecticut, then crossed over the New York border, where its pilots made the requested turn away from American Flight 11.

50 Saracini: Unbylined story, "9/11 Widow Recalls Last Words With Pilot Husband, 'I love you,'" CBS, September 11, 2013.

50 Michael Horrocks: Schuyler Knopf, "C of C Athlete Lost Her Dad, A Co-Pilot, During 9/11," *The Post and Courier*, September 7, 2010.

51 something nagged: As explained later in the chapter, pilots Saracini and Horrocks later said they heard a suspicious transmission "on our departure out of Boston" but purposely waited until 8:41:33 a.m. to report it to air traffic control.

51 fifty-six passengers: Four Flights Monograph, p. 19.

51 David Brandhorst: 9/11 Memorial Blog. www.911memorial.org/blog/honoring-life-david-reed-gamboa-brandhorst.

51 Daniel Brandhorst: Unbylined story, "Daniel Brandhorst," *Los Angeles Times*, September 18, 2001.

51   smuggled aboard: Interview with Ron Clifford, February 9, 2017.

52   Reverend Francis Grogan: "Portraits of Grief: Rev. Francis Grogan," *New York Times*, December 27, 2001.

52   Garnet "Ace" Bailey . . . Mark Bavis: Kevin Paul Dupont, "Widow Still Holds Ace in Her Hand," *Boston Globe*, September 11, 2011; "Portraits of Grief: Ace of 'Bailey-Baisse,'" *New York Times*, November 25, 2001.

52   eleven hours: Bailey earned 633 penalty minutes in the regular season, plus 28 minutes during playoffs, in his NHL career, according to the Internet Hockey Database, www.hockeydb.com/ihdb/stats/pdisplay.php?pid=169.

52   fear of flying: Jackie L. Larson, "'Ace' Bailey, Gretzky's Friend and Mentor, Among the 9/11 Dead," *Edmonton Sun*, September 10, 2011.

52   Touri Bolourchi: Unbylined story, "Muslim-American Family Mourns Loss of Loved One," *CNN*, September 18, 2001.

52   Michael Tarrou . . . Amy King: Lizbeth Hall, "Living Memorial to 9/11 Victims," *Hartford Courant*, August 23, 2002.

52   Alfred Marchand: Heather Clark, "Religious Faith Helps Family Cope with 9/11," The Associated Press, December 17, 2001.

53   Robert Fangman: "Portraits of Grief: Robert Fangman," *New York Times*, November 18, 2001.

53   "I have a ten-year plan": Interviews with Andrea LeBlanc, January 31 and February 1, 2017. Information on Robert LeBlanc also comes from an archive of remembrances and readings from a memorial service held at the University of New Hampshire on September 21, 2001.

55   almost missed the exit: Carolyn LeBlanc comments at memorial service, September 21, 2001.

56   before they saw stars: Reminiscence of Alan Sweeney, posted on Boston University 9/11 Memorial website, September 17, 2001. "I remember his eyes, because I was an offensive back and we . . . met quite often at the line of scrimmage."

56   F-14 fighter jets: FBI interview with Louise Sweeney, March 25, 2004.

56   top of his class: FBI interview with Louise Sweeney, March 25, 2004.

56   "heart of a warrior": FBI interview with Louise Sweeney, March 25, 2004.

56   "That's the kind of guy": Interview with Julie Sweeney Roth, February 3, 2017.

57   "You celebrate life": Interview with Julie Sweeney Roth, February 3, 2017.

58   bought a multitool: Four Flights Monograph, p. 17.

58   commercial pilot certificate: Staff Memo No. 4, p. 5.

58   "cousin": Kate Zernike and Don Van Natta Jr., "The Plot: Hijackers' Meticulous Strategy of Brains, Muscle and Practice," *New York Times*, November 4, 2001.

58   two short-bladed knives: Four Flights Monograph, p. 17.

58   final confirmation: Four Flights Monograph, p. 18. Although it is unknown who placed the call to Atta, the timing and the location of the pay phone, combined with al-Shehhi's relationship with Atta and his status as leader of the Flight 175 hijackers, led authorities to believe the caller was most likely al-Shehhi. The exact content of the call is unknown.

59   seemed confused: Four Flights Monograph, p. 17. Also FBI interviews with
Gail Jawahir, Sept. 21, 2001, and September 28, 2001.

59   additional security screenings: Four Flights Monograph, p. 18.

59   overbearing fragrance: 9/11 Commission staff memorandum for the record,
Boston, MA, Summary, February 2, 2004, p. 8.

59   fifteen-cent tip: Ibid.

CHAPTER 4: "I THINK WE'RE BEING HIJACKED"

60   in the tail: FBI interview with American flight service manager Nancy Wyatt,
September 15, 2001.

61   The call: Betty Ong's phone call was a pivotal initial piece of evidence in the
hijacking of Flight 11. It is detailed in numerous official reports, including the 9/11
Commission Report and the Four Flights Monograph, pp. 8–10. For a transcript
with audio, see: www.nytimes.com/interactive/2011/09/08/nyregion/911-tapes
.html?_r=0.

61   "I think we're being hijacked": FBI interview with Vanessa Minter, September 12, 2001.

62   a wonderful job: 9/11 Commission interview with Craig Marquis and others,
November 19, 2003.

63   "pray for us": FBI interview with Winston Courtney Sadler, September 12, 2001.

63   No one . . . relayed information: 9/11 Commission interview with Craig Marquis and others, November 19, 2003.

63   almost spitting: 9/11 Commission Staff Visit to Boston Center, New England
Region, FAA, p. 6.

64   "We have some planes": 9/11 Commission Report, p. 19. The version of
the Atta quote used here is from the Full Audio Transcript published in 2011 by
the Rutgers University Law Review, at http://www.rutgerslawreview.com/2011
/full-audio-transcript/, which corrected, clarified, or reinterpreted earlier transcriptions of key FAA and other recorded communications on 9/11. (The 9/11 Commission transcribed this passage, in part, "Just stay quiet, and you'll be OK.") It
bears mentioning that 911 Commission investigator Miles Kara does not subscribe
to the belief that Atta mistakenly keyed the mic and "accidentally" broadcast his
message beyond the confines of Flight 11. Rather, Kara maintains that Atta was,
in part, attempting to sow confusion within the FAA. He also suggests that Atta
was delivering a message to his collaborator pilot on United Flight 175, Marwan
al-Shehhi. In Kara's view, explained more deeply at his oredigger61.org website, the
hijackings were more sophisticated than many observers believe, and the terrorists
would have known that passengers likely could monitor cockpit communications
on Channel 9 of United's onboard entertainment system. Because both American
Flight 11 and United Flight 175 departed on cross-country routes from Boston at
nearly the same time, Kara explains, the terrorists could feel confident that the two
cockpits would be using the same radio frequency during the first minutes after
takeoff. Under that scenario, Atta's "We have some planes" remark could be viewed
as a signal to Marwan al-Shehhi that their plan was working and that the United
Flight 175 group should execute its piece of the attack. Although it is unknown

whether al-Shehhi heard Atta's comment or was listening to Channel 9, Kara considers it likely. One piece of evidence he cites is the fact that al-Shehhi waited to initiate the hijacking until after Flight 175 had crossed into the airspace of a different air traffic control center. Kara believes that al-Shehhi knew the crossover took place because he heard the legitimate United Flight 175 pilots say so. If that was the case, Kara maintains, al-Shehhi also would have heard the earlier transmissions from Atta that were picked up in the cockpit of United Flight 175 and reported later to air traffic control. Separately, John Farmer, senior counsel to the 9/11 Commission, raised questions to the author about whether the sequence of the hijackings, in which two United flights were hijacked after American flights, might have been influenced by the terrorists' hope to use United Channel 9 to gather real-time intelligence on the other hijackings.

65   screamed for his supervisor: Zalewski, 9/11 Commission, pp. 3–4.

65   verge of panic: During his interview with 9/11 Commission staff, Zalewski said he "freaked out" on Jones. He also said other controllers looked at him as if he were crazy, and was "shouting and screaming."

65   Flight 11 had been hijacked: Four Flights Monograph, p. 11.

65   terror: Zalewski 9/11 Commission interview and NBC Skies Report.

66   set pattern: Anti-hijacking training for flight crews was laid out in the Air Carrier Standard Security Program. See: "The Four Flights, Staff Memo No. 4," http://govinfo.library.unt.edu/911/staff_statements/staff_statement_4.pdf. Hereafter: Staff Memo No. 4.

66   dedicated messaging system: Four Flights Monograph, pp. 8–10.

67   calling card: Interview with Mike Low, January 18, 2017.

67   third try: Four Flights Monograph, p. 10.

67   "What, what, what?": 9/11 Commission interview with Michael Woodward, January 25, 2004, p. 2.

68   At 8:29 a.m.: NTSB Flight 11, p. 3.

68   fell silent: Zalewski, 9/11 Commission, p. 6.

68   take too long: Four Flights Monograph, p. 12; 9/11 Commission interview with Terry Biggio, September 23, 2003.

68   correct protocol: The issue of FAA notification of the military is discussed at length in Spencer, *Touching History*, p. 22.

68   transponder turned off: Priscilla D. Jones, *The First 109 Minutes: 9/11 and the U.S. Air Force*, Monograph published by the Air Force History and Museums Program, Washington, D.C., 2011.

69   called the Otis Air National Guard base: Four Flights Monograph, p. 13; 102nd Fighter Wing Historian's Report for September 11, 2001, by TSgt. Bruce Vittner.

69   first direct notification: 9/11 Commission Report, p. 20. Four Flights Monograph, p. 13.

69   "[W]e have a problem here": 9/11 Commission Report, p. 20.

69   "needed in Ops, pronto": Interview with Kevin Nasypany, March 20, 2017. Also transcript from Voice Recorder, September 11, 2001, Northeast Air Defense Sector, Rome, New York. Quotes from NEADS tapes also derive from the Full Audio Transcript created by the *Rutgers Law Review*, at http://www.rutgerslaw

review.com/2011/full-audio-transcript/. The actions and statements of Major Kevin "Nasty" Nasypany have been detailed elsewhere, including the impressive "9-11 Revisited" blog maintained by former 9/11 Commission investigator Miles Kara, at www.oredigger61.org. Also see Spencer, *Touching History*; Michael Bronner, "9/11 Live: The Norad Tapes," *Vanity Fair*, August 2006; Priscilla D. Jones, *The First 109 Minutes: 9/11 and the U.S. Air Force*, a monograph published in 2011 by the Air Force History and Museums Program; and 9/11 Commission Staff Statement No. 17. A great deal of research has explored the faulty timelines of military response to 9/11 provided in the initial days, months, and even years after the attacks. One clear conclusion is that Nasypany acted with extraordinary grace under pressure and played no role in the erroneous reports. This account relies on the generally accepted timeline that emerged after release of the 9/11 Commission report, relying also on interviews with Nasypany, the transcripts of the NORAD tapes, and the work of Farmer, Kara, Bronner, Spencer, Summers, and Swan and many others. Among the most significant conclusions was that initial claims by Vice President Dick Cheney and others about a hot pursuit of a hijacked plane by fighter jets with shoot-to-kill orders was not supported by the evidence. Earlier research and the 9/11 Commission also demonstrated falsehoods or errors in claims by the military and the FAA about situational awareness. These issues are examined in depth by Farmer, Summers, and Swan and others.

70	Nasypany thought: Interview with Kevin Nasypany, March 20, 2017; Michael Bronner, "9/11 Live: The Norad Tapes," *Vanity Fair*, August 2006.

70	"another hour": Michael Bronner, "9/11 Live: The Norad Tapes," *Vanity Fair*, August 2006.

70	weren't a top priority: 9/11 Commission Report, p. 17: "NORAD perceived the dominant threat to be from cruise missiles."

70	"on the shitter": Transcript from Voice Recorder, September 11, 2001, Northeast Air Defense Sector, Rome, New York.

71	locker room: Matt Viser, "Two Pilots Revisit Their 9/11," *Boston Globe*, September 11, 2005.

71	fourteen years old: Matt Viser, "Two Pilots Revisit Their 9/11," *Boston Globe*, September 11, 2005.

72	"follow the flight": 9/11 Commission Report, pp. 17–18; Farmer, p. 116.

72	five miles: Staff Memo No. 17, p. 3.

72	"tracking coast": Farmer, pp. 134–35.

72	"I don't know": 9/11 Commission Report, p. 20.

73	"scramble the airplanes": 9/11 Commission Staff Statement No. 17, "Improvising a Homeland Defense," p. 6. Hereafter, Staff Memo No. 17.

74	more than one hijacking: 9/11 Commission Report, p. 10.

74	"sleeper" comrades: Interview with Peg Ogonowski, January 24, 2017.

74	a program known as the Common Strategy: 9/11 Commission Staff, "The Four Flights, Staff Memo No. 4," http://govinfo.library.unt.edu/911/staff_statements/staff_statement_4.pdf, p. 1. Hereafter, Staff Memo No. 4.

75	"suicide wasn't in the game plan": Staff Memo No. 4, p. 2.

75	attempt to disappear: 9/11 Commission Report, p. 18.

75	a run for Cuba: NBC Skies Report, quoting Dave Bottiglia.

75   when the pilots turned off the Fasten Seatbelt signs: Staff Memo No. 4, p. 3.

75   keys to the cockpit: 9/11 Commission Report, p. 453. Multiple interviews with American Airlines employees confirmed this fact.

76   weren't strong enough: Staff Memo No. 4, p. 3.

76   one key: Four Flights Monograph, p. 81.

76   loud arguing: FBI interview with Craig Marquis, September 16, 2001.

76   first, brief call: FBI interview with Evelyn Nunez, September 12, 2001.

76   "Tom Sukani": Four Flights Monograph, p. 12.

77   Another possibility: Callers from Flight 93 also reported that one passenger was killed in first class. Based on the fact that Mark "Mickey" Rothenberg was the only first-class passenger who didn't make a phone call, it's reasonable to conclude that he was the victim. Rothenberg was seated directly in front of one of the Flight 93 hijackers, essentially in the same seating arrangement as Lewin and al-Suqami. That makes it at least possible that part of Atta's plan called for the hijacker seated farthest back to attack a passenger, even unprovoked, to frighten other passengers and crew into compliance. Understandably, Lewin's family and friends remain certain that he acted heroically, and no evidence contradicts that.

78   heard Amy scream: Four Flights Monograph, p. 14, based on transcript of call from Nancy Wyatt to Ray Howland.

79   440 miles per hour: An FAA report put the speed of Flight 11 at 494 miles per hour, while a report by a professor at MIT calculated it at 429 mph. The NIST report put the speed at about 440 miles per hour. NIST NC STAR 1, WTC Investigation, p. 20. Also see Brian Dakss, "Speed Likely Factor in WTC Collapse," CBS News, February 5, 2002. www.cbsnews.com/news/speed-likely-factor-in-wtc-collapse-25-02-2002/.

80   normally controlled: David Perry, "There's No Answer You Can Understand," The Lowell Sun, September 11, 2006.

CHAPTER 5: "DON'T WORRY, DAD"

81   "suspicious transmission": NTSB Flight Path Study, United Airlines Flight 175, p. 6.

82   hadn't yet heard anything: Spencer, Touching History, pp. 36–37. Also see NBC Skies Report.

82   "Oh, okay": NTSB Flight Path Study, United Airlines Flight 175, p. 7.

82   Under normal circumstances: Four Flights Monograph, p. 20.

82   "You see this target here?": NBC Skies Report.

83   didn't notice: Four Flights Monograph, p. 21.

83   "smoke in Lower Manhattan": Four Flights Monograph, p. 21; also FAA memo "Full Transcript; Aircraft Accident; UAL 175," dated May 2, 2002.

83   climbing several thousand feet: NTSB Flight Path Study, United Flight 175, p. 4.

84   starting to shake: NBC Skies Report.

84   "being hijacked": Interview with Lee and Eunice Hanson, February 22, 2017. This account of Peter Hanson's two phone calls also relies on Lee Hanson's interviews with the FBI on September 11, 2001, and June 20, 2002; and a transcript of Lee Hanson's testimony at the Moussaoui trial.

84   prank call: Brian McGrory, "Up From the Ashes," *Boston Globe*, September 11, 2001.

85   "This just in": CNN Transcript from September 11, 2001, 8:48 a.m., http://transcripts.cnn.com/TRANSCRIPTS/0109/11/bn.01.html.

86   refused to be disturbed: Staff Memo No. 17, p. 7.

87   dial "f-i-x": FBI interview with Derek Price, January 24, 2002.

87   reported details: 9/11 Commission Report, p. 8.

87   "a reported incident": Four Flights Monograph, p. 22.

87   tried four times: Four Flights Monograph, p. 21. FBI 302 on phone calls reveals the caller was Garnet "Ace" Bailey. See FBI report, "Documents relating to any report, including but not limited to FD-302S, electronic communications, and inserts, by any passenger on board any of the four flights hijacked on 9/11/2001 following the hijacking of such flight (including voice mail, messages and calls to the Somerset County, PA, Emergency 911 Operator Service)." Found at Scribd.com, uploaded by the 9/11 Document Archive, Feb. 1, 2011.

88   fishhook turn: NTSB Flight Path Study, United Airlines Flight 175, p. 3.

88   "Traffic, two o'clock": Spencer, *Touching History*, p. 48. Also see FAA memo "Full Transcript; Aircraft Accident; UAL 175," dated May 2, 2002.

88   collision course: Charles Lane, Don Phillips, and David Snyder, "A Sky Filled with Chaos, Uncertainty and True Heroism," *Washington Post*, September 17, 2001.

89   run errands: Louise Sweeney interviews with FBI, September 11, 2001, and March 24, 2004.

89   "It's getting bad, Dad": 9/11 Commission Report, p. 8; Interview with Lee and Eunice Hanson, February 22, 2017.

90   "several situations": Four Flights Monograph, p. 23.

90   beset by a routine: "FAA Summary of Air Traffic Hijack Events, September 11, 2001." Radio communication timeline dated September 17, 2001, p. 13. http://nsarchive.gwu.edu/NSAEBB/NSAEBB165/faa7.pdf.

91   "He's going in!": David Maraniss, "September 11, 2001: Portrait of a Day That Began in Routine and Ended in Ashes," *Washington Post*, September 16, 2001, p. 1.

91   9:03:11 a.m.: Four Flights Monograph, p. 24. There are various estimates of the speed, with 540 mph coming from NIST NCSTAR 1, WTC Investigation, p. 38, and 587 mph coming from the NTSB Flight Path Study, United Airlines Flight 175.

91   "Oh my God": Video of ABC *Good Morning America* coverage of 9/11, found at http://abcnews.go.com/WNT/video/reporting-911-attacks-14491835. In *The Eleventh Day*, Summers and Swan use a slightly different transcript, pp. 33–34.

93   same time: Four Flights Monograph, p. 15. Also p. 23.

93   previous ten minutes: "FAA Summary of Air Traffic Hijack Events, September 11, 2001." Radio communication timeline dated September 17, 2001. http://nsarchive.gwu.edu/NSAEBB/NSAEBB165/faa7.pdf.

CHAPTER 6: "THE START OF WORLD WAR III"

94   "I love you": Toby Harnden, "She Asked Me How to Stop the Plane," *The Telegraph*, March 5, 2002.

95   "Oh God": Summers and Swan, p. 129; Transcript from Voice Recorder, September 11, 2001, Northeast Air Defense Sector, Rome, New York.

95   "Send 'em": Transcript from Voice Recorder, September 11, 2001, Northeast Air Defense Sector, Rome, New York, p. 10. Also Michael Bronner, "9/11 Live: The Norad Tapes," *Vanity Fair*, August 2006.

95   The call: Transcript from Voice Recorder, September 11, 2001, Northeast Air Defense Sector, Rome, New York, p. 10. Also Michael Bronner, "9/11 Live: The Norad Tapes," *Vanity Fair*, August 2006.

96   didn't learn: 9/11 Commission Staff Statement No. 17, p. 9.

96   watched it live: Michael Bronner, "9/11 Live: The Norad Tapes," *Vanity Fair*, August 2006.

97   "We've already had two": Ibid.

97   Bernard C. Brown II: Lynette Clemetson, "Washington School Still Feels Pain of 9/11," *New York Times*, September 9, 2006; 9/11 Memorial Biography; Bernard Curtis Brown II "Profile in Grief," *New York Times*, October 12, 2001.

98   fifty-eight passengers: Details of Flight 77 passengers come from the Four Flights Monograph, p. 28.

98   Mari-Rae Sopper: Sylvia Moreno, "Mari-Rae Sopper," *Washington Post*, "Remembering the Pentagon Victims," undated; National Pentagon 9/11 Memorial Biography.

98   Leslie Whittington, Charles Falkenberg, and their daughters: National Pentagon 9/11 Memorial Biographies, http://pentagonmemorial.org/explore/biographies; *Washington Post*, "Remembering the Pentagon Victims."

99   Yugang Zheng and Shuying Yang: Testimony of Rui Zheng at Moussaoui trial, transcript pp. 3442–49. (These names are spelled differently in the National Pentagon 9/11 Memorial Biographies, but this is how they were spelled by their daughter in trial testimony.)

99   Retired Rear Admiral Wilson "Bud" Flagg and Darlene "Dee" Flagg: Adam Bernstein, "Wilson 'Bud' Flagg and Darlene 'Dee' Flagg," *Washington Post*, "Remembering the Pentagon Victims," undated; National Pentagon 9/11 Memorial Biographies.

100   Dr. Yeneneh Betru: National Pentagon 9/11 Memorial Biography of Dr. Yeneneh Betru; Erin Chan, "Keeping a 9/11 Victim's Dream Alive," *Los Angeles Times*, September 3, 2002.

100   Eddie Dillard: National Pentagon 9/11 Memorial Biography of Eddie A. Dillard. Also Rosemary Dillard, "Local 9/11 Widow Reflects 10 Years Later," *Bloomfield Patch*, September 8, 2011.

100   Zandra and Robert Riis Ploger III: Sylvia Moreno, "Zandra and Robert Riis Ploger III," *Washington Post*, "Remembering the Pentagon Victims," undated; National Pentagon 9/11 Memorial Biographies.

100   connections to the government: National Pentagon 9/11 Memorial Biographies, http://pentagonmemorial.org/explore/biographies.

101   Laminated prayer card: Interview with Debra Burlingame, July 26, 2018.

101   was seven weeks pregnant: FBI interview with Ronald and Nancy May, June 5, 2002, p. 2. Also Emily Ngo, "9/11 Memorial Honors Unborn Babies," *Newsday*, September 1, 2011.

101   hop a quick flight: Henry Breanlas, "Las Vegas Still Feels 9/11 Aftershocks," *(Las Vegas) Review Journal*, September 11, 2011.

101   Ken and Jennifer Lewis: Ian Shapira, "Kenneth and Jennifer Lewis," *Washington Post*, September 15, 2001.

102   gaze at the stars: Mike Roberts et al., "I Hope and Pray They Were in Each Other's Arms," *(Vancouver) Province*, September 16, 2001.

102   airport security: Four Flights Monograph, pp. 27-28. This account from the 9/11 Commission staff monograph relies on reviews of the airport security video and the commission's interviews with more than forty screeners.

102   multitool knives: Four Flights Monograph, p. 27; the purchase was made by Nawaf al-Hazmi on August 27, 2001.

102   anxious or excited: FBI interview with Vaughn Allex, July 13, 2004.

103   Hindu option: FBI interview with Jorge C. Villasenor, September 11, 2001.

103   issued an order: 9/11 Commission Staff Statement No. 17, p. 9.

104   safety notice nationwide: 9/11 Commission Staff Statement No. 17, p. 9.

104   "a new type of war": Rutgers Law Review audio transcript: www.rutgerslaw review.com/2011/full-audio-transcript/.

104   "smart terrorists today": Priscilla D. Jones, *The First 109 Minutes: 9/11 and the U.S. Air Force*, a monograph published in 2011 by the Air Force History and Museums Program, Washington, p. 22; confirmed by NEADS transcripts.

104   "talk to FAA": 9/11 Commission Staff Statement No. 17, p. 9.

105   "do more than fuck with this": Farmer, p. 160.

105   "My recommendation": Transcript from Voice Recorder, September 11, 2001, Northeast Air Defense Sector, Rome, New York, p. 61.

105   Sidewinder: U.S. Navy Fact File Data Sheet, www.navy.mil/navydata/fact _display.asp?cid=2200&tid=1000&ct=2.

105   Sparrows: 9/11 Commission staff interview with Major James Fox, October 29, 2003, p. 9.

106   "better call the president": Transcript from Voice Recorder, September 11, 2001, Northeast Air Defense Sector, Rome, New York, time-stamped 9:22 a.m., p. 14. On a separate transcript the speaker is identified as female, on p. 12.

106   shoot down: 102nd Fighter Wing Historian's Report for September 11, 2001, by TSgt. Bruce Vittner.

106   "World War III": Brian MacQuarrie, "War Without End, On Both Fronts, There and Here," *Boston Globe*, September 10, 2011; Kevin Dennehy, "I Thought It Was the Start of World War III," *Cape Cod Times*, August 21, 2002.

106   "Good luck": ABC News 20/20, "Get These Planes on the Ground," October 24, 2001.

106   73,000 square miles: FAA Facility Orientation Guide, Indianapolis ARTCC, www.air-traffic-control.org/pdf-files/indianapolis.pdf.

107   fourteen other planes: Farmer, p. 161.

107   8:54 a.m.: Four Flights Monograph, p. 29.

107   turned off its transponder: Four Flights Monograph, pp. 29-30.

107   8:56 a.m.: National Transportation Safety Board, Flight Path Study—American Airlines Flight 77, p. 7. The first set of calls began at 8:56:32 a.m. and continued to 8:58:16 a.m. Radio calls to Flight 77 continued from Indianapolis Center until 9:03:06 a.m.

107   text message: FBI Interview with Donald A. Robinson, September 11, 2001.

107   "sterilize the airspace": Four Flights Monograph, p. 30.

108   Still in the dark: Handwritten notes of 9/11 Commission staff interview with Chuck Thomas, September 24, 2003.

108   they suspected: 9/11 Commission staff interviews with Chuck Thomas, John A. Thomas, and Linda Povinelli, September 24, 2003.

108   the possible crash: Indianapolis Center first learned of the other hijackings and suicide crashes at 9:11 a.m. Rutgers Law Review audio transcript: www.rutgers lawreview.com/2011/full-audio-transcript/.

108   Its autopilot: 9/11 Commission hearing transcript, testimony of commission staff member Sam Brinkley, January 27, 2004, https://9-11commission.gov/archive /hearing7/9-11Commission_Hearing_2004-01-27.htm.

CHAPTER 7: "BEWARE ANY COCKPIT INTRUSION"

109   wake her up: Flight 93 Memorial Oral History Transcript, interview with Lorne Lyles, October 25, 2006, conducted by Kathie Shaffer.

109   phone conversation: FBI interview with Lorne Lyles, September 12, 2001.

109   "miss that shuttle": Linda Shrieves, "CeeCee Lyles Was Soaring Through Life, Then Destiny Came Calling," *Orlando Sentinel*, September 29, 2001.

109   bills and chores: Jere Longman, *Among the Heroes*, New York: HarperCollins, 2002, p. 9.

110   ten passengers: During one of these calls, CeeCee Lyles told her husband there were eleven passengers in first class and twenty-four passengers in coach. Later it was reported that first class had ten passengers. This contributed to confusion and controversy over the correct number of people aboard Flight 93, particularly since there were four hijackers as opposed to five on each of the other affected flights. Ultimately it was determined there were forty legitimate passengers and crew members and four hijackers. Ten passengers sat in first class, including all the hijackers, and twenty-seven sat in coach. Part of the confusion resulted from passenger Marion Britton's purchase of two seats. See 9/11 Commission Report, p. 10; Longman, p. xiii; Tom McMillan, *Flight 93: The Story, the Aftermath and the Legacy of American Courage on 9/11* (Guilford, CT: Lyons Press, 2014), p. vi. (A few notes on McMillan's excellent book are also included below, with regard to telephone calls from Flight 93.)

110   gaping weaknesses: The conclusion reached by the 9/11 Commission puts it most succinctly: "By 8:00 a.m. on the morning of Tuesday, September 11, 2001, they had defeated all the security layers that America's civil aviation security system then had in place to prevent a hijacking." 9/11 Commission Report, p. 4.

111   supposed to leave the gate: 9/11 Commission Report, Chapter 1 footnote No. 63, p. 455.

112   always flew first class: "Passenger Mark 'Mickey' Rothenberg," *Pittsburgh*

Post-Gazette, October 28, 2001; Flight 93 National Memorial biography of Rothenberg, http://www.honorflight93.org/remember/?fa=passengers-crew.

112   Thomas E. Burnett Jr.: Deena Burnett, *Fighting Back: Living Beyond Ourselves*, Altamonte Springs, FL: Advantage Inspirational, 2006. Flight 93 National Memorial biography of Burnett, http://www.honorflight93.org/remember/?fa=passengers-crew.

112   Mark Bingham: Jon Barrett, "Person of the Year, This Is the Mark Bingham Story," *The Advocate*, January 22, 2002, at www.markbingham.com/legend.html; "Passenger Mark Bingham," *Pittsburgh Post-Gazette*, October 28, 2001; Flight 93 National Memorial biography of Bingham, http://www.honorflight93.org/remember/?fa=passengers-crew.

112   Six years earlier: Longman, pp. 137–38.

113   Todd Beamer: Lisa Beamer, *Let's Roll! Ordinary People, Extraordinary Courage* (Carol Stream, IL: Tyndale House Publishers, 2006); "Passenger Todd Beamer," *Pittsburgh Post-Gazette*, October 28, 2001; Flight 93 National Memorial biography of Beamer, http://www.honorflight93.org/remember/?fa=passengers-crew.

113   Jeremy Glick: "Passenger Jeremy Glick," *Pittsburgh Post-Gazette*, October 28, 2001; Flight 93 National Memorial biography of Glick, http://www.honorflight93.org/remember/?fa=passengers-crew; McMillan, p. 77.

113   Louis "Joey" Nacke II: "Passenger Louis J. Nacke II," *Pittsburgh Post-Gazette*, October 28, 2001; Flight 93 National Memorial biography of Nacke, http://www.honorflight93.org/remember/?fa=passengers-crew.

114   Toshiya Kuge: "Passenger Toshiya Kuge," *Pittsburgh Post-Gazette*, October 28, 2001; Flight 93 National Memorial biography of Kuge, http://www.honorflight93.org/remember/?fa=passengers-crew.

114   William Cashman . . . Patrick "Joe" Driscoll: Flight 93 National Memorial biographies of Cashman and Driscoll, http://www.honorflight93.org/remember/?fa=passengers-crew.

114   Hilda Marcin et al.: Flight 93 National Memorial biographies, http://www.honorflight93.org/remember/?fa=passengers-crew; passenger biographies in special section of *Pittsburgh Post-Gazette*, October 28, 2001.

115   "Courageous Challenge": FBI interview with Claudette Greene, September 15, 2001.

115   commandeered a paratransit bus: Longman, p. 62.

116   Jason Dahl: Flight 93 National Memorial biographies, http://www.honorflight93.org/remember/?fa=passengers-crew; pilot biographies in special section of *Pittsburgh Post-Gazette*, October 28, 2001; Longman, p. 2.

116   LeRoy Homer Jr.: Flight 93 National Memorial biographies, http://www.honorflight93.org/remember/?fa=passengers-crew; pilot biographies in special section of *Pittsburgh Post-Gazette*, October 28, 2001.

116   inside his wedding band: Melodie Homer, *From Where I Stand: Flight #93 Pilot's Widow Sets the Record Straight* (Minneapolis, MN: Langdon Street Press, 2012). Kindle location 524.

116   without incident: Four Flights Monograph, p. 35.

117   five telephone calls: Timeline, Flight 93, September 11, 2001, National Park

Service Flight 93 Memorial site, www.nps.gov/flni/learn/historyculture/upload /timeline_flight_93.pdf; FBI Chronology of Hijackers, Part 2 of 2, p. 293.

117   Aysel Sengün: FBI Translation of an interview conducted by German authorities, September 18, 2001; Dirk Laabs, "Testimony Offers Intimate Look at a Sept. 11 Hijacker's Life," *Los Angeles Times*, November 21, 2002.

117   "The Last Night": Attorney General John Ashcroft revealed on September 28, 2001, that three copies of the instruction sheet had been found, including one at the Flight 93 crash site. Bob Woodward, "In Hijackers' Bags, a Call to Planning, Prayer and Death," *Washington Post*, September 28, 2001.

118   A Saudi: 9/11 Commission Report, p. 11. The man was identified as Mohamed al-Kahtani, although it was sometimes spelled Mohammed al-Qahtani. He was captured in 2002 and subsequently held at the U.S. detention center at Guantanamo Bay, Cuba. The senior Pentagon official in the Bush administration who oversaw prosecutions of detainees later told journalist Bob Woodward that she concluded he had been tortured during interrogations.

118   Mohamed Atta: Philip Shenon, "Panel Says a Deported Saudi Was Likely '20th Hijacker,'" *New York Times*, January 27, 2004. During a hearing, members of 9/11 Commission hailed the inspector, Jose E. Melendez-Perez, as a hero.

118   "hit man": 9/11 and Terrorist Travel Monograph, 9/11 Commission, Staff Report, August 21, 2004, p. 30.

118   Waiting in vain: Philip Shenon, "Panel Says a Deported Saudi Was Likely '20th Hijacker,'" *New York Times*, January 27, 2004.

118   perhaps fifteen other planes: Longman, p. 44.

118   drank juice: Matt Hall, who drove Mark Bingham to the airport, said Mark called him at 7:40 a.m. to say he'd made the flight and was drinking orange juice. Jon Barrett, "Person of the Year: This is the Mark Bingham Life Story," *The Advocate*, January 22, 2002.

118   "cleared for takeoff": NTSB Flight Path Study, United Flight 93, February 19, 2002, p. 5.

119   another thirty minutes: Four Flights Monograph, p. 22.

119   9:03 a.m.: Four Flights Monograph, p. 36.

119   a ground stop: Four Flights Monograph, p. 37.

120   more than ten minutes: NTSB Flight Path Study, United Flight 93, February 19, 2002, p. 4.

120   engaged the 757's autopilot: FBI Summary of PENTTBOM Investigation, p. 70.

120   biggest worry: At 9:21 a.m., Flight 93's pilots sent a routine ACARS message to dispatcher Ed Ballinger that said: "Good mornin' . . . Nice clb [climb] outta EWR [Newark Airport] . . . at 350 [35,000 feet] occl [occasional] lt [light] chop. Wind 290/50 ain't helping. J." The "J" apparently meant it came from Jason Dahl, who knew Ballinger. Four Flights Monograph, p. 37.

120   heard her alarm: Melodie Homer, *From Where I Stand: Flight #93 Pilot's Widow Sets the Record Straight* (Minneapolis: Langdon Street Press, 2012), p. 20.

120   dress silently in the bathroom: Flight 93 Memorial Oral History Transcript, interview with Melodie Homer, February 2, 2008, conducted by Kathie Shaffer, p. 34.

121   "I promise you": Homer, p. 21.

121   As sent: FBI interview with Tara Campbell, January 22, 2002; FBI interview with an unnamed Special Investigator and Firearms Instructor who familiarized FBI agents with ACARS, autopilot, and other cockpit technology, December 19, 2001, p. 4.

121   reached Flight 93: Four Flights Monograph, p. 37.

121   arrive . . . two ways: 9/11 Commission interview with Ed Ballinger, April 24, 2004, p. 3; FBI interview with an unnamed Special Investigator and Firearms Instructor who familiarized FBI agents with ACARS, autopilot, and other cockpit technology, December 19, 2001, p. 4. It was widely reported that when an ACARS message reached the cockpit, a chime sounded. However, that more likely referred to a different communication method, called a "SELCAL," in which a bell rings in the cockpit to alert pilots of an incoming radio call on that system. The interview subject told the FBI that method was not believed to be used on Flights 175 or 93 on 9/11. See 9/11 Commission staff interview with Dave Knerr, United Airlines technical expert on cockpit communications, et al., May 27, 2004, p. 1. However, Ed Ballinger told 9/11 Commission staff members he sent his ACARS messages two ways, digitally with a bell, and as a printout.

121   Personal messages: FBI interview with an unnamed Special Investigator and Firearms Instructor who familiarized FBI agents with ACARS, autopilot, and other cockpit technology, December 19, 2001, p. 4.

122   a second time: FBI interview with Tara Campbell, January 22, 2002.

122   started working . . . in 1958: 9/11 Commission staff interview with Ed Ballinger, April 24, 2004, p. 1.

122   Ballinger's job: 9/11 Commission staff interview with Ed Ballinger, April 24, 2004, p. 2.

122   Rule of Five: Flight 93 Oral History Transcript, interview with Edward and Sally Ballinger, April 23, 2007, conducted by Kathie Shaffer, p. 3.

122   perfect weather: 9/11 Commission interview with Ed Ballinger, April 24, 2004, p. 3.

123   "How is the ride": FBI document labeled "ACARS Messages from Dispatch; Messages from Aircraft to Ed Ballinger and Chad McCurdy; and Messages to Flight 93."

123   sent that message: Four Flights Monograph, p. 23.

124   "Beware any cockpit intrusion [sic]": 9/11 Commission interview with Ed Ballinger, April 24, 2004, p. 5; FBI document labeled "ACARS Messages from Dispatch; Messages from Aircraft to Ed Ballinger and Chad McCurdy; and Messages to Flight 93."

124   Flight 175: FBI document labeled "ACARS Messages from Dispatch; Messages from Aircraft to Ed Ballinger and Chad McCurdy; and Messages to Flight 93."

124   the first direct warnings: Four Flights Monograph, p. 37.

124   sent them: 9/11 Commission interview with Ed Ballinger, April 24, 2004, p. 6.

124   fire ax: 9/11 Commission interview with Ed Ballinger, April 24, 2004, p. 9.

124   "Good morning": Four Flights Monograph, p. 37.

124   "hijackings in progress": Four Flights Monograph, p. 37.

124   didn't notice the message: 9/11 Commission interview with Ed Ballinger, April 24, 2004, p. 6.

124   "beware": Ballinger sent the message at 9:23 a.m., and it arrived one minute later. Four Flights Monograph, p. 38.

125   "accident at New York": 9/11 Commission interview with Ed Ballinger, April 24, 2004, p. 6.

125   "Morning Cleveland": Four Flights Monograph, p. 37.

125   "cofirm latest mssg": Four Flights Monograph, p. 38; FBI-released printout of ACARS messages from Flight 93, p. 103.

125   routine radio call: NTSB Flight Path Study, United Flight 93, February 19, 2002, p. 6; Four Flights Monograph, p. 38.

CHAPTER 8: "AMERICA IS UNDER ATTACK"

127   phone call at 9:21 a.m.: 9/11 Commission Staff Statement No. 17, p. 12.

127   "Another hijack!": Michael Bronner, "9/11 Live: The Norad Tapes," Vanity Fair, August 2006.

128   "chase this guy down": 9/11 Commission Staff Statement No. 17, p. 13.

128   Around 8:51 a.m.: This time is based primarily on circumstantial evidence, particularly the loss of radio and transponder and the unauthorized turn. See Four Flights Monograph, p. 29.

128   getting ready for work: FBI interview with Ronald and Nancy May, September 11, 2001.

128   sounded happy: FBI interview with Ronald and Nancy May, June 5, 2002, p. 2.

128   told her mother: FBI interview with Ronald and Nancy May, September 11, 2001; Four Flights Monograph, p. 31.

129   "I love you, Mom": Henry Breanlas, "Las Vegas Still Feels 9/11 Aftershocks," (Las Vegas) Review Journal, September 11, 2011.

129   Patty Carson: 9/11 Commission staff interview, November 19, 2003.

129   "hijacked and held hostage": 9/11 Commission staff interview with Patty Carson, November 19, 2003.

129   Rosemary Dillard stumbled backward: Tom Murphy, Reclaiming the Sky: 9/11 and the Untold Story of the Men and Women Who Kept America Flying (New York: AMACOM Books, 2007), pp. 57–58. Also Rosemary Dillard, "Local 9/11 Widow Reflects 10 Years Later," Bloomfield Patch, September 8, 2011.

130   Lori Keyton: FBI interview with Lori Lynn Keyton, September 11, 2001.

130   "Barbara is on the line": FBI interview with Helen Voss, September 11, 2001. The statement is not in quotation marks in the FBI notes, but it is presented there as dialogue, which is why it is quoted here.

130   viewing a replay: FBI interview with Ted Olson, September 11, 2001.

130   first thought: Ted Olson interview at Hudson Union Society, posted August 8, 2014, https://www.youtube.com/watch?v=5ppFvUc10nc.

130   "my plane's been hijacked": Ted Olson interview with Alan Colmes, Sean Hannity, and Brit Hume, Hannity and Colmes, Fox News, September 14, 2001.

130   "the pilot": FBI interview with Ted Olson, September 11, 2001; Four Flights Monograph, p. 32.

131   "tell the pilot": Toby Harden, "She Asked Me How to Stop the Plane," *The Telegraph*, March 5, 2002.

131   reassured the other: Ibid.

131   9:09 a.m.: 9/11 Commission Staff Statement No. 17, p. 10.

131   9:20 a.m.: 9/11 Commission Staff Statement No. 17, pp. 10–11.

132   As late as 9:20 a.m.: Four Flights Monograph, p. 25.

132   false sense of security: This issue is explored at some length in the Four Flights Monograph, pp. 53–59.

132   presidential commission: 9/11 Commission Report, pp. 82–83. Also see *Final Report of the White House Commission on Safety and Security* (Gore Commission), February 12, 1997.

132   twelve names: Farmer, p. 101.

132   didn't even know that . . . list existed: Farmer, p. 101.

133   had written a memo: The so-called "Phoenix EC" memo is discussed in the Four Flights Monograph, p. 64; also see 9/11 Commission staff interview with Kenneth Williams, October 22, 2003; 9/11 Commission Report, p. 272.

133   "an Islamic extremist": 9/11 Commission Report, p. 273.

133   didn't mention the agent's belief: 9/11 Commission Report, p. 274.

133   "suicide in a spectacular explosion": Four Flights Monograph, p. 59, based on FAA, 2001 CD-ROM Terrorism Threat Presentation to Aviation Security Personnel at Airports and Air Carriers, slide 24.

134   never saw it: 9/11 Commission Staff Statement No. 17, p. 11.

134   undetected for thirty-six minutes: 9/11 Commission Staff Statement No. 17, p. 11.

134   disappearance of Flight 77: Four Flights Monograph, Footnote 277, p. 95, based on Benedict Sliney interview, May 21, 2004. Sliney described seeing Flight 175 hit the South Tower in an interview with the BBC that aired September 12, 2011.

134   haunted by the question: Alan Levin, Marilyn Adams, and Blake Morrison, "Clear the Skies," *USA Today*, August 12, 2002, p. A1.

134   "nationwide ground stop": Four Flights Monograph, p. 33.

135   first day in his new job: David Germain, "FAA Official Plays Self in 'United 93,'" The Associated Press, April 24, 2006.

135   dropped from an altitude of 25,000 feet: National Transportation Safety Board, Flight Path Study—American Airlines Flight 77, p. 4; Four Flights Monograph, p. 33.

135   "We have two reports": Rutgers Law Review audio transcript: www.rutgerslawreview.com/2011/full-audio-transcript/.

136   military jet: ABC News 20/20, "Get These Planes on the Ground," October 24, 2001.

136   her fiancé: Miles Kara, "9-11: The Andrews Fighters; an expeditionary force, not an air defense force," at http://www.oredigger61.org/?p=154.

136   "Oh my God": ABC News 20/20, "Get These Planes on the Ground," October 24, 2001.

136 "unidentified, very fast-moving aircraft": Ibid.

137 Russian planes or cruise missiles: 9/11 Commission interview with Major Dean Eckmann, December 1, 2003.

See also Michael Bronner, "9/11 Live: The Norad Tapes," *Vanity Fair*, August 2006, and Priscilla D. Jones, *The First 109 Minutes: 9/11 and the U.S. Air Force*, Monograph published by the Air Force History and Museums Program, Washington, D.C., 2011, pp. 37–41.

137 "gonna shoot him down": Michael Bronner, "9/11 Live: The Norad Tapes," *Vanity Fair*, August 2006.

137 heightened terrorism risk: Accounts of the first hours of Bush's day come from numerous sources, including the 9/11 Commission Report; the 9/11 Commission Staff Statement No. 17; and an excellent oral history, Garrett M. Graff, "We're the Only Plane in the Sky," *Politico*, September 9, 2016. Also William Langley, "Revealed: What Really Went on During Bush's 'Missing Hours,'" *The Telegraph*, December 16, 2001; Dan Balz and Bob Woodward, "America's Chaotic Road to War," *Washington Post*, January 27, 2002; Bill Sammon, "Suddenly, a Time to Lead," *Washington Times*, October 7, 2002; Summers and Swan, pp. 32–40; Farmer, p. 155.

137 "Bin Laden Determined": 9/11 Commission Report, p. 261.

137 wasn't a word about terrorism: Garrett M. Graff, "We're the Only Plane in the Sky," *Politico*, September 9, 2016, quoting Mike Morell: "There was nothing in the briefing about terrorism."

138 "so off course": Bill Sammon, "Suddenly, a Time to Lead," *Washington Times*, October 7, 2002.

138 "America is under attack.": Card has described this moment in numerous interviews, and it is memorialized in the 9/11 Commission Report, p. 38. Other reports have suggested he told the president the more prosaic, "Captain Loewer says it's terrorism," referring to Navy Captain Deborah Loewer, director of the White House Situation Room.

139 purposely stepped back: MSNBC interview with Andy Card, September 10, 2016, www.youtube.com/watch?v=7fs2duxjpE4.

139 project calm: 9/11 Commission Staff Statement No. 17, p. 22.

139 "Ladies and gentlemen": Bush's full remarks can be found at numerous sources, including: "Remarks in Sarasota, Florida, on the Terrorist Attack on New York City's World Trade Center," September 11, 2001. Online by Gerhard Peters and John T. Woolley, *The American Presidency Project*. http://www.presidency.ucsb.edu/ws/?pid=58055.

140 Lieutenant Colonel Steven O'Brien: 9/11 Commission staff interview with Lt. Colonel Steven O'Brien, May 6, 2004; Bob Von Sternberg, "How We've Changed, 9-11-01 to 9-11-02," *(Minneapolis) Star Tribune*, September 11, 2002; 1st Lt. Sheree Savage, "Witnessing, Now Remembering, the 9/11 Attacks," *Northstar Guardian*, September 11, 2006.

140 "let them land safely": Bill Catlin, "Museum Features Air Guard's History and Role in the War on Terror," *Minnesota Public Radio*, May 31, 2004.

140 a short time earlier: 9/11 Commission staff interview with Lt. Colonel Steven O'Brien, May 6, 2004; investigator Miles Kara told O'Brien that he had calculated that if Gofer 06 had not been delayed three minutes by the takeoff of a 747, "you would have been in the flight path of AAL 77."

141	called the FAA's Washington Center: Farmer, pp. 172–83; Four Flights Monograph, p. 33; 9/11 Commission Staff Statement No. 17, p. 13; Rutgers Law Review audio transcript: www.rutgerslawreview.com/2011/full-audio-transcript/.

141	"three aircraft": Farmer, p. 173.

142	seized control of the airspace: 9/11 Commission Staff Statement No. 17, p. 13.

142	wanted them in a position: Interview with Kevin Nasypany, March 17, 2017.

143	"run 'em to the White House": Four Flights Monograph, p. 33.

143	mistake on coordinates: Miles Kara, "9-11 NEADS Mission Crew Commander; A Valiant Effort, Ultimately Futile, Part V," 9-11 Revisited, www.oredigger 61.org/?p=5042.

143	Father Stephen McGraw: Interview with Father Stephen McGraw on June 9, 2017. Also see Angela E. Pometto, "Fairfax Priest Remembers Pentagon Scene on 9/11," Arlington Catholic Herald, September 7, 2006; James Graves, "Grief and Grace From the Ashes," OSV Newsweekly, September 9, 2016.

144	"miss the graveside service": Interview with Father Stephen McGraw on June 9, 2017.

144	maximum power: 9/11 Commission Report, p. 9; NTSB Flight Path Study, American Airlines Flight 77, p. 2.

145	just below the second floor: American Society of Civil Engineers, The Pentagon Building Performance Report, pp. 12–20. Cited in Goldberg et al., Pentagon 9/11 (Washington: Historical Office of the Secretary of Defense, 2007).

145	fought to the death: Interview with Debra Burlingame, July 26, 2018.

146	"that aircraft crashed": 9/11 Commission Staff Statement No. 17, p. 12.

146	almost ten minutes: 9/11 Commission staff interview with Major Kevin Nasypany, January 22 and 23, 2004, p. 5.

146	top NEADS officers: 9/11 Commission staff interview with Major James Fox, October 29, 2003, p. 11. Notes say: "Fox speculates that if they had five or more minutes and had gotten the position in [a] timely manner they would have had time to intercept [Flight 77 before it hit the Pentagon]; but that orders would still have taken time."

146	"Goddammit!": Michael Bronner, "9/11 Live: The Norad Tapes," Vanity Fair, August 2006.

146	"Talk to me": Taken from the transcript on the "9-11 Revisited" blog maintained by former 9/11 Commission investigator Miles Kara, at www.oredigger61.org.

147	ride to the airport: 9/11 Commission Report, p. 39.

147	several minutes earlier: 9/11 Commission Report, p. 40, says the vice president reached the entrance to the tunnel at 9:37 a.m. and spoke to the president from a phone there.

147	"a minor war": 9/11 Commission Staff Statement No. 17, p. 23.

147	no set destination: 9/11 Commission Staff Statement No. 17, p. 23.

148	parking lot of the Pentagon: 9/11 Commission Staff Statement No. 17, p. 27.

148	"I'm landing everyone!": Transcript of oral history interview of Ben Sliney at the Flight 93 Memorial, conducted by Kathie Shaffer, October 28, 2007.

148   "Regardless of destination!": Alan Levin, Marilyn Adams, and Blake Morrison, "Clear the Skies," *USA Today*, August 12, 2002, p. A1.

148   4,546 planes: "FAA Chronology of the September 11 Attacks," http://ns archive.gwu.edu/NSAEBB/NSAEBB165/faa4.pdf. Also David Bond, "The Other Aircraft of Sept. 11 – Crisis at Herndon," *Aviation Week*, December 17, 2001.

CHAPTER 9: "MAKE HIM BRAVE"

149   abruptly dropped: Four Flights Monograph, p. 38; NTSB Flight Path Study, Flight 93, p. 6.

149   "Mayday!": FBI evidence for Moussaoui trial; NTSB Flight Path Study, United Flight 93, February 19, 2002, p. 6; Four Flights Monograph, p. 38. Although the NTSB report lists only one "Mayday," an account of the voice recording played at the Moussaoui trial indicated the word was repeated three times, which is standard practice among pilots and boat captains, to be sure the emergency is clearly communicated. See David Stout, "Recording from Flight 93 Played at Trial," *New York Times*, April 12, 2006.

150   calling pilots: 9/11 Commission staff interview with John Werth, October 1, 2003, p. 1.

150   he might panic them: 9/11 Commission staff interview with John Werth, October 1, 2003, pp. 3–4.

150   "guttural sounds": Transcript of oral history interview of John Werth at the Flight 93 Memorial, conducted by Kathie Shaffer.

150   "Somebody call Cleveland?": Four Flights Monograph, p. 38.

150   "I think we've got one!": 9/11 Commission staff interview with John Werth, October 1, 2003, p. 3.

150   "Tell Washington": 9/11 Commission staff interview with John Werth, October 1, 2003, p. 3.

151   "not another one": 9/11 Commission staff interview with Mark Barnik, October 2, 2003, p. 3.

151   moved the passengers: Four Flights Monograph, p. 38.

151   breathed heavily: "FAA Summary of Air Traffic Hijack Events, September 11, 2001."

151   "Here the captain": Four Flights Monograph, p. 39.

151   never before flown: Four Flights Monograph, pp. 49–50.

151   the voice of a second hijacker: In the United Airlines Flight #93 Cockpit Voice Recorder Transcript, introduced as evidence in the Moussaoui trial, twice the pilot hijacker addresses his companion as "Saeed."

151   bomb on board: Four Flights Monograph, p. 39.

152   Jarrah: In light of conclusions by investigators about who was flying the plane and speaking, recorded statements from the hijacker pilot are attributed to Jarrah.

152   using a key: Four Flights Monograph, p. 36.

152   cockpit voice recorder: United Flight #93 Cockpit Voice Recorder Transcript, introduced at Moussaoui trial. A more complete version of the transcript was obtained by INTELWIRE.com, at http://intelfiles.egoplex.com/2003-12-04-FBI-cock

pit-recorder-93.pdf, hereafter Intelwire CVR Flight 93. That document identifies the location of the speaker, based on four distinct microphone positions. All statements coming from the person in the "left seat cockpit," who initially identified himself as the "captain," are attributed to Jarrah. The recording itself was not released publicly, but was played at the trial. See David Stout, "Recording from Flight 93 Played at Trial," *New York Times*, April 12, 2006. Stout wrote: "In some instances, it is hard to tell if a speaker is American or Arab, man or woman. In a transcript of the recording, the hijackers' words in Arabic are accompanied in some cases by printed English translations, which also appeared on a screen in the courtroom while the recordings were being played." The FBI document obtained by intelwire.com goes a long way to distinguish the voices.

153 Someone pleaded: In the *Washington Post*'s account of the recording played at trial, this comment is attributed to a victim, although in the full transcript it comes from the left seat cockpit position occupied by Jarrah. Timothy Dwyer, Jerry Markon, and William Branigin, "Flight 93 Recording Played at Moussaoui Trial," *Washington Post*, April 12, 2006.

153 a native English-speaking woman: Intelwire CVR, Flight 93.

153 she moaned: Neil A. Lewis, "Final Struggles on 9/11 Plane Fill Courtroom," *New York Times*, April 13, 2006.

156 40,700 feet: Four Flights Monograph, p. 39.

156 tossed like a rag doll: 9/11 Commission staff interview with John Werth, October 1, 2003, p. 4.

156 "squawk 'trip,' please": NTSB Flight Path Study, Flight 93, p. 6.

156 "a different level": Four Flights Monograph, p. 40, citing FAA memo "Full transcription; Air Traffic Control System Command Center, National Traffic Management Officer, East position; September 11, 2001," October 31, 2003, pp. 10, 13. Also see *Rutgers Law Review* audio transcript: www.rutgerslawreview.com/2011/full-audio-transcript/.

157 "Ah, here's the captain.": There are several different transcribed versions of this message, with slight variations in wording. This version is from the FBI transcript of the cockpit voice recorder.

157 at a rate of 4,000 feet per minute: NTSB Flight Path Study, Flight 93, p. 2.

158 seatback Airfones: A great deal of discussion and a certain amount of controversy has surrounded the phone calls from Flight 93, including disputes over whether callers used cellphones, Airfones, or both. Reports about the calls in the immediate aftermath of 9/11, by the government, the media, and some family members, contributed to the disputes by including inaccuracies. The author has no interest in rehashing those issues, only in presenting an accurate account of the phone calls based on the most credible information available. This account relies on interviews by the author; summaries of FBI interviews immediately after the attacks; sworn testimony and exhibits from the Moussaoui trial; a memorandum from a May 13, 2004, Justice Department briefing of the 9/11 Commission staff about Flight 93 phone calls; the Four Flights Monograph and the 9/11 Commission Report; published and oral accounts by family members and others who received the calls; interviews for the Oral History Project at the Flight 93 Memorial; and a summary titled "Phone Calls from the Passengers and Crew of Flight 93"

compiled by the National Park Service for the Flight 93 Memorial, which relies on FBI information, at www.nps.gov/flni/learn/historyculture/upload/phone_calls _formatted_web_fall2013.pdf. Tom McMillan's book, *Flight 93: The Story, the Aftermath, and the Legacy of American Courage on 9/11*, was also a source of information and inspiration regarding the calls and all the events of Flight 93.

158   eight outgoing calls: 9/11 Commission staff memorandum on Justice Department briefing about cell and phone calls from Flight 93, May 13, 2004, p. 1.

158   wearing the robe: Deena Burnett with Anthony Giombetti, *Fighting Back: Living Life Beyond Ourselves* (Altamonte Springs, FL: Advantage Inspirational, 2006), p. 59.

159   "an airplane that's been hijacked": Tom Burnett made three confirmed Airfone calls, starting at 9:30, according to the FBI. His wife, Deena, has written and said that he first called on his cellphone, but there was no record of that call on his bill. Dialogue of the call comes from Burnett and Giombetti, p. 61.

159   other planes: This account relies on the FBI interview with Deena Burnett on the day of the attacks, September 11, 2001.

159   9:35 a.m.: Four Flights Monograph, p. 40. An FBI report on calls from Flight 93 indicated that a flight attendant attempted to call United's "Starfix" maintenance center in San Francisco as early as 9:31 a.m., but 9/11 Commission staff interviewed workers there and concluded that only one call went through, which they concluded was Sandy Bradshaw's call at 9:35 a.m.

160   giving birth to triplets: Jane Ganahl, "A Steely Resolve, Born of Anguish," *San Francisco Chronicle*, September 10, 2003.

160   "this is Mark Bingham": Notes of FBI interview with Alice Hoglan, quoted in "Phone Calls from the Passengers and Crew of Flight 93" compiled by the National Park Service for the Flight 93 Memorial. Alice has spelled her last name differently over the years, including Hoglan and Hoaglan. She gave details of their conversation on the day of the attacks to ABC News: https://www.youtube.com /watch?v=qQTs5uuq0II.

160   call Mark's cellphone: Longman, p. 135. In the notes to his engrossing book, written in 2002, Jere Longman wrote, "Alice Hoaglan (she later changed the spelling of her last name) played for me a tape of the calls that she made to her son Mark Bingham's cellphone."

161   "We're so worried": Dennis Roddy, "Flight 93: Forty Lives, One Destiny," *Pittsburgh Post-Gazette*, October 28, 2001.

161   "bad men on this plane": This account of Jeremy Glick's phone call comes from the FBI transcript of a 9-1-1 call, with a New York State Police dispatcher patched in, from Glick's mother-in-law, Joanne Makely. That call was made simultaneous to the lengthy Airfone call between Jeremy and Lyzbeth Glick, www .scribd.com/document/18886268/T7-B12-Flight-93-Calls-Jeremy-Glick-Fdr-FBI-302 -Transcript-Joanne-Makely-911-Call-412. Also the call is described in great detail in Lyz Glick and Dan Zegart, *Your Father's Voice: Letters for Emmy About Life with Jeremy—and Without Him after 9/11* (New York: St. Martin's Press, 2004). Also see notes of FBI interviews with Lyzbeth Glick, September 12, 2001, and Richard Makely, September 12, 2001; 9/11 Commission staff interview with Lyzbeth Glick, April 22, 2004.

161　didn't participate: The 9/11 Commission and its staff considered this question and concluded, "It is reasonable to expect that the hijackers would take all precautions necessary to protect the one among them required to fly the plane. Given their unwillingness to risk his death or injury during the takeover of the aircraft, it made operational sense for the pilot hijacker to remain seated and inconspicuous until he was needed, most likely after the cockpit had been seized." Four Flights Monograph, pp. 40–41.

164　"I'll finish the call": Lisa Jefferson and Felicia Middlebrooks, *Called* (Chicago: Northfield Publishing, 2006), p. 32; 9/11 Commission staff interview with Lisa Jefferson, May 11, 2004.

164　"we're going down": Jefferson, p. 44.

165　"having a little problem": Longman, p. 128.

165　"Mostly I just love you": FBI transcript of Linda Grondlund call, https://vault.fbi.gov/9-11%20Commission%20Report/9-11-interviews-2001-09-sep-05-of-08.

165　"Hi, baby": CeeCee Lyles' voice message is quoted in "Phone Calls from the Passengers and Crew of Flight 93" compiled by the National Park Service for the Flight 93 Memorial, based on FBI reports.

166　land in another country: FBI interview with Fred Fiumano, September 20, 2001. Fiumano's name is redacted, but he later was revealed to have been on the call with Marion Britton.

166　"Hello, Mom": Longman, p. 167.

166　"be in the present": Longman, p. 168.

166　"arms around you": Longman, p. 168.

166　"harder this is going to be on you": Longman, p. 171. A similar version of this quote is contained in "Phone Calls from the Passengers and Crew of Flight 93" compiled by the National Park Service for the Flight 93 Memorial, based on FBI reports, trial evidence, and a family statement.

167　told her husband, Phil: Flight 93 National Memorial Oral History Transcript, interview with Philip Bradshaw conducted by Barbara Black, June 12, 2006, p. 9.

167　his voice shaking: Steve Levin, "'It hurt to listen': A wife describes pain of hearing 911 call from Flight 93," *Pittsburgh Post-Gazette*, April 21, 2002. Levin describes Felt's voice as "quivering."

167　"My plane has been hijacked": Testimony of Lorne Lyles at Moussaoui trial, pp. 3510–11.

168　heard people yelling: Flight 93 National Memorial Oral History Transcript, interview with Lorne Lyles conducted by Kathie Shaffer, October 25, 2006, p. 30.

168　"We've got a plan": Testimony of Lorne Lyles at Moussaoui trial, p. 3511.

CHAPTER 10: "LET'S ROLL"

169　"How about we let them in": Intelwire CVR, Flight 93.

169　"talk to the pilot": Intelwire CVR, Flight 93.

170　"Evacuate the facility": Flight 93 National Memorial Oral History Transcript, interview with Mahlon Fuller conducted by Kathie Shaffer, November 3, 2005, p. 8.

170   "There's a plane coming": The Associated Press, "In Washington, chaos and fear as national capital tries to empty after attacks," September 11, 2001.

170   Cold War–era bunker: U.S. House history interview with Brian Gunderson, Chief of Staff to Representative Richard K. Armey of Texas. https://www.youtube.com/watch?v=iytb4-3CdkY.

170   jerked the plane's nose: NTSB Flight Path Study, United Airlines Flight 93, p. 2.

171   he'd impressed: Flight 93 National Memorial Oral History Transcript, interview with Arne Kruithof, conducted by Kathie Shaffer, April 20, 2009, p. 11.

171   one hundred hours: Four Flights Monograph, p. 50.

171   "do you hear Cleveland Center?": NTSB Air Traffic Control Recording, Flight 93, dated December 21, 2001, p. 10.

171   "scrambling aircraft": Rutgers University Law Review, Rutgers Law Review audio transcript: www.rutgerslawreview.com/2011/full-audio-transcript/.

172   wholly unaware: Four Flights Monograph, p. 45.

172   "that's the hijack": Michael Bronner, "9/11 Live: The Norad Tapes," Vanity Fair, August 2006; also see Rutgers University Law Review audio, "Another Mistaken Report, Delta Flight 1989," http://www.rutgerslawreview.com/2011/7-another-mistaken-report-delta-flight-1989/#_ednref.

172   diverted two unarmed fighters: Rutgers University Law Review audio, "Another Mistaken Report, Delta Flight 1989," http://www.rutgerslawreview.com/2011/7-another-mistaken-report-delta-flight-1989/#_ednref.

172   offers of more fighters: Priscilla D. Jones, The First 109 Minutes: 9/11 and the U.S. Air Force, Monograph published by the Air Force History and Museums Program, Washington, D.C., 2011, p. 44.

172   "in the face": Transcript from Voice Recorder, September 11, 2001, Northeast Air Defense Sector, Rome, New York, p. 61.

172   shootdown order: Interview with Kevin Nasypany, March 20, 2017.

173   "What are we gonna do": Priscilla D. Jones, The First 109 Minutes: 9/11 and the U.S. Air Force, Monograph published by the Air Force History and Museums Program, Washington, D.C., 2011, p. 44.

173   Trolley Problem: Thomas Cathcart, The Trolley Problem, or Would You Throw the Fat Guy Off the Bridge? A Philosophical Conundrum (New York: Workman Publishing Company, 2013).

173   whatever needed to be done: Interview with Kevin Nasypany, March 20, 2017.

173   it reached 19,000 feet: NTSB Flight Path Study, United Airlines Flight 93, p. 4.

173   "The best thing": Intelwire CVR, Flight 93.

174   dialed a navigational code: Four Flights Monograph, p. 45.

174   preferred the White House: 9/11 Commission Report, p. 244 and p. 248.

174   "High Security Alert": FBI document labeled "ACARS Messages from Dispatch; Messages from Aircraft to Ed Ballinger and Chad McCurdy; and Messages to Flight 93."

174   could only wonder: 9/11 Commission staff interview with Ed Ballinger, April 24, 2004, p. 6; Drew Griffin, Kathleen Johnston, and Brian Rokus, "The

Footnotes of 9/11," *CNN*, September 9, 2011, www.cnn.com/2011/US/09/06/sep tember.11.footnotes/. In this tearful video interview, aboard his boat after his retirement, Ballinger asks the interviewer, "Should I have said something else? What's more to the point than 'beware of cockpit intrusion'? Should I have said 'lock the so-and-so door'? Should I have said 'hijacking alert, hijacking'? . . . I wake up at night, thinking about it."

175　"passengers are getting together": Jefferson, p. 53.

175　fly the plane: 9/11 Commission staff interview with Lisa Jefferson, May 11, 2004.

176　Several minutes before ten: The Four Flights Monograph places the time of the revolt at 9:57 a.m. On the cockpit voice recorder, a hijacker asks in Arabic, "Is there something?" at 9:57:55 a.m.

176　Fearsome yells: Intelwire CVR, Flight 93. Dialogue and descriptions of fight noises and yells are from this document.

176　rock the plane: Four Flights Monograph, p. 45.

176　"In the cockpit!": The transcript notes that these yells came from "a native English-speaking male."

178　streaking toward him at 8,000 feet: Four Flights Monograph, p. 45.

178　landing gear down: "FAA Summary of Air Traffic Hijack Events, September 11, 2001," p. 23.

178　two most likely targets: A great deal of speculation and investigation has focused on the hijackers' thwarted destination for Flight 93. Based on intelligence reports and interrogations of Khalid Sheikh Mohammed and Ramzi Binalshibh, U.S. investigators concluded that the fourth flight targeted either the U.S. Capitol or the White House. Investigators said Osama bin Laden specifically wanted a plane to strike the White House, but Mohamed Atta worried that the president's home would be more difficult to reach than the U.S. Capitol. According to the 9/11 Commission Report: "When Binalshibh persisted, Atta agreed to include the White House but suggested they keep the Capitol as an alternate target in case the White House proved too difficult." 9/11 Commission Report, p. 248.

179　crash the plane: 9/11 Commission Report, p. 244: "If any pilot could not reach his intended target, he was to crash the plane."

179　hard to the right: Four Flights Monograph, p. 46.

180　563 miles per hour: Four Flights Monograph, p. 46.

180　fifteen minutes' flight time away: 9/11 Commission Staff Statement No. 17, "Improvising a Homeland Defense," p. 28. "Had it not crashed in Pennsylvania at 10:03, we estimate that United 93 could not have reached Washington, D.C., any earlier than 10:13, and most probably would have arrived before 10:23."

180　Steve O'Brien asked his crew: 9/11 Commission staff interview with Lt. Colonel Steven O'Brien, May 6, 2004, conducted by Miles Kara and Lisa Sullivan.

181　reached some five thousand feet: Audio tape of Lt. Colonel Steven O'Brien reporting black smoke to the Cleveland Center, found at http://www.oredigger61 .org/wp-content/uploads/2014/02/1005-Gofer-06-reports-black-smoke.mp3.

181　wasn't a hijacking: Rutgers University Law Review, Rutgers Law Review audio transcript: www.rutgerslawreview.com/2011/full-audio-transcript/.

181   first learned: This timeline was confirmed by the 9/11 Commission (See Staff Statement No. 17, "Improvising a Homeland Defense," pp. 17–18), which contradicted claims by numerous government and military officials in the immediate aftermath of 9/11 that fighters were pursuing Flight 93 and would have prevented it from reaching Washington, had the plane not crashed. Controversy also erupted over whether President Bush or Vice President Cheney authorized a shoot-down at roughly 10:15 a.m. In his outstanding book, 9/11 Commission senior counsel John Farmer concludes that the authorization came from the vice president, based largely on the absence of any record of a phone call between Bush and Cheney in which the issue was settled. Farmer, pp. 226–32.

182   "Negative clearance": Rutgers University Law Review, Rutgers Law Review audio transcript: www.rutgerslawreview.com/2011/full-audio-transcript/; Michael Bronner, "9/11 Live: The Norad Tapes," *Vanity Fair*, August 2006.

182   "Are they loaded?": Rutgers University Law Review, Rutgers Law Review audio transcript: www.rutgerslawreview.com/2011/full-audio-transcript/.

182   "United Ninety-Three": Rutgers University Law Review, Rutgers Law Review audio transcript: www.rutgerslawreview.com/2011/full-audio-transcript/.

183   eighty miles from the capital: Farmer, p. 226.

183   permissive rules of engagement: 9/11 Commission Staff Statement No. 17, "Improvising a Homeland Defense," p. 17. Farmer, pp. 230–31. Although the 9/11 Commission indicated that the Andrews jets launched "weapons free," based on a statement by General David Wherley, Sasseville told the commission and interviewers that he knew he had wide latitude but didn't formally receive "weapons free" ROE until later.

184   "shoot them down": 9/11 Commission Staff Statement No. 17, "Improvising a Homeland Defense," pp. 25–26.

184   "bastards snuck one by us": 9/11 Commission Staff Statement No. 17, "Improvising a Homeland Defense," p. 28.

184   prudent to wait: 9/11 Commission Staff Statement No. 17, "Improvising a Homeland Defense," pp. 25–26. As described in the statement, Marr told the 9/11 Commission "he did not pass along the order because he was unaware of its ramifications." Nasypany and Fox "indicated they did not pass the order to the fighters circling Washington and New York City because they were unsure how the pilots would, or should, proceed with this guidance."

184   At 10:39 a.m., Cheney updated: 9/11 Commission Report, p. 43.

185   depths of confusion, misinformation, and chaos: This is captured clearly in the declassified transcript of the Air Threat Conference call, available at https://www.archives.gov/files/declassification/iscap/pdf/2012-076-doc1.pdf.

185   seventh in line for takeoff: Matthew L. Wald, "An Inquiry: FBI Asks if Hijacking Plot Included Plane at Kennedy," *New York Times*, October 20, 2001.

185   told a similar story: 9/11 Commission staff interview with Ed Ballinger, April 14, 2004, p. 5. Also see Spencer, *Touching History*, pp. 102–105. The 9/11 Commission Report made no mention of Flight 23.

186   shortly after noon: "FAA Chronology of the September 11 Attacks," http://nsarchive.gwu.edu/NSAEBB/NSAEBB165/faa4.pdf.

186   Gander, Newfoundland: Jim DeFede. *The Day the World Came to Town: 9/11 in Gander, Newfoundland* (New York: Regan Books, 2003).

186   "any of the four hijacked flights": Farmer, p. 240.

186   bold claims: The clearest and most authoritative explanation of the false claims can be found in Farmer, pp. 241–90.

187   "It's happening now!": Flight 93 National Memorial Oral History Transcript, interview with Lorne Lyles conducted by Kathie Shaffer, October 25, 2006, p. 30.

188   "The mightiest building in the world": Garrett M. Graff, "We're the Only Plane in the Sky," *Politico*, September 9, 2016.

189   "keep that light from shining": Statement by the President in His Address to the Nation, September 11, 2001, https://georgewbush-whitehouse.archives.gov /news/releases/2001/09/20010911-16.html.

189   wrote in his diary: Dan Balz and Bob Woodward, "America's Chaotic Road to War," *Washington Post*, January 27, 2008.

189   one hundred fifty fighter jets: Spencer, *Touching History*, p. 281.

CHAPTER 11: "WE NEED YOU"

193   hemorrhaging money: Interview with Ron Clifford, January 25, 2017, with follow-ups for fact-checking. In addition to interviews with Clifford, information about his experiences comes from his testimony at Moussaoui trial, transcript pp. 3154–60; Niall O'Dowd, *Fire in the Morning* (Kerry, Ireland: Brandon, 2002), pp. 13–23; and Hampton Sides, *Americana: Dispatches from the New Frontier* (New York: Anchor Books, 2002), pp. 369–401.

195   Elaine Duch swam leisurely: Interview with Elaine Duch and Janet Cardwell, September 27, 2017.

197   agreement to lease: July 24, 2001, press release from Port Authority of New York & New Jersey on privatization lease of the World Trade Center. See www .panynj.gov/press-room/press-item.cfm?headLine_id=81.

197   Jay Jonas's mind: Interview with FDNY Deputy Chief John A. (Jay) Jonas, October 13, 2017, with email and phone follow-ups for fact-checking.

198   a burning tenement: Leslie Gevirtz, "Smoke-Eater Saves a Life": *New York Post*, May 4, 1982, p. 13. Jay Jonas wrote an account of the fire in the Division 7 Training and Safety Newsletter from September 2016.

198   "We need you": Interview with Deputy Chief John A. (Jay) Jonas, October 13, 2017.

199   11,336 firefighters: U.S. Centers for Disease Control Report, "Injuries and Illnesses Among New York City Fire Department Rescue Workers After Responding to the World Trade Center Attacks," dated September 11, 2002. www.cdc.gov /mmwr/preview/mmwrhtml/mm51SPa1.htm.

200   To pay his rent: Interview with Christopher Briggs Young, November 15, 2017.

200   "a pizza bagel": "Portraits of Grief: Dominique Pandolfo," *New York Times*, July 14, 2002.

201   Carlos and Cecilia Lillo: Interviews with Cecilia Lillo, January 6 and 8, 2018, with phone and email follow-ups.

204   Brian Clark and Stan Praimnath: Interviews with Brian Clark on October 5, 2017, and Stan Praimnath on October 4, 2017, with follow-up phone and email interviews. Throughout the text, quotes and descriptions from both men come from those interviews as well as the accounts they each gave to the 9/11 Commission. Some details also came from videos of their public talks and previous media accounts of their survival, subsequently reconfirmed by both men.

205   men in shirtsleeves: A video posted on YouTube captures the atmosphere of Euro Brokers in 1988, including images of Brian Clark at work: https://www.youtube.com/watch?v=aOsaYsUo_x4.

205   a small village: Interview with Stan Praimnath, October 4, 2017. Also see Stanley Praimnath and William Hennessey, "Plucked from the Fire," (Pittsburgh: Rose Dog Books, 2004), pp. 1–3.

206   Alayne Gentul: Interview with Jack Gentul, November 13, 2017, and follow-up emails and phone calls.

208   Harry rode to the top: Interview with Harry Friedenreich, November 24, 2017, with email follow-ups.

209   plunged toward earth: Interview with Dr. David Tarantino, July 12, 2017, with telephone follow-ups. Also David Tarantino Jr., "A Letter to Stanford Crew from a 9/11 Hero," 2003 Stanford Crew Yearbook, p. 3.

211   Marilyn Wills: Interview with Lieutenant Colonel Marilyn Wills (Ret.), June 8, 2017.

213   Jack Punches: Interviews with Janice Punches, Jennifer Punches-Botta, and Jeremy Punches, August 4, 2017. Video of Jack Punches's 2000 retirement ceremony provided by his family. Also see Tom Philpott, "Remember This Name," *Washingtonian*, November 2001.

216   Terry Shaffer was in trouble: Interviews with Terry and Kathie Shaffer, May 8–10, 2018, with phone and email follow-ups.

217   Christian Shank: Yvonne Brett, "Reflections of Stonycreek, 1776–1976," Stonycreek-Shanksville-Indian Lake Bicentennial Committee (1976), p. 7.

217   "You've got to remember": Flight 93 Memorial Oral History Transcript, interview with Wallace Miller, July 19, 2007, conducted by Kathie Shaffer.

CHAPTER 12: "HOW LUCKY AM I?"

220   "human scale": The comment attributed to Yamasaki reportedly came in response to a press conference question about why he designed two 110-story buildings and not one 220-story building. If he did say it, most considered the statement to be tongue-in-cheek. See Ada Louise Huxtable, "Big, But Not So Bold, Trade Center Towers are Tallest, But Architecture is Smaller Scale," *New York Times*, April 5, 1973, p. 34.

220   Monica Rodriguez Smith: Family Tribute to Monica Rodriguez Smith at the 9/11 Memorial. www.911memorial.org/family-tribute-monica-rodriguez-smith, accessed February 4, 2017.

221   large display ad: James Glanz and Eric Lipton, "The Height of Ambition," *New York Times*, Sept. 8, 2002. Glanz and Lipton's book, *City in the Sky: The Rise and Fall of the World Trade Center* (New York: Times Books, 2003), recounts this episode as well, and does a masterful job of telling the political, economic, architectural,

and engineering history of the complex. Although the display ad appeared in May 1968, Wien had been talking about the B-25 crash and raising similar doubts about the towers for years.

221 "only local damage": Glanz and Lipton, pp. 135–36. Also see Thomas W. Ennis, "Critics Impugned on Trade Center," *New York Times*, Feb. 15, 1964.

221 no such detailed analysis: NIST NCSTAR 1, WTC Investigation, p. 13. Also see Glanz and Lipton, pp. 135–39.

221 oversized filing cabinets: Pioneering city planner and philosopher Lewis Mumford is widely credited with calling them "glass and metal filing cabinets," although he used that phrase generally to describe modern skyscrapers in the late 1960s. He did, however, call the plans for the World Trade Center "a sheer disaster." See Alden Whitman, "Mumford Finds City Strangled by Excess Cars and People," *New York Times*, March 22, 1967.

221 "On balance": Ada Louise Huxtable, "Who's Afraid of the Big Bad Buildings?" *New York Times*, May 29, 1966.

222 structural steel: NIST NCSTAR 1, WTC Investigation, p. 3. The NIST report offers a detailed description of the towers' design and engineering.

222 thirty thousand square feet: NIST NCSTAR 1, WTC Investigation, p. 57.

222 afraid of heights: Glanz and Lipton, p. 109.

223 wasn't required to comply: NIST NCSTAR 1, WTC Investigation, pp. 52–60.

223 at least a fourth stairwell: This was a significant finding of the NIST report, although the Port Authority disputed it. It is worth noting that the issue was not whether the three stairwells had the capacity to evacuate everyone in the towers. Based on occupancy on September 11, they did. Rather, the question surrounding a fourth stairwell was whether it might have been accessible to some of the nearly two thousand people trapped above the impact zones of the two buildings. As explained on NIST p. 59, "It is conceivable that such a fourth stairwell, depending on its location and the effects of aircraft impact on its functional integrity, could have remained passable, allowing evacuation by an unknown number of additional occupants from above the floors of impact." Also see Dwyer and Flynn, pp. 105–109.

224 sprayed-on fire retardants: NIST NCSTAR 1, WTC Investigation, p. 56; Glanz and Lipton, pp. 325–27.

224 Port Authority officials essentially guessed: NIST NCSTAR 1, WTC Investigation, p. 56. The NIST report uses more bureaucratic language: "NIST was unable to find any indication that such tests were performed, nor any technical basis for the specification of the particular SFRM (sprayed fire-resistive material) selected or its application thickness."

224 more than four hundred companies: Tenant list for the World Trade Center on September 11, 2001, http://interactive.wsj.com/public/resources/documents/Tenant-List.htm and www.cnn.com/SPECIALS/2001/trade.center/tenants1.html.

226 upgraded and replaced fireproofing: NIST NCSTAR 1, WTC Investigation, pp. 73–74. Dwyer and Flynn, p. 10.

226 $2 million a year: David W. Dunlap, "Parking That Yacht—For $2.25 Million," *New York Times*, June 19, 1989.

227 ordered her morning coffee: Interview with Elaine Duch, September 27, 2017.

Elaine Duch's experience, thoughts, and quotes come from this interview in her home and several subsequent phone calls for elaboration and fact-checking.

228   bowl of Wheaties: Interview with FDNY Deputy Chief John A. (Jay) Jonas, October 13, 2017, with email and phone follow-ups for fact-checking. Although each quote from Jonas is not cited individually in the notes, all quotes from him come from these interviews. Some, confirmed independently with Jonas, originally appeared in an oral history interview he contributed to *Report from Ground Zero* by Dennis Smith (New York: Viking, 2002). Also valuable was a transcript of a talk Jonas gave in 2002 called "The Entombed Man's Tale," published September 8, 2002, in the *Times Herald-Record* of Orange County, NY.

228   Chris Young arrived by subway: Interview with Christopher Briggs Young, November 15, 2017.

229   Cecilia Lillo was hungry: Interviews with Cecilia Lillo, January 6 and 8, 2018, with phone and email follow-ups. Her account of the entire day comes from those interviews and follow-ups.

229   the usual awful time: Some background on Moussa Diaz, including certain phrases, comes from a story published five days after 9/11: Mitchell Zuckoff, "Six Lives: Reliving the Morning of Death," *Boston Globe*, September 16, 2001, p. A1. Most information about the experiences of Diaz, Paul Adams, Carlos Lillo, and other Battalion 49 EMTs comes from multiple interviews with Diaz and Adams in January 2017, along with transcribed interviews they and fellow EMTs Roberto Abril, Kevin Barrett, and Alwish Moncherry gave to the World Trade Center Task Force: Diaz and Barrett on January 17, 2002; Adams on November 1, 2001; Abril on January 27, 2002; and Moncherry on October 22, 2001. In cases of minor disagreements involving sequences or dialogue, this account relies on the author's interviews with participants.

231   a million or more: It is difficult to determine the exact number of people in the entire Financial District on September 11, 2001, or any other day. However, see the March 2012 paper titled "The Dynamic Population of Manhattan," by Mitchell L. Moss and Carson Qing of the Rudin Center at New York University. As they write, "An analysis of tract-to-tract worker flow data from the 2000 Census Transportation Planning Package indicates that census tracts in Midtown and Financial District (typically less than one-tenth of a square mile) have up to 70,000 commuters and residents in skyscrapers and office buildings during the day with a population density of up to 980,000 people per square mile. If visitors staying in hotels or touring nearby neighborhoods were included, the number of people per square mile could even exceed 1 million in several of these tracts." https://wagner.nyu.edu/files/rudincenter/dynamic_pop_manhattan.pdf, accessed February 4, 2017.

232   five feet one: Interview with Andrea Maffeo, February 1, 2017.

232   Wai-ching Chung: *New York Times* "Portraits of Grief" tribute, found at www.legacy.com/sept11/Story.aspx?PersonID=94791&location=2, accessed February 6, 2017.

233   the sight of rainbows: Emily Gold Boutilier, "Three in 2,996," *Amherst (College) Magazine*, Summer issue 2011, www.amherst.edu/amherst-story/magazine/issues/2011summer/safronoff/3in2996/node/332776, accessed February 8, 2017.

233   Trader: Archived website of destroyed World Trade Center Marriott Hotel:

http://web.archive.org/web/20010302170701/http://www.marriotthotels.com/NYCWT/meeting.asp, accessed February 7, 2017.

234 "bus sooooooooo crowded": Copy of email sent by Elaine Duch at 8:41 a.m., September 11, 2001, provided by Janet Duch Cardwell.

235 Roughly 8,900 people: As discussed elsewhere, there remains disagreement about the exact number of people in each tower at the time of the attacks. This figure comes from NIST NCSTAR 1, WTC Investigation, p. 26.

CHAPTER 13: "GOD SAVE ME!"

236 283,600-pound guided missile: NIST NCSTAR 1, WTC Investigation, pp. 20–30. The complete description of initial damage to the North Tower relies upon the NIST report findings unless otherwise noted.

237 1,355 people: NIST NCSTAR 1, WTC Investigation, p. xxxviii.

238 more than seventy-five hundred: NIST NCSTAR 1, WTC Investigation, p. 26. NIST puts the estimated number of people on the 91st floor and below at 7,545, while acknowledging rounding errors.

238 could have remained standing: This is a significant finding of NIST, found on p. 23 of the NIST NCSTAR 1 report.

239 Vaswald George Hall: Unbylined story, "Can't You Ever Say No?" *New York Times*, December 3, 2001. Elaine Duch didn't know Hall's name. His identity was established with help from the 9/11 Memorial and Museum's chief curator, Jan Seidler Ramirez, who reached out to former Port Authority officials. Elaine Duch's former boss, Jim Connors, recalled that the messenger was employed by a company that worked for the law firm Skadden Arps. That memory led to Martha Feltenstein, a retired Skadden Arps lawyer, who confirmed Hall's name, which appears on the 9/11 Memorial.

241 her boss standing over her: Interview with Janet Duch Cardwell, September 27, 2017.

242 battalion chief: World Trade Center Task Force Interview with Chief Joseph Pfeifer, October 23, 2001. Manhattan Dispatch Tape 432 transcript, Side A, 8:46 a.m. to 9:31 a.m. www.nytimes.com/packages/pdf/nyregion/wtctape1.1.pdf. Also see Joseph Pfeifer, "First Chief on the Scene," *Fire Engineering*, September 1, 2002. www.fireengineering.com/articles/print/volume-155/issue-9/world-trade-center-disaster/volume-i-initial-response/first-chief-on-the-scene.html.

242 "A plane just crashed!": Interview with FDNY Deputy Chief John A. (Jay) Jonas, October 13, 2017, with email and phone follow-ups for fact-checking.

243 "a number of floors on fire": Manhattan Dispatch Tape 432 transcript, Side A, 8:46 a.m. to 9:31 a.m. www.nytimes.com/packages/pdf/nyregion/wtctape1.1.pdf.

243 more than two hundred fire units: McKinsey Report: FDNY 9/11 Response, 2002, p. 9. Also see 9/11 Commission Report, p. 289.

244 "You show up": Michael Daly, *The Book of Mychal* (New York: Thomas Dunne Books, 2008), p. 324. Judge's homily also is captured on numerous YouTube videos, including: https://www.youtube.com/watch?v=mpI6oRgHeNU.

247 "A plane hit the towers": Interview with Moussa "Moose" Diaz, January 17, 2017.

249   "How many people": Interview with FDNY Deputy Chief John A. (Jay) Jonas, October 13, 2017, with email and phone follow-ups for fact-checking.

249   "could be a terror attack": Time stamped is 8:49 a.m., from Squad 1-8. Manhattan Dispatch Tape 432 transcript, Side A, 8:46 a.m. to 9:31 a.m. www.nytimes.com/packages/pdf/nyregion/wtctape1.1.pdf.

250   "The best-kept secret": Vincent Dunn, *Command and Control of Fires and Emergencies* (New York: Fire Engineering Books, 1999), pp. 145–46.

251   "People trapped": Dunn, *Command and Control*, p. 145.

253   "Listen up, everybody": Genelle Guzman-McMillan with William Croyle, *Angel in the Rubble* (New York: Howard Books, 2011), p. 23.

253   on street level: Dwyer and Flynn, p. 4.

253   about two dozen: Jim Dwyer, "The Port Authority Tapes: Overview; Fresh Glimpse in 9/11 Files of the Struggle for Survival," *New York Times*, August 29, 2003.

254   "This building was designed": *World Trade Center, in Memoriam*, History Channel, produced by Actuality Productions Inc. This version of the documentary was revised in 2002, after the attacks.

254   architect Gerry Gaeta: Interview with Gerry Gaeta, October 23, 2017. Also see Dean E. Murphy, *September 11: An Oral History* (New York: Doubleday, 2002), pp. 50–52.

254   "Uh, we're on the eighty-eighth floor": Transcripts of World Trade Center radio transmissions, WTC Ch. 25, Radio Channel B. Also see Dwyer and Flynn, p. 84.

256   charred remains: Several firefighters reported seeing this early victim in the North Tower lobby. See the account of Firefighter Peter Blaich of Ladder 123 in *Firehouse Magazine*, September 9, 2002. www.firehouse.com/article/10579453/ff-peter-blaich-ladder-123.

256   nineteen commendations: City of New York press release, "Mayor Bloomberg, Former Mayor Giuliani, and Elizabeth Petrone-Hatton Rename West 43rd Street Between 10th and 11th Avenue After FDNY Captain Terence S. Hatton," June 4, 2005.

257   as many as fifty thousand people: 9/11 Commission Report, pp. 290–91.

257   between 14,000 and 17,400: Dwyer and Flynn, p. 2. The *New York Times* and *USA Today* conducted extensive research into the number of people in the trade center buildings and reached rough agreement on the lower number, while NIST reported the higher estimate in July 2004.

258   born on the Fourth of July: *New York Times*, "Portraits of Grief: Christine Olender," October 20, 2001. Transcripts of Christine Olender's calls: WTC CH. 10, PAPD Police Desk, 3541 Right.

258   He suspected: Interview with PAPD Officer Ray Murray, June 23, 2018.

258   "do whatever you have to": Transcripts of Christine Olender's calls: WTC CH. 10, PAPD Police Desk, 3541 Right.

258   more than fifty emergency calls: Complete transcripts of WTC CH. 10, PAPD Police Desk, 3541 Right. Times of calls are approximate.

259   sixteenth floor: 9/11 Commission Report, p. 298.

259   critical problems: The communication issue was examined closely by the 9/11 Commission and others, including a five-month review by McKinsey & Co. consultants commissioned by the FDNY. See McKinsey Report: FDNY 9/11 Response, 2002. See World Trade Center Task Force Interview with Battalion Chief Joseph Pfeifer, October 23, 2001, p. 5. Also see Dwyer and Flynn, pp. 60–61. Although this was squarely disputed by Chief Pfeifer, the 9/11 Commission concluded (p. 297) that only one of two necessary buttons on the repeater system was activated on 9/11: "The activation of *transmission* on the master handset required, however, that a second button be pressed. That second button was never activated on the morning of September 11."

259   ultra-high-frequency Motorola radios: McKinsey Report: FDNY 9/11 Response, 2002, p. 13. Also Dwyer and Flynn, pp. 55–56.

260   rooftop rescues would be impossible: 9/11 Commission report, p. 291.

260   Within six minutes of the crash: NIST NCSTAR 1, WTC Investigation, p. 26.

261   a fire at the Triangle Shirtwaist Factory: David Von Drehle, *Triangle: The Fire That Changed America* (New York: Grove Press, 2004). Also NIST NCSTAR 1-7, WTC Investigation, p. 1.

261   "Please don't jump": Dwyer and Flynn, p. 62.

262   the unaffected South Tower: 9/11 Commission Report, p. 290.

262   not to evacuate: 9/11 Commission Report, p. 318.

262   "We need to know": Transcripts of World Trade Center radio transmissions, Channel 8 Police Desk, 3541 Left, pp. 8–10.

262   different advice: 9/11 Commission Report, p. 287. This is borne out by hundreds of pages of transcripts of PAPD dispatch calls.

263   refused orders by their chiefs: 9/11 Commission Report, p. 303.

263   "Maybe they got one by us": Smith, p. 88.

263   "I am not dead": Dean E. Murphy, *September 11: An Oral History* (New York: Random House, 2002), pp. 152–53. The author located Mr. Armstead and attempted to speak with him for this book, but he did not reply to several interview requests.

264   "Jay, don't even bother": Interview with FDNY Deputy Chief John A. (Jay) Jonas, October 13, 2017, with email and phone follow-ups for fact-checking.

265   "There's a second plane": Transcript from ABC News, "Why No Rooftop Rescues on Sept. 11?" November 8, 2001.

CHAPTER 14: "WE'LL BE BROTHERS FOR LIFE"

266   sat silently at his desk: Interview with Brian Clark, October 5, 2017, with email and phone follow-ups.

267   Stan Praimnath rode a local elevator: Interview with Stan Praimnath, October 4, 2017, with email and phone follow-ups. Also see Stanley Praimnath and William Hennessey, *Plucked from the Fire* (Pittsburgh: Rose Dog Books, 2004).

267   eighty-six hundred people: NIST NCSTAR 1, WTC Investigation, p. 37. This official number is higher than several careful counts by media organizations, notably the *New York Times* and *USA Today*. Overall, the NIST report put the total population of the Towers at 17,400, while the *Times* used turnstile counts and other methods to conclude that 14,154 people were in the Towers plus another 940 reg-

istered at the Marriott Hotel between them. Despite the difference, the NIST and *NYT/USA Today* numbers are far more credible than those contained in earlier reports, which consistently overstated the Towers' population.

267 roughly half heard the sound: NIST NCSTAR 1, WTC Investigation, p. 37.

268 about half of the people: NIST NCSTAR 1, WTC Investigation, p. 37.

269 "Your attention, please": Brian Clark's account of the instructions from Port Authority officials was confirmed in the 9/11 Commission Report, pp. 287–89. Clark's quote is from interviews with the author and also from his interview with an investigator for the 9/11 Commission, Madeleine Blot, March 2, 2004.

271 fond of Brooks Brothers clothes: *New York Times*, "Portraits of Grief: Hideya Kawauchi," December 6, 2001.

271 often returned home: *New York Times*, "Portraits of Grief: Alisha Levin," October 12, 2001.

271 a new announcement: NIST NCSTAR 1, WTC Investigation, p. 37, and Jim Dwyer, "9/11 Tape Has Late Change on Evacuation," *New York Times*, May 17, 2004.

271 more than six hundred people: NIST NCSTAR 1, WTC Investigation, pp. 37 and 42. Also see Jim Dwyer, "9/11 Tape Has Late Change on Evacuation," *New York Times*, May 17, 2004.

272 thirty-two hundred: NIST NCSTAR 1, WTC Investigation, p. 42.

272 "the fuck out of here": James B. Stewart, "The Real Heroes Are Dead," *New Yorker*, February 3, 2002. Also see Michael Grunwald, "A Tower of Courage," *Washington Post*, October 28, 2001, and James B. Stewart, *Heart of a Soldier* (New York: Simon and Schuster, 2003).

273 bright red blazer: Interview with Jack Gentul, November 13, 2017.

273 group of ten people: Dwyer and Flynn, p. 76. Early reporting about Alayne Gentul suggested that her father, Harry Friedenreich, was project manager for Otis Elevator Company at the World Trade Center; although Otis did build those elevators, Friedenreich was not directly involved in the World Trade Center work.

273 resembled a teacher: Jim Dwyer, Eric Lipton, et al., "Fighting to Live as the Towers Died," *New York Times*, May 26, 2002. While Nora Hutton was quoted as saying Alayne pointed toward the elevators, Mona Dunn described her leading others to the stairs.

273 like a family: Interview with Anne Foodim, January 4, 2018.

275 a 38-degree angle: NIST NCSTAR 1, WTC Investigation, p. 38. Additional details of Flight 175's impact in this passage also come from the NIST report.

277 packed lobby-bound express elevators: 9/11 Commission Report, pp. 293–94.

277 return to their offices: NIST NCSTAR 1, WTC Investigation, p. 41.

277 A handful of bloodied survivors: Jane Lerner, "Bandanna Links Acts of Courage," *The Journal News*, June 10, 2002. Also see Tom Rinaldi, *The Red Bandanna: A life. A choice. A legacy* (New York: Penguin Press, 2016), pp. 123–26; and Jim Dwyer, Eric Lipton, et al., "Fighting to Live as the Towers Died," *New York Times*, May 26, 2002.

277 she'd sat motionless: The timing of Ling Young's account, as described in media reports, contributes to the credibility of the identification of "the man in the red bandanna" as Welles Crowther. Young told author Tom Rinaldi that she'd

sat motionless for ten minutes, maybe longer, before the man arrived on the 78th floor. Welles Crowther left a phone message for his mother at 9:12 a.m., saying that he was all right. If he then rushed to Stairwell A, it would have taken him several minutes to get from the 104th floor to the 78th floor.

279 "84th floor": John Breunig, "Father's Note Changes Family's 9/11 Account," *Stamford Advocate*, December 28, 2012, p. 1. Also see *New York Times*, "Portraits of Grief: Randolph Scott," November 17, 2001.

280 Melissa Doi: A recording of Melissa Doi's 911 call was played at the 2006 trial of Zacarias Moussaoui. Also see *New York Times*, "Portraits of Grief: Melissa Doi," February 12, 2002. Dispatcher Vanessa Barnes was identified in the Moussaoui trial transcript.

285 Ron climbed: Andrew Duffy, "Tower of Pain for Canadian Who Survived 9/11," *Ottawa Citizen*, June 5, 2005.

287 called her husband, Jack: Interview with Jack Gentul, November 13, 2017, and follow-up emails and phone calls.

CHAPTER 15: "THEY'RE TRYING TO KILL US, BOYS"

289 "We may not live through today": Interview with FDNY Deputy Chief John A. (Jay) Jonas, October 13, 2017, with email and phone follow-ups for fact-checking. Another reported version of Nevins's quote is "We're going to be lucky if we survive this."

290 "I love you, brother": World Trade Center Task Force Interview with Firefighter Timothy Brown, January 15, 2002, p. 6.

290 Assistant Chief Donald Burns: World Trade Center Task Force Interview with Battalion Chief Joseph Pfeifer, October 23, 2001. Also Joseph Pfeifer, "First Chief on the Scene," *Fire Engineering*, September 1, 2002, and unbylined story, "Fire Chief's Remains Are among Thirteen Found," *New York Times*, March 22, 2002.

290 missiles being fired: Transcript from NBC News, "The Miracle of Ladder Company 6," September 28, 2001.

294 a girl's foot: Several EMTs and paramedics mentioned this, most vividly EMT Lonnie Penn, interviewed by the World Trade Center Task Force on November 9, 2001, p. x.

294 "have you seen my wife?": World Trade Center Task Force interview with Kevin Barrett, January 17, 2002, p. 4.

295 "I shall impersonate a man": From the 1965 Broadway musical *Man of La Mancha*, book by Dale Wasserman, lyrics by Joe Darion, music by Mitch Leigh.

296 loudly quoted scripture: Renea Henry, "9/11 Survivor Gil Weinstein Speaks to Community Synagogue," *New York Patch*, September 10, 2011.

296 "Let's go for it!": Interview with Gerry Gaeta, October 23, 2017. Also Murphy, p. 52.

296 trapped on higher floors: The heroism of Frank De Martini, Pablo Ortiz, and others is covered at length by Dwyer and Flynn in *102 Minutes*. Their account is based in part on a narrative record assembled by Roberta Gordon, an attorney and friend of the De Martini family. Their story also is featured in the Discovery Channel documentary *Heroes of the 88th Floor*.

298   "We are going to die": Interview with Gerry Gaeta, October 23, 2017. Also Murphy, p. 53.

298   56-inch-wide: NIST NCSTAR 1-7 report: Occupant Behavior, Egress and Emergency Communications, http://ws680.nist.gov/publication/get_pdf.cfm?pub_id=101046, p. 27.

299   "Only another ten floors": Interview with Gerry Gaeta, October 23, 2017. Also Murphy, p. 53.

299   "What do you suggest?": Transcripts of World Trade Center radio transmissions, Channel 24, Phone Clerk, pp. 6–7. Also see Dwyer and Flynn, pp. 123–25.

301   accountant John Abruzzo: September 11 Memorial Memo Blog, "Colleagues Use Special Chair to Save Quadriplegic on 9/11," www.911memorial.org/blog/colleagues-use-special-chair-save-quadriplegic-911.

301   "Oh my god . . . Jennieann is there": Interview with Andrea Maffeo, February 1, 2017.

303   belonged to Dave Bobbitt: Dennis Cauchon and Martha T. Moore, "Elevators Were Disaster Within Disaster," USA Today, September 11, 2002. The investigation by USA Today provided a complex and authoritative look at the failure of elevators in the World Trade Center after the planes hit the Twin Towers. Chris Young was among the scores of people interviewed by Cauchon and Moore.

303   blowing out doors and walls: NIST NCSTAR 1, WTC Investigation, p. 24.

303   Jan Demczur: Jim Dwyer, "A NATION CHALLENGED: OBJECTS; Fighting for Life 50 Floors Up, with One Tool and Ingenuity," New York Times, October 9, 2001. Also see Victoria Dawson, "Handed Down to History," Smithsonian, July 2002, Vol. 33, Issue 4, p. 40.

304   never systematically checked: Dennis Cauchon and Martha T. Moore, "Elevators Were Disaster Within Disaster," USA Today, September 11, 2002.

304   four men and three women: Young didn't recall their names, but their story is included in Dwyer and Flynn, pp. 179–81.

306   "lock release order": 9/11 Commission Report, p. 294.

306   returned to the lobby: 9/11 Commission Report, p. 299.

CHAPTER 16: "THEY DONE BLOWED UP THE PENTAGON"

307   sixty years to the day: Steve Vogel, The Pentagon: A History (New York: Random House, 2007), p. 126.

307   71 feet tall: Details about the building's design and construction come from numerous sources, primarily Vogel. Also see Goldberg et al., pp. 2–4.

307   6.6 million square feet: Goldberg et al., Pentagon 9/11 (Washington, D.C.: Historical Office of the Secretary of Defense, 2007), pp. 2–4.

307   Pentagon lore: Steven Donald Smith, "Pentagon Hot Dog Stand, Cold War Legend, to Be Torn Down," American Forces Press Service, September 20, 2006.

308   salute their superiors: Official Pentagon tour brochure, https://pentagontours.osd.mil/Tours/documents/Self.Guided.Tour.pdf.

308   puttering around: Details about Dr. David Tarantino's actions come primarily from an interview on July 12, 2017, with telephone follow-ups, as well as from

transcripts of accounts he gave shortly after 9/11, confirmed with him for accuracy. The accounts of David Tarantino and David Thomas differ slightly in terms of sequence, but overall matched closely and were confirmed by Jerry Henson in an interview July 14, 2017. This account is based on a combination of those accounts, confirmed by the participants, as well as on contemporaneous news stories and official Navy accounts of the aftermath of the Pentagon strike, cited where appropriate.

309   a distracted Jerry Henson: Interviews with Jerry and Kathleen Henson, July 14 and July 31, 2017. Also see Tom Philpott, "Remember This Name," *Washingtonian*, November 2001.

311   Lieutenant Colonel Marilyn Wills: Interview with Lieutenant Colonel Marilyn Wills (Ret.), June 8, 2017. Details about the events in ODCSPER also come from multiple interviews conducted by the U.S. Army Center of Military History for the Noble Eagle oral history project, including an interview conducted with Marilyn Wills by ACMH Historian Steve Lofgren on July 12, 2002. Also see Robert Rossow III, *Uncommon Strength: The Story of the U.S. Army Office of the Deputy Chief of Staff for Personnel during the Attack on the Pentagon, 11 September 2001* (Washington, DC: Department of the Army, Office of the Deputy Chief of Staff, 2003).

312   Dilbertville was humming: The actions and locations of OCDSPER staff are detailed by Rossow in *Uncommon Strength* and also in oral histories conducted by the U.S. Army Center of Military History.

312   Max Beilke: Biographical listing of Master Sergeant Max J. Beilke, Retired, Pentagon 9/11 Memorial, http://pentagonmemorial.org/explore/biographies/msg-max-j-beilke-usa-retired.

312   last American combat soldier: Robert A. Rosenblatt and Richard T. Cooper, "Last Soldier to Leave Vietnam Is Feared Dead," *Los Angeles Times*, September 16, 2001.

313   "Quick meeting": This dialogue, confirmed in interviews with Marilyn Wills, appeared in a newspaper series: Earl Swift, "Out of Nowhere: Sept. 11 at the Pentagon, Part 1," *Virginian-Pilot*, September 7, 2002.

313   "Honey, do me a favor": Interview with John Yates, U.S. Army Center of Military History, oral history project, March 21, 2002.

313   startled facilities manager George Aman: U.S. Army Center of Military History interview with George P. Aman, part of Noble Eagle oral history project, December 12, 2001.

314   530 miles per hour: 9/11 Commission Report, p. 9; NTSB Flight Path Study, American Airlines Flight 77, p. 2.

314   5,300 gallons of fuel: This is an estimate from the American Society of Civil Engineers, *Pentagon Building Performance Report*, p. 12.

315   twenty-five feet up the outer wall: Vogel, p. 432.

315   superheated, superdeadly shrapnel: Arlington County After-Action Report on the Response to the September 11 Terrorist Attack on the Pentagon, p. A-8. www.floridadisaster.org/publications/Arl_Co_AAR.pdf.

316   an angled path: American Society of Civil Engineers, *Pentagon Building Performance Report*, p. 12. Also see Goldberg et al., pp. 17–19.

321 Grunewald . . . disobeyed: Interview with Lieutenant Colonel Robert Grunewald (Ret.), June 8, 2017. Grunewald's account is also included in his oral history interview with the U.S. Army Center of Military History and in *Uncommon Strength*. Also see interview with Martha Carden with the U.S. Army Center of Military History, October 29, 2001.

322 about twenty people: Goldberg, p. 20.

322 Navy Captain Dave Thomas: Interview with Retired Admiral David M. Thomas Jr., August 2, 2017. Elements of Thomas's account, confirmed with him, also came from Captain David M. Thomas Jr., "Everyone helps in some capacity. It's automatic," *National Journal*, August 30, 2002; transcript of CSPAN interview, August 10, 2001, www.c-span.org/video/?300993-1/rear-admiral-david-thomas-september-11 -2001; and Tom Philpott, "Remember This Name," *Washingtonian*, November 2001.

322 Wounded and blinded by a bomb: Obituary, "Capt. Donald K. Ross, 81; Won Medal of Honor," *New York Times*, June 1, 1992.

325 Both perished: Goldberg et al., p. 42.

326 John Yates: Noble Eagle oral history interview with John Yates, March 21, 2002. Also see Earl Swift, "Out of Nowhere: Sept. 11 at the Pentagon, Parts 1–4," *Virginian-Pilot*, September 7–10, 2002.

326 blown out of a cannon: Interview with Major Regina Grant, U.S. Army Center of Military History, oral history project, February 26, 2002.

326 snatched her by the skirt: There exist relatively minor differences in the various accounts of the escape from the OCDSPER cubicle area and conference room, as described in interviews and Noble Eagle oral history transcripts from Marilyn Wills, Robert Grunewald, Martha Carden, Lois Stevens, Regina Grant, John Yates, and others. The differences mainly involve the sequence of events and discussions along the route from the conference room to the C Ring windows or the central courtyard. This account relies primarily on Marilyn Wills' version. There is no disagreement on the most essential elements of their accounts, particularly in terms of the heroism of Wills, Grunewald, and others.

327 "I am too tired": Interview with Lois Stevens, U.S. Army Center of Military History, oral history project, December 13, 2001.

CHAPTER 17: "I THINK THOSE BUILDINGS ARE GOING DOWN"

329 "I think we should break the window": Dwyer and Flynn, pp. 185–86. The call is logged on an EMS document obtained by *New York Times*, http://www.nytimes.com/packages/pdf/nyregion/EMSLog.pdf.

331 three minutes and seventeen seconds: Brian Clark timed his recorded call to 911 when it was played for him three months later by a 9/11 Commission investigator. In the investigator's notes, she reports that "staff has located Mr. Clark's 9-1-1 call and found it to be almost precisely as he described it."

332 seek out Stairwell A: NIST NCSTAR 1, WTC Investigation, p. 44.

332 "How can you have a big building": FDNY 911 Calls Transcript, EMS Part 2.

332 rooftop rescues: 9/11 Commission Report, p. 292.

332 "It's just me and the sky": Keith Elliot Greenberg, *Risky Business: Window Washer at Work above the Clouds* (Woodridge, CT: Blackbirch Press, 1995), p. 18.

332   called his wife: Jenny Pachucki, "Remembering WTC Window Washer Roko Camaj," The Memo Blog, National September 11 Memorial & Museum. www .911memorial.org/blog/remembering-wtc-window-washer-roko-camaj.

332   "big smoke": Transcript of WTC Radio Channel 28, Radio Channel Y, p. 24.

333   When Brian Clark made his 9-1-1 call: These evacuation numbers are esti-mated for 9:37 a.m., NIST NCSTAR 1, WTC Investigation, p. 42.

333   not to leave the floor: 9/11 Commission Report, p. 295.

333   "You have to wait": Jim Dwyer, "City Releases Tapes of 911 Calls from Sept. 11 Attack," New York Times, Nov. 30, 2005. Numerous calls to 911 operators were released in response to a lawsuit filed by the New York Times. Also see 9/11 Commis-sion Report, p. 295.

333   first fatality: 9/11 Commission Report, p. 300. Also see "He Kept Everyone Safe," New York Times, September 20, 2001.

333   "Let's make this quick": World Trade Center Task Force Interview with Cap-tain Paul Conlon, January 26, 2002, p. 7.

333   "small glimmer of hope": World Trade Center Task Force Interview with EMT Richard Erdey, October 10, 2001, p. 9.

334   bursts of smoke: NIST NCSTAR 1, WTC Investigation, p. 43.

335   department's fitness award: FDNY Division 7 Training and Safety Newslet-ter, September 2016, courtesy of Jay Jonas.

336   functioning freight elevator: 9/11 Commission Report, p. 299, among other sources and radio transcripts of Battalion Chief Orio Palmer in the South Tower.

336   FDNY's radio repeater channel: As discussed elsewhere, questions remain whether the repeater was properly activated at the North Tower command desk. It isn't known how Chief Palmer learned that it worked inside the South Tower, and it's notable that he was considered an expert in the system's use and was among the FDNY officials who tried to activate it when he first arrived in the North Tower.

336   "a lot of bodies": FDNY 9/11 dispatch transcript. Radio transmissions from Battalion Chief Orio Palmer and other firefighters in this section also come from this tape. Also see Dwyer and Flynn, pp. 196–199, 204–206, and 208–209.

337   the Flying Fireman: Michael Daly, "The Flying New York Fireman Who Shined on 9/11," The Daily Beast, September 11, 2014.

338   "We need EMS": World Trade Center Channel 15, EMS Direct line, tran-script p. 49. Robert Gabriel Martinez's identity was confirmed by Dwyer and Flynn, pp. 209–10.

339   Kevin Cosgrove: A recording of Kevin Cosgrove's 911 call was played at the 2006 trial of Zacarias Moussaoui. Also see the trial testimony of his widow, Wendy Cosgrove, pp. 3283–88, and New York Times, "Portraits of Grief: Kevin Cosgrove," October 17, 2001.

339   down to the 79th floor: Dwyer and Flynn, p. 207.

341   another person he didn't name: Although Kevin Cosgrove said he and the others were in the office of Jon "Ostaru," it was in fact the office of Jon Ostrau, who wasn't there.

341   The forces of catastrophe: Although criticized by some structural engineers, the NIST NCSTAR 1, WTC Investigation Report remains the authoritative account

of the towers' collapse. For the South Tower, see pp. 37–46. http://ws680.nist.gov/publication/get_pdf.cfm?pub_id=909017.

341   a minor earthquake: Kevin Krajick, "A Morning That Shook the World: The Seismology of 9/11," *State of the Planet: General Earth Institute, Columbia University*, September 6, 2016. http://blogs.ei.columbia.edu/2016/09/06/a-morning-that-shook-the-world/.

342   eight thousand people: NIST NCSTAR 1, WTC Investigation, p. 42.

342   Welles Crowther: Tom Rinaldi, *The Red Bandanna: A Life. A Choice. A Legacy* (New York: Penguin Press, 2016). Also see Jane Lerner, "Bandanna Links Acts of Courage," *Journal News*, June 10, 2002; and Jim Dwyer, Eric Lipton, et al., "Fighting to Live as the Towers Died," *New York Times*, May 26, 2002, which was the first media report that mentioned the mysterious man in the red bandanna on the 78th floor. The story of how the survivors came to identify Welles as their rescuer, a remarkable effort initiated by his parents, was first reported by Lerner and is told elegantly in Rinaldi's book.

342   Police Officer Moira Smith: Susan Hagen and Mary Carouba, *Women at Ground Zero* (Indianapolis: Alpha Books, 2002), pp. 311–13.

343   John O'Neill: Lawrence Wright, "The Counter-Terrorist," *New Yorker*, January 14, 2002. Also see Robert Kolker, "O'Neill Versus Osama," *New York* magazine, December 17, 2001, and Murray Weiss, *The Man Who Warned America: The Life and Death of John O'Neill, The FBI's Embattled Counterterrorism Warrior* (New York: William Morrow, 2003).

343   "A lot of these groups": Robin Pogrebin, "John O'Neill is Dead at 49; Trade Center Security Chief," *New York Times*, September 23, 2001.

344   Brian and Stan became separated: This is one area where their accounts diverge. In Stan's account, he gave Brian his business card as they ran, and then separated.

CHAPTER 18: "TO RUN, WHERE THE BRAVE DARE NOT GO"

346   "Burn victim!": Interview with Gerry Gaeta, October 23, 2017. Also Murphy, p. 54.

350   By one estimate: NIST NCSTAR 1, WTC Investigation, p. 26.

350   "Emptiness is the only way": Andrew Fredericks, "Father's Day," *Fire Nuggets*, August-November issue, 2001.

351   To dream the impossible dream: From the 1965 Broadway musical, "Man of La Mancha," book by Dale Wasserman, lyrics by Joe Darion, music by Mitch Leigh.

356   "I'm missing my partner": World Trade Center Task Force interview with Roberto Abril, January 27, 2002, p. 7.

356   "Listen, man": World Trade Center Task Force interview with Manuel Delgado, October 2, 2001, p. 12.

359   blinding dust and debris: McKinsey Report: FDNY 9/11 Response, 2002, p. 39.

359   Chief Hayden stumbled: Daly, pp. 338–39.

359   "evacuate the building": 9/11 Commission Report, p. 306.

360   roughly eight hundred people: NIST NCSTAR 1, WTC Investigation, p. 32.

361   "Start your way down": World Trade Center Task Force Interview with Firefighter Michael Byrne, January 21, 2002, p. 4.

361   to help two other men: Dwyer and Flynn, pp. 219–20. Also Sean Kirst, "Capt. Billy Burke and a Choice Made on Sept. 11: 'This Is What I Do,'" *(Syracuse) Post-Standard,* September 11, 2011. Also see Kirst, "Fourteen Years After September 11, 2001: A Burning Tower, a Friend He Loved, 'The Way He Always Was,'" *(Syracuse) Post-Standard,* September 11, 2015.

CHAPTER 19: "REMEMBER THIS NAME"

362   possibility for deliverance: Tom Philpott, "Remember This Name," *Washingtonian,* November 2001.

363   spit fire: Interview with Lois Stevens, U.S. Army Center of Military History, oral history project, December 13, 2001.

363   She rose to her knees and hung on: In multiple interviews, including discussions with the author in the summer of 2017, Marilyn Wills maintained that Lois Stevens accepted her offer to carry her on her back. In Lois Stevens's oral history interview with the U.S. Army Center of Military History, she acknowledged that she gave up hope and said Wills offered to carry her on her back, but she said she declined. In her account, Stevens said she instead hung onto Wills's belt as they crawled toward the AE Drive windows. However it occurred, Lois Stevens has consistently credited Marilyn Wills with guiding her through the wreckage and saving her life.

367   Carlton stepped up: Dean E. Murphy, *September 11: An Oral History* (New York: Doubleday, 2002), p. 217.

367   Powell stood: Interview with CDR Craig Powell, Naval Historical Center, conducted by Capt. Michael McDaniel, October 29, 2001.

367   "Hey, ceiling's going": Ibid.

367   General Carlton: Condensed transcript of Lt. Gen. Paul K. Carlton Jr., "Medical Response to 9/11," from an interview by the Air Force History Support office, December 4, 2001.

369   "jump or get toasted": Interview with Dalisay Olaes, U.S. Army Center of Military History, oral history project, September 13, 2001.

373   "I want reimbursement": Interview with Martha Carden, U.S. Army Center of Military History, oral history project, October 29, 2001.

373   "needs to get out of here": Interview with John Yates, U.S. Army Center of Military History, oral history project, March 21, 2002.

375   Captain Darrell Oliver: Interview with Captain Darrell Oliver, U.S. Army Center of Military History, oral history project, October 31, 2001, and Goldberg et al., pp. 52–53. Also see Ron Kampeas, "Discipline, Training Saved Lives After Pentagon Attack," The Associated Press, September 17, 2001. Oliver's personal account was recorded October 17, 2002, http://hereisnewyorkv911.org/2011/darrell-oliver/.

376   rolled on the floor: Lieutenant Kevin P. Shaeffer (Ret.), "Never Forget," *Proceedings* magazine, September 2011, Vol. 137/9/1,303. Also see Goldberg et al., pp. 32–33.

376   ran toward the carnage: Interview with Father Stephen McGraw, June 9, 2017.

377   Major John Thurman: Interview with Major John Thurman, U.S. Army Cen-

ter of Military History, oral history project, September 20, 2001; and Thurman's testimony at Moussaoui trial, transcript pp. 3397–3414. Also see John Spong, "Karen Wagner's Life," *Texas Monthly*, September 2011.

379   he hadn't recognized: Goldberg et al., p. 58.

379   surveyed the scene: U.S. Army Center of Military History interview with George P. Aman, part of Noble Eagle oral history project, December 12, 2001.

380   "huge heaps of rubble": Arlington County After-Action Report on the Response to the September 11 Terrorist Attack on the Pentagon, p. A-8. www.florida disaster.org/publications/Arl_Co_AAR.pdf.

380   ninety-five feet: Goldberg et al., pp. 80–81. Also see American Society of Civil Engineers, *Pentagon Building Performance Report*, pp. 12–20.

381   "The Pentagon is functioning": Nicholas M. Horrock and Pamela Hess, "Rumsfeld: The Pentagon Is Still Functioning," UPI, September 11, 2001.

381   Dave Tarantino hoped: Interview with Dr. David Tarantino, July 12, 2017.

CHAPTER 20: "THIS IS *YOUR* PLANE CRASH"

383   Linda Shepley watched her television in shock: Interview with Linda and Jim Shepley, May 9, 2018.

383   "You say that emergency vehicles are there": NBC News live coverage of September 11, 2001, (Complete *Today Show*), on YouTube, https://www.youtube.com /watch?v=89G749GrtBQ.

385   "No, no, no, no": Interview with Linda Shepley, May 9, 2018. Also see Flight 93 Memorial Oral History Transcript, interview with Linda Shepley, January 28, 2006, conducted by Kathie Shaffer.

385   The first 9-1-1 call: Flight 93 Memorial Oral History Transcript, interview with Paula Pluta, June 15, 2008, conducted by Kathie Shaffer. Also see Transcript of Somerset County Emergency Management Agency September 11, 2001, communications. Call No. 27301, Paula Pluta, transcribed by Donna Glessner.

386   "Oh, my God!": Transcript of Somerset County Emergency Management Agency September 11, 2001, communications. Call No. 27301, Paula Pluta, transcribed by Donna Glessner.

386   "There was an airplane": Transcript of Somerset County Emergency Management Agency September 11, 2001, communications. Call No. 27303, Daniel Meyers, transcribed by Donna Glessner.

387   At the family medical practice: Interviews with Terry and Kathie Shaffer, May 8–10, 2018, with phone and email follow-ups.

388   between his house and his country store: Flight 93 Memorial Oral History Transcript, interview with Rick King, July 5 and December 13, 2006, conducted by Kathie Shaffer.

389   "What stations are dispatched?": Somerset County Emergency Management Agency 9-1-1 Call Center, Transcript of September 11, 2001, communications, transcribed by Donna Glessner.

389   every available emergency unit: Shanksville Volunteer Fire Company, Station 627 Timeline, Calls from 9-1-1 Center, Recorder 3 transcript, call number 588086.

389   "Guys, prepare yourself": Flight 93 Memorial Oral History Transcript, interview with Keith Custer, March 24, 2006, conducted by Kathie Shaffer.

390   "how'd you like to be the coroner": McMillan, p. 7.

390   The county job: Sara Rimer, "Public Lives: Where Death Mostly Tiptoes, It Rushed Violently In," *New York Times*, Sept. 22, 2001.

390   A car had struck a deer: The Associated Press, "Tax Collector Killed When Deer Hits Truck," *(Sharon, PA) Herald*, August 21, 2001.

390   "Could you ever think": Flight 93 Memorial Oral History Transcript, interview with Wallace Miller, July 19, 2007, conducted by Kathie Shaffer.

391   "possibly a hijacked aircraft": Transcript of conversation between Wally Miller and 9-1-1 dispatcher and EMA Director Rick Lohr, September 11, 2001.

393   "What do you want us to do?": Flight 93 Memorial Oral History Transcript, interview with Rick King, July 5 and December 13, 2006, conducted by Kathie Shaffer.

393   a wave of relief: Flight 93 Memorial Oral History Transcript, interview with Rick King, July 5 and December 13, 2006, conducted by Kathie Shaffer.

393   He knew the Diamond T site: McMillan, pp. 113–14. Also see Flight 93 Memorial Oral History Transcript, interview with Wallace Miller, July 19, 2007, conducted by Kathie Shaffer.

394   "If there's four hundred people": Flight 93 Memorial Oral History Transcript, interview with Wallace Miller, July 19, 2007, conducted by Kathie Shaffer.

394   Terry walked through a blue haze: This section comes from interviews with Terry and Kathie Shaffer, May 8–10, 2018, with phone and email follow-ups, and Flight 93 Memorial Oral History Transcript, interview with Terry Shaffer, June 25, 2007, and April 5, 2017, conducted by Kathie Shaffer.

395   Andrew "Sonny" Garcia's wedding ring: Steve Levin, "Flight 93 Victims' Effects Go Back to Families," *(Pittsburgh) Post-Gazette*, December 30, 2001.

395   Sandy Bradshaw's flight log: Flight 93 Memorial Oral History Transcript, interview with Trooper James Broderick, May 11, 2005, conducted by Barbara Black.

396   tree-climbing arborists: Flight 93 Memorial Oral History Transcript, interview with Mark Trautman, August 25, 2006, conducted by Kathie Shaffer.

396   Superman logo tattoo: Flight 93 Memorial Oral History Transcript, interview with Trooper James Broderick, May 11, 2005, conducted by Barbara Black.

CHAPTER 21: "MAYDAY, MAYDAY, MAYDAY!"

398   "door restrictors": Dennis Cauchon and Martha T. Moore, "Elevators Were Disaster Within Disaster," *USA Today*, September 11, 2002. Cauchon and Moore explained the use of "door restrictors" in their reporting. It is the author's belief that Chris Young's ability to force open the doors indicated that his elevator was not yet equipped with one. Also see NIST NCSTAR 1-7, p. 124.

399   more than ninety minutes: This estimate is based on Chris Young's estimate that he left the elevator about five minutes before the North Tower collapsed. That would be about 10:23 a.m., or about ninety-seven minutes after the crash of Flight 11. He also spent about one minute in the elevator before it stalled.

401   "What's your name?": Transcript from NBC News, "The Miracle of Ladder Company 6," September 28, 2001.

403 "We gotta keep moving": Ibid.

404 "All FDNY, get the fuck out!": 9/11 Commission Report, p. 307. Jonas and others credited Picciotto with using his bullhorn to order firefighters to leave. However, the men of Ladder 6 vehemently disputed claims Picciotto made in a book, television appearances, and elsewhere about his actions regarding the rescue of Josephine Harris. See Alice McQuillan, "9/11 Tale Doubted: Some Bravest Say Chief's Book Exaggerated His Role," (New York) Daily News, November 26, 2002.

404 as many as one hundred men: Jim Dwyer and Michelle O'Donnell, "9/11 Firefighters Told of Isolation Amid Disaster," New York Times, September 9, 2005. Also see 9/11 Commission Report, p. 310.

404 "Didn't you hear the Mayday?": World Trade Center Task Force Interview with Lieutenant William Walsh, January 11, 2002, p. 18.

404 "We're not fucking coming out!": 9/11 Commission Report, p. 310.

404 pulse of air pressure: NIST WTC Investigation, Final Report, p. 32.

405 "About fifteen floors down": Kevin Dwyer, Jim Flynn, and Ford Fessenden, "FATAL CONFUSION: A Troubled Emergency Response; 9/11 Exposed Deadly Flaws in Rescue Plan," New York Times, July 7, 2002.

405 "I'm in the Trade Center, Tower One": Transcripts of World Trade Center radio transmissions, Channel 23, Sergeant's Desk, p. 42. Also see Dwyer and Flynn, p. 224.

406 "That's okay": Jonas provided the dialogue from the fallen men he encountered, including Firefighter Faustino Apostol and Lieutenant Michael Warchola.

406 "Dispatch, Captain Brown": Michael Daly, The Book of Mychal: The Surprising Life and Heroic Death of Father Mychal Judge (New York: Thomas Dunne Books, 2008), p. 330. Transcript of dispatch tape released August 16, 2006.

408 upper floors sagged: The definitive report on the causes of the collapse, by the National Institutes of Standards and Technology, or NIST, is known as NC-STAR-1, "Federal Building and Fire Safety Investigation of the World Trade Center Disaster." Issued October 26, 2005, it was based on "some 200 technical experts—including about 85 career NIST experts and 125 leading experts from the private sector and academia—[who] reviewed tens of thousands of documents, interviewed more than 1,000 people, reviewed 7,000 segments of video footage and 7,000 photographs, analyzed 236 pieces of steel from the wreckage, performed laboratory tests, and created sophisticated computer simulations of the sequence of events that occurred from the moment the aircraft struck the towers until they began to collapse." An excellent summary of the NIST findings appears here: www.nist.gov/el/faqs-nist-wtc-towers-investigation. The complete report can be found here: http://ws680.nist.gov/publication/get_pdf.cfm?pub_id=909017.

408 five hundred million fiery, falling pounds: Steven Ashley, "When the Twin Towers Fell," Scientific American, October 9, 2001. The NIST study also estimated that each tower weighed approximately 250,000 tons.

411 lower sky lobby: Smith, p. 90.

412 "Mayday, Mayday, Mayday!": Interview with FDNY Deputy Chief John A. (Jay) Jonas, October 13, 2017, with email and phone follow-ups for fact-checking. Numerous firefighters heard Warchola's calls.

412   didn't have the heart: Testimony of Deputy Chief Nicholas Visconti before the House Committee on Homeland Security, September 20, 2007, p. 1.

413   "Tell my wife and kids": Smith, p. 87.

413   evacuation of Lower Manhattan: Jessica DuLong, *Dust to Deliverance: Untold Stories from the Maritime Evacuation on September 11* (New York: McGraw-Hill Education, 2017), p. 87.

415   didn't want to hang up: Smith, p. 91.

416   more than twenty-seven hundred people: In the years since 9/11, reports have claimed that the official death toll in New York does not include an unknown number of undocumented immigrants who worked in the towers.

417   forty-six rounds: Smith, p. 91.

417   WTC 7 collapsed: The fall of what has become known as Building 7 is at the center of numerous conspiracy theories built on claims that it was a controlled demotion. They point out that the building was not hit by a plane, no other steel-frame high-rise ever collapsed as a result of fire, and its tenants included offices of the Department of Defense, the Secret Service, the IRS, and the Securities and Exchange Commission. Those theories persisted despite an official 2008 report by the National Institute of Standards and Technology report about the WTC 7 collapse. See http://ws680.nist.gov/publication/get_pdf.cfm?pub_id=861610.

417   "We'll come back": Smith, p. 106.

419   "a million planes flying": Interview with Allan Hackel, January 28, 2017.

CHAPTER 22: "YOUR SISTER AND NIECE WILL NEVER BE LONELY"

427   officially ending in 2011: Barbara Salazar Torreon, "U.S. Periods of War and Dates of Recent Conflicts," *Congressional Research Service,* October 11, 2017, pp. 8–9.

428   "The number of casualties": Dan Barry, "A DAY OF TERROR: HOSPITALS; Pictures of Medical Readiness, Waiting and Hoping for Survivors to Fill Their Wards," *New York Times,* September 12, 2001, p. A9. Also see CNN transcript of press conference by Mayor Rudy Giuliani and Governor George Pataki, September 11, 2001, http://transcripts.cnn.com/TRANSCRIPTS/0109/11/bn.42.html.

428   "Dave died in the line of duty": FBI Press Release, dated May 26, 2018, from the Atlanta Field Office.

428   178th member: Announcement on FDNY's official Facebook page, June 23, 2018: https://www.facebook.com/304603755728/posts/10156296133045729/.

428   more people will have died: Noah Goldberg and Thomas Tracy, "So many deaths from 9/11-related illnesses, victims' fund may run out of money," *(New York) Daily News,* September 10, 2018.

429   three thousand children: Carol Polsky, "Children of 9/11: Life with a Parent Missing," *Newsday,* September 10, 2011.

429   "We're close to my house": Interviews with John and Gerry Creamer, January 27, 2017. Details also come from John Creamer's testimony at the Moussaoui trial, transcript pp. 3113–17.

430   "I need to talk to you about something": Interviews with John, Colin, and Julie Creamer, January 27, 2017.

431 "life-safety systems": One World Trade Center website, https://www.wtc
.com/about/buildings/1-world-trade-center, accessed September 9, 2017.

432 "Hi, baby": CeeCee Lyles' voice message is quoted in "Phone Calls from the
Passengers and Crew of Flight 93" compiled by the National Park Service for the
Flight 93 Memorial, based on FBI reports.

432 "I love you": Toby Harnden, "She Asked Me How to Stop the Plane," *The
Telegraph,* March 5, 2002.

433 a minor miracle: Interview with Sonia Tita Puopolo, January 20, 2017.

433 "I love Vamsi": Testimony of Chandra Shakara Kalahasti at Moussaoui trial,
transcript pp. 3087–88.

433 "all I have": Interview with Lee and Eunice Hanson, February 22, 2017; Lee
Hanson testimony at Moussaoui trial.

434 walked among the graduates: "Westling Tells BU Graduates to Build on Nation's Democratic Institutions," *BU Bridge,* May 31, 2002, p. 1. www.bu.edu/bridge
/archive/2002/05-31/westling.htm.

434 "We miss you so much": Eunice Hanson, "Dear Son, You Remain in Our
Hearts," *(Dubuque, Iowa) Telegraph Herald,* September 11, 2002, p. e7.

434 complicated feelings: Brian Charles, "Mother of 9/11 Victim Reacts to News
of Bin Laden's Death," *San Gabriel Valley (Calif.) Tribune,* May 2, 2011.

435 "I didn't want terrorists": Interview with Peg Ogonowski Hatch, January 27,
2017.

436 "took their loved ones' lives": September Eleventh Families for Peaceful Tomorrows website, http://peacefultomorrows.org/history/, accessed February 2, 2017.

436 "It was obvious pretty quickly": Interview with Andrea LeBlanc, January 21,
2017.

436 "We're here today": Text of Donald Rumsfeld's speech at the Pentagon rededi-
cation, *New York Times,* September 11, 2002, http://www.nytimes.com/2002/09/11
/national/text-rumsfelds-speech-at-pentagon.html.

437 One hundred twenty-five people: Goldberg et al., p. 23.

437 laminated prayer card: Interview with Debra Burlingame, July 26, 2018.

438 Her group's mission: From the group's website, at https://911familiesfor
america.org/about/.

438 replaced by plans: Ronda Kaysen, "Condo Tower to Rise Where Muslim
Community Center Was Proposed," *New York Times,* May 12, 2017.

438 For rescuing Jerry Henson: Interviews with David Tarantino, MD, July 12,
2017; Jerry Henson, July 14 and 31, 2017; and Dave Thomas, August 2, 2017.

440 Janice Punches: Interviews with Janice Punches, Jennifer Punches-Botta, and
Jeremy Punches, August 4, 2017.

441 Soldier's Medal: Interview with Lieutenant Colonel Marilyn Wills (Ret.), June
8, 2017. Also see Denise Steele, "Valor, Pain and Tears," Association of the United
States Army, December 1, 2001, www.ausa.org/articles/valor-pain-and-tears.

443 "the other shoe to drop": Interview with Kevin Nasypany, March 20, 2017.
Also see Flight 93 Memorial Oral History Transcript, interview with Kevin Nasypany, July 23, 2013, conducted by Donna Glessner.

444 "They did my job": Flight 93 Memorial Oral History Transcript, interview with Kevin Nasypany, July 23, 2013, conducted by Donna Glessner.

444 "The Americans lost some people": Flight 93 Memorial Oral History Transcript, interview with Wallace Miller, July 19, 2007, conducted by Kathie Shaffer.

444 "All you saw was debris": Sara Rimer, "Public Lives: Where Death Mostly Tiptoes, It Rushed Violently In," *New York Times*, Sept. 22, 2001.

444 a single tooth and a piece of skull: Flight 93 Memorial Oral History Transcript, interview with Wallace Miller, July 19, 2007, conducted by Kathie Shaffer.

444 Forty-eight samples: Eve Conant, "Terror: The Remains of 9/11 Hijackers," *Newsweek*, January 2, 2009.

446 "You're seeing grass": Interview with Terry Shaffer, May 9, 2018.

447 "You have to call Janet": Interview with Elaine Duch and Janet Cardwell, September 27, 2017.

448 killed by the fireball: Until late 2017, Elaine Duch had never seen the form that nominated her for a civilian Medal of Honor from the Port Authority of New York and New Jersey. The form stated: "The messenger she was to meet perished."

449 had just begun classes: Marsh & McLennan Companies, Tributes to Our Colleagues, Dominique Pandolfo, http://memorial.mmc.com/P/dominique-pandolfo.html.

449 the families of Brian Clark and Stan Praimnath: Interviews with Brian Clark, October 5, 2017, and Stan Praimnath, October 4, 2017.

450 eighteen people: NIST NCSTAR 1, WTC Investigation, pp. xxxiv and 105.

451 "She may have jumped": Interview with Jack Gentul, November 13, 2017. Gentul made similar comments to the makers of a 2006 documentary, *The Falling Man*.

451 Jack sat weeping: Interview with Jack Gentul, November 13, 2017, and follow-up emails and phone calls.

453 Edith Torres: Cecilia Lillo suffered another blow in 2017 when Lieutenant Edith Torres died from what FDNY and union officials described as 9/11-related cervical cancer. See Thomas Tracy and John Annese, "FDNY EMS Lieutenant, Who Helped Sept. 11 Victims While Off Duty, Dies of 9/11-Linked Cervical Cancer," *(New York) Daily News*, Feb. 9, 2017.

453 special edition of *Newsweek*: "America Under Attack," *Newsweek Extra Edition*, September 13, 2001, pp. 16–17. NYPD Officer Moira Smith is visible in the background of the photo, by Jason Szenes of Corbis Sygma, although her face is obscured by red hazard tape.

455 Bryan Craig Bennett: Unbylined story, "Boston College Remembers its 9/11 Victims," *The Heights*, September 10, 2001, p. A4.

459 previous worst day: David J. Krajicek, "Deadly 1966 Blaze Kills 12 Firefighters in Manhattan Building, Changes NYC Policy," *(New York) Daily News*, October 23, 2016.

459 the flag-draped remains of "Little Jimmy": FDNY Deputy Chief (ret.) Jim Riches, "Firefighter Jimmy Riches: A Smiling Face That We Miss So Much," *Fire Rescue1.com*, September 8, 2011.

460 "strong, brave, caring, kind people": Transcript from NBC News, "The Miracle of Ladder Company 6," September 28, 2001.

SELECT BIBLIOGRAPHY

In light of the sheer number of books about 9/11 and the continuing flow of new works, a bibliography related to the attacks feels incomplete by necessity and outdated the moment it's published. Consider this a partial and somewhat idiosyncratic collection that ranges from essential materials read deeply and repeatedly during research on this book (e.g. *9/11 Commission Report, 102 Minutes*) to more obscure works consulted out of a reporter's healthy neurosis about not leaving too many stones unturned.

Atkins, Stephen E. *The 9/11 Encyclopedia*. Westport: Praeger Security International, 2008.

Aronson, Jay D. *Who Owns the Dead?: The Science and Politics of Death at Ground Zero*. Cambridge, MA: Harvard University Press, 2016.

Baer, Ulrich, ed. *110 Stories: New York Writes After September 11*. New York: New York University Press, 2002.

Bauer, Nona Kilgore. *Dog Heroes of September 11th: A Tribute to America's Search and Rescue Dogs*. Allenhurst: Kennel Club Books, 2011.

Barbash, Tom. *On Top of the World: Cantor Fitzgerald, Howard Lutnick, & 9/11: A Story of Loss & Renewal*. New York: Harper, 2003.

Beamer, Lisa. *Let's Roll!: Ordinary People, Extraordinary Courage*. Carol Stream, IL: Tyndale House, 2006.

Benfante, Michael, and Dave Hollander. *Reluctant Hero: A 9/11 Survivor Speaks Out About That Unthinkable Day, What He's Learned, How He's Struggled, and What No One Should Ever Forget*. New York: Skyhorse Publishing, 2011.

Bernstein, Richard. *Out of the Blue: The Story of September 11, 2001, from Jihad to Ground Zero*. New York: Times Books, 2002.

Boulden, Jane, and Thomas G. Weiss, eds. *Terrorism and the UN: Before and After September 11*. Bloomington: Indiana University Press, 2004.

Braiker, Harriet B. *The September 11 Syndrome: Anxious Days and Sleepless Nights. Seven Steps to Getting a Grip in Uncertain Times*. New York: McGraw-Hill, 2002.

Brewer, Paul. *September 11 and Radical Islamic Terrorism*. Milwaukee: World Almanac Library, 2006.

Brill, Steven. *After: How America Confronted the September 12 Era*. New York: Simon and Schuster, 2003.

Bull, Chris, and Sam Erman, eds. *At Ground Zero: 25 Stories From Young Reporters Who Were There*. New York: Thunder's Mouth Press, 2002.

Burnett, Deena, and Anthony Giombetti. *Fighting Back: Living Beyond Ourselves*. Altamonte Springs, FL: Advantage Inspirational, 2006.

Calhoun, Craig, Paul Price, and Ashley Timmer. *Understanding September 11*. New York: New Press–W. W. Norton, 2002.

Chomsky, Noam. *Imperial Ambitions: Conversations with Noam Chomsky on the Post-9/11 World*. New York: Metropolitan Books, 2005.

———. *9-11: Was There an Alternative?* New York: Seven Stories Press, 2011.

Coll, Steve. *Ghost Wars: The Secret History of the CIA, Afghanistan, and Bin Laden, from the Soviet Invasion to September 10, 2001*. New York: Penguin Press, 2004.

Condon-Rall, Mary Ellen, *Attack on the Pentagon: The Medical Response to 9/11*. Fort Detrick, MD: Borden Institute, 2011.

Creed, Patrick, and Rick Newman. *Firefight: Inside the Battle to Save the Pentagon on 9/11*. New York: Ballantine Books, 2008.

Crockatt, Richard. *America Embattled: September 11, Anti-Americanism and the Global Order*. London: Routledge, 2003.

Crotty, William J. *The Politics of Terror: The U.S. Response to 9/11*. Boston: Northeastern University Press, 2004.

Daly, Michael. *The Book of Mychal: The Surprising Life and Heroic Death of Father Mychal Judge*. New York: Thomas Dunne Books, 2008.

Darling, Robert J. *24 Hours Inside the President's Bunker: 9-11-01; The White House*. Bloomington, IN: iUniverse, 2010.

DeFede, Jim. *The Day the World Came to Town: 9/11 in Gander, Newfoundland*. New York: Regan Books, 2003.

Didion, Joan. *Fixed Ideas: America since 9/11*. New York: New York Review of Books, 2006.

DiMarco, Damon, and Thomas Kean. *Tower Stories*. Santa Monica: Santa Monica Press, 2007.

Downey, Tom. *The Last Men Out: Life on the Edge at Rescue 2 Firehouse*. New York: Holt Paperbacks, 2004.

Dudziak, Mary L., ed. *September 11 in History: A Watershed Moment?* Durham and London: Duke University Press, 2003.

DuLong, Jessica. *Dust to Deliverance: Untold Stories from the Maritime Evacuation on September 11*. New York: McGraw-Hill, 2017.

Dupré, Judith. *One World Trade Center: Biography of the Building*. New York: Little, Brown, 2016.

Dwyer, Jim, and Kevin Flynn. *102 Minutes: The Untold Story of the Fight to Survive Inside the Twin Towers*. New York: Times Books, 2005.

Engelhardt, Tom. *The United States of Fear*. Chicago: Haymarket Books, 2011.

Ensalaco, Mark. *Middle Eastern Terrorism: From Black September to September 11*. Philadelphia: University of Pennsylvania Press, 2008.

Faludi, Susan. *The Terror Dream: What 9/11 Revealed about America*. London: Atlantic, 2008.

Farmer, John. *The Ground Truth: The Untold Story of America Under Attack on 9/11*. New York: Riverhead-Penguin, 2010.

Feinberg, Kenneth R. *What Is Life Worth? The Inside Story of the 9/11 Fund and Its Effort to Compensate the Victims of September 11th*. New York: PublicAffairs, 2006.

Filson, Leslie. *Air War Over America: September 11 Alters Face of Air Defense Mission*. Tyndall Air Force Base, FL: Headquarters, 1st Air Force, 2003.

Fink, Mitchell. *Never Forget: An Oral History of September 11, 2001*. New York: Regan Books, 2002.

Fontana, Marian. *A Widow's Walk: A Memoir of 9/11*. New York: Simon and Schuster Paperbacks, 2005.

Frank, Mitch. *Understanding September 11th: Answering Questions about the Attacks on America*. New York: Viking Books for Young Readers, 2002.

Friedman, Thomas L. *Longitudes and Attitudes: Exploring the World After September 11*. New York: Farrar, Straus and Giroux, 2002.

Friend, David. *Watching the World Change: The Stories Behind the Images of 9/11*. New York: Farrar, Straus and Giroux, 2006.

Gehring, Verna V. *War after September 11*. Oxford: Rowman and Littlefield, 2003.

Gertz, Bill. *Breakdown: How America's Intelligence Failures Led to September 11*. Washington, DC: Regnery, 2002.

Gilbert, Allison, Phil Hirschkorn, Melinda Murphy, Robyn Walensky, and Mitchell Stephens, eds. *Covering Catastrophe: Broadcast Journalists Report September 11*. Chicago: Bonus Books, 2002.

Glanz, James, and Eric Lipton. *City in the Sky: The Rise and Fall of the World Trade Center*. New York: Times Books, 2003.

Glick, Lyz, and Dan Zegart. *Your Father's Voice: Letters for Emmy About Life with Jeremy—and Without Him After 9/11*. New York: St. Martin's Press, 2004.

Goldberg, Alfred, Sarandis Papadopoulos, Diane Putney, Nancy Berlage, and Rebecca Welch. *Pentagon 9/11*. Washington, DC: Historical Office of the Secretary of Defense, 2007.

Good, Jennifer. *Photography and September 11th: Spectacle, Memory, Trauma*. New York: Bloomsbury Academic, 2016.

Greenwald, Alice M. *The Stories They Tell: Artifacts from the National September 11 Memorial Museum*. New York: Skira Rizzoli, 2013.

———. *No Day Shall Erase You: The Story of 9/11 as Told at the September 11 Museum*. New York: Skira Rizzoli, 2016.

Guzman-McMillan, Genelle. *Angel in the Rubble: The Miraculous Rescue of 9/11's Last Survivor*. New York: Howard Books, 2011.

Hagen, Susan and Mary Carouba. *Women at Ground Zero: Stories of Courage and Compassion*. Indianapolis: Alpha Books, 2002.

Halberstam, David. *Firehouse*. New York: Hyperion, 2003.

Hampton, Wilborn. *September 11, 2001: Attack on New York City*. Cambridge, MA: Candlewick Press, 2003.

Harun, Abdul Hakeem. *Before and Beyond September 11*. Bloomington: New Era Institute for Islamic Thought and Heritage, 2002.

Hauerwas, Stanley, and Frank Lentricchia, eds. *Dissent From the Homeland: Essays after September 11*. Durham and London: Duke University Press, 2003.

Heller, D., ed. *The Selling of 9/11*. London: Palgrave Macmillan, 2005.

Hoge, James Jr., and Gideon Rose. *How Did This Happen? Terrorism and The New War*. New York: Public Affairs, 2001.

Homer, Melodie. *From Where I Stand—Flight #93 Pilot's Widow Sets the Record Straight*. Minneapolis: Langdon Street Press, 2012.

Hovitz, Helaina. *After 9/11: A Young Girl's Journey Through Darkness to a New Beginning*. New York: Skyhorse Publishing, 2016.

Jefferson, Lisa, and Felicia Middlebrooks. *Called: Hello, My Name Is Mrs. Jefferson. I Understand Your Plane Is Being Hijacked. 9:45 am, Flight 93, September 11, 2001*. Chicago: Northfield Publishing, 2006.

Jones, Priscilla D. *The First 109 Minutes: 9/11 and the U.S. Air Force*. Washington DC: Air Force History and Museums Program, 2011.

Kean, Thomas H., and Lee H. Hamilton. *Without Precedent: The Inside Story of the 9/11 Commission*. New York: Vintage, 2007.

Kellner, Douglas. *From 9/11 to Terror War: The Dangers of the Bush Legacy*. Oxford: Rowman and Littlefield, 2003.

Keniston, Ann, and Jeanne Follansbee Quinn, eds. *Literature After 9/11*. London: Routledge, 2008.

Kendra, James M., and Tricia Wachtendorf. *American Dunkirk: The Waterborne Evacuation of Manhattan on 9/11*. Philadelphia: Temple University Press, 2016.

Ketcham, Christopher. *Notes from September 11: Poems and Stories*. Petaluma, CA: Wordrunner, Chapbooks, 2004.

Langewiesche, William. *American Ground: Unbuilding the World Trade Center*. New York: North Point Press, 2002.

Life Magazine, eds. *One Nation: America Remembers September 11, 2001, 10 Years Later*. New York: Little, Brown, 2011.

Life Magazine, eds., and George Bush. *The American Spirit: Meeting the Challenge of September 11*. New York: Time Home Entertainment, 2002.

Lincoln, Bruce. *Holy Terrors: Thinking About Religion After September 11, 2nd edition*. Chicago: University of Chicago Press, 2006.

Lofgren, Stephen J. *Then Came the Fire: Personal Accounts from the Pentagon, 11 September 2001*. Washington, DC: U.S. Army Center of Military History, 2011.

Longman, Jere. *Among the Heroes: United Flight 93 and the Passengers and Crew Who Fought Back*. New York: HarperCollins, 2002.

Lord, Walter. *A Night to Remember*. New York: St. Martin's Griffin, 1955, 1983.

———. *Day of Infamy*. New York: Henry Holt, 60th anniversary edition, 2001.

Luft, Benjamin J. *We're Not Leaving: 9/11 Responders Tell Their Stories of Courage, Sacrifice, and Renewal*. New York: Greenpoint Press, 2011.

Lutnick, Edie. *An Unbroken Bond: The Untold Story of How the 658 Cantor Fitzgerald Families Faced the Tragedy of 9/11*. New York: Emergence Press, 2011.

Lyon, David. *Surveillance After September 11*. Cambridge: Polity Press, 2003.

Malinek, Judy, and T. J. Mitchell. *Working Stiff: Two Years, 262 Bodies, and the Making of a Medical Examiner*. New York: Scribner, 2015.

Manning, Lauren. *Unmeasured Strength*. New York: Henry Holt, 2011.

Marra, Frank, and Maria Bellia Abbate. *From Landfill to Hallowed Ground: The Largest Crime Scene in America*. Dallas: Brown Books Publishing Group, 2015.

Marsoobian, Armen T., Tom Rockmore, and Joseph Margolis, eds. *The Philosophical Challenge of September 11*. Oxford: Blackwell, 2004.

Mayer, Jane. *The Dark Side: The Inside Story of How the War on Terror Turned Into a War on American Ideals*. New York: Anchor Books, 2009.

McDermott, Terry. *Perfect Soldiers: The 9/11 Hijackers: Who They Were, Why They Did It*. New York: Harper Perennial, 2006.

McEneaney, Bonnie. *Messages: Signs, Visits, and Premonitions from Loved Ones Lost on 9/11*. New York: HarperCollins Publishers, 2011.

McMillan, Tom. *Flight 93: The Story, the Aftermath, and the Legacy of American Courage on 9/11*. Guilford: Lyons Press, 2015.

Miller, John C., Michael Stone, and Chris Mitchell. *The Cell: Inside the 9/11 Plot, and Why the FBI and CIA Failed to Stop It*. New York: Hachette Books, 2003.

Murphy, Dean E. *September 11: An oral history*. New York: Doubleday, 2002.

Murphy, Tom. *Reclaiming the Sky: 9/11 and the Untold Story of the Men and Women Who Kept America Flying*. New York: Amacom Books, 2007.

National Commission on Terrorist Attacks Upon the United States. *The 9/11 Commission Report: The Final Report of the National Commission on Terrorist Attacks Upon the United States*. New York: W. W. Norton and Co., 2004.

New York City Police Department, Christopher Sweet, David Fitzpatrick, and Gregory Semendinger. *Above Hallowed Ground: A Photographic Record of September 11*. New York: Avery-Penguin Books, 2002.

New York Times Staff. *Portraits: 9/11/01: The Collected "Portraits of Grief" from The New York Times*. New York: Times Books, 2002.

Noll, Michael A., ed. *Crisis Communications: Lessons from September 11*. Oxford: Rowman & Littlefield Publishers, Inc., 2003.

O'Dowd, Niall. *Fire in the Morning: The Story of the Irish and the Twin Towers on September 11*. Dublin: Brandon Books, 2002.

Peek, Lori A. *Behind the Backlash: Muslim Americans After 9/11*. Philadelphia: Temple University Press, 2011.

Picciotto, Richard and Daniel Paisner. *Last Man Down: A Firefighter's Story of Survival and Escape from the World Trade Center*. New York: Berkley Publishing Group, 2003.

Popular Mechanics, David Dunbar, ed., Brad Reagan, ed. *Debunking 9/11 Myths: Why Conspiracy Theories Can't Stand Up to the Facts*. New York: Hearst Books, 2011.

Potorti, David. *September 11th Families for Peaceful Tomorrows*. New York: Akashic Books, 2003.

Puopolo, Sonia Tita. *Sonia's Ring: 11 Ways to Heal Your Heart*. Minneapolis: Publish Green, 2010.

Quay, Sara E., and Amy M. Damico, eds. *September 11 in Popular Culture: A Guide*. Santa Barbara: Greenwood-ABC-CLIO, LLC, 2010.

Randall, Martin. *9/11 and the Literature of Terror*. Edinburgh: Edinburgh University Press, 2014.

Raskin, Molly Knight. *No Better Time: The Brief, Remarkable Life of Danny Lewin, The Genius Who Transformed the Internet*. New York: Da Capo Press, 2013.

Redfield, Marc. *The Rhetoric of Terror: Reflections on 9/11 and the War on Terror*. New York: Fordham University Press, 2009.

Rinaldi, Tom. *The Red Bandanna: A Life. A Choice. A Legacy*. New York: Penguin, 2016.

Roach, Kent. *The 9/11 Effect: Comparative Counter-Terrorism*. New York: Cambridge University Press, 2011.

Ronningen, Erik O. *From the Inside Out: Harrowing Escapes from the Twin Towers of the World Trade Center*. New York: Welcome Rain Publishers, 2013.

Sagalyn, Lynne B. *Power at Ground Zero: Politics, Money, and the Remaking of Lower Manhattan*. New York: Oxford University Press, 2016.

Salon.com, eds. *Afterwords: Stories and Reports from 9/11 and Beyond*. New York: Washington Square Press, 2002.

Schopp, Andrew, and Matthew B. Hill. *The War on Terror and American Popular Culture: September 11 and Beyond*. Madison: Fairleigh Dickinson University Press, 2009.

Scott, Peter Dale. *The Road to 9/11: Wealth, Empire, and the Future of America*. Berkeley and Los Angeles: University of California Press, 2008.

Scraton, Phil, ed. *Beyond September 11: An Anthology of Dissent*. London: Pluto, 2002.

Shaler, Robert C. *Who They Were: Inside the World Trade Center DNA Story: The Unprecedented Effort to Identify the Missing.* New York: Free Press–Simon & Schuster, 2005.

Shenon, Philip. *The Commission: The Uncensored History of the 9/11 Investigation.* New York: Twelve, 2008.

Sides, Hampton. *Americana: Dispatches From the New Frontier.* New York: Anchor Books, 2004.

Silberstein, Sandra. *War of Words: Language, Politics and 9/11.* London: Routledge, 2002.

Simpson, David. *9/11: The Culture of Commemoration.* Chicago: University of Chicago Press, 2006.

Smith, Dennis. *Report from Engine Co. 82.* New York: Grand Central Publishing–Warner Books, 1999.

——. *Firefighters: Their Lives in Their Own Words.* New York: Broadway Books, 2002.

——. *Report from Ground Zero.* New York: Plume Books, 2003.

——. *A Decade of Hope: Stories of Grief and Endurance from 9/11 Families and Friends.* New York: Viking, 2011.

Soufan, Ali. *The Black Banners: The Inside Story of 9/11 and the War Against al-Qaeda.* New York: W. W. Norton & Company, 2011.

Stewart, James B. *Heart of a Soldier: A Story of Love, Heroism, and September 11th.* New York: Simon & Schuster, 2002.

Stout, Glenn, Charles Vitchers, Robert Gray, and Joel Meyerowitz. *Nine Months at Ground Zero: The Story of the Brotherhood of Workers Who Took on a Job Like No Other.* New York: Scribner, 2006.

Summers, Anthony and Robbyn Swan. *The Eleventh Day: The Full Story of 9/11.* New York: Ballantine Books, 2012.

Taibbi, Matt. *The Great Derangement: A Terrifying True Story of War, Politics, and Religion at the Twilight of the American Empire.* New York: Spiegel & Grau, 2008.

Talbott, Strobe, and Nayan Chanda, eds. *The Age Of Terror: America and the World After September 11.* New York: Basic Books, 2001.

Tarshis, Lauren. *I Survived the Attacks of September 11th, 2001.* New York: Scholastic, 2012.

Thompson, Paul. *The Terror Timeline: Year by Year, Day by Day, Minute by Minute: A Comprehensive Chronicle of the Road to 9/11—And America's Response.* New York: Harper Paperbacks, 2004.

The Poynter Institute. *September 11, 2001: A Collection of Newspaper Front Pages Selected by The Poynter Institute.* Kansas City: Andrews McMeel Publishing, 2001.

Trulson, Jennifer Gardner. *Where You Left Me.* New York: Gallery Books, 2011.

U.S. Department of Justice and Kurtis Toppert, ed. *Responding to September 11 Victims: Lessons Learned from the States.* Damascus, MD: Penny Hill Press, 2010.

Virilio, Paul. *Ground Zero*, trans. Chris Turner. London: Verso, 2002.

Vogel, Steve. *The Pentagon: A History; The Untold Story of the Wartime Race to Build the Pentagon—and to Restore it Sixty Years Later*. New York: Random House, 2007.

Von Drehle, David. *Triangle: The Fire That Changed America*. New York: Grove Press, 2004.

Weiss, Murray. *The Man Who Warned America: The Life and Death of John O'Neill, the FBI's Embattled Counterterror Warrior*. New York: William Morrow, 2003.

Welch, Michael. *Scapegoats of September 11th: Hate Crimes & State Crimes in the War on Terror (Critical Issues in Crime and Society)*. New Brunswick, NJ: Rutgers University Press, 2006.

Winston Dixon, Wheeler, ed. *Film and Television after 9/11*. Carbondale: Southern Illinois University Press, 2004.

Wright, Lawrence. *The Looming Tower: Al-Qaeda and the Road to 9/11*. New York: Vintage, 2007.

Zegart, Amy B. *Spying Blind: The CIA, the FBI, and the Origins of 9/11*. Princeton: Princeton University Press, 2009.

Zelizer, Barbie, and Stuart Allan, eds. *Journalism After September 11*. London and New York: Routledge, 2002.

INDEX

ABOUT THE AUTHOR

MITCHELL ZUCKOFF is the Sumner M. Redstone Professor of Narrative Studies at Boston University. He covered 9/11 for the *Boston Globe* and wrote the lead news story on the day of the attacks. Zuckoff is the author of seven previous works of nonfiction, including the number one *New York Times* bestseller *13 Hours: The Inside Account of What Really Happened in Benghazi*, which became the basis of the Paramount Pictures movie of the same name. As a member of the *Boston Globe* Spotlight Team, he was a finalist for the Pulitzer Prize in investigative reporting and the winner of numerous national awards. He lives outside Boston.

ALSO BY MITCHELL ZUCKOFF

FROZEN IN TIME
An Epic Story of Survival and a Modern Quest for Lost Heroes of World War II

"One of those epic adventure stories that will hold you in its grip from beginning to end."
—David Grann, author of *Killers of the Flower Moon*

A breathtaking blend of mystery and adventure in the vast Arctic wilderness during World War II, and a poignant reminder of the sacrifices of our military personnel and the everyday heroism of the U.S. Coast Guard.

LOST IN SHANGRI-LA
A True Story of Survival, Adventure, and the Most Incredible Rescue Mission of World War II

"A truly incredible adventure."
—*New York Times Book Review*

The exhilarating, untold story of an extraordinary rescue mission during the final days of WWII, where a plane crash in the South Pacific plunged a trio of U.S. military personnel into a land that time forgot, and the ancient indigenous tribe members who aided those stranded on the ground.

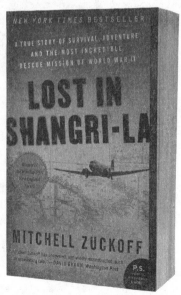

HarperCollins*Publishers* | HARPER PERENNIAL
DISCOVER GREAT AUTHORS, EXCLUSIVE OFFERS, AND MORE AT HC.COM.